Generalized Linear Models

WILEY SERIES IN PROBABILITY AND STATISTICS

Established by WALTER A. SHEWHART and SAMUEL S. WILKS

A complete list of the titles in this series appears at the end of this volume.

Generalized Linear Models

With Applications in Engineering and the Sciences

Second Edition

RAYMOND H. MYERS

Virginia Polytechnic Institute and State University
Blacksburg, Virginia

DOUGLAS C. MONTGOMERY

Arizona State University
Tempe, Arizona

G. GEOFFREY VINING

Virginia Polytechnic Institute and State University
Blacksburg, Virginia

TIMOTHY J. ROBINSON

University of Wyoming
Laramie, Wyoming

A JOHN WILEY & SONS, INC., PUBLICATION

Published by John Wiley & Sons, Inc., Hoboken, New Jersey
Published simultaneously in Canada

Library of Congress Cataloging-in-Publication Data:

Generalized linear models : with applications in engineering and the sciences / Raymond H. Myers ... [et al.]. — 2nd ed.
p. cm.
Rev. ed. of: Generalized linear models / Raymond H. Myers, Douglas C. Montgomery, G. Geoffrey Vining. c2002.
Includes bibliographical references and index.
ISBN 978-0-470-45463-3 (cloth)
1. Linear models (Statistics) I. Myers, Raymond H. Generalized linear models.
QA276.M94 2010
519.5′35—dc22

2009049310

Printed in the United States of America

10 9 8 7 6 5 4 3 2 1

Contents

Preface

This is an introductory textbook on the generalized linear model (GLM). We intend this book for anyone who has completed a course in regression analysis that covers basic model-fitting and statistical inference at the level of an upper-division or first-year graduate course. Some previous background in maximum likelihood estimation for linear regression and some exposure to nonlinear regression is helpful. However, Chapters 2 and 3 cover these topics in sufficient detail.

This book has several unique features. First, we give a thorough treatment of logistic and Poisson regression. Although these are special cases of the GLM, they are very important in their own right, and they deserve special attention. More importantly, the treatment of these two cases provides a solid foundation for the GLM. Second, we provide an introduction to generalized estimating equations, which is a topic closely related to GLM. Third, this text provides an introduction to the generalized linear mixed model (GLMM). Both generalized estimating equations and GLMMs are of increasing importance to many practitioners. Important application areas include biology, to analyze longitudinal data, and the physical sciences and engineering, to analyze correlated observations.

Another useful feature of this book is the many examples of the GLM, in settings ranging from the "classical" applications in biology and biopharmaceuticals, to engineering, quality assurance, and designed experiments. We use real data throughout to illustrate all of the methodologies and analyses. We think that the range of topics and applications gives the book broad appeal both to practicing professionals in a variety of fields and as a textbook for a second course in regression. Finally, we provide considerable guidance on computing. This text illustrates how to use R, SAS, SAS-JMP, and MINITAB to fit GLMs, to perform inference, and to do diagnostic checking.

We fully intend this book as an introductory text that makes this material accessible to a wide audience. There are many other excellent works on GLM, but many of these are written at a much higher level and are aimed primarily at theoretical statisticians. There are other texts that are primarily monographs and much narrower in scope than our book. We intend this book to be an excellent medium for a broad group of engineers, scientists, and statisticians to learn about the GLM, including both the underlying theory and the proper application of these methods. The website

ftp://ftp.wiley.com/public/sci_tech_med/generalized_linear

contains electronic versions of all data sets found in this text.

Chapter 1 is an introduction designed to give the reader some insight into what types of problems support the use of GLM. It also establishes a tone that we use throughout this book of motivating GLMs from classical, normal-theory linear models and regression.

Chapter 2 provides an overview of many of the basic concepts of multiple linear regression. We discuss such fundamental concepts as the least squares and the maximum likelihood estimation procedures. We present confidence interval estimation, hypothesis testing procedures, and model diagnostic checking techniques such as residual plots and influence diagnostics. We also discuss procedures for dealing with nonhomogenous variance through transformations and through weighted least squares estimation. The chapter concludes with a discussion of experimental design. We expect that many readers will find much of this material as review. Nonetheless, this material serves as our starting point for appreciating GLMs, which make extensive use of maximum likelihood estimation and the basic concepts underlying weighted least squares estimation and transformations.

Chapter 3 is a concise presentation of the fundamental ideas of nonlinear regression. It begins by drawing a clear distinction between linear and nonlinear models. It also points out that many nonlinear models are the results of solutions to differential equations. We then discuss both least squares and maximum likelihood estimation procedures. It concludes with a discussion on experimental design for nonlinear models. Chapter 3 is extremely important for our discussion of GLMs, which often are nonlinear in nature.

In Chapter 4 we develop and illustrate both logistic and Poisson regression, which are two special cases of the GLM. Poisson and logistic regression find many applications in the biological, medical, and social sciences. Many physical scientists and engineers are discovering their value as well. This chapter's theoretical development provides a foundation that makes the GLM much easier to grasp. We present examples from a wide array of fields. These examples include both observational studies and designed experiments.

Chapter 5 introduces the exponential family of distributions and motivates the model-building process for a GLM through the use of a link function. We fully develop the connection between weighted least squares and maximum likelihood estimation of the GLM. We make extensive use of the "gamma family" to illustrate the proper analysis of GLMs. This chapter discusses appropriate residual plots to check the adequacy of our model assumptions.

Chapter 6 presents generalized estimating equations (GEEs). GEEs are extensions of the GLM involving more general correlation structures. Typically, these correlation structures are the result of repeated observations within large experimental units or across time within subjects. We discuss the use of GEEs with certain types of industrial experimental designs and in biomedical applications.

In Chapter 7 we extend the GLM to mixed models, where some of the regressors are fixed (nonrandom) effects and some of the regressors are random effects. The generalized linear mixed model (GLMM) has many applications. This chapter begins by reviewing the classical, normal-theory linear mixed model and its analysis. It then extends these ideas to other distributions from the exponential family. The dependence of the response's variance upon the response's mean complicates the estimation, analysis, and interpretation of the results. This chapter fully discusses all of these issues. It concludes with a brief discussion of the Bayesian approach to the GLMM.

Chapter 8 is a collection of topics on experimental designs for the GLM. It discusses optimal design for the GLM. It also examines the relationship between the choice of the link function and design orthogonality. It presents results on the impact of using standard experimental designs. It also considers screening experiments.

We thank the many students at Arizona State University, the University of Wyoming, and Virginia Tech who have relayed their comments to us from the courses we have taught. Their feedback and criticism have proved invaluable. We also thank Dr. Katina Skinner, Dr. Duangporn Jearkpaporn, and Dr. Sharon Lewis for providing valuable computing assistance for the first edition.

Raymond H. Myers
Douglas C. Montgomery
G. Geoffrey Vining
Timothy J. Robinson

January 2010

CHAPTER 1

Introduction to Generalized Linear Models

1.1 LINEAR MODELS

A model is just a simple abstraction of reality in that it provides an approximation of some relatively more complex phenomenon. Models may be broadly classified as **deterministic** or **probabilistic**. In a deterministic model, the system outcomes and responses are precisely defined, often by a set of equations. Deterministic models abound in the sciences and engineering; examples include Ohm's law ($E = IR$), the ideal gas law ($PV = nRT$), and the first law of thermodynamics ($\oint dW = J\oint dQ$). In probabilistic models, the system outcomes or responses exhibit variability, because the model either contains random elements or is impacted in some way by random forces.

There is certainly no more important class of probabilistic model than the probabilistic **linear model**

$$y = \beta_0 + \beta_1 x_1 + \beta_2 x_2 + \cdots + \beta_k x_k + \varepsilon \tag{1.1}$$

where y is the outcome or response variable, $x_1, x_2, \ldots, x_k$ are a set of predictor or regressor variables, $\beta_0, \beta_1, \ldots, \beta_k$ are a set of unknown parameters, and ε is the random error term. Sometimes the regressor variables $x_1, x_2, \ldots, x_k$ are called **covariates**. Often Equation (1.1) is called the **linear regression model**. We typically assume that the error term ε has expected value zero. Therefore the expected value (or expectation function) of the response y in our linear regression model is

$$E(y) = \beta_0 + \beta_1 x_1 + \beta_2 x_2 + \cdots + \beta_k x_k \tag{1.2}$$

Generalized Linear Models, Second Edition, by Myers, Montgomery, Vining, and Robinson

Equation (1.1) is called a **linear** model because the mean response is a linear function of the unknown parameters $\beta_0, \beta_1, \ldots, \beta_k$. That means that an interaction model in, for example, k=2 variables

$$y = \beta_0 + \beta_1 x_1 + \beta_2 x_2 + \beta_{12} x_1 x_2 + \varepsilon \tag{1.3}$$

or a second-order polynomial in k=2 variables

$$y = \beta_0 + \beta_1 x_1 + \beta_2 x_2 + \beta_{12} x_1 x_2 + \beta_{11} x_1^2 + \beta_{22} x_2^2 + \varepsilon \tag{1.4}$$

or even a model with transcendental terms such as

$$y = \beta_0 + \beta_1 \sin\left(\frac{2\pi x}{12}\right) + \beta_2 \cos\left(\frac{2\pi x}{12}\right) + \varepsilon$$

are all linear models. The linear regression model in Equation (1.1) is usually called the first-order linear model. When it is used in conjunction with the analysis of data from a designed experiment, Equation (1.1) is typically called a main-effects model. The interaction model (1.3) and the second-order model (1.4) also arise often in the field of designed experiments.

Linear regression models are used widely for several reasons. First, they are natural approximating polynomials for more complex functional relationships. That is, if the expectation function $E(y) = f(x)$ is the true relationship between the response and a predictor variable x, then the first-order Taylor series approximation to this relationship at some point of interest x_0 is

$$\begin{aligned} E(y) &\cong f(x_0) + \left.\frac{df(x)}{dx}\right|_{x=x_0} (x - x_0) + R \\ &\cong \beta_0 + \beta_1 (x - x_0) \end{aligned}$$

which, ignoring the remainder R (and apart from the error term), is a linear regression model in one variable. When k predictor variables are involved, the first-order Taylor series approximation leads directly to a first-order linear regression model in k variables. Use of a second-order Taylor series results in the second-order model Equation (1.4) or if the pure second-order derivatives are omitted we get an interaction model, of which Equation (1.3) is an example. Because linear regression models are used so often (and so successfully) as approximating polynomials, we sometimes refer to them as **empirical models**.

The second reason for the popularity of linear regression models is that it is straightforward to estimate the unknown parameters $\beta_0, \beta_1, \ldots, \beta_k$. The **method of least squares** is a parameter estimation technique that dates from the early part of the nineteenth century. When this method is applied to a linear model, the analyst has only to solve a set of $p=k+1$ simultaneous linear equations in the $p=k+1$ unknowns β_0, β_1, ..., β_k. There are many statistical software packages (and hand calculators and spreadsheet programs) that have implemented the method of least squares for linear models so regression model fitting is very easy.

Finally, there is a really nice, elegant, and well-developed **statistical theory** for the linear model. If we assume that the errors (ε) in the linear model are normally and independently distributed with constant variance, then statistical tests on the model parameters, confidence intervals on the parameters, and confidence and prediction intervals for the mean response can easily be obtained. Furthermore, these procedures have been implemented nicely in many statistics software packages and so they are easy to perform. In our view, there is no more elegant theory in statistics than that of the linear model, because it not only contains mathematical aesthetics, but it actually works easily and effectively in practical problems as well.

On the way to learning about **generalized linear models** (the main objective of this book), we have to be comfortable with some of the theory and practical aspects of using the linear regression model. Chapter 2 covers the essentials.

1.2 NONLINEAR MODELS

Linear regression models often arise as empirical models for more complex, and generally unknown phenomena. However, there are situations where the phenomenon is well understood and can be described by a mathematical relationship. For example, consider Newton's law of cooling, which states that the rate of change of temperature of an object is proportional to the difference between the object's current temperature and the temperature of the surrounding environment. Thus if f is the current temperature and T_A is the ambient or environmental temperature, then

$$\frac{df}{dt} = -\beta(f - T_A) \tag{1.5}$$

where β is the constant of proportionality. The value of β depends on the thermal conductivity of the object and other factors. Now the actual temperature of the object at time t is the solution to Equation (1.5), or

$$f(t, \beta) = T_A + (T_I - T_A)e^{\beta t} \tag{1.6}$$

where T_I is the initial temperature of the object. In practice, a person measures the temperature at time t with an instrument, and both the person and the instrument are potential sources of variability not accounted for in Equation (1.6). Combining these and all other sources of variability into an error term ε, we may write the actual observed value of temperature at time t as

$$
\begin{aligned}
y &= f(t, \beta) + \varepsilon \\
&= T_{\mathrm{A}} + (T_{\mathrm{I}} - T_{\mathrm{A}})\, e^{\beta t} + \varepsilon
\end{aligned}
\tag{1.7}
$$

Equation (1.7) is an example of a **nonlinear model**, because the response is not a linear function of the unknown parameter β. Sometimes we say that Equation (1.7) is an example of a **nonlinear regression model**. Nonlinear regression models play a very important role in science and engineering, and they usually arise from our knowledge of the underlying **mechanism** that describes the phenomena. Indeed, nonlinear models are sometimes called **mechanistic models**, distinguishing them from linear models, which are typically thought of as **empirical models**. Many nonlinear models are developed directly from the solution to differential equations, as was illustrated in Equation (1.7).

Just as in the case of the linear regression model, it is necessary to estimate the parameters in a nonlinear regression model, and to test hypotheses and construct confidence intervals. There is a statistical theory supporting inference for the nonlinear model. This theory makes use of the normal distribution, and typically assumes that observations are independent with constant variance. The essential elements of nonlinear regression models are summarized in Chapter 3.

1.3 THE GENERALIZED LINEAR MODEL

It should be clear that in dealing with the linear and nonlinear regression models of the previous two sections, the normal distribution played a central role. Inference procedures for both linear and nonlinear regression models in fact assume that the response variable y follows the normal distribution. There are a lot of practical situations where this assumption is not going to be even approximately satisfied. For example, suppose that the response variable is a discrete variable, such as a **count**. We often encounter counts of defects or other rare events, such as injuries, patients with particular diseases, and even the occurrence of natural phenomena including earthquakes and Atlantic hurricanes. Another possibility is a **binary response** variable. Situations where the response variable is either success or failure (i.e., 0 or 1) are fairly common in nearly all areas of science and engineering.

As an example, consider the space shuttle *Challenger* accident, which occurred on January 28, 1986. The space shuttle was made up of the *Challenger* orbiter, an external liquid fuel tank containing liquid hydrogen fuel and liquid oxygen oxidizer, and two solid rocket boosters. At 11:39 EST about 73 seconds after launch, the space shuttle exploded and crashed into the Atlantic Ocean off the coast of Florida, killing all seven astronauts aboard. The cause of the accident was eventually traced to the failure of O-rings on the solid rocket booster. The O-rings failed because they lost flexibility at low temperatures, and the temperature that morning was 31°F, far below the lowest temperature recorded for previous launches.

Table 1.1 Temperature and O-Ring Failure Data from the *Challenger* Accident

Temperature at Launch (°F)	At Least One O-Ring Failure	Temperature at Launch (°F)	At Least One O-Ring Failure
53	1	70	1
56	1	70	1
57	1	72	0
63	0	73	0
66	0	75	0
67	0	75	1
67	0	76	0
67	0	76	0
68	0	78	0
69	0	79	0
70	0	80	0
70	1	81	0

Table 1.1 presents the temperature at the 24 launches or static tests preceding the *Challenger* launch along with an indicator variable denoting whether or not O-ring failure or damage had occurred (0 indicates no failure, 1 indicates failure). A scatter diagram of the data is shown in Figure 1.1. There does appear to be some relationship between failure and temperature, with a higher likelihood of failure at lower temperatures, but it is not immediately obvious what kind of model might describe this relationship. A linear regression model does not seem appropriate, because there are likely some temperatures for which the fitted or predicted value of failure would either be greater than unity or less than zero, clearly impossible values. This is a situation where some type of generalized linear model is more appropriate than an ordinary linear regression model.

There are also situations where the response variable is continuous, but the assumption of normality is not reasonable. Examples include the distribution of stresses in mechanical components and the failure times of systems or components. These types of responses are nonnegative and typically have a highly right-skewed behavior. GLMs are often better models for these situations than ordinary linear regression models.

The **generalized linear model** or (GLM) allows us to fit regression models for univariate response data that follow a very general distribution called the **exponential family**. The exponential family includes the normal, binomial, Poisson, geometric, negative binomial, exponential, gamma, and inverse normal distributions. Furthermore, if the y_i, $i = 1, 2, \ldots, n$, represent the response values, then the GLM is

$$g(\mu_i) = g[E(y_i)] = \mathbf{x}_i'\boldsymbol{\beta}$$

where $\mathbf{x}_i$ is a vector of regressor variables or covariates for the ith observation and $\boldsymbol{\beta}$ is the vector of parameters or regression coefficients. Every generalized

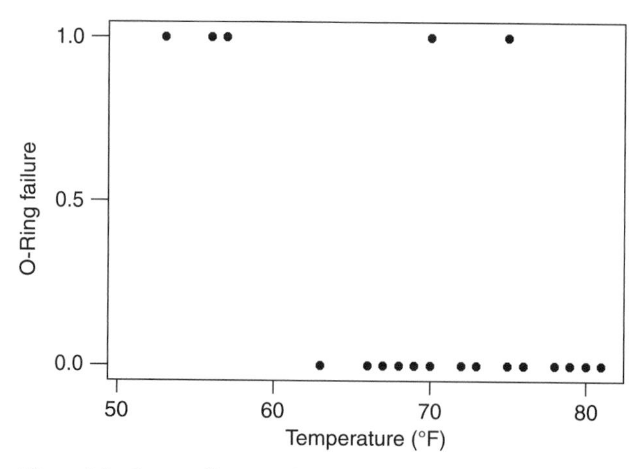

Figure 1.1 Scatter diagram of O-ring failures versus temperature.

linear model has three components: a **response variable distribution** (sometimes called the error structure), a **linear predictor** that involves the regressor variables or covariates, and a **link function** g that connects the linear predictor to the natural mean of the response variable.

For example, consider the linear regression model in Equation (1.1). The response distribution is normal, the linear predictor is

$$\mathbf{x}'\boldsymbol{\beta} = \beta_0 + \beta_1 x_1 + \beta_2 x_2 + \cdots + \beta_k x_k$$

and the link function is an **identity link,** $g(a) = a$, or

$$\begin{aligned} E(y) &= \mu \\ &= \beta_0 + \beta_1 x_1 + \beta_2 x_2 + \cdots + \beta_k x_k \end{aligned}$$

Thus the standard linear regression model in Equation (1.1) is a GLM. Depending on the choice of the link function g, a GLM can include a nonlinear model. For example, if we use a **log link**, $g(a) = \ln(a)$, then

$$\begin{aligned} E(y) &= \mu \\ &= e^{\beta_0 + \beta_1 x_1 + \beta_2 x_2 + \cdots + \beta_k x_k} \end{aligned}$$

For the case of a binomial distribution, a fairly standard choice of link function is the **logit link**. For the *Challenger* data, where there is a single regressor variable, this leads to the model

$$E(y) = \frac{\exp(\beta_0 + \beta_1 x)}{1 + \exp(\beta_0 + \beta_1 x)}$$

We show in Chapter 4 that the estimates of the model parameters in this equation, β_0 and β_1, are $b_0 = 10.875$ and $b_1 = -0.17132$. Therefore the fitted function is

$$\hat{y} = \frac{\exp(10.875 - 0.17132x)}{1 + \exp(10.875 - 0.17132x)}$$

This is called a **logistic regression model**, and it is a very common way to model binomial response data. A graph of the fitted function is shown in Figure 1.2. Notice that the model will not result in fitted values outside the 0–1 range, regardless of the value of temperature.

The generalized linear model may be viewed as a **unification** of linear and nonlinear regression models that incorporates a rich family of normal and nonnormal response distributions. Model fitting and inference can all be performed under the same framework. Furthermore, computer software that supports this unified approach has become widely available and easy to use. Thus while the earliest use of GLMs was confined to the life sciences and the biopharmaceutical industries, applications to other areas of science and engineering have been growing rapidly. Chapters 4 and 5 give a detailed presentation of generalized linear models along with examples from several areas of science and engineering.

The usual GLM assumes that the observations are independent. There are situations where this assumption is inappropriate; examples include data where there are multiple measurements on the same subject or experimental unit, split-plot and other types of experiments that have restrictions on randomization, and experiments involving both random and fixed factors (the mixed

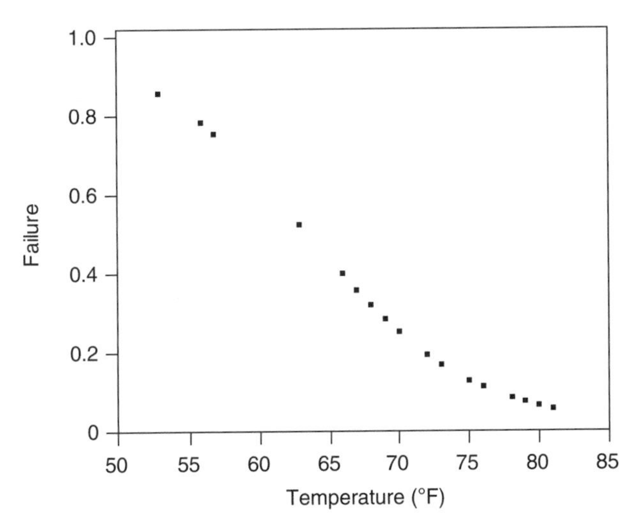

Figure 1.2 Graph of the fitted logistic regression model (from JMP) for the *Challenger* O-ring failure data.

model). **Generalized estimating equations (GEEs)** are introduced to account for a correlation structure between observations in the generalized linear model. Chapter 6 discusses GEEs and presents several applications. GLMs can also include random effects, just as in linear models. Chapter 7 presents methods for including random factors in GLMs. Designed experiments are widely used in fitting linear models and can also be used with GLMs. Some aspects of this are discussed in Chapter 8.

As we will see in subsequent chapters, GLMs, GLMMs, and GEEs are powerful extensions to the familiar linear and nonlinear regression models that have proved to be so useful in many scientific and engineering fields. There are many other books devoted to GLMs and related techniques, including Agresti (1990), Collett (1991), Dobson (1990), Fahrmeir and Tutz (1994), Lindsey (1994), McCullagh and Nelder (1987), Hosmer and Lemeshow (2002), and Kleinbaum (1994). These books are either specialized texts devoted to special subjects of GLM methodology, or they are higher-level works targeted at technical specialists or researchers. Our objective is to provide an introduction to GLMs and GEEs for a broad audience of potential users, including statisticians, engineers, physical, chemical, medical, and life scientists, and other readers with a background in linear regression methods. We also illustrate how the implementation of these techniques in modern computer software facilitates their application.

CHAPTER 2

Linear Regression Models

2.1 THE LINEAR REGRESSION MODEL AND ITS APPLICATION

Regression analysis is a collection of statistical techniques for modeling and investigating the relationship between a **response variable** of interest y and a set of **regressor** or predictor variables $x_1, x_2, \ldots, x_k$. Applications of regression are numerous and occur in almost every applied field including engineering and the chemical/physical sciences, life and biological sciences, the social sciences, management and economics. A very important type of regression model is the **linear regression model**

$$y = \beta_0 + \beta_1 x_1 + \beta_2 x_2 + \cdots + \beta_k x_k + \varepsilon \tag{2.1}$$

in which the response is a linear function of the unknown model parameters or regression coefficients $\beta_0, \beta_1, \ldots, \beta_k$. Linear regression models are widely used as **empirical models** to approximate some more complex and usually unknown functional relationship between the response and the regressor variables.

In this chapter we summarize the techniques for estimating the parameters in multiple regression models. We also present the standard methods for testing hypotheses and constructing confidence intervals for these models, as well as methods for checking model adequacy and quality of fit. We discuss two important parameter estimation techniques for the linear regression model: the method of least squares and the method of maximum likelihood. The important role of the normal distribution in linear regression is discussed. This chapter ends with a discussion of designing experiments.

Generalized Linear Models, Second Edition, by Myers, Montgomery, Vining, and Robinson

2.2 MULTIPLE REGRESSION MODELS

2.2.1 Parameter Estimation with Ordinary Least Squares

The method of least squares is typically used to estimate the regression coefficients in a multiple linear regression model. Suppose that $n > k$ observations on the response variable are available, say, $y_1, y_2, \ldots, y_n$. Along with each observed response y_i, we have an observation on each regressor variable, and let x_{ij} denote the ith observation or level of variable x_j. Table 2.1 summarizes the structure of our dataset. We assume that the error term ε in the model has mean zero and constant variance σ^2, that is, $E(\varepsilon) = 0$ and $\text{Var}(\varepsilon) = \sigma^2$, and that the $\{\varepsilon_i\}$ are uncorrelated random variables.

We may write the model equation (Equation (2.1)) in terms of the observations in Table 2.1 as

$$\begin{aligned} y_i &= \beta_0 + \beta_1 x_{i1} + \beta_2 x_{i2} + \cdots + \beta_k x_{ik} + \varepsilon_i \\ &= \beta_0 + \sum_{j=1}^{k} \beta_j x_{ij} + \varepsilon_i, \quad i = 1, 2, \ldots, n \end{aligned} \tag{2.2}$$

The **method of least squares** chooses the β's in Equation (2.2) so that the sum of the squares of the errors, the ε_i, is minimized. The least squares function is

$$\begin{aligned} S &= \sum_{i=1}^{n} \varepsilon_i^2 \\ &= \sum_{i=1}^{n} \left(y_i - \beta_0 - \sum_{j=1}^{k} \beta_j x_{ij} \right)^2 \end{aligned} \tag{2.3}$$

The function S is to be minimized with respect to $\beta_0, \beta_1, \ldots, \beta_k$. The least squares estimators, say, $b_0, b_1, \ldots, b_k$, must satisfy

$$\left.\frac{\partial S}{\partial \beta_0}\right|_{b_0, b_1, \ldots, b_k} = -2 \sum_{i=1}^{n} \left(y_i - b_0 - \sum_{j=1}^{k} b_j x_{ij} \right) = 0 \tag{2.4a}$$

Table 2.1 Data for Multiple Linear Regression

y	x_1	x_2	$\ldots$	x_k
y_1	x_{11}	x_{12}	$\ldots$	x_{1k}
y_2	x_{21}	x_{22}	$\ldots$	x_{2k}
$\vdots$	$\vdots$	$\vdots$		$\vdots$
y_n	x_{n1}	x_{n2}	$\ldots$	x_{nk}

and

$$\left.\frac{\partial S}{\partial \beta_j}\right|_{b_0, b_1, \ldots, b_k} = -2\sum_{i=1}^{n}\left(y_i - b_0 - \sum_{j=1}^{k} b_j x_{ij}\right)x_{ij} = 0, \quad j = 1, 2, \ldots, k \qquad (2.4b)$$

Simplifying Equation (2.4), we obtain

$$\begin{aligned}
nb_0 + b_1\sum_{i=1}^{n} x_{i1} = b_2\sum_{i=1}^{n} x_{i2} + \cdots + b_k\sum_{i=1}^{n} x_{ik} &= \sum_{i=1}^{n} y_i \\
b_0\sum_{i=1}^{n} x_{i1} + b_1\sum_{i=1}^{n} x_{i1}^2 + b_2\sum_{i=1}^{n} x_{i1}x_{i2} + \cdots + b_k\sum_{i=1}^{n} x_{i1}x_{ik} &= \sum_{i=1}^{n} x_{i1}y_i \\
\vdots \qquad\qquad \vdots \qquad\qquad \vdots \qquad\qquad\qquad \vdots \qquad & \qquad \vdots \\
b_0\sum_{i=1}^{n} x_{ik} + b_1\sum_{i=1}^{n} x_{ik}x_{i1} + b_2\sum_{i=1}^{n} x_{ik}x_{i2} + \cdots + b_k\sum_{i=1}^{n} x_{ik}^2 &= \sum_{i=1}^{n} x_{ik}y_i \qquad (2.5)
\end{aligned}$$

These equations are called the **least squares normal equations**. Note that there are $p = k + 1$ normal equations, one for each of the unknown regression coefficients. The solution to the normal equations will be the least squares estimators of the regression coefficients b_0, b_1, ..., b_k.

It is simpler to solve the normal equations if they are expressed in matrix notation. We now give a matrix development of the normal equations that parallels the development of Equation (2.5). The model in terms of the observations, Equation (2.2), may be written in matrix notation as

$$\mathbf{y} = \mathbf{X}\boldsymbol{\beta} + \boldsymbol{\varepsilon}$$

where

$$\mathbf{y} = \begin{bmatrix} y_1 \\ y_2 \\ \vdots \\ y_n \end{bmatrix}, \quad \mathbf{X} = \begin{bmatrix} 1 & x_{11} & x_{12} & \cdots & x_{1k} \\ 1 & x_{21} & x_{22} & \cdots & x_{2k} \\ \vdots & \vdots & \vdots & & \vdots \\ 1 & x_{n1} & x_{n2} & \cdots & x_{nk} \end{bmatrix}$$

$$\boldsymbol{\beta} = \begin{bmatrix} \beta_0 \\ \beta_1 \\ \vdots \\ \beta_k \end{bmatrix}, \quad \text{and} \quad \boldsymbol{\varepsilon} = \begin{bmatrix} \varepsilon_1 \\ \varepsilon_2 \\ \vdots \\ \varepsilon_n \end{bmatrix}$$

In general, $\mathbf{y}$ is an $(n \times 1)$ vector of the observations, $\mathbf{X}$ is an $(n \times p)$ matrix of the levels of the independent variables expanded to the form of the model (which in this case includes an intercept, leading to the first column containing all elements of unity), $\boldsymbol{\beta}$ is a $(p \times 1)$ vector of the regression coefficients, and $\boldsymbol{\varepsilon}$ is an $(n \times 1)$ vector of random errors. The $\mathbf{X}$ matrix is often called the **model matrix**.

We wish to find the vector of least squares estimators, $\mathbf{b}$, that minimizes

$$S(\boldsymbol{\beta}) = \sum_{i=1}^{n} \varepsilon_i^2 = \boldsymbol{\varepsilon}'\boldsymbol{\varepsilon} = (\mathbf{y} - \mathbf{X}\boldsymbol{\beta})'(\mathbf{y} - \mathbf{X}\boldsymbol{\beta})$$

Note that $S(\boldsymbol{\beta})$ may be expressed as

$$\begin{aligned} S(\boldsymbol{\beta}) &= \mathbf{y}'\mathbf{y} - \boldsymbol{\beta}'\mathbf{X}'\mathbf{y} - \mathbf{y}'\mathbf{X}\boldsymbol{\beta} + \boldsymbol{\beta}'\mathbf{X}'\mathbf{X}\boldsymbol{\beta} \\ &= \mathbf{y}'\mathbf{y} - 2\boldsymbol{\beta}'\mathbf{X}'\mathbf{y} + \boldsymbol{\beta}'\mathbf{X}'\mathbf{X}\boldsymbol{\beta} \end{aligned} \tag{2.6}$$

since $\boldsymbol{\beta}'\mathbf{X}'\mathbf{y}$ is a (1×1) matrix, or a scalar, and its transpose $(\boldsymbol{\beta}'\mathbf{X}'\mathbf{y})' = \mathbf{y}'\mathbf{X}\boldsymbol{\beta}$ is the same scalar. The least squares estimators must satisfy

$$\left.\frac{\partial S}{\partial \boldsymbol{\beta}}\right|_{\mathbf{b}} = -2\mathbf{X}'\mathbf{y} + 2\mathbf{X}'\mathbf{X}\mathbf{b} = \mathbf{0}$$

which simplifies to

$$\mathbf{X}'\mathbf{X}\mathbf{b} = \mathbf{X}'\mathbf{y} \tag{2.7}$$

Equation (2.7) gives the least squares normal equations in matrix form. It is identical to Equation (2.5). As long as the columns of the model matrix $\mathbf{X}$ are not collinear, $\mathbf{X}'\mathbf{X}$ is positive definite; thus to solve the normal equations, multiply both sides of Equation (2.7) by the inverse of $\mathbf{X}'\mathbf{X}$. Consequently, the **least squares estimator** of $\boldsymbol{\beta}$ is

$$\mathbf{b} = (\mathbf{X}'\mathbf{X})^{-1}\mathbf{X}'\mathbf{y} \tag{2.8}$$

We also call $\mathbf{b}$ the **ordinary least squares estimator** of $\boldsymbol{\beta}$ to distinguish it from other estimators based on the least squares idea. It is easy to see that the matrix form of the normal equations is identical to the scalar form. Writing out Equation (2.7) in detail, we obtain

$$\begin{bmatrix} n & \sum_{i=1}^{n} x_{i1} & \sum_{i=1}^{n} x_{12} & \cdots & \sum_{i=1}^{n} x_{ik} \\ \sum_{i=1}^{n} x_{i1} & \sum_{i=1}^{n} x_{i1}^2 & \sum_{i=1}^{n} x_{i1}x_{i2} & \cdots & \sum_{i=1}^{n} x_{i1}x_{ik} \\ \vdots & \vdots & \vdots & & \vdots \\ \sum_{i=1}^{n} x_{ik} & \sum_{i=1}^{n} x_{ik}x_{i1} & \sum_{i=1}^{n} x_{ik}x_{i2} & \cdots & \sum_{i=1}^{n} x_{ik}^2 \end{bmatrix} \begin{bmatrix} b_0 \\ b_1 \\ \vdots \\ b_k \end{bmatrix} = \begin{bmatrix} \sum_{i=1}^{n} y_i \\ \sum_{i=1}^{n} x_{i1}y_i \\ \vdots \\ \sum_{i=1}^{n} x_{ik}y_i \end{bmatrix}$$

If the indicated matrix multiplication is performed, the scalar forms of the normal equations (i.e., Equation (2.5)) result. In this form it is easy to see that $\mathbf{X}'\mathbf{X}$ is a $(p \times p)$ symmetric matrix and $\mathbf{X}'\mathbf{y}$ is a $(p \times 1)$ column vector. Note the special structure of the $\mathbf{X}'\mathbf{X}$ matrix. The diagonal elements of $\mathbf{X}'\mathbf{X}$ are the sums of squares of the elements in the columns of $\mathbf{X}$, and the off-diagonal elements are the sums of cross-products of the elements in the columns of $\mathbf{X}$. Furthermore, note that the elements of $\mathbf{X}'\mathbf{y}$ are the sums of cross-products of the columns of $\mathbf{X}$ and the observations $\{y_i\}$.

The fitted regression model is

$$\hat{\mathbf{y}} = \mathbf{X}\mathbf{b} \tag{2.9}$$

In scalar notion, the fitted model is

$$\hat{y}_i = b_0 + \sum_{j=1}^{k} b_j x_{ij}, \quad i = 1, 2, \ldots, n$$

The difference between the observation y_i and the fitted value $\hat{y}_i$ is a **residual,** say, $e_i = y_i - \hat{y}_i$. The $(n \times 1)$ vector of residuals is denoted by

$$\mathbf{e} = \mathbf{y} - \hat{\mathbf{y}} \tag{2.10}$$

Example 2.1. The Transistor Gain Data. Myers, Montgomery, and Anderson-Cook (2009) describe a study in which the transistor gain in an integrated circuit device between emitter and collector (*hFE*) is reported along

Table 2.2 Data on Transistor Gain (y) for Example 2.1

Observation	x_1 (drive-in time, minutes)	x_2 (dose, ions $\times$ 10^{14})	y (gain or *hFE*)
1	195	4.00	1004
2	255	4.00	1636
3	195	4.60	852
4	255	4.60	1506
5	225	4.20	1272
6	225	4.10	1270
7	225	4.60	1269
8	195	4.30	903
9	255	4.30	1555
10	225	4.00	1260
11	225	4.70	1146
12	225	4.30	1276
13	225	4.72	1225
14	230	4.30	1321

with two variables that can be controlled at the deposition process—emitter drive-in time (x_1, in minutes) and emitter dose (x_2, in ions $\times 10^{14}$). Fourteen observations were obtained following deposition, and the resulting data are shown in Table 2.2. We fit a linear regression model using gain as the response and emitter drive-in time and emitter dose as the regressor variables.

We fit the model

$$y = \beta_0 + \beta_1 x_1 + \beta_2 x_2 + \varepsilon$$

The $\mathbf{X}$ matrix and $\mathbf{y}$ vector are

$$\mathbf{X} = \begin{bmatrix} 1 & 195 & 4.00 \\ 1 & 255 & 4.00 \\ 1 & 195 & 4.60 \\ 1 & 255 & 4.60 \\ 1 & 225 & 4.20 \\ 1 & 225 & 4.10 \\ 1 & 225 & 4.60 \\ 1 & 195 & 4.30 \\ 1 & 255 & 4.30 \\ 1 & 225 & 4.00 \\ 1 & 225 & 4.70 \\ 1 & 225 & 4.30 \\ 1 & 225 & 4.72 \\ 1 & 230 & 4.30 \end{bmatrix}, \quad \mathbf{y} = \begin{bmatrix} 1004 \\ 1636 \\ 852 \\ 1506 \\ 1272 \\ 1270 \\ 1269 \\ 903 \\ 1555 \\ 1260 \\ 1146 \\ 1276 \\ 1225 \\ 1321 \end{bmatrix}$$

The $\mathbf{X}'\mathbf{X}$ matrix is

$$\mathbf{X}'\mathbf{X} = \begin{bmatrix} 1 & 1 & \cdots & 1 \\ 195 & 255 & \cdots & 230 \\ 4.00 & 4.00 & \cdots & 4.30 \end{bmatrix} \begin{bmatrix} 1 & 195 & 4.00 \\ 1 & 255 & 4.00 \\ \vdots & \vdots & \vdots \\ 1 & 230 & 4.30 \end{bmatrix}$$

$$= \begin{bmatrix} 14 & 3155 & 60.72 \\ 3155 & 716{,}425 & 13{,}683.5 \\ 60.72 & 13{,}683.5 & 264.2584 \end{bmatrix}$$

The $\mathbf{X}'\mathbf{y}$ vector is

$$\mathbf{X}'\mathbf{y} = \begin{bmatrix} 1 & 1 & \cdots & 1 \\ 195 & 255 & \cdots & 230 \\ 4.00 & 4.00 & \cdots & 4.30 \end{bmatrix} \begin{bmatrix} 1004 \\ 1636 \\ \vdots \\ 1321 \end{bmatrix}$$

$$= \begin{bmatrix} 7{,}495 \\ 4{,}001{,}120 \\ 75{,}738.30 \end{bmatrix}$$

The least squares estimate of $\boldsymbol{\beta}$ is

$$\mathbf{b} = (\mathbf{X}'\mathbf{X})^{-1}\mathbf{X}'\mathbf{y}$$

or

$$\mathbf{b} = \begin{bmatrix} 30.247596 & -0.041712 & -4.789945 \\ -0.041712 & 0.000184 & 0.000038 \\ -4.789945 & 0.000038 & 1.102439 \end{bmatrix} \begin{bmatrix} 17{,}495 \\ 4{,}001{,}120 \\ 75{,}738.30 \end{bmatrix}$$

$$= \begin{bmatrix} -520.1 \\ 10.7812 \\ -152.15 \end{bmatrix}$$

The fitted regression model is

$$\hat{y} = -520.1 + 10.7812x_1 - 152.15x_2$$

Table 2.3 shows the observed values of y_i, the corresponding fitted values $\hat{y}_i$, and the residuals from this model. There are several other quantities given in this table that will be defined and discussed later. Figure 2.1 shows the fitted regression model **response surface** and the **contour plot** for this model. The regression model for gain is a plane laying above the time–dose space. □

2.2.2 Properties of the Least Squares Estimator and Estimation of σ^2

The method of least squares produces an **unbiased estimator** of the parameter $\boldsymbol{\beta}$ in the multiple linear regression model. This property may easily be demonstrated by finding the expected value of **b** as follows:

$$\begin{aligned}E(\mathbf{b}) &= E[(\mathbf{X}'\mathbf{X})^{-1}\mathbf{X}'\mathbf{y}] \\ &= E[(\mathbf{X}'\mathbf{X})^{-1}\mathbf{X}'(\mathbf{X}\boldsymbol{\beta} + \boldsymbol{\varepsilon})] \\ &= E[(\mathbf{X}'\mathbf{X})^{-1}\mathbf{X}'\mathbf{X}\boldsymbol{\beta} + (\mathbf{X}'\mathbf{X})^{-1}\mathbf{X}'\boldsymbol{\varepsilon}] \\ &= \boldsymbol{\beta}\end{aligned}$$

because $E(\boldsymbol{\varepsilon}) = \mathbf{0}$ and $(\mathbf{X}'\mathbf{X})^{-1}\mathbf{X}'\mathbf{X} = \mathbf{I}$. Thus $\mathbf{b}$ is an unbiased estimator of $\boldsymbol{\beta}$. The variance property of $\mathbf{b}$ is expressed by the covariance matrix

$$\text{Cov}(\mathbf{b}) = E\{[\mathbf{b} - E(\mathbf{b})][\mathbf{b} - E(\mathbf{b})]'\}$$

The covariance matrix of $\mathbf{b}$ is a $(p \times p)$ symmetric matrix whose jjth element is the variance of (b_j) and whose (i, j)th element is the covariance between b_i and b_j. The **covariance matrix** of $\mathbf{b}$ is

$$\begin{aligned}\text{Cov}(\mathbf{b}) &= \text{Var}(\mathbf{b}) \\ &= \text{Var}[(\mathbf{X}'\mathbf{X})^{-1}\mathbf{X}'\mathbf{y}] \\ &= (\mathbf{X}'\mathbf{X})^{-1}\mathbf{X}'\text{Var}(\mathbf{y})\mathbf{X}(\mathbf{X}'\mathbf{X})^{-1} \\ &= \sigma^2(\mathbf{X}'\mathbf{X})^{-1}\mathbf{X}'\mathbf{X}(\mathbf{X}'\mathbf{X})^{-1} \\ &= \sigma^2(\mathbf{X}'\mathbf{X})^{-1}\end{aligned} \tag{2.11}$$

Table 2.3 Observations, Fitted Values, Residuals, and Other Summary Information for Example 2.1

Observation	y_i	$\hat{y}_i$	e_i	h_{ij}	r_i	t_i	D_i
1	1004.0	973.7	30.3	0.367	1.092	1.103	0.231
2	1636.0	1620.5	15.5	0.358	0.553	0.535	0.057
3	852.0	882.4	−30.4	0.317	−1.052	−1.057	0.171
4	1506.0	1529.2	−23.2	0.310	−0.801	−0.787	0.096
5	1272.0	1266.7	5.3	0.092	0.160	0.153	0.001
6	1270.0	1281.9	−11.9	0.133	−0.365	−0.350	0.007
7	1269.0	1205.8	63.2	0.148	1.960	2.316	0.222
8	903.0	928.0	−25.0	0.243	−0.823	−0.810	0.072
9	1555.0	1574.9	−19.9	0.235	−0.651	−0.633	0.043
10	1260.0	1297.1	−37.1	0.197	−1.185	−1.209	0.115
11	1146.0	1190.6	−44.6	0.217	−1.442	−1.527	0.192
12	1276.0	1251.4	24.6	0.073	0.730	0.714	0.014
13	1225.0	1187.5	37.5	0.233	1.225	1.256	0.152
14	1321.0	1305.3	15.7	0.077	0.466	0.449	0.006

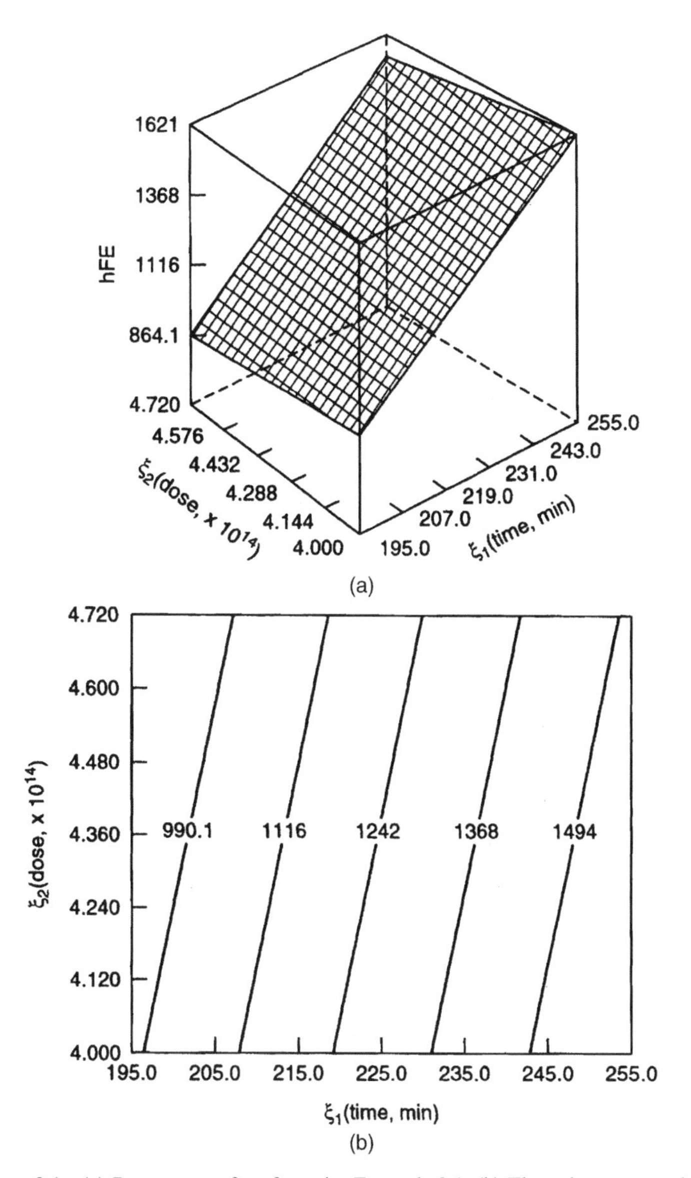

Figure 2.1 (a) Response surface for gain, Example 2.1. (b) The gain contour plot.

Appendices A.1 and A.2 provide some useful background material on the statistical properties of **b**. Appendix A.3 demonstrates that **b** is the **best linear unbiased estimator**. The **best linear unbiased estimator** means that, of all linear functions of the observations, **b** is the best in the sense that it is an unbiased estimator with minimum variance.

It is also usually necessary to estimate σ^2. To develop an estimator of this parameter, consider the sum of squares of the residuals, say,

$$\begin{aligned} SS_{\text{res}} &= \sum_{i=1}^{n} (y_i - \hat{y}_i)^2 \\ &= \sum_{i=1}^{n} e_i^2 \\ &= \mathbf{e}'\mathbf{e} \end{aligned}$$

Substituting $\mathbf{e} = \mathbf{y} - \hat{\mathbf{y}} = \mathbf{y} - \mathbf{Xb}$, we have

$$\begin{aligned} SS_{\text{res}} &= (\mathbf{y} - \mathbf{Xb})'(\mathbf{y} - \mathbf{Xb}) \\ &= \mathbf{y}'\mathbf{y} - \mathbf{b}'\mathbf{X}'\mathbf{y} - \mathbf{y}'\mathbf{Xb} + \mathbf{b}'\mathbf{X}'\mathbf{Xb} \\ &= \mathbf{y}'\mathbf{y} - 2\mathbf{b}'\mathbf{X}'\mathbf{y} + \mathbf{b}'\mathbf{X}'\mathbf{Xb} \end{aligned}$$

Because $\mathbf{X}'\mathbf{Xb} = \mathbf{X}'\mathbf{y}$, this last equation becomes

$$SS_{\text{res}} = \mathbf{y}'\mathbf{y} - \mathbf{b}'\mathbf{X}'\mathbf{y} \tag{2.12}$$

Equation (2.12) is called the **error** or **residual sum of squares,** and it has $n-p$ degrees of freedom associated with it. It can be shown that

$$\mathrm{E}[SS_{\text{res}}] = \sigma^2(n - p)$$

so an unbiased estimator of σ^2 is given by

$$\hat{\sigma}^2 = \frac{SS_{\text{res}}}{n - p} \tag{2.13}$$

Example 2.2. The Transistor Gain Data. We estimate σ^2 for the regression model from Example 2.1. Because

$$\mathbf{y}'\mathbf{y} = \sum_{i=1}^{14} y_i^2 = 22{,}527{,}889.0$$

and

$$\mathbf{b}'\mathbf{X}'\mathbf{y} = \begin{bmatrix} -520.1 & 10.7812 & -152.15 \end{bmatrix} \begin{bmatrix} 17{,}495 \\ 4{,}001{,}120 \\ 75{,}738.30 \end{bmatrix} = 22{,}514{,}467.9$$

the residual sum of squares is

$$\begin{aligned} SS_{\text{res}} &= \mathbf{y}'\mathbf{y} - \mathbf{b}'\mathbf{X}'\mathbf{y} \\ &= 22{,}527{,}889.0 - 22{,}514{,}467.9 \\ &= 13{,}421.1 \end{aligned}$$

Therefore the estimate of σ^2 is computed from Equation (2.13) as follows:

$$\hat{\sigma}^2 = \frac{SS_{\text{res}}}{n-p} = \frac{13{,}421.1}{14-3} = 1220.1$$

The estimate of σ^2 produced by Equation (2.13) is **model dependent**. That is, it depends on the form of the model that is fit to the data. To illustrate this point, suppose that we fit a quadratic model to the gain data, say,

$$y = \beta_0 + \beta_1 x_1 + \beta_2 x_2 + \beta_{11} x_1^2 + \beta_{22} x_2^2 + \beta_{12} x_1 x_2 + \varepsilon$$

In this model it can be shown that $SS_{\text{res}} = 12{,}479.8$. Because the number of model parameters, p, equals 6, the estimate of σ^2 based on this model is

$$\hat{\sigma}^2 = \frac{12{,}479.8}{14-6} = 1559.975$$

This estimate of σ^2 is actually larger than the estimate obtained from the first-order model, suggesting that the first-order model is superior to the quadratic in that there is less unexplained variability resulting from the first-order fit. If replicate runs are available (i.e., more than one observation on y at the same x-levels), then a **model-independent** estimate of σ^2 can be obtained. □

2.2.3 Hypothesis Testing in Multiple Regression

In multiple linear regression problems, certain tests of hypotheses about the model parameters are helpful in measuring the usefulness of the model. In this section we describe several important hypothesis-testing procedures. These procedures require that the errors ε_i in the model be normally and independently distributed with mean zero and constant but unknown variance σ^2, abbreviated $\varepsilon \sim \text{NID}(0, \sigma^2)$. By this assumption, the observations y_i are normally and independently distributed and variance σ^2. If we assume that the correct model is first-order, then the mean of these responses is

$$\beta_0 + \sum_{j=1}^{k} \beta_j x_{ij}$$

Test for Significance of Regression

The test for significance of regression is a test to determine if there is a linear relationship between the response variable y and a subset of the regressor variables $x_1, x_2, \ldots, x_k$. The appropriate hypotheses are

$$\begin{aligned} &H_0 : \beta_1 = \beta_2 = \cdots = \beta_k = 0 \\ &H_1 : \beta_j \neq 0 \text{ for at least one } j \end{aligned} \tag{2.14}$$

Rejection of H_0: in (2.14) implies that at least one of the regressor variables x_1, $x_2, \ldots, x_k$ contributes significantly to the model. The test procedure involves partitioning the corrected total sum of squares, SS_{T}, into a sum of squares due to the model (or to regression) and a sum of squares due to residual, say,

$$SS_{\mathrm{T}} = SS_{\mathrm{R}} + SS_{\mathrm{res}} \tag{2.15}$$

"Corrected" means that the sum of squares is adjusted for the presence of the intercept.

Now if the null hypothesis $H_0 : \beta_1 = \beta_2 = \cdots = \beta_k = 0$ is true, then SS_{R}/σ^2 is distributed as χ_k^2, where the number of degrees of freedom for χ^2, the number of regressor variables in the model. Also, we can show that $SS_{\mathrm{res}}/\sigma^2$ is distributed as χ^2_{n-k-1} and that SS_{res} and SS_{R} are independent. The test procedure for $H_0 : \beta_1 = \beta_2 = \cdots = \beta_k = 0$ is to compute

$$F_0 = \frac{SS_{\mathrm{R}}/k}{SS_{\mathrm{res}}/(n-k-1)} = \frac{MS_{\mathrm{R}}}{MS_{\mathrm{res}}} \tag{2.16}$$

and to reject H_0 if F_0 exceeds $F_{\alpha,\ k,\ n-k-1}$. Alternatively, one could use the P-value approach to hypothesis testing and thus reject H_0 if the P-value for the statistic F_0 is less than α. The test is usually summarized in a table such as Table 2.4. This test procedure is called an **analysis of variance** because it is based on a decomposition of the total variability in the response variable y.

A computational formula for SS_{R} may be found easily. We have derived a computational formula for SS_{res} in Equation (2.12), that is,

Table 2.4 Analysis of Variance for Significance of Regression in Multiple Regression

Source of Variation	Sum of Squares	Degrees of Freedom	Mean Square	F_0
Regression	SS_{R}	k	MS_{R}	$MS_{\mathrm{R}}/MS_{\mathrm{res}}$
Residual	SS_{res}	$n-k-1$	MS_{res}	
Total	SS_{T}	$n-1$		

$$SS_{\text{res}} = \mathbf{y}'\mathbf{y} - \mathbf{b}'\mathbf{X}'\mathbf{y}$$

Now, because $SS_T = \sum_{i=1}^{n} y_i^2 - (\sum_{i=1}^{n} y_i)^2/n = \mathbf{y}'\mathbf{y} - (\sum_{i=1}^{n} y_i)^2/n$, we may rewrite the foregoing equation as

$$SS_{\text{res}} = \mathbf{y}'\mathbf{y} - \frac{\left(\sum_{i=1}^{n} y_i\right)^2}{n} - \left[\mathbf{b}'\mathbf{X}'\mathbf{y} - \frac{\left(\sum_{i=1}^{n} y_i\right)^2}{n}\right]$$

or

$$SS_{\text{res}} = \text{SS}_\text{T} - \text{SS}_\text{R}$$

Therefore the regression sum of squares is

$$SS_R = \mathbf{b}'\mathbf{X}'\mathbf{y} - \frac{\left(\sum_{i=1}^{n} y_i\right)^2}{n} \tag{2.17}$$

the residual sum of squares is

$$SS_{\text{res}} = \mathbf{y}'\mathbf{y} - \mathbf{b}'\mathbf{X}'\mathbf{y} \tag{2.18}$$

and the corrected total sum of squares is

$$SS_T = \mathbf{y}'\mathbf{y} - \frac{\left(\sum_{i=1}^{n} y\right)^2}{n} \tag{2.19}$$

Example 2.3. The Transistor Gain Data. We test for significance of regression using the model fit to the transistor gain data for Example 2.1. Note that

$$\begin{aligned} SS_T &= \mathbf{y}'\mathbf{y} - \frac{\left(\sum_{i=1}^{14} y_i\right)^2}{14} \\ &= 22{,}527{,}889.0 - \frac{(17{,}495)^2}{14} \\ &= 665{,}387.2 \end{aligned}$$

$$\begin{aligned} SS_{\mathrm{R}} &= \mathbf{b}'\mathbf{X}'\mathbf{y} - \frac{\left(\sum_{i=1}^{14} y_i\right)^2}{14} \\ &= 22{,}514{,}467.9 - 21{,}862{,}501.8 \\ &= 651{,}966.1 \end{aligned}$$

and

$$\begin{aligned} SS_{\mathrm{res}} &= SS_{\mathrm{T}} - SS_{\mathrm{R}} \\ &= 665{,}387.2 - 651{,}966.1 \\ &= 13{,}421.1 \end{aligned}$$

The analysis of variance is shown in Table 2.5. If we select $\alpha = 0.05$, then we would reject $H_0 : \beta_1 = \beta_2 = 0$ because $F_0 = 267.2 > F_{0.05,2,11} = 3.98$. Also note that the P-value for F_0 (shown in Table 2.5) is considerably smaller than $\alpha = 0.05$. □

The coefficient of multiple determination R^2 is defined as

$$R^2 = \frac{SS_{\mathrm{R}}}{\mathrm{SS_T}} = 1 - \frac{SS_{\mathrm{res}}}{\mathrm{SS_T}} \tag{2.20}$$

R^2 is a measure of the amount of variability of y explained by using the regressor variables $x_1, x_2, \ldots, x_k$ in the model. From inspection of the analysis of variance identity equation (Equation (2.15)), we see that $0 \leq R^2 \leq 1$. However, a large value of R^2 does not necessarily imply that the regression model is a good one. Adding a variable to the model cannot decrease R^2, regardless of whether the additional variable is statistically significant or not. Thus it is possible for models that have large values of R^2 to yield poor predictions of new observations or estimates of the mean response.

To illustrate, consider the first-order model for the transistor gain data. The value of R^2 for this model is

$$R^2 = \frac{SS_{\mathrm{R}}}{SS_{\mathrm{T}}} = \frac{651{,}966.1}{665{,}387.2} = 0.9798$$

Table 2.5 Test for Significance of Regression, Example 2.3

Source of Variation	Sum of Squares	Degrees of Freedom	Mean Square	F_0	P-Value
Regression	651,996.1	2	325,983.0	267.2	4.74×10^{-10}
Residual	13,421.1	11	1220.1		
Total	665,387.2	13			

That is, the first-order model explains about 97.98% of the variability observed in gain. Now, if we add quadratic terms to this model, we can show that the value of R^2 increases to 0.9812. This increase in R^2 is relatively small, implying that the quadratic terms do not really improve the model.

Because R^2 cannot decrease as we add terms to the model, some regression model builders prefer to use an **adjusted** R^2 statistic defined as

$$R^2_{\text{adj}} = 1 - \frac{SS_{\text{res}}/(n-p)}{SS_{\text{T}}/(n-1)} = 1 - \left(\frac{n-1}{n-p}\right)(1-R^2) \tag{2.21}$$

In general, the adjusted R^2 statistic may not increase as variables are added to the model. In fact, if unnecessary terms are added, the value of R^2_{adj} will often decrease.

For example, consider the transistor gain data. The adjusted R^2 for the first-order model is

$$\begin{aligned} R^2_{\text{adj}} &= 1 - \left(\frac{n-1}{n-p}\right)(1-R^2) \\ &= 1 - \left(\frac{13}{11}\right)(1-0.9798) \\ &= 0.9762 \end{aligned}$$

which is very close to the ordinary R^2 for the first-order model. When R^2 and R^2_{adj} differ dramatically, there is a good chance that nonsignificant terms have been included in the model. Now, when the quadratic terms are added to the first-order model, we can show that $R^2_{\text{adj}} = 0.9695$; that is, the adjusted R^2 actually decreases when the quadratic terms are included in the model. This is a strong indication that the quadratic terms are unnecessary.

Tests on Individual Regression Coefficients and Groups of Coefficients

We are frequently interested in testing hypotheses on the individual regression coefficients. Such tests would be useful in determining the value of each of the regressor variables in the regression model. For example, the model might be more effective with the inclusion of additional variables, or perhaps with the deletion of one or more of the variables already in the model.

Adding a variable to the regression model cannot cause the sum of squares for regression to decrease and the error sum of squares cannot increase. We must decide whether any increase in the regression sum of squares is sufficient to warrant using the additional variable in the model. Furthermore, adding an unimportant variable to the model can actually increase the mean square error, thereby decreasing the usefulness of the model.

The hypotheses for testing the significance of any individual regression coefficient, say, β_j, are

$$H_0 : \beta_j = 0$$
$$H_1 : \beta_j \neq 0$$

If $H_0 : \beta_j = 0$ is not rejected, then this indicates that x_j can be deleted from the model. The test statistic for this hypothesis is

$$t_0 = \frac{b_j}{\sqrt{\hat{\sigma}^2 C_{jj}}} \tag{2.22}$$

where C_{jj} is the diagonal element of $(\mathbf{X}'\mathbf{X})^{-1}$ corresponding to b_j. The null hypothesis $H_0 : \beta_j = 0$ is rejected if $|t_0| > t_{\alpha/2,\ n-k-1}$. Note that this is really a partial or marginal test, because the regression coefficient b_j depends on all the other regressor variables x_i $(i \neq j)$ that are in the model.

The denominator of Equation (2.22), $\sqrt{\hat{\sigma}^2 C_{jj}}$, is often called the estimated **standard error** of the regression coefficient b_j. That is,

$$\widehat{se}(b_j) = \sqrt{\hat{\sigma}^2 C_{jj}} \tag{2.23}$$

Therefore an equivalent way to write the test statistic in Equation (2.22) is

$$t_0 = \frac{b_j}{\widehat{se}(b_j)} \tag{2.24}$$

Example 2.4. The Transistor Gain Data. To illustrate the use of the t-test, consider the regression model for the transistor gain data. We construct the t-statistic for the hypotheses $H_0 : \beta_1 = 0$ and $H_0 : \beta_2 = 0$. The main diagonal elements of $(\mathbf{X}'\mathbf{X})^{-1}$ corresponding to β_1 and β_2 are $C_{11} = 0.000184$ and $C_{22} = 1.102439$, respectively, so the two t-statistics are computed as follows.

For $H_0 : \beta_1 = 0$,

$$t_0 = \frac{b_1}{\sqrt{\hat{\sigma}^2 C_{11}}}$$
$$= \frac{10.7812}{\sqrt{(1220.1)(0.000184)}} = \frac{10.7812}{0.4743} = 22.73$$

For $H_0 : \beta_2 = 0$,

$$t_0 = \frac{b_2}{\sqrt{\hat{\sigma}^2 C_{22}}} = \frac{-152.5}{\sqrt{(1220.1)(1.102439)}}$$
$$= \frac{-152.15}{36.68} = -4.15$$

The absolute values of these t-statistics would be compared to $t_{0.025,\ 11} = 2.201$ (assuming that we select $\alpha = 0.05$). Both t-statistics are larger than this criterion. Consequently, we would conclude that $\beta_1 \neq 0$, which implies that x_1 contributes significantly to the model given that x_2 is included, and that $\beta_2 \neq 0$, which implies that x_2 contributes significantly to the model given that x_1 is included. □

We may also directly examine the contribution to the regression sum of squares for a particular variable, say, x_j, given that other variables x_i $(i \neq j)$ are included in the model. The procedure used to do this is called the **extra sum of squares method**. This procedure can also be used to investigate the contribution of a **subset** of the regressor variables to the model. Consider the regression model with k regressor variables:

$$\mathbf{y} = \mathbf{X}\boldsymbol{\beta} + \boldsymbol{\varepsilon}$$

where $\mathbf{y}$ is $(n \times 1)$, $\mathbf{X}$ is $(n \times p)$, $\boldsymbol{\beta}$ is $(p \times 1)$, $\boldsymbol{\varepsilon}$ is $(n \times 1)$, and $p = k + 1$. We would like to determine if the subset of regressor variables $x_1, x_2, \ldots, x_r$ $(r < k)$ contributes significantly to the regression model. Let the vector of regression coefficients be partitioned as follows:

$$\boldsymbol{\beta} = \begin{bmatrix} \boldsymbol{\beta}_1 \\ \boldsymbol{\beta}_2 \end{bmatrix}$$

where $\boldsymbol{\beta}_1$ is $(r \times 1)$ and $\boldsymbol{\beta}_2$ is $[(p - r) \times 1]$. We wish to test the hypotheses

$$\begin{aligned} H_0 &: \boldsymbol{\beta}_1 = \mathbf{0} \\ H_1 &: \boldsymbol{\beta}_1 \neq \mathbf{0} \end{aligned} \tag{2.25}$$

The model may be written as

$$\mathbf{y} = \mathbf{X}\boldsymbol{\beta} + \boldsymbol{\varepsilon} = \mathbf{X}_1\boldsymbol{\beta}_1 + \mathbf{X}_2\boldsymbol{\beta}_2 + \boldsymbol{\varepsilon} \tag{2.26}$$

where $\mathbf{X}_1$ represents the columns of $\mathbf{X}$ associated with $\boldsymbol{\beta}_1$, and $\mathbf{X}_2$ represents the columns of $\mathbf{X}$ associated with $\boldsymbol{\beta}_2$.

For the **full model** (including both $\boldsymbol{\beta}_1$, and $\boldsymbol{\beta}_2$), we know that $\mathbf{b} = (\mathbf{X}'\mathbf{X})^{-1}\,\mathbf{X}'\mathbf{y}$. Also, the regression sum of squares for all variables including the intercept is

$$SS_{\mathrm{R}}(\boldsymbol{\beta}) = \mathbf{b}'\mathbf{X}'\mathbf{y} \quad (p \text{ degrees of freedom})$$

and the mean square residual for the full model is

$$MS_{\mathrm{res}} = \frac{\mathbf{y}'\mathbf{y} - \mathbf{b}'\mathbf{X}'\mathbf{y}}{n - p}$$

$SS_R(\boldsymbol{\beta})$ refers to the uncorrected regression sum of squares due to $\boldsymbol{\beta}$. To find the contribution of the terms in $\boldsymbol{\beta}_1$ to the regression, fit the model assuming the null hypothesis $H_0 : \boldsymbol{\beta}_1 = \mathbf{0}$ to be true. The **reduced model** is found from Equation (2.26) with $\boldsymbol{\beta}_1 = \mathbf{0}$:

$$\mathbf{y} = \mathbf{X}_2\boldsymbol{\beta}_2 + \boldsymbol{\varepsilon} \tag{2.27}$$

The least squares estimator of $\boldsymbol{\beta}_2$ is $\mathbf{b}_2 = (\mathbf{X'}_2\mathbf{X}_2)^{-1}\ \mathbf{X'}_2\mathbf{y}$, and

$$SS_R(\boldsymbol{\beta}_2) = \mathbf{b}_2'\mathbf{X}_2'\mathbf{y} \quad (p - r \text{ degrees of freedom}) \tag{2.28}$$

The regression sum of squares due to $\boldsymbol{\beta}_1$ given that $\boldsymbol{\beta}_2$ is already in the model is

$$SS_R(\boldsymbol{\beta}_1|\boldsymbol{\beta}_2) = SS_R(\boldsymbol{\beta}) - SS_R(\boldsymbol{\beta}_2) \tag{2.29}$$

This sum of squares has r degrees of freedom. It is the **extra sum of squares** due to $\boldsymbol{\beta}_1$. Note that $SS_R(\boldsymbol{\beta}_1|\boldsymbol{\beta}_2)$ is the increase in the regression sum of squares due to including the variables $x_1, x_2, \ldots, x_r$ in the model. Now $\mathbf{SS}_R(\boldsymbol{\beta}_1|\boldsymbol{\beta}_2)$ is independent of MS_{res}, and the null hypothesis $\boldsymbol{\beta}_1 = \mathbf{0}$ may be tested by the statistic

$$F_0 = \frac{SS_R(\boldsymbol{\beta}_1|\boldsymbol{\beta}_2)/r}{MS_{res}} \tag{2.30}$$

If $F_0 > F_{\alpha, r, n-p}$, we reject H_0, concluding that at least one of the parameters in $\boldsymbol{\beta}_1$ is not zero, and consequently at least one of the variables $x_1, x_2, \ldots, x_r$ in $\mathbf{X}_1$ contributes significantly to the regression model. Some authors call the test in Equation (2.30) a **partial** F-test.

The partial F-test is very useful. We can use it to measure the contribution of x_j as if it were the last variable added to the model by computing

$$SS_R(\beta_j|\beta_0, \beta_1, \ldots, \beta_{j-1}, \beta_{j+1}, \ldots, \beta_k)$$

This is the increase in the regression sum of squares due to adding x_j to a model that already includes $x_1, \ldots, x_{j-1}, x_{j+1}, \ldots, x_k$. Note that the partial F-test on a single variable x_j is equivalent to the t-test in Equation (2.22). However, the partial F-test is a more general procedure in that we can measure the effect of sets of variables. This procedure is used often in response surface work. For example, suppose that we are considering fitting the second-order model

$$y = \beta_0 + \beta_1 x_1 + \beta_2 x_2 + \beta_{11} x_1^2 + \beta_{22} x_2^2 + \beta_{12} x_1 x_2 + \varepsilon$$

and we wish to test the contribution of the second-order terms over and above the contribution from the first-order model. Therefore the hypotheses of interest are

$$H_0 : \beta_{11} = \beta_{22} = \beta_{12} = 0$$
$$H_1 : \beta_{11} \neq 0 \text{ and/or } \beta_{22} \neq 0 \text{ and/or } \beta_{12} \neq 0$$

In the notation of this section, $\boldsymbol{\beta}_1' = [\boldsymbol{\beta}_{11}, \boldsymbol{\beta}_{22}, \boldsymbol{\beta}_{12}]$ and $\boldsymbol{\beta}_2' = [\boldsymbol{\beta}_0, \boldsymbol{\beta}_1, \boldsymbol{\beta}_2]$, and the columns of $\mathbf{X}_1$ and $\mathbf{X}_2$ are the columns of the original $\mathbf{X}$ matrix associated with the second order and linear terms in the model, respectively.

Example 2.5. The Extra Sum of Square Procedure. Consider the transistor gain data in Example 2.1. We have previously fit a first-order model to these data, but now suppose that we want to consider using a higher-order model, the full quadratic:

$$y = \beta_0 + \beta_1 x_1 + \beta_2 + \beta_{12} x_1 x_2 + \beta_{11} x_1^2 + \beta_{22} x_2^2 + \varepsilon$$

We can use the extra sum of squares method to investigate the contribution of the second-order terms to this model. We need to calculate

$$\begin{aligned} &SS_R(\beta_{12}, \beta_{11}, \beta_{22} | \beta_0, \beta_1, \beta_2) \\ &= SS_R(\beta_0, \beta_1, \beta_2, \beta_{12}, \beta_{11}, \beta_{22}) - SS_R(\beta_0, \beta_1, \beta_2) \\ &= SS_R(\beta_1, \beta_2, \beta_{12}, \beta_{11}, \beta_{22} | \beta_0) - SS_R(\beta_1, \beta_2 | \beta_0) \end{aligned}$$

The regression sum of squares $SS_R(\beta_1,\beta_2|\beta_0)$ is calculated in Example 2.3, where we were testing for significance of regression as $SS_R(\beta_1,\beta_2|\beta_0) = 651{,}966.1$. If we fit the complete second-order model to these data, the regression sum of squares for this model is $SS_R(\beta_1,\beta_2,\beta_{12},\beta_{11},\beta_{22}|\beta_0) = 652{,}907.4$. The error mean square for this full model is $MS_{\text{res}} = 1560$. Therefore the extra sum of squares for the quadratic terms is

$$\begin{aligned} &SS_R(\beta_{12}, \beta_{11}, \beta_{22} | \beta_1, \beta_1, \beta_0) \\ &= SS_R(\beta_1, \beta_2, \beta_{12}, \beta_{11}, \beta_{22} | \beta_0) - SS_R(\beta_1, \beta_2 | \beta_0) \\ &= 652{,}907.4 - 651{,}966.1 \\ &= 941.3 \end{aligned}$$

with $r = 3$ degrees of freedom. To test the hypotheses

$$H_0 : \beta_{12} = \beta_{11} = \beta_{22} = 0$$
$$H_1 : \text{At least one } \beta \neq 0$$

we use the F-statistic in Equation (2.30):

$$F = \frac{SS_{\mathrm{R}}(\beta_{12}, \beta_{11}, \beta_{22}|\beta_1, \beta_1, \beta_0)}{MS_{\mathrm{res}}}$$
$$= \frac{941.3}{1560}$$
$$= 0.60$$

This F-statistic has three numerator and eight denominator degrees of freedom. Because $F_{0.05,\ 3,\ 8} = 4.07$, there is insufficient evidence to reject the null hypothesis, so we conclude that the first-order model is adequate and we do not need the higher-order terms. □

This extra sum of squares procedure can also be applied to testing the contribution of a single variable to the regression model. For example, suppose that we wish to investigate the contribution of the variable x_2 = dose to the original first-order model. That is, the hypotheses we want to test are

$$H_0 : \beta_2 = 0$$
$$H_1 : \beta_2 \neq 0$$

This requires the extra sum of squares due to β_2, or

$$\begin{aligned} SS_{\mathrm{R}}(\beta_2|\beta_1, \beta_0) &= SS_{\mathrm{R}}(\beta_0, \beta_1, \beta_2) - SS_{\mathrm{R}}(\beta_0, \beta_1) \\ &= SS_{\mathrm{R}}(\beta_1, \beta_2|\beta_0) - SS_{\mathrm{R}}(\beta_1|\beta_0) \end{aligned}$$

Now from Example 2.3, where we tested for significance of regression, we have (from Table 2.5)

$$SS_{\mathrm{R}}(\beta_1, \beta_2|\beta_0) = 651{,}966.1$$

This sum of squares has 2 degrees of freedom. The reduced model is

$$y = \beta_0 + \beta_1 x_1 + \varepsilon$$

The least squares fit for this model is

$$\hat{y} = -1181.1 + 10.7864x_1$$

and the regression sum of squares for this model (with 1 degree of freedom) is

$$SS_{\mathrm{R}}(\beta_1|\beta_0) = 630{,}967.9$$

Therefore

$$\begin{aligned} SS_{\mathrm{R}}(\beta_2|\beta_0, \beta_1) &= 651{,}966.1 - 630{,}967.9 \\ &= 20{,}998.2 \end{aligned}$$

with 2−1 = 1 degree of freedom. This is the increase in the regression sum of squares that results from adding x_2 to a model already containing x_1. To test $H_0: \beta_2 = 0$, from the test statistic in Equation (2.30) we obtain

$$F_0 = \frac{SS_R(\beta_2|\beta_0,\beta_1)/1}{MS_{\text{res}}} = \frac{20,998.2/1}{1220.1} = 17.21$$

Note that MS_{res} from the full model (Table 2.5) is used in the denominator of F_0. Now, because $F_{0.05,\,1,\,11} = 4.84$, we would reject $H_0: \beta_2 = 0$ and conclude that x_2 (dose) contributes significantly to the model.

Because this partial F-test involves only a single regressor, it is equivalent to the t-test introduced earlier, because the square of a t random variable with υ degrees of freedom is an F random variable with 1 and υ degrees of freedom. To see this, recall that the t-statistic for $H_0: \beta_2 = 0$ resulted in $t_0 = -4.15$ and that $t_0^2 = (-4.15)^2 = 17.22 \simeq F_0$.

2.2.4 Confidence Intervals in Multiple Regression

It is often necessary to construct confidence interval estimates for the regression coefficients $\{\beta_j\}$ and for other quantities of interest from the regression model. The development of a procedure for obtaining these confidence intervals requires that we assume the errors $\{\varepsilon_i\}$ to be normally and independently distributed with mean zero and variance σ^2, the same assumption made in the section on hypothesis testing (Section 2.2.3).

Confidence Intervals on the Individual Regression Coefficients

Because the least squares estimator $\mathbf{b}$ is a linear combination of the observations, it follows that $\mathbf{b}$ is normally distributed with mean vector $\boldsymbol{\beta}$ and covariance matrix $\sigma^2(\mathbf{X}'\mathbf{X})^{-1}$. Then each of the statistics

$$\frac{b_j - \beta_j}{\sqrt{\hat{\sigma}^2 C_{jj}}}, \quad i = 0, 1, \ldots, k \tag{2.31}$$

is distributed as t with n–p degrees of freedom, where C_{jj} is the jjth element of the $(\mathbf{X}'\mathbf{X})^{-1}$ matrix, and $\hat{\sigma}^2$ is the estimate of the error variance, obtained from Equation (2.13). Therefore a 100(1–α)% confidence interval for the regression coefficient β_j, $j = 0, 1, \ldots, k$, is

$$b_j - t_{\alpha/2,n-p}\sqrt{\hat{\sigma}^2 C_{jj}} \leq \beta_j \leq b_j + t_{\alpha/2,n-p}\sqrt{\hat{\sigma}^2 C_{jj}} \tag{2.32}$$

Note that this confidence interval could also be written as

$$b_j - t_{\alpha/2,n-p}\widehat{se}(b_j) \leq \beta_j \leq b_j + t_{\alpha/2,n-p}\widehat{se}(b_j)$$

because $\widehat{se}(b_j) = \sqrt{\hat{\sigma}^2 C_{jj}}$.

Example 2.6. We construct a 95% confidence interval for the parameter β_1 in Example 2.1. Now $b_1 = 10.7812$, and because $\hat{\sigma}^2 = 1220.1$ and $se(b_1) = 0.4743$, we find that

$$\begin{aligned} b_1 - t_{0.025,11}\widehat{se}(b_i) \leq \beta_1 &\leq b_1 + t_{0.025,11}\widehat{se}(b_1) \\ 10,7812 - 2.201(0.4743) \leq \beta_1 &\leq 10.7813 + 2.201(0.4743) \\ 10.7812 - 1.0439 \leq \beta_1 &\leq 10.7812 + 1.0439 \end{aligned}$$

and the 95% confidence interval on β_1 is

$$9.7373 \leq \beta_1 \leq 11.8251$$

□

A Joint Confidence Region on the Regression Coefficients β

The confidence intervals in the previous section should be thought of as one-at-a-time intervals; that is, the confidence coefficient 1–α applies only to one such interval. Some problems require that several confidence intervals be constructed from the same data. In such cases, the analyst is usually interested in specifying a confidence coefficient that applies to the entire set of confidence intervals. Such intervals are called **simultaneous confidence intervals**.

It is relatively easy to specify a **joint confidence region** for the parameters **β** in a multiple regression model. We may show that

$$\frac{(\mathbf{b} - \boldsymbol{\beta})'\mathbf{X}'\mathbf{X}(\mathbf{b} - \boldsymbol{\beta})}{pMS_{\text{res}}}$$

has an F distribution with p numerator and n–p denominator degrees of freedom, and this implies that

$$P\left\{\frac{(\mathbf{b} - \boldsymbol{\beta})'\mathbf{X}'\mathbf{X}(\mathbf{b} - \boldsymbol{\beta})}{pMS_{\text{res}}} \leq F_{\alpha,p,n-p}\right\} = 1 - \alpha$$

Consequently, a 100(1–α)% joint confidence region for all the parameters in **β** is

$$\frac{(\mathbf{b} - \boldsymbol{\beta})'\mathbf{X}'\mathbf{X}(\mathbf{b} - \boldsymbol{\beta})}{pMS_{\text{res}}} \leq F_{\alpha,p,n-p} \tag{2.33}$$

This inequality describes an **elliptically shaped** region. Montgomery, Peck, and Vining (2006) and Myers (1990) demonstrate the construction of this region for $p = 2$. When there are only two parameters, finding this region is relatively simple; however, when more than two parameters are involved, the construction problem is considerably harder.

There are other methods for finding joint simultaneous intervals on regression coefficients. Montgomery, Peck and Vining (2006) and Myers (1990) discuss and illustrate some of these methods. They also present methods for finding several other types of interval estimates.

Confidence Interval on the Mean Response

We may also obtain a confidence interval on the mean response at a particular point, say, $x_{01}, x_{02}, \ldots, x_{0k}$. Define the vector

$$\mathbf{x}_0 = \begin{bmatrix} 1 \\ x_{01} \\ x_{02} \\ \vdots \\ x_{0k} \end{bmatrix}$$

The mean response at this point is

$$\mu_{y|\mathbf{x}_0} = \beta_0 + \beta_1 x_{01} + \beta_2 x_{02} + \cdots + \beta_k x_{0k} = \mathbf{x}_0'\boldsymbol{\beta}$$

The estimated mean response at this point is

$$\hat{y}(\mathbf{x}_0) = \mathbf{x}_0'\mathbf{b} \tag{2.34}$$

This estimator is unbiased, because $E[\hat{y}(\mathbf{x}_0)] = E(\mathbf{x}_0'\mathbf{b}) = \mathbf{x}_0'\boldsymbol{\beta} = \mu_{y|\mathbf{x}_0}$, and the variance of $\hat{y}(\mathbf{x}_0)$ is

$$\text{Var}[\hat{y}(\mathbf{x}_0)] = \sigma^2 \mathbf{x}_0'(\mathbf{X}'\mathbf{X})^{-1}\mathbf{x}_0 \tag{2.35}$$

Therefore a 100(1–α)% confidence interval on the mean response at the point $x_{01}, x_{02}, \ldots, x_{0k}$ is

$$\hat{y}(\mathbf{x}_0) - t_{\alpha/2,n-p}\sqrt{\hat{\sigma}^2\mathbf{x}_0'(\mathbf{X}'\mathbf{X})^{-1}\mathbf{x}_0} \\ \leq \mu_{y|\mathbf{x}_0} \leq \hat{y}(\mathbf{x}_0) + t_{\alpha/2,n-p}\sqrt{\hat{\sigma}^2\mathbf{x}_0'(\mathbf{X}'\mathbf{X})^{-1}\mathbf{x}_0} \tag{2.36}$$

Example 2.7. Suppose that we wish to find a 95% confidence interval on the mean response for the transistor gain problem for the point $x_{01} = 225$ min and $x_{02} = 4.36 \times 10^{14}$ ions, so that

$$\mathbf{x}_0 = \begin{bmatrix} 1 \\ 225 \\ 4.36 \end{bmatrix}$$

The estimate of the mean response at this point is computed from Equation (2.34) as

$$\hat{y}(\mathbf{x}_0) = \mathbf{x}_0'\mathbf{b} = [1, 225, 4.36]\begin{bmatrix} -520.1 \\ 10.7812 \\ -152.15 \end{bmatrix} = 1242.3$$

Now from Equation (2.35), we find $\text{Var}[\hat{y}(\mathbf{x}_0)]$ as

$$\begin{aligned}\text{Var}[\hat{y}(\mathbf{x}_0)] &= \sigma^2\mathbf{x}_0'(\mathbf{X}'\mathbf{X})^{-1}\mathbf{x}_0 \\ &= \sigma^2[1, 225, 4.36]\begin{bmatrix} 30.247596 & -0.041712 & -4.789945 \\ -0.041712 & 0.000184 & 0.000038 \\ -4.789945 & 0.000038 & 1.102439 \end{bmatrix}\begin{bmatrix} 1 \\ 225 \\ 4.36 \end{bmatrix} \\ &= \sigma^2(0.072027)\end{aligned}$$

Using $\hat{\sigma}^2 = MS_{\text{res}} = 1220.1$ and Equation (2.36), we find the confidence interval as

$$\begin{aligned}&\hat{y}(\mathbf{x}_0) - t_{\alpha/2,n-p}\sqrt{\hat{\sigma}^2\mathbf{x}_0'(\mathbf{X}'\mathbf{X})^{-1}\mathbf{x}_0} \\ &\qquad \leq \mu_{y|x_0} \leq \hat{y}(\mathbf{x}_0) + t_{\alpha/2,n-p}\sqrt{\hat{\sigma}^2\mathbf{x}_0'(\mathbf{X}'\mathbf{X})^{-1}\mathbf{x}_0} \\ &1242.3 - 2.201\sqrt{1220.1(0.072027)} \\ &\qquad \leq \mu_{y|x_0} \leq 1242.3 + 2.201\sqrt{1220.1(0.072027)} \\ &1242.3 - 20.6 \leq \mu_{y|x_0} \leq 1242.3 + 20.6\end{aligned}$$

or

$$1221.7 \leq \mu_{y|x_0} \leq 1262.9$$

□

2.2.5 Prediction of New Response Observations

A regression model can be used to predict future observations on the response y corresponding to particular values of the regressor variables, say, x_{01}, x_{02}, ..., x_{0k}. If $\mathbf{x}_0' = [1, x_{01}, x_{02}, \ldots, x_{0k}]$, then a point estimate for the future observation y_0 at the point x_{01}, x_{02}, ..., x_{0k} is computed from Equation (2.34):

$$\hat{y}(\mathbf{x}_0) = \mathbf{x}_0'\mathbf{b}$$

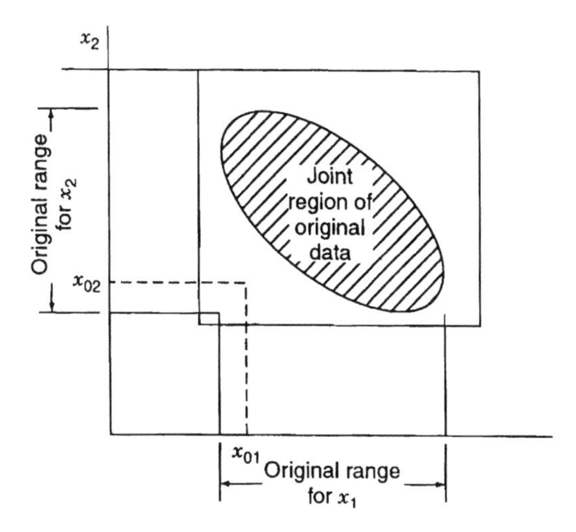

Figure 2.2 An example of extrapolation in multiple regression.

A 100(1–α)% **prediction interval** for this future observation is

$$\hat{y}(\mathbf{x}_0) - t_{\alpha/2,n-p}\sqrt{\hat{\sigma}^2(1 + \mathbf{x}_0'(\mathbf{X}'\mathbf{X})^{-1}x_0)}$$
$$\leq y_0 \leq \hat{y}(\mathbf{x}_0) + t_{\alpha/2,n-p}\sqrt{\hat{\sigma}^2(1 + \mathbf{x}_0'(\mathbf{X}'\mathbf{X})^{-1}\mathbf{x}_0} \quad (2.37)$$

In predicting new observations and in estimating the mean response at a given point $x_{01}, x_{02}, \ldots, x_{0k}$, one must be careful about **extrapolating** beyond the region containing the original observations. It is very possible that a model that fits well in the region of the original data will no longer fit well outside that region. In multiple regression it is often easy to **inadvertently extrapolate,** since the levels of the variables $(x_{i1}, x_{i2}, \ldots, x_{ik})$, $i = 1, 2, \ldots, n$, jointly define the region containing the data. As an example, consider Figure 2.2, which illustrates the region containing the observations for a two-variable regression model. Note that the point (x_{01}, x_{02}) lies within the ranges of both regressor variables x_1, and x_2, but it is outside the region of the original observations. Thus either predicting the value of a new observation or estimating the mean response at this point is an extrapolation of the original regression model.

Example 2.8. Suppose that we wish to find a 95% prediction interval on the next observation on transistor gain at the point $x_{01} = 225$ min and $x_{02} = 4.36 \times 10^{14}$ ions. The predicted value of gain at this point is $\hat{y}(\mathbf{x}_0) = \mathbf{x}_0'\mathbf{b} = 1242.3$. From Example 2.7 we know that

$$\mathbf{x}_0'(\mathbf{X}'\mathbf{X})^{-1}\mathbf{x}_0 = 0.072027$$

Therefore, using Equation (2.37), we can find the 95% prediction interval on y_0 as follows:

$$\hat{y}(\mathbf{x}_0) - t_{\alpha/2,\, n-p}\sqrt{\hat{\sigma}^2(1 + \mathbf{x}_0'(\mathbf{X}'\mathbf{X})^{-1}\mathbf{x}_0)} \le y_0 \le \hat{y}(\mathbf{x}_0) + t_{\alpha/2,\, n-p}\sqrt{\hat{\sigma}^2(1 + \mathbf{x}_0'(\mathbf{X}'\mathbf{X})^{-1}\mathbf{x}_0)}$$

$$1242.3 - 2.201\sqrt{1220.1(1 + 0.072027)} \le y_0 \le 1242.3 + 2.201\sqrt{1220.1(1 + 0.072027)}$$

$$1242.3 - 79.6 \le y_0 \le 1242.3 + 79.6$$

or

$$1162.7 \le y_0 \le 1321.9$$

If we compare the width of the prediction interval at this point with the width of the confidence interval on the mean gain at the same point from Example 2.7, we observe that the prediction interval is much wider. This reflects the fact that it is much more difficult to predict an individual future value of a random variable than it is to estimate the mean of the probability distribution from which that future observation will be drawn. □

2.2.6 Linear Regression Computer Output

There are many statistics software packages with excellent linear regression modeling capability. The output for the linear regression model for the transistor gain data from JMP is shown in Table 2.6. Notice that some of the computer results are slightly different from those reported earlier. This is the result of either rounding intermediate calculations or truncation of values reported by JMP.

Most of the quantities in the output have been explained previously. The PRESS statistic is discussed in Section 2.4.3. Notice that JMP computes the sequential sums of squares $SS_R(\beta_1 \mid \beta_0)$ and $SS_R(\beta_2 \mid \beta_1, \beta_0)$ using the extra sum of squares method from Section 2.2.3. There is also a plot of actual gain versus predicted gain, a plot of residuals versus the predicted gain (this is discussed in Section 2.4.1), and a prediction profile, an interactive tool for predicting the response at different levels of the x's.

2.3 PARAMETER ESTIMATION USING MAXIMUM LIKELIHOOD

2.3.1 Parameter Estimation Under the Normal-Theory Assumptions

The method of least squares can be used to estimate the parameters in a linear regression model regardless of the form of the distribution of the response variable y. Appendix A.3 shows that least squares produces **best linear unbiased**

Table 2.6 JMP Multiple Regression Output for the Transistor Gain Data

Response y Gain
Whole Model
Actual by Predicted Plot

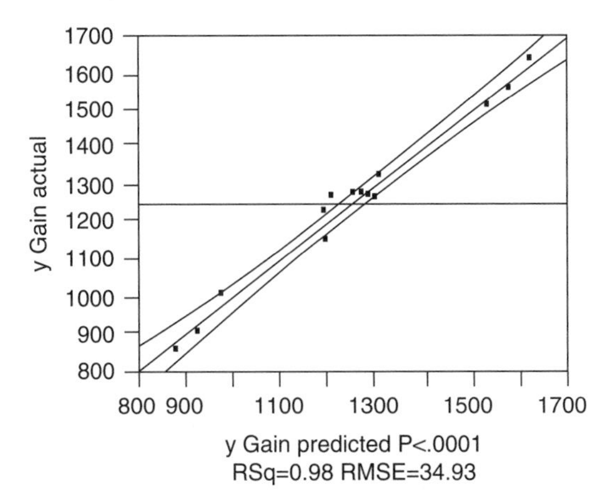

Summary of Fit

RSquare	0.97983
RSquare Adj	0.976162
Root Mean Square Error	34.92995
Mean of Response	1249.643
Observations (or Sum Wgts)	14

Analysis of Variance

Source	DF	Sum of Squares	Mean Square	F Ratio
Model	2	651966.10	325983	267.1770
Error	11	13421.12	1220	Prob > F
C. Total	13	665387.21		<.0001

Parameter Estimates

Term	Estimate	Std Error	t Ratio	Prob>|t|
Intercept	-520.0767	192.1071	-2.71	0.0204
x1 Drive-in time	10.781158	0.47432	22.73	<.0001
x2 Dose	-152.1489	36.67544	-4.15	0.0016

(Continued)

estimators (BLUEs) of the model parameters. However, other inference procedures such as hypothesis tests and confidence intervals assume that the response is normally distributed. If the form of the response distribution is known, another method of parameter estimation, the method of **maximum likelihood**, can be used.

Table 2.6 *Continued*

Residual by Predicted Plot

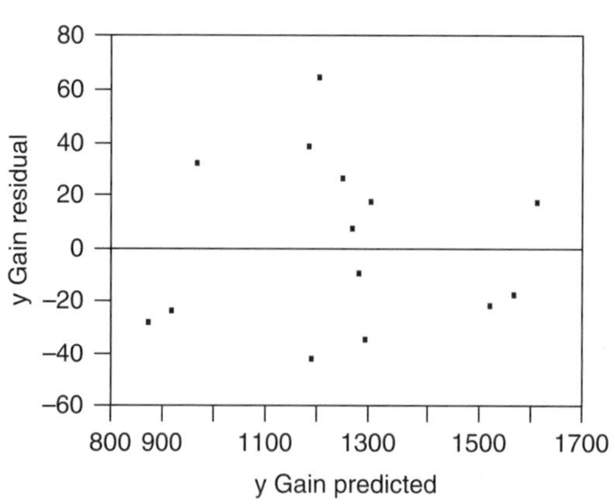

Press

Press	Press RMSE
22225.010037	39.8434526

Prediction Profiler

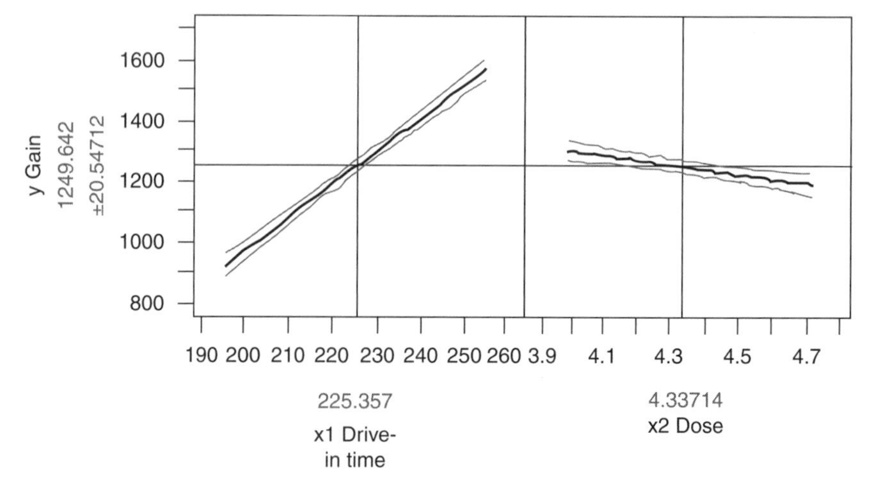

Sequential (Type 1) Tests

Source	Nparm	DF	Seq SS	F Ratio	Prob > F
x1 Drive-in time	1	1	630967.86	517.1438	<.0001
x2 Dose	1	1	20998.23	17.2102	0.0016

Consider the linear regression model

$$\mathbf{y} = \mathbf{X}\boldsymbol{\beta} + \boldsymbol{\varepsilon}$$

Suppose that the errors in this model are normally and independently distributed with mean zero and constant variance σ^2 (NID(0, σ^2)). Then the observations $\mathbf{Y}$ are normally and independently distributed with mean $\mathbf{X\beta}$ and variance $\sigma\mathbf{I}$. The **likelihood function** is found from the joint probability distribution of the observations. If we consider this joint distribution with the observations given and the parameters as unknown we have the likelihood function. For the linear regression model the likelihood function is

$$\mathscr{L}(\mathbf{y}|\boldsymbol{\beta},\sigma^2) = \frac{1}{(2\pi\sigma^2)^{n/2}} e^{(-1/2\sigma^2)(\mathbf{y}-\mathbf{X\beta})'(\mathbf{y}-\mathbf{X\beta})} \quad (2.38)$$

The maximum likelihood estimators are the values of the parameters $\boldsymbol{\beta}$ and σ^2 that maximize the likelihood function.

Maximizing the likelihood function $\mathscr{L}$ is equivalent to maximizing the log-likelihood, $\ln(\mathscr{L})$. The log-likelihood is

$$\ln[\mathscr{L}(\mathbf{y}|\boldsymbol{\beta},\sigma^2)] = -\frac{n}{2}\ln(2\pi) - \frac{n}{2}\ln(\sigma^2) - \frac{1}{2\sigma^2}(\mathbf{y}-\mathbf{X\beta})'(\mathbf{y}-\mathbf{X\beta}) \quad (2.39)$$

The derivative of the log-likelihood is called the **score function**. Taking the partial derivatives of the log-likelihood with respect to the parameters $\boldsymbol{\beta}$ and equating to zero yields

$$\frac{\partial \ln(\mathscr{L})}{\partial \boldsymbol{\beta}} = -\frac{1}{2\sigma^2}(-2\mathbf{b}'\mathbf{X}'\mathbf{y} + \mathbf{y}'\mathbf{X}'\mathbf{X}\mathbf{b}) = \mathbf{0}$$

or

$$\frac{1}{\sigma^2}\mathbf{X}'(\mathbf{y} - \mathbf{Xb}) = 0 \quad (2.40)$$

The $p \times p$ system of Equations (2.40) are called the **maximum likelihood score equations**. The solution to the score equations is the **maximum likelihood estimator** (MLE)

$$\mathbf{b} = (\mathbf{X}'\mathbf{X})^{-1}\mathbf{X}'\mathbf{y} \quad (2.41)$$

The maximum likelihood estimator of σ^2 is the solution to

$$\frac{\partial \ln(\mathscr{L})}{\partial \sigma^2} = -\frac{n}{2\tilde{\sigma}^2} - \frac{1}{2\tilde{\sigma}^4}(\mathbf{y} - \mathbf{Xb})'(\mathbf{y} - \mathbf{Xb}) = 0$$

which is

$$\tilde{\sigma}^2 = \frac{1}{n}(\mathbf{y} - \mathbf{Xb})'(\mathbf{y} - \mathbf{Xb}) \quad (2.42)$$

Notice that the MLE $\mathbf{b}$ for the normal-theory linear regression model is identical to the ordinary least squares estimator. It is clear from examining the likelihood function in Equation (2.38) or the log-likelihood, that maximizing the likelihood function involves minimizing the quantity in the exponent, which is the least squares function $S(\boldsymbol{\beta})$. Furthermore, the maximum likelihood score equations are identical to the least squares normal equations, since $\mathbf{X}'(\mathbf{y}-\mathbf{X}\mathbf{b}) = \mathbf{X}'\mathbf{y}-\mathbf{X}'\mathbf{X}\mathbf{b} = \mathbf{0}$ or $\mathbf{X}'\mathbf{X}\mathbf{b} = \mathbf{X}'\mathbf{y}$.

2.3.2 Properties of the Maximum Likelihood Estimators

In general, maximum likelihood estimators have better **statistical properties** than estimators obtained by least squares. This occurs at the expense of additional assumptions, however, as the MLE requires that the observations be normally distributed while the least squares procedure does not. Maximum likelihood estimators are either **unbiased** or **asymptotically unbiased,** that is, unbiased as n becomes large. For example, the MLE $\mathbf{b}$ is an unbiased estimator, and since

$$\tilde{\sigma}^2 = \frac{n-p}{n}\hat{\sigma}^2$$

the MLE of σ is unbiased as n becomes large. In the linear regression model, the MLEs have minimum variance when compared to *all* other unbiased estimators.

The second derivative of the log-likelihood function is called the **Hessian**, and for the linear regression model the Hessian is

$$\frac{\partial}{\partial \boldsymbol{\beta}}\left(\frac{\partial \ln(\mathscr{L})}{\partial \boldsymbol{\beta}}\right) = \frac{\partial^2 \ln(\mathscr{L})}{\partial \boldsymbol{\beta}^2} = -\frac{\mathbf{X}'\mathbf{X}}{\sigma^2}$$

The negative of the Hessian, or $\mathbf{X}'\mathbf{X}/\sigma^2$, is sometimes called the **information matrix**. Another way to find the information matrix is as the variance of the score. That is, the variance of the left-hand side of (2.40) is

$$\text{Var}\left[\frac{1}{\sigma^2}\mathbf{X}'(\mathbf{y}-\mathbf{X}\mathbf{b})\right] = \frac{1}{\sigma^2}\mathbf{X}'\mathbf{X}$$

The inverse of the negative of the Hessian is the covariance matrix of the MLEs. In the case of the linear regression model, this is just the familiar

$$\text{Var}(\mathbf{b}) = \sigma^2(\mathbf{X}'\mathbf{X})^{-1}$$

Strictly speaking, these are large-sample properties of the MLE, but they happen to hold exactly in the case of the linear regression model.

The MLE is also a **consistent** estimator (consistency is another large-sample property indicating that the estimator differs from the true value of the parameter by a very small amount as n becomes large). Furthermore, the MLEs form a set of **sufficient statistics**, which implies that the estimators contain all of the information about the parameters contained in the original sample of size n.

2.4 MODEL ADEQUACY CHECKING

It is always necessary to (1) examine a fitted regression model to ensure that it provides an adequate approximation to the true system and (2) verify that none of the least squares regression assumptions are violated. The model for prediction or estimation will give poor or misleading results unless the model is an adequate fit. In this section we present several techniques for checking model adequacy.

2.4.1 Residual Analysis

The residuals from the least squares fit, defined by $e_i = y_i - \hat{y}_i$, $i = 1, 2, \ldots, n$, play an important role in judging model adequacy. The residuals from Example 2.1 are shown in column 3 of Table 2.3.

A check of the normality assumption may be made by constructing a normal probability plot of the residuals, as in Figure 2.3. If the residuals plot approximately along a straight line, then the normality assumption is satisfied. Figure 2.3 reveals no apparent problem with normality. The straight line in this normal probability plot was determined by eye, concentrating on the central portion of the data. When this plot indicates problems with the normality assumption, we often **transform** the response variable as a remedial

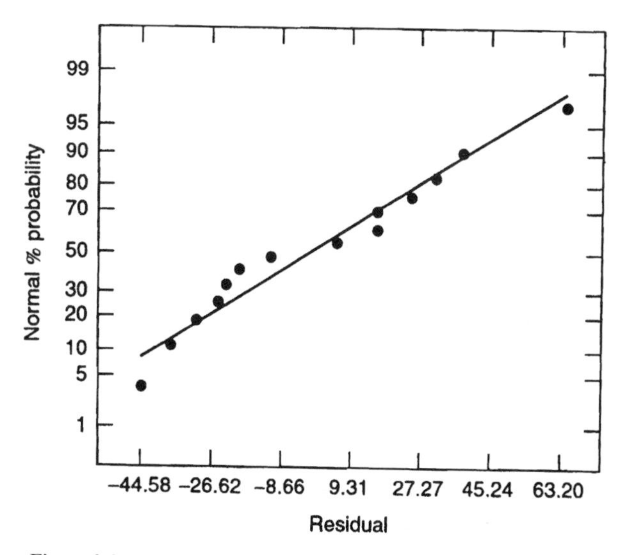

Figure 2.3 Normal probability plot of residuals, Example 2.1.

measure. For more details, see Montgomery, Peck, and Vining (2006) and Myers (1990).

Figure 2.4 presents a plot of residuals e_i versus the predicted response $\hat{y}_i$. The general impression is that the residuals scatter randomly on the display, suggesting that the variance of the original observations is constant for all values of y. If the variance of the response depends on the mean level of y, then this plot will often exhibit a funnel-shaped pattern. This is also suggestive of the need for transformation of the response variable y.

It is also useful to plot the residuals in time or run order and versus each of the individual regressors. Nonrandom patterns on these plots would indicate model inadequacy. In some of these cases, transformations may stabilize the situation. See Montgomery, Peck, and Vining (2006) and Myers (1990) for more details.

Example 2.9. The Worsted Yarn Data. The data in Table 2.7 (taken from Box and Draper, 1987) show the number of cycles to failure of worsted yarn (y) and three factors defined as follows:

$$\text{Length of test specimen (mm):} \quad x_1 = \frac{\text{length} - 300}{50}$$

$$\text{Amplitude of load cycle (mm):} \quad x_2 = \text{amplitude} - 9$$

$$\text{Load(grams):} \quad x_3 = \frac{\text{load} - 45}{5}$$

These factors form a 3^3 factorial experiment. This experiment will support a complete second-order polynomial. The least squares fit is

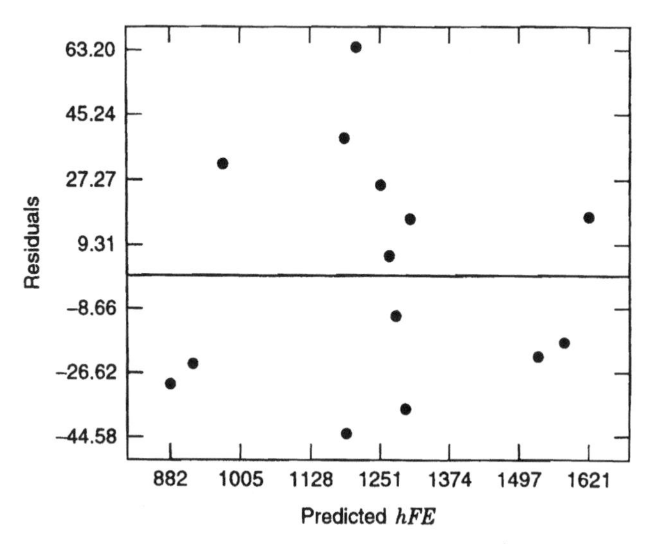

Figure 2.4 Plot of residuals versus predicted response $\hat{y}_i$, Example 2.1.

Table 2.7 The Worsted Yarn Data

Run Number	Length, x_1	Amplitude, x_2	Load, x_3	Cycles to Failure, y
1	−1	−1	−1	674
2	0	−1	−1	1414
3	1	−1	−1	3636
4	−1	0	−1	338
5	0	0	−1	1022
6	1	0	−1	1368
7	−1	1	−1	170
8	0	1	−1	442
9	1	1	−1	1140
10	−1	−1	0	370
11	0	−1	0	1198
12	1	−1	0	3184
13	−1	0	0	266
14	0	0	0	620
15	1	0	0	1070
16	−1	1	0	118
17	0	1	0	332
18	1	1	0	884
19	−1	−1	1	292
20	0	−1	1	634
21	1	−1	1	2000
22	−1	0	1	210
23	0	0	1	438
24	1	0	1	566
25	−1	1	1	90
26	0	1	1	220
27	1	1	1	360

$$\hat{y} = 550.7 + 660x_1 - 535.9x_2 - 310.8x_3 + 238.7x_1^2 + 275.7x_2^2 - 48.3x_3^2 - 456.5x_1x_2 - 235.7x_1x_3 + 143.0x_2x_3$$

The R^2 value is 0.975. An analysis of variance is given in Table 2.8. The fit appears to be reasonable and both the first- and second-order terms appear to be necessary.

Figure 2.5 is a plot of residuals versus the predicted cycles to failure $\hat{y}$ for this model. There is an indication of an outward-opening funnel in this plot, implying possible inequality of variance.

When a natural log transformation is used for y, we obtain the following model:

$$\ln \hat{y} = 6.33 + 0.82x_1 - 0.63x_2 - 0.38x_3$$

or

$$\hat{y} = e^{6.33+0.82x_1-0.63x_2-0.38x_3}$$

Table 2.8 Analysis of Variance for the Quadratic Model for the Worsted Yarn Data

Source of Variability	Sum of Squares ($\times 10^{-3}$)	Degrees of Freedom	Mean Square ($\times 10^{-3}$)	F_0
First-order terms	14,748.5	3	4,916.2	70
Added second-order terms	4,224.3	6	704.1	9.5
Residual	1,256.6	17	73.9	
Total	20,229.4	26		

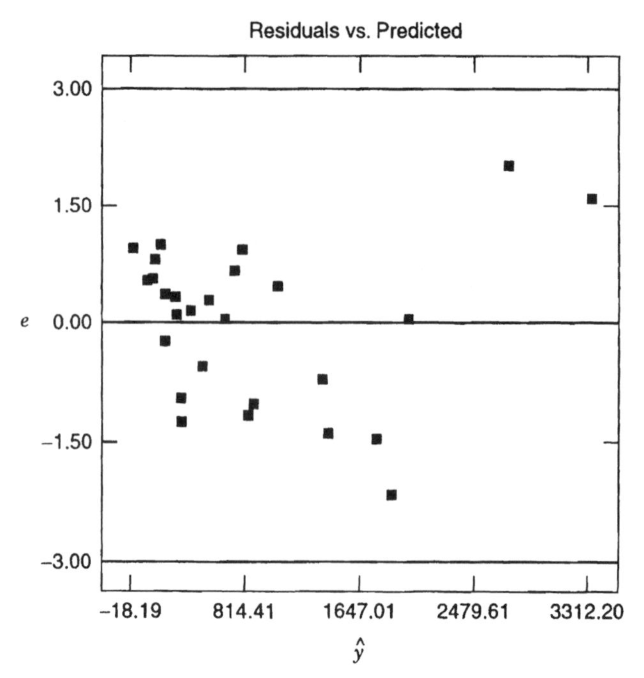

Figure 2.5 Plot of residuals versus predicted cycles of failure for the worsted yarn data, quadratic mode.

This model has $R^2 = 0.963$ and has only three model terms (apart from the intercept). None of the second-order terms are significant. Here, as in most modeling exercises, simplicity is of vital importance. The elimination of the quadratic terms and interaction terms with the change in response metric not only allows a better fit than the second-order model with the natural metric, but the impact of the design variables x_1, x_2, and x_3 on the response is clear.

Figure 2.6 is a plot of residuals versus the predicted response for the log model. There is still some indication of inequality of variance, but the log model, overall, is an improvement on the original quadratic fit. □

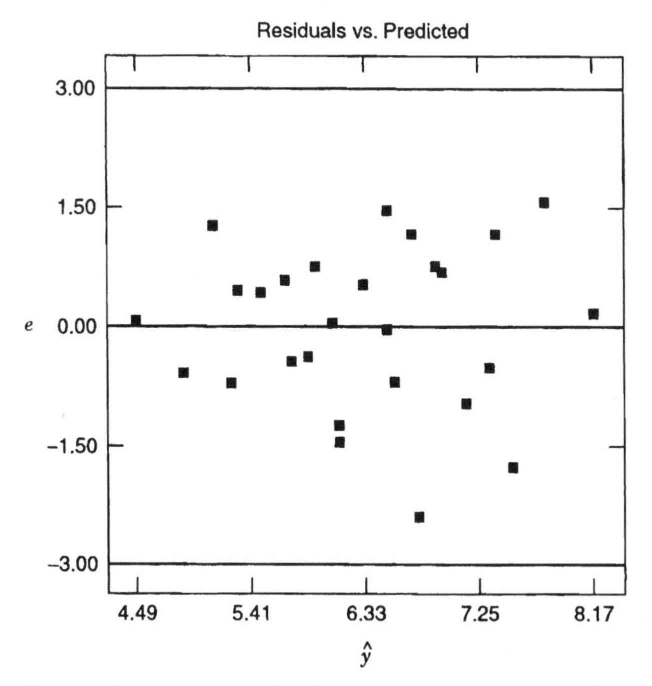

Figure 2.6 Plot of residuals versus predicted response for the worsted yarn data, log model.

2.4.2 Transformation of the Response Variable Using the Box–Cox Method

In the previous section we discussed the problem of nonconstant variance in the response variable y in linear regression and noted that this is a departure from the standard least squares assumptions. This inequality of variance problem occurs relatively often in practice, frequently in conjunction with a nonnormal response variable. Examples would include a count of defects or particles, proportion data such as yield or fraction defective, or a response variable that follows some skewed distribution (one tail of the response distribution is longer than the other). We introduced and illustrated **transformation of the response variable** as an appropriate method for stabilizing the variance of the response. In our example we selected a log transformation empirically, by noting that it greatly improved the appearance of the residual plots.

Generally, transformations are used for three purposes: stabilizing response variance, making the distribution of the response variable closer to the normal distribution, and improving the fit of the model to the data. This last objective could include model simplification, say, by eliminating interaction or higher-order polynomial terms. Sometimes a transformation will be reasonably effective in simultaneously accomplishing more than one of these objectives.

We often find that the **power family** of transformations $y^* = y^\lambda$ is very useful, where λ is the parameter of the transformation to be determined (e.g., $\lambda = \frac{1}{2}$

means use the square root of the original response). Box and Cox (1964) have shown how the transformation parameter λ may be estimated simultaneously with the other model parameters (overall mean and treatment effects). The theory underlying their method uses the method of maximum likelihood. The actual computational procedure consists of performing, for various values of λ, a standard analysis of variance on

$$y^{(\lambda)} = \begin{cases} \dfrac{y^{\lambda} - 1}{\lambda \dot{y}^{\lambda-1}}, & \lambda \neq 0 \\ \dot{y} \ln y, & \lambda = 0 \end{cases} \tag{2.43}$$

where $\dot{y} = \ln^{-1}[(1/n)\Sigma \ln y]$ is the geometric mean of the observations. The maximum likelihood estimate of λ is the value for which the error sum of squares, say, $SS_{\text{res}}(\lambda)$, is a minimum. This value of λ is usually found by plotting a graph of $SS_{\text{res}}(\lambda)$ versus λ and then reading the value of λ that minimizes in each case $SS_{\text{res}}(\lambda)$ from the graph. Usually between 10 and 20 values of λ are sufficient for estimation of the optimum value. A second iteration using a finer mesh of values could be performed if a more accurate estimate of λ is necessary.

Notice that we *cannot* select the value of λ by *directly* comparing residual sums of squares from analyses of variance on y^{λ} because for each value of λ the residual sum of squares is measured on a different scale. Furthermore, a problem arises in y when $\lambda = 0$; namely, as λ approaches zero, y^{λ} approaches unity. That is, when $\lambda = 0$, all the response values are a constant. The component $(y^{\lambda}-1)/\lambda$ of Equation (2.43) alleviates this problem because as λ tends to zero, $(y^{\lambda}-1)/\lambda$ goes to a limit of $\ln y$. The divisor component $\dot{y}^{\lambda-1}$ in Equation (2.43) rescales the responses so that the residual sums of squares are directly comparable.

In applying the Box–Cox method, we recommend using simple choices for λ because the practical difference between $\lambda = 0.5$ and $\lambda = 0.58$ is likely to be small, but the square root transformation ($\lambda = 0.5$) is much easier to interpret. Obviously, values of λ close to unity would suggest that no transformation is necessary.

Once a value of λ is selected by the Box–Cox method, the experimenter can analyze the data using y^{λ} as the response, unless of course $\lambda = 0$, in which he/she can use $\ln y$. It is perfectly acceptable to use $y^{(\lambda)}$ as the actual response, although the model parameter estimates will have a scale difference and origin shift in comparison to the results obtained using y^{λ} (or $\ln y$).

An approximate $100(1-\alpha)\%$ confidence interval for λ can be found by computing

$$SS^{*} = SS_{\text{res}}(\lambda)\left(1 + \frac{t^2_{\alpha/2,v}}{v}\right) \tag{2.44}$$

where v is the number of degrees of freedom, and plotting a line parallel to the λ axis at height SS^{*} on the graph of $SS_{\text{res}}(\lambda)$ versus λ. Then by locating the points on the λ axis where SS^{*} cuts the curve $SS_{\text{res}}(\lambda)$, we can read confidence limits on

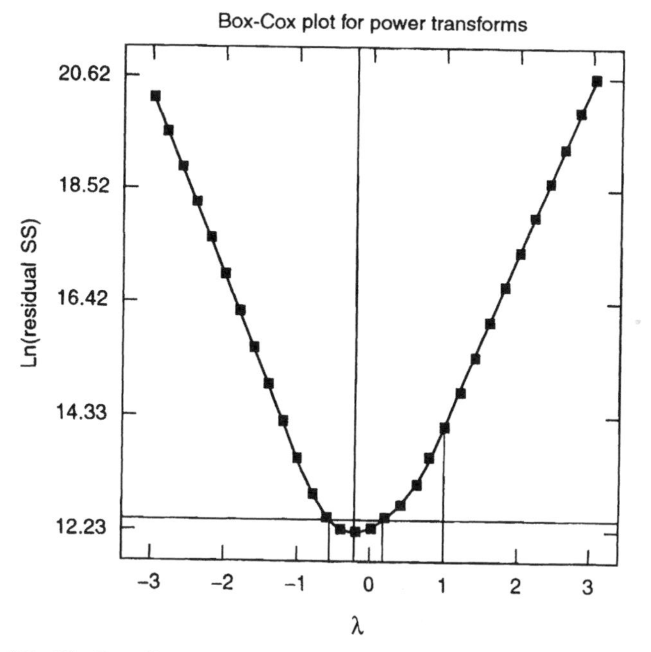

Figure 2.7 The Box–Cox procedure applied to the worsted yarn data in Table 2.7.

λ directly from the graph. If this confidence interval includes the value $\lambda = 1$, this implies (as noted above) that the data do not support the need for transformation.

Several software packages have implemented the Box–Cox procedure. Figure 2.7 shows the output graphics from Design-Expert when the Box–Cox procedure is applied to the worsted yarn data in Table 2.7. The optimum value of λ is −0.24, and the 95% confidence interval for λ contains zero, so the use of a log transformation is indicated.

2.4.3 Scaling Residuals

Standardized and Studentized Residuals

Many analysts prefer to work with **scaled residuals**, in contrast to the ordinary least squares residuals. These scaled residuals often convey more information than do the ordinary residuals.

One type of scaled residual is the **standardized residual**:

$$d_i = \frac{e_i}{\hat{\sigma}}, \quad i = 1, 2, \ldots, n \tag{2.45}$$

where we generally use $\hat{\sigma} = \sqrt{MS_{\text{res}}}$ in the computation. These standardized residuals have mean zero and approximately unit variance; consequently, they are useful in looking for **outliers**. Most of the standardized residuals should lie in the interval $-3 \leq d_i \leq 3$, and any observation with a standardized residual outside this interval is potentially unusual with respect to its observed response. These outliers should be examined carefully, because they may represent something as simple as a data recording error or something of more serious concern, such as a region of the regressor variable space where the fitted model is a poor approximation to the true response surface.

The standardizing process in Equation (2.45) scales the residuals by dividing them by their average standard deviation. In some data sets, residuals may have standard deviations that differ greatly. We now present a scaling that takes this into account.

The vector of fitted values $\hat{\mathbf{y}}$, corresponding to the observed values $\mathbf{y}$ is

$$\begin{aligned} \mathbf{y} &= \mathbf{Xb} \\ &= \mathbf{X}(\mathbf{X}'\mathbf{X})^{-1}\mathbf{X}'\mathbf{y} \\ &= \mathbf{Hy} \end{aligned} \tag{2.46}$$

The $n \times n$ matrix $\mathbf{H} = \mathbf{X}(\mathbf{X}'\mathbf{X})^{-1}\mathbf{X}'$ is usually called the hat matrix because it maps the vector of observed values into a vector of fitted values. The hat matrix and its properties play a central role in regression analysis.

The residuals from the fitted model may conveniently be written in matrix notation as

$$\mathbf{e} = \mathbf{y} - \hat{\mathbf{y}} \tag{2.47}$$

There are several other ways to express the vector of residuals $\mathbf{e}$ that will prove useful, including

$$\mathbf{e} = \mathbf{y} - \mathbf{Xb} \tag{2.48}$$

$$\begin{aligned} &= \mathbf{y} - \mathbf{Hy} \\ &= (\mathbf{I} - \mathbf{H})\mathbf{y} \end{aligned} \tag{2.49}$$

The hat matrix has several useful properties. It is symmetric ($\mathbf{H}' = \mathbf{H}$) and idempotent ($\mathbf{HH} = \mathbf{H}$). Similarly, the matrix $\mathbf{I}$–$\mathbf{H}$ is symmetric and idempotent.

The covariance matrix of the residuals is

$$\begin{aligned} \mathrm{Var}(e) &= \mathrm{Var}[(\mathbf{I} - \mathbf{H})\mathbf{y}] \\ &= (\mathbf{I} - \mathbf{H})\mathrm{Var}(\mathbf{y})(\mathbf{I} - \mathbf{H})' \\ &= \sigma^2(\mathbf{I} - \mathbf{H}) \end{aligned} \tag{2.50}$$

because $\mathrm{Var}(\mathbf{y}) = \sigma^2\mathbf{I}$ and $\mathbf{I}$–$\mathbf{H}$ is symmetric and idempotent. The matrix $\mathbf{I}$–$\mathbf{H}$ is generally not diagonal, so the residuals have different variances and they are correlated.

The variance of the ith residual is

$$\text{Var}(e_i) = \sigma^2(1 - h_{ii}) \tag{2.51}$$

where h_{ii} is the ith diagonal element of $\mathbf{H}$. Because $0 \leq h_{ii} \leq 1$, using the residual mean square MS_{res} to estimate the variance of the residuals actually overestimates Var(e_i). Furthermore, because h_{ii} is a measure of the location of the ith point in x-space, the variance of e_i depends on where the point $\mathbf{x}_i$ lies. Generally, residuals near the center of the x-space have larger variance than do residuals at more remote locations. Violations of model assumptions are more likely at remote points, and these violations may be hard to detect from inspection of e_i, (or d_i) because their residuals will usually be smaller.

We recommend taking this inequality of variance into account when scaling the residuals. We suggest plotting the **studentized residuals**:

$$r_i = \frac{e_i}{\sqrt{\hat{\sigma}^2(1 - h_{ii})}}, \quad i = 1, 2, \ldots, n \tag{2.52}$$

with $\hat{\sigma}^2 = MS_{\text{res}}$ instead of e_i (or d_i). The studentized residuals have constant variance Var(r_i) = 1 regardless of the location of $\mathbf{x}_i$, when the form of the model is correct. In many situations the variance of the residuals stabilizes, particularly for large data sets. In these cases there may be little difference between the standardized and studentized residuals. Thus standardized and studentized residuals often convey equivalent information. However, because any point with a large residual and a large h_{ii} is potentially highly influential on the least squares fit, examination of the studentized residuals is generally recommended.

PRESS Residuals and the PRESS Statistic

The prediction error sum of squares (PRESS) proposed by Allen (1971, 1974) provides a useful residual scaling. To calculate PRESS, select an observation—for example, y_i. Fit the regression model to the remaining n–1 observations and use this equation to predict the withheld observation y_i. Denoting this predicted value $\hat{y}_{(i)}$, we may find the **prediction error** for point i as $e_{(i)} = y_i - \hat{y}_{(i)}$. The prediction error is often called the ith **PRESS residual**. This procedure is repeated for each observation i = 1, 2,…, n, producing a set of n PRESS residuals $e_{(1)}, e_{(2)}, \ldots, e_{(n)}$.

The **PRESS statistic** is defined as the sum of squares of the n PRESS residuals:

$$\text{PRESS} = \sum_{i=1}^{n} e_{(i)}^2 = \sum_{i=1}^{n} [y_i - \hat{y}_{(i)}]^2 \tag{2.53}$$

Thus PRESS uses each possible subset of n–1 observations as an estimation data set, and every observation in turn is used to form a prediction data set.

It would initially seem that calculating PRESS requires fitting n different regressions. However, it is possible to calculate PRESS from the results of a single least squares fit to all n observations. It turns out that the ith PRESS residual is

$$e_{(i)} = \frac{e_i}{1 - h_{ii}} \tag{2.54}$$

Thus, because PRESS is just the sum of the squares of the PRESS residuals, a simple computing formula is

$$\text{PRESS} = \sum_{i=1}^{n} \left(\frac{e_i}{1 - h_{ii}} \right)^2 \tag{2.55}$$

From Equation (2.54) it is easy to see that the PRESS residual is just the ordinary residual weighted according to the diagonal elements of the hat matrix h_{ii}. Data points for which h_{ii} are large will have large PRESS residuals. These observations will generally be **high-influence** points. Generally, a large difference between the ordinary residual and the PRESS residual will indicate a point where the model fits the data well, but a model built without that point predicts poorly. In the next section we discuss some other measures of influence.

The variance of the ith PRESS residual is

$$\begin{aligned}
\text{Var}[e_{(i)}] &= \text{Var}\left[\frac{e_i}{1 - h_{ii}}\right] \\
&= \frac{1}{(1 - h_{ii})^2}[\sigma^2(1 - h_{ii})] \\
&= \frac{\sigma^2}{1 - h_{ii}}
\end{aligned}$$

so that the standardized PRESS residual is

$$\begin{aligned}
\frac{e_{(i)}}{\sqrt{\text{Var}[e_{(i)}]}} &= \frac{e_i/(1 - h_{ii})}{\sqrt{\sigma^2/(1 - h_{ii})}} \\
&= \frac{e_i}{\sqrt{\sigma^2(1 - h_{ii})}}
\end{aligned}$$

which if we use MS_{res} to estimate σ^2 is just the studentized residual discussed previously.

Finally, we note that PRESS can be used to compute an approximate R^2 for prediction, say,

$$R^2_{\text{prediction}} = 1 - \frac{\text{PRESS}}{SS_{\text{T}}} \tag{2.56}$$

This statistic gives some indication of the predictive capability of the regression model.

For the transistor gain model we can compute the PRESS residuals using the ordinary residuals and the values of h_{ii} found in Table 2.3. The resulting value is PRESS = 22,225.0. This was also computed by JMP and reported in Table 2.6. Then

$$\begin{aligned} R^2_{\text{prediction}} &= 1 - \frac{\text{PRESS}}{SS_{\text{T}}} \\ &= 1 - \frac{22,225.0}{665,387.2} \\ &= 0.9666 \end{aligned}$$

Therefore we could expect this model to explain about 96.66% of the variability in predicting new observations, as compared to the approximately 97.98% of the variability in the original data explained by the least squares fit. The overall predictive capability of the model based on this criterion seems very satisfactory.

R-Student

The studentized residual r_i discussed above is often considered an **outlier diagnostic**. It is customary to use MS_{res} as an estimate of σ^2 in computing r_i. This is referred to as internal scaling of the residual because MS_{res} is an internally generated estimate of σ^2 obtained from fitting the model to all n observations. Another approach would be to use an estimate of σ^2 based on a data set with the ith observation removed. Denote the estimate of σ^2 so obtained by $S^2_{(i)}$. We can show that

$$S^2_{(i)} = \frac{(n-p)MS_{\text{res}} - e_i^2/(1-h_{ii})}{n-p-1} \tag{2.57}$$

The estimate of σ^2 in Equation (2.57) is used instead of MS_{res} to produce an externally studentized residual, usually called R-student, given by

$$t_i = \frac{e_i}{\sqrt{S^2_{(i)}(1-h_{ii})}}, \quad i = 1, 2, \ldots, n \tag{2.58}$$

In many situations t_i will differ little from the studentized residual r_i. However, if the ith observation is influential, then $S^2_{(i)}$ can differ significantly from MS_{res}, and thus the R-student will be more sensitive to this point. Furthermore, under the standard assumptions t_i has t_{n-p-1} distribution. Thus

R-student offers a more formal procedure for outlier detection via hypothesis testing. One could use a simultaneous inference procedure called the **Bonferroni approach** and compare all n values of $|t_i|$ to $t_{\alpha/(2n),-p-1}$, to provide guidance regarding outliers. However, it is our view that a formal approach is usually not necessary and that only relatively crude cutoff values need be considered. In general, a diagnostic view as opposed to a strict statistical hypothesis-testing view is best. Furthermore, detection of outliers needs to be considered simultaneously with detection of influential observations.

Example 2.10. The Transistor Gain Data. Table 2.3 presents the studentized residuals r_i and the R-student values t_i defined in Equations (2.52) and (2.58) for the transistor gain data. None of these values are large enough to cause any concern regarding outliers.

Figure 2.8 is a normal probability plot of the studentized residuals. It conveys exactly the same information as the normal probability plot of the ordinary residuals e_i in Figure 2.3. This is because most of the h_{ii} values are relatively similar and there are no unusually large residuals. In some applications, however, the h_{ii} can differ considerably, and in those cases plotting the studentized residuals is the best approach. □

2.4.4 Influence Diagnostics

We occasionally find that a small subset of the data exerts a disproportionate influence on the fitted regression model. That is, parameter estimates or predictions may depend more on the influential subset than on the majority of the data. We would like to locate these influential points and assess their

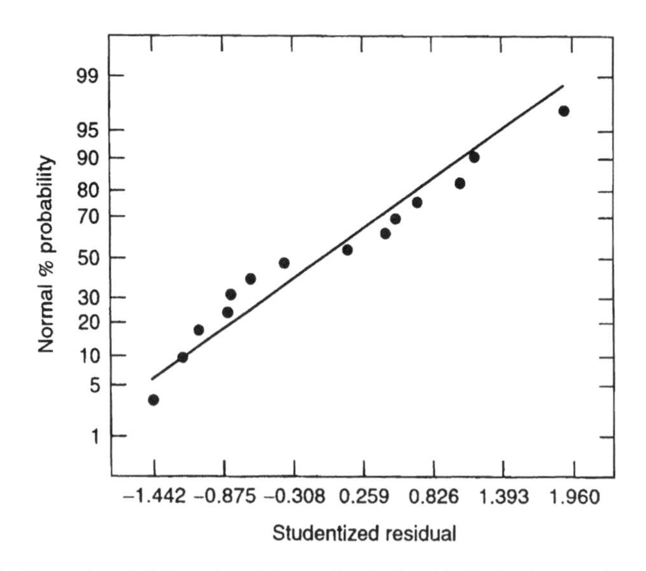

Figure 2.8 Normal probability plot of the studentized residuals for the transistor gain data.

impact on the model. If these influential points are bad values, then they should be eliminated. On the other hand, there may be nothing wrong with these points, but if they control key model properties, we would like to know it because it could affect the use of the model. In this section we describe and illustrate several useful measure of influence.

Leverage Points

The disposition of points in x-space is important in determining model properties. In particular, remote observations potentially have disproportionate leverage on the parameter estimates, predicted values, and the usual summary statistics.

The hat matrix $\mathbf{H} = \mathbf{X}(\mathbf{X}'\mathbf{X})^{-1}\mathbf{X}'$ is very useful in identifying influential observations. As noted earlier, $\mathbf{H}$ determines the variances and covariances of $\hat{\mathbf{y}}$ and $\mathbf{e}$ because $\text{Var}(\hat{\mathbf{y}}) = \sigma^2\mathbf{H}$ and $\text{Var}(\mathbf{e}) = \sigma^2(\mathbf{I}\text{–}\mathbf{H})$. The elements h_{ij} of $\mathbf{H}$ may be interpreted as the amount of leverage exerted by y_j on $\hat{y}_i$. Thus inspection of the elements of $\mathbf{H}$ can reveal points that are potentially influential by virtue of their location in x-space. Attention is usually focused on the diagonal elements h_{ii}. Because $\sum_{i=1}^{n} h_{ii} = \text{rank}(\mathbf{H}) = \text{rank}(\mathbf{X}) = p$, the average size of the diagonal element of the $\mathbf{H}$ matrix is p/n. As a rough guideline, then, if a diagonal element h_{ii} is greater than $2p/n$, observation i is a high-leverage point. To apply this to the transistor gain data in Example 2.1, note that $2p/n = 2(3)/14 = 0.43$.

Table 2.3 gives the hat diagonals h_{ii} for the first-order model; and because none of the h_{ii} exceed 0.43, we would conclude that there are no leverage points in these data. For more information about the hat matrix, see Myers (1990) and Montgomery, Peck, and Vining (2006).

Influence on Regression Coefficients

The hat diagonals identify points that are potentially influential due to their location in x-space. It is desirable to consider both the location of the point and the response variable in measuring influence. Cook (1977, 1979) has suggested using a measure of the squared distance between the least squares estimate based on all n points $\mathbf{b}$, and the estimate obtained by deleting the ith point, say, $\mathbf{b}_{(i)}$. This distance measure can be expressed in a general form as

$$D_i(\mathbf{M}, c) = \frac{(\mathbf{b}_{(i)} - \mathbf{b})'\mathbf{M}(\mathbf{b}_{(i)} - \mathbf{b})}{c}, \quad i = 1, 2, \ldots, n \tag{2.59}$$

The usual choices of $\mathbf{M}$ and c are $\mathbf{M} = \mathbf{X}'\mathbf{X}$ and $c = pMS_{\text{res}}$, so that Equation (2.59) becomes

$$D_i(\mathbf{M}, c) \equiv D_i = \frac{(\mathbf{b}_{(i)} - \mathbf{b})'\mathbf{X}'\mathbf{X}(\mathbf{b}_{(i)} - \mathbf{b})}{pMS_{\text{res}}}, \quad i = 1, 2, \ldots, n \tag{2.60}$$

Points with large values of D_i have considerable influence on the least squares estimates $\mathbf{b}$. The magnitude of D_i may be assessed by comparing it to $F_{\alpha,p,n-p}$. If $D_i \simeq F_{0.5,p,n-p}$, then deleting point i would move $\mathbf{b}$ to the boundary of a 50%

confidence region for $\boldsymbol{\beta}$ based on the complete data set. This is a large displacement and indicates that the least squares estimate is sensitive to the ith data point. Because $F_{0.5,p,n-p} \simeq 1$, we usually consider points for which $D_i > 1$ to be influential. Practical experience has shown the cutoff value of 1 works well in identifying influential points.

The D_i statistic may be rewritten as

$$D_i = \frac{r_1^2}{p} \frac{\text{Var}[\hat{y}(\mathbf{x}_i)]}{\text{Var}(e_i)} = \frac{r_i^2}{p} \frac{h_{ii}}{(1 - h_{ii})}, \quad i = 1, 2, \ldots, n \tag{2.61}$$

Thus we see that apart from the constant p, D_i is the product of the square of the ith studentized residual and $h_{ii}/(1-h_{ii})$. This ratio can be shown to be the distance from the vector $\mathbf{x}_i$ to the centroid of the remaining data. Thus D_i is made up of a component that reflects how well the model fits the ith observation y_i and a component that measures how far that point is from the rest of the data. Either component (or both) may contribute to a larger value of D_i.

Table 2.3 presents the values of D_i for the first-order model fit to the transistor gain data in Example 2.1. None of these values of D_i exceed 1, so there is no strong evidence of influential observations in these data.

2.5 USING R TO PERFORM LINEAR REGRESSION ANALYSIS

R is a popular statistical software package, primarily because it is freely available at www.r-project.org. An easier but limited version of R is R Commander. According to the project's webpage:

> The R Foundation is a not for profit organization working in the public interest. It has been founded by the members of the R Development Core Team in order to
>
> - Provide support for the R project and other innovations in statistical computing. We believe that R has become a mature and valuable tool and we would like to ensure its continued development and the development of future innovations in software for statistical and computational research.
> - Provide a reference point for individuals, institutions or commercial enterprises that want to support or interact with the R development community.
> - Hold and administer the copyright of R software and documentation.
>
> R is an official part of the Free Software Foundation's GNU project, and the R Foundation has similar goals to other open source software foundations like the Apache Foundation or the GNOME Foundation.
>
> Among the goals of the R Foundation are the support of continued development of R, the exploration of new methodology, teaching and training of statistical computing and the organization of meetings and conferences with a statistical computing orientation.

R is a very sophisticated statistical software environment, even though it is freely available. The contributors include many of the top researchers in statistical computing. In many ways, it reflects the very latest statistical methodologies.

R itself is a high-level programming language. Most of its commands are prewritten functions. It does have the ability to run loops and call other routines, for example, in C. Since it is primarily a programming language, it often presents challenges to novice users. The purpose of this section is to introduce the reader as to how to use R to analyze multiple linear regression data sets.

The first step is to create the data set. One method inputs the data into a text file using spaces for delimiters. Each row of the data file is a record. The top row should give the names for each variable. All other rows are the actual data records. For example, consider the transistor gain data from Example 2.1 given in Table 2.2. Let transistor.txt be the name of the data file. The first row of the text file gives the variable names:

```
x1  x2  y
```

The next row is the first data record, with spaces delimiting each data item:

```
195 4.00 1004
```

The R code to read the data into the package is

```
trans <- read.table("transistor.txt",header=TRUE, sep=" ")
```

The object trans is the R data set, and "transistor.txt" is the original data file. The phrase header = TRUE tells R that the first row is the variable names. The phrase sep = " " tells R that the data are space delimited.
The commands

```
trans.model <- lm(y~x1+x2, data=trans)
summary(trans.model)
print (influence.measures(trans.model))
```

tell R

- to estimate the model,
- to print the analysis of variance, the estimated coefficients, and their tests, and
- to print the influence measures.

The commands

```
yhat <- trans.model$fit
t <- rstudent(trans.model)
```

```
qqnorm (t)
plot(yhat,t)
plot(trans$x1,t)
plot(trans$x2,t)
```

set up and then create the appropriate residual plots based on R-student. Generally, R does not produce satisfactory plots of data. The commands

```
trans2 <- cbind(trans,yhat,t)
write.table (trans2,"transistor_output.txt")
```

create a file "transistor_output.txt" which the user can import into his/her favorite package for doing graphics.

R Commander is an add-on package to R. It also is freely available. It provides an easy-to-use user interface, much like MINITAB and JMP, to the parent R product. R Commander makes it much more convenient to use R; however, it does not provide much flexibility in its analysis. For example, R Commander does not allow the user to use R-student for the residual plots. R Commander is a good way for users to get familiar with R. Ultimately, however, we recommend the use of the parent R product.

2.6 PARAMETER ESTIMATION BY WEIGHTED LEAST SQUARES

2.6.1 The Constant Variance Assumption

In both the ordinary least squares and maximum likelihood approaches to parameter estimation, we made the assumption of constant variance. By this assumption we mean that the variance of an observation is the same regardless of the values of the predictor or regressor variables associated with it. Since the regressor variables determine the mean or expected value of the observation, we are in effect assuming that the variance of the observation is unrelated to its mean.

There are a lot of practical situations where the constant variance assumption turns out to be inappropriate. Perhaps the observation y_i is an average of several (e.g., n_i) observations at a point in the predictor variable space. If this is the case and the original observations have constant variance, then the actual observations have variance $\text{Var}(y_i) = \sigma^2/n_i$. An important source of variability may be the measurement system used, and it is not unusual to find that in many such systems the size of the measurement error is proportional to the size of the measured quantity (i.e., the measurement error as a percentage of the mean is constant). Also, if the underlying distribution of the response variable is not normal, but is a continuous, skewed distribution, such as the lognormal, gamma, or Weibull distribution, we often find that the constant variance assumption is violated.

In this case the inequality of variance is directly related to the nonnormality of the response variable distribution, as for a skewed distribution, the variance is a function of the mean. An important point here is that **the constant variance assumption is often linked to the assumption of a normal distribution for the response**.

When the constant variance assumption is not satisfied, one approach to the problem is through data transformation, as illustrated in Example 2.9. Another approach is based on **generalized** or **weighted least squares,** a variation of the least squares procedure that takes the inequality of variance in the observations into account. Weighted least squares play a very important role in parameter estimation for generalized linear models.

2.6.2 Generalized and Weighted Least Squares

Generalized Least Squares

We now consider what modifications to the ordinary least squares procedure are necessary when Var $(\mathbf{y}) = \sigma^2\ \mathbf{V}$, where $\mathbf{V}$ is a known $n \times n$ matrix. This situation has an easy interpretation; if $\mathbf{V}$ is diagonal but with unequal diagonal elements, then the observations $\mathbf{y}$ are **uncorrelated** but have **unequal variances**, while if some of the off-diagonal elements of $\mathbf{V}$ are nonzero, then the observations are **correlated**.

When the model is

$$\begin{gathered} y = \mathbf{X}\boldsymbol{\beta} + \boldsymbol{\varepsilon} \\ \mathbf{E}(\boldsymbol{\varepsilon}) = \mathbf{0},\ \mathrm{Var}(\boldsymbol{\varepsilon}) = \sigma^2\mathbf{V} \end{gathered} \tag{2.62}$$

the ordinary least squares estimator $\mathbf{b} = (\mathbf{X}'\mathbf{X})^{-1}\mathbf{X}'\mathbf{y}$ is no longer optimal. We approach this problem by transforming the model to a new set of observations that satisfy the standard least squares assumptions. Then we use ordinary least squares on the transformed data. Since $\sigma^2\mathbf{V}$ is the covariance matrix of the errors, $\mathbf{V}$ must be nonsingular and positive definite, so there exists an $n \times n$ nonsingular symmetric matrix $\mathbf{K}$, where $\mathbf{K}'\mathbf{K} = \mathbf{K}\mathbf{K} = \mathbf{V}$. The matrix $\mathbf{K}$ is often called the **square root** of $\mathbf{V}$.

Define the new variables

$$\mathbf{z} = \mathbf{K}^{-1}\mathbf{y},\ \ \mathbf{B} = \mathbf{K}^{-1}\mathbf{X},\ \ \mathbf{g} = \mathbf{K}^{-1}\boldsymbol{\varepsilon} \tag{2.63}$$

so that the regression model $\mathbf{y} = \mathbf{X}\boldsymbol{\beta} + \boldsymbol{\varepsilon}$ becomes $\mathbf{K}^{-1}\mathbf{y} = \mathbf{K}^{-1}\mathbf{X}\boldsymbol{\beta} + \mathbf{K}^{-1}\boldsymbol{\varepsilon}$, or

$$\mathbf{z} = \mathbf{B}\boldsymbol{\beta} + \mathbf{g} \tag{2.64}$$

The errors in this transformed model have zero expectation; that is, $E(\mathbf{g}) = \mathbf{K}^{-1}E(\boldsymbol{\varepsilon}) = \mathbf{0}$. Furthermore, the covariance matrix of $\mathbf{g}$ is

$$\begin{aligned}\text{Var}(\mathbf{g}) &= E\{[\mathbf{g} - E(\mathbf{g})][\mathbf{g} - E(\mathbf{g})]'\} \\ &= E(\mathbf{g}\mathbf{g}') \\ &= E(\mathbf{K}^{-1}\boldsymbol{\varepsilon}\boldsymbol{\varepsilon}'\mathbf{K}^{-1}) \\ &= \mathbf{K}^{-1}E(\boldsymbol{\varepsilon}\boldsymbol{\varepsilon}')\mathbf{K}^{-1} \\ &= \sigma^2\mathbf{K}^{-1}\mathbf{V}\mathbf{K}^{-1} \\ &= \sigma^2\mathbf{K}^{-1}\mathbf{K}\mathbf{K}\mathbf{K}^{-1} \\ &= \sigma^2\mathbf{I} \end{aligned} \tag{2.65}$$

Thus the elements of **g** have mean zero and constant variance and are uncorrelated. Since the errors **g** in the model (2.64) satisfy the usual assumptions, we may apply ordinary least squares. The least squares function is

$$\begin{aligned} S(\boldsymbol{\beta}) &= \mathbf{g}'\mathbf{g} = \boldsymbol{\varepsilon}'\mathbf{V}^{-1}\boldsymbol{\varepsilon} \\ &= (\mathbf{y} - \mathbf{X}\boldsymbol{\beta})'\mathbf{V}^{-1}(\mathbf{y} - \mathbf{X}\boldsymbol{\beta}) \end{aligned} \tag{2.66}$$

The least squares normal equations are

$$= (\mathbf{X}'\mathbf{V}^{-1}\mathbf{X})\mathbf{b} = (\mathbf{X}'\mathbf{V}^{-1}\mathbf{y}) \tag{2.67}$$

and the solution to these equations is

$$\mathbf{b} = (\mathbf{X}'\mathbf{V}^{-1}\mathbf{X})^{-1}\mathbf{X}'\mathbf{V}^{-1}\mathbf{y} \tag{2.68}$$

Here **b** is called the **generalized least squares estimator** of $\boldsymbol{\beta}$.

It is not difficult to show that **b** is an unbiased estimator of $\boldsymbol{\beta}$. The covariance matrix of **b** is

$$\text{Var}(\mathbf{b}) = \sigma^2(\mathbf{B}'\mathbf{B})^{-1} = \sigma^2(\mathbf{X}'\mathbf{V}^{-1}\mathbf{X})^{1} \tag{2.69}$$

Furthermore, under the assumption of normal errors in (2.62), **b** is the best linear unbiased estimator of $\boldsymbol{\beta}$. The analysis of variance in terms of generalized least squares is summarized in Table 2.9.

Weighted Least Squares

When the errors $\boldsymbol{\varepsilon}$ are uncorrelated but have unequal variances so that the covariance matrix of $\boldsymbol{\varepsilon}$ is

$$\sigma^2\mathbf{V} = \sigma^2\begin{bmatrix} \frac{1}{w_1} & & & 0 \\ & \frac{1}{w_1} & & \\ & & \ddots & \\ 0 & & & \frac{1}{w_n} \end{bmatrix}$$

Table 2.9 Analysis of Variance for Generalized Least Squares

Source	Sum of Squares	Degrees of Freedom	Mean Square	F_0
Regression	$SS_R = \mathbf{b}'\mathbf{B}'z$ $= \mathbf{y}'\mathbf{V}^{-1}\mathbf{X}(\mathbf{X}'\mathbf{V}^{-1}\mathbf{X})^{-1}\mathbf{X}'\mathbf{V}^{-1}y$	p	SS_R/p	MS_R/MS_{res}
Residual	$SS_{res} = \mathbf{z}'\mathbf{z} - \mathbf{b}'\mathbf{B}'z$ $= \mathbf{y}'\mathbf{V}^{-1}\mathbf{y} - \mathbf{y}'\mathbf{V}^{-1}\mathbf{X}(\mathbf{X}'\mathbf{V}^{-1}\mathbf{X})^{-1}\mathbf{X}'\mathbf{V}^{-1}\mathbf{y}$	$n-p$	$SS_{res}/(n-p)$	
Total	$\mathbf{z}'\mathbf{z} = \mathbf{y}'\mathbf{V}'\mathbf{y}$	n		

the estimation procedure is usually called **weighted least squares**. Let $\mathbf{W} = \mathbf{V}^{-1}$. Since $\mathbf{V}$ is a diagonal matrix, $\mathbf{W}$ is also diagonal with diagonal elements or **weights** $w_1, w_2, \ldots, w_n$. From (2.65), the weighted least squares normal equations are

$$(\mathbf{X}'\mathbf{W}\mathbf{X})\mathbf{b} = \mathbf{X}'\mathbf{W}\mathbf{y} \tag{2.70}$$

and

$$\mathbf{b} = (\mathbf{X}'\mathbf{W}\mathbf{X})^{-1}\mathbf{X}'\mathbf{W}\mathbf{y} \tag{2.71}$$

which is the **weighted least squares estimator**. Note that observations with large variances have smaller weights than observations with small variances.

To use weighted least squares in a practical sense, we must know the weights $w_1, w_2, \ldots, w_n$. Sometimes prior knowledge or experience or information based on underlying theoretical considerations can be used to determine the weights. In other situations we may find empirically that the variability in the response is a function of one or more regressors, and so a model may be fit to predict the variance of each observation and hence determine the weights. In some cases we may have to estimate the weights, perform the analysis, reestimate a new set of weights based on these results, and then perform the analysis again. Several iterations may be necessary to obtain both reasonable weights and a reasonable solution to the overall model-fitting problem. We call a procedure such as this **iteratively reweighted least squares (IRLS)**. There are several applications of IRLS in regression, including generalized linear models.

When weights are known, the inference procedures associated with weighted and generalized least squares are exact, as long as the response distribution is normal. However, when the weights are estimated, the inference procedures are only approximate. This is less of an issue when the sample size used is not too small.

2.6.3 Generalized Least Squares and Maximum Likelihood

Recall that in Section 2.3 we showed that, under the normality assumption for the linear regression model, the method of maximum likelihood results in the same estimator as ordinary least squares. Exactly the same thing happens in the generalized (or weighted) least squares case.

Suppose that $\mathbf{y}$ is a vector of n observations and $\mathbf{X}$ is the corresponding matrix of predictor variables. Let the joint distribution of $\mathbf{y}$ be the multivariate normal distribution with mean vector $\mathbf{X\beta}$ and covariance matrix $\sigma^2\mathbf{V}$. The likelihood function for the n observations is

$$\mathscr{L}(\mathbf{y}, \boldsymbol{\beta}, \sigma^2\mathbf{V}) = \frac{1}{(2\pi)^{n/2}\sigma^2|\mathbf{V}|^{1/2}} e^{-(1/2\sigma^2)(\mathbf{y}-\mathbf{X\beta})'\mathbf{V}^{-1}(\mathbf{y}-\mathbf{X\beta})} \tag{2.72}$$

It is clear from examining Equation (2.72) that to maximize the likelihood function, we must minimize the expression in the exponent, namely, $[(\mathbf{y}-\mathbf{X\beta})']\mathbf{V}^{-1}(\mathbf{y}-\mathbf{X\beta})$. However, this is exactly the same as the generalized least squares criterion used in Equation (2.66). Therefore generalized and weighted least squares estimators are maximum likelihood estimators under normal theory.

2.7 DESIGNS FOR REGRESSION MODELS

Typically, regression analysis uses either historical data or data from an observational study. However, there are many cases where we apply regression analysis to designed experiments. The question then becomes: If we could choose the levels of the regressor variables, what values should we use? The experimental design problem in regression consists of selecting the elements of a design matrix, which in turn determines the elements of the model matrix, $\mathbf{X}$.

Classical experimentation often focuses on factorial designs. Complete factorial experiments use as their distinct design runs all of the possible combinations of the levels for the factors involved. Classical second-order experimental designs include the central composite design (Box and Wilson, 1951) and the Box–Behnken design (Box and Behnken, 1960). Myers, Montgomery, and Anderson-Cook (2009) summarize the classical designs typically used in response surface methodology.

For example, suppose that we are fitting a first-order linear regression model. A factorial experiment often is a very good design for fitting this model. To illustrate, consider the following first-order model with the two-factor interaction in two variables:

$$y_i = \beta_0 + \beta_1 x_{i1} + \beta_2 x_{i2} + \beta_{12} x_{i1} x_{i2} + \varepsilon_i$$

A reasonable choice of factorial design for fitting this model is a design with both factors at two levels, say, ± 1. This results in a design with four runs, each run located at the corner of the square, and is commonly called a 2^2 factorial design. If this design is replicated twice to give eight runs, the model matrix in the coded units is

$$\mathbf{X} = \begin{bmatrix} 1 & -1 & -1 & 1 \\ 1 & 1 & -1 & -1 \\ 1 & -1 & 1 & -1 \\ 1 & 1 & 1 & 1 \\ 1 & -1 & -1 & 1 \\ 1 & 1 & -1 & -1 \\ 1 & -1 & 1 & -1 \\ 1 & 1 & 1 & 1 \end{bmatrix}$$

The resulting $\mathbf{X}'\mathbf{X}$ matrix is

$$\mathbf{X}'\mathbf{X} = \begin{bmatrix} 8 & 0 & 0 & 0 \\ 0 & 8 & 0 & 0 \\ 0 & 0 & 8 & 0 \\ 0 & 0 & 0 & 8 \end{bmatrix}$$

The 2^2 factorial design is an orthogonal design for fitting this model. This particular structure greatly simplifies the model fitting, statistical inference, and practical interpretation of the fitted model. In general, the 2^k factorial design is an orthogonal design for fitting the first-order model with interactions. Orthogonality is an important property for many experimental designs. Good experimental designs often are either orthogonal or nearly orthogonal.

Optimal design theory provides one approach for developing experimental designs. Most optimal design criteria are variance based. The popular criteria are: **D-optimality**, **G-optimality**, **A-optimality**, **E-optimality**, and **I-optimality**. D-, A-, and E-optimalities are examples of Φ -optimal designs, which is an attempt to unify several of the basic variance-based optimality criteria.

D-optimality is the single most popular approach, probably because it was the first programmed (Mitchell, 1974). Under normal theory, the D-optimal design minimizes the volume of the confidence ellipsoid around the vector of coefficient estimates. Another way to express this concept is that the D-optimal design minimizes the generalized variance of the estimated coefficients. This approach maximizes the determinant of $\mathbf{X}'\mathbf{X}$, which in turn minimizes the determinant of $(\mathbf{X}'\mathbf{X})^{-1}$. This criterion's focus on the determinant leads to its name. Another way to view D-optimality is that the D-optimal design maximizes the product of the eigenvalues of $\mathbf{X}'\mathbf{X}$. This focus on the eigenvalues of $\mathbf{X}'\mathbf{X}$ underlies much of optimal design theory.

Under normal theory, G-optimality minimizes the maximum value of the variance for a predicted value over the region of interest. The G refers to **global**. Operationally, the G-optimal design minimizes the maximum of the function $\mathbf{x}'(\mathbf{X}'\mathbf{X})^{-1}\mathbf{x}$ over the experimental region. For a G-optimal design the largest prediction variance occurs at the design points and is $\sigma^2 \cdot p/n$, where p is the number of parameters in the model and n is the total number of runs in the design. In general, it is difficult to construct G-optimal designs, but it is easy to confirm weather a specific design is G-optimal. Asymptotically, D- and G-optimalities are equivalent.

A-optimality minimizes the sum of the variances of the estimated coefficients. It thus minimizes trace $[(\mathbf{X}'\mathbf{X})]^{-1}]$. The trace is the sum of the diagonal elements of a matrix. It also is the sum of the eigenvalues. As a result, this criterion minimizes the sum of the eigenvalues of $(\mathbf{X}'\mathbf{X})^{-1}$. If $\lambda_1, \lambda_2, \ldots, \lambda_p$ are the eigenvalues of $\mathbf{X}'\mathbf{X}$, then $1/\lambda_1, 1/\lambda_2, \ldots, 1/\lambda_p$ are the eigenvalues of $(\mathbf{X}'\mathbf{X})^{-1}$.

E-optimality minimizes the maximum value of $\mathbf{z}'(\mathbf{X}'\mathbf{X})^{-1}\mathbf{z}$ subject to the constraint $\mathbf{z}'\mathbf{z} = 1$. It thus minimizes the largest eigenvalue of $(\mathbf{X}'\mathbf{X})^{-1}$, which is equivalent to maximizing the smallest eigenvalue of $\mathbf{X}'\mathbf{X}$. It is closely related to G-optimality, but not identical. Most people who use E-optimality design experiments involving categorical factors. As a result, they generally focus on contrasts, which are of limited interest in regression analysis.

Φ-optimality is a unifying concept often seen in the optimal design literature. The Φ_0 design is D-optimal; the Φ_1 design is A-optimal, and the Φ_∞ is E-optimal.

I-optimality seeks to choose the design to minimize the average prediction variance over the experimental region. Actually, Box and Draper (1959) first mentioned this criterion, which they called V-optimality. The average prediction variance over the region is defined by

$$\frac{1}{\Psi}\int_R \text{Var}(\hat{y})d\mathbf{x}$$

$$\frac{1}{\Psi}\int_R \sigma^2\mathbf{x}'(\mathbf{X}'\mathbf{X})^{-1}\mathbf{x}\ d\mathbf{x}$$

where R is the design region and $\Psi = \int_R d\mathbf{x}$ is the volume of the region. Clearly, *I*- and *G*-optimalities are related; however, they are not identical.

The designs recommended as being optimal depend heavily upon

- the assumed form of the model and
- the shape of the experimental region of interest.

In addition, these designs are optimal only in terms of a very specific and narrow criterion. These designs are very useful for nonstandard regions, usually due to constraints, situations requiring a total number of runs not covered by classical designs, and for augmenting a set of additional runs to a previously run experiment. An example of a very irregular design region is the mixture design problem as a result of necessary constraints on the factor levels. See Cornell (2002), Montgomery (2009), and Myers, Montgomery, and Anderson-Cook (2009) for more information.

The original computer codes for generating optimal designs required a list of candidate points. Such codes created an initial design based on the desired model form and number of experimental runs. It then used a point exchange algorithm to improve the design's performance in terms of the specific criterion, almost always D-optimality. One advantage to this approach was that all levels in the recommended design were convenient, for example, ± 1.

Many current software packages use a coordinate exchange algorithm (Meyer and Nachtsheim, 1995) to select the design. This algorithm finds designs with better values for the specific criterion; however, there is no guarantee that all of the resulting design levels are convenient.

Box (1982) outlines what he considers to be vital in the appropriate selection of an experimental design. Fundamental to Box's perspective are two points. First, "all models are wrong; some models are useful." Second, all scientific inquiry involves experimentation; and all scientific experimentation involves a series of experiments. Each experimental phase must build on what is learned from the previous phases. Both of these points stand in stark contrast to much of optimal design theory, which typically assumes a single, one-shot experiment to estimate a known model form. In this paper, Box discusses his fourteen points. According to Box, a good experiment design

1. provides a satisfactory distribution of information,
2. gives a fitted value as close as possible to the true,
3. provides the ability to test for lack of fit,
4. allows transformations,
5. allows for blocking,
6. allows sequential assembly (design augmentation),
7. provides an internal estimate of error,
8. is insensitive to the presence of outliers,
9. uses a near minimum number of runs,
10. provides data patterns that allow visual appreciation of the information in the data,
11. ensures simplicity of calculation,

12. behaves well when there are errors in the factors,

13. requires only a few levels for the factors, and

14. provides a check of the constancy of variance assumption.

Underlying many of these points is the concept of projection properties, which consider the design's structure if one or more of the experimental factors proves insignificant in the analysis. Good projection properties ensure that the experimental design in the remaining factors maintains a good structure.

Box's major point is that most classical experimental designs perform well in terms of these fourteen points in addition to performing well, although not necessarily best, in terms of the various optimality criteria. In this light, he strongly recommends the use of such classical experimental designs as the 2^k factorial system and the central composite design. For example, recall the simple, replicated 2^2 factorial designed that we introduced at the beginning of this section. If the levels ± 1 are at the extremes of the design region, then this design is optimal in terms of all the criteria we have outlined. In addition, it meets almost all of the criteria outlined by Box. It is an excellent example of an optimal design that is also a good design.

One may or may not agree with all of Box's fourteen points; however, they do underscore an extremely important issue. The final choice of any experimental design involves a complex compromise across many competing and often contradictory criteria. For example, the ability to detect lack of fit requires a certain number of experimental runs that may provide no meaningful information for estimating the presumed model. In a similar manner, pure error estimates of the σ^2 involve replication that often provides no additional information for estimating the model. Most statistical software packages that produce optimal designs try to take these issues into consideration. In most cases, the packages outline the specifics in their help manuals.

Example 2.11. Optimal Designs for a Second-Order Model. To illustrate constructing optimal designs, suppose that we want to fit a second-order model in $k = 4$ variables. A central composite design has at least 25 runs (16 factorial runs, 8 axial runs, and at least one center point) and probably closer to 30 runs, because the center point would typically be replicated several times. The second-order model only has 15 parameters, so we could logically be interested in a smaller experiment. We use JMP to construct both a D-optimal and an I-optimal design with 18 runs where the runs are restricted to a cube.

Table 2.10 shows some of the output from JMP for the D-optimal design. The prediction variance profiler is set to the place in the design space where the prediction variance is maximized. The output also contains a fraction of

Table 2.10 A D-Optimal Design from JMP for a Second-Order Model in k = 4 Factors

Design

Run	X1	X2	X3	X4
1	0	-1	0	0
2	1	-1	-1	-1
3	-1	-1	-1	1
4	-1	1	-1	-1
5	-1	-1	1	-1
6	-1	-1	1	1
7	0	0	-1	-1
8	1	-1	-1	1
9	1	0	1	-1
10	-1	-1	-1	-1
11	1	-1	1	0
12	0	1	1	-1
13	-1	1	1	0
14	1	1	-1	0
15	-1	0	0	0
16	-1	1	-1	1
17	1	1	1	1
18	1	1	0	-1

Prediction Variance Profile

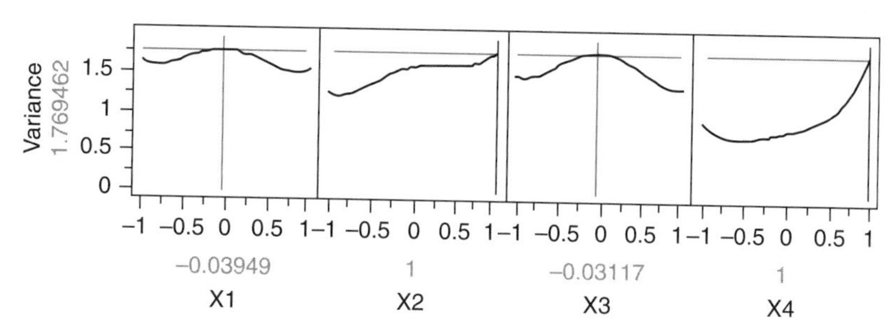

Fraction of Design Space Plot

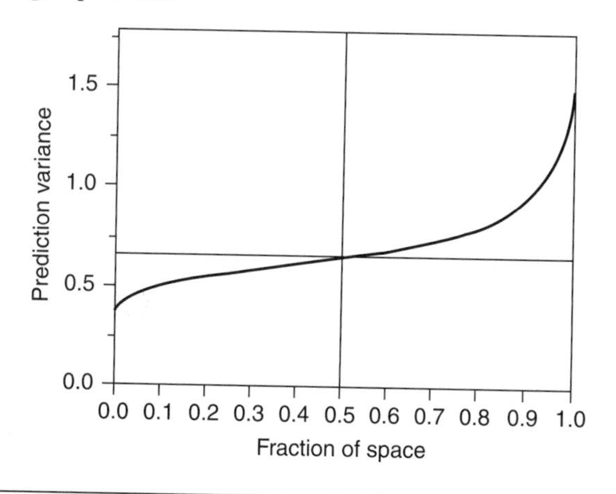

Table 2.11 An *I*-Optimal Design from JMP a Second-Order Model in $k = 4$ Factors

Run	X1	X2	X3	X4	
1	1	-1	0	1	
2	0	1	0	0	
3	-1	0	0	-1	
4	1	1	1	1	
5	1	-1	-1	-1	
6	-1	-1	-1	0	
7	-1	1	0	1	
8	0	1	-1	-1	
9	1	1	1	-1	
1	0	-1	-1	1	1
1	1	-1	1	1	-1
1	2	0	0	1	0
1	3	0	-1	0	-1
1	4	0	0	-1	1
1	5	0	0	0	0
1	6	1	1	-1	1
1	7	1	-1	1	0
1	8	1	0	0	0

Prediction Variance Profile

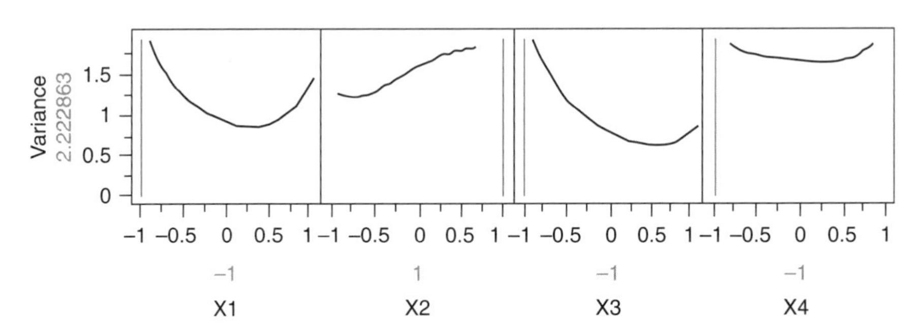

(Continued)

design (FDS) space plot, which is a plot of the prediction variance (apart from σ^2) versus the fraction of the design space. The cross-hairs on the FDS plot show that over about 50% of the design space the prediction variance will be less that about $0.7\sigma^2$. Table 2.11 shows the I-optimal design from JMP. Notice from the prediction variance profiler that the maximum prediction variance occurs at the boundary of the experimental region and is greater than the maximum prediction variance for the D-optimal design ($2.22\sigma^2$ versus $1.77\sigma^2$). However, the FDS plot reveals that the prediction

Table 2.11 *Continued*

Fraction of Design Space Plot

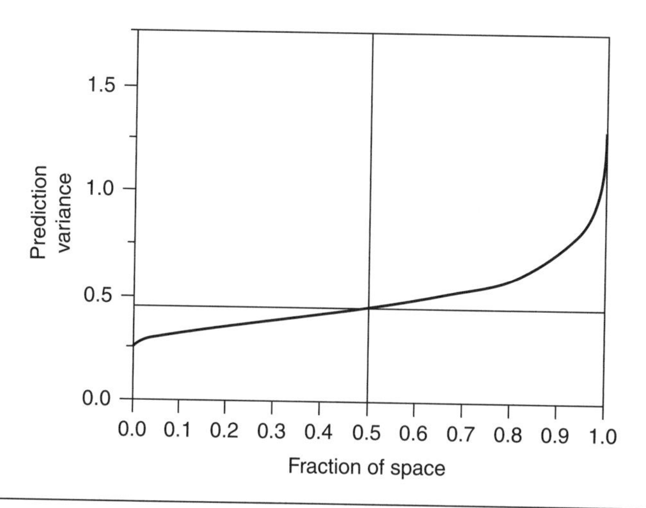

variance is much smaller over most of the deign region than in the D-optimal design. The I-optimal design basically trades smaller prediction variance on the average over the design space for a larger variance at the extremes of the region.

EXERCISES

2.1 The following data were collected on the wear of a bearing y, the oil viscosity x_1, and load x_2.

y	x_1	x_2
193	1.6	851
230	15.5	816
172	22.0	1058
91	43.0	1201
113	33.0	1357
125	40.0	1115

(a) Fit a multiple linear regression model to these data.

(b) Test for significance of regression.

(c) Compute t-statistics for each of the regression coefficients and provide an interpretation.

2.2 Reconsider the bearing data from Exercise 2.1. Expand the multiple regression model to include an interaction term.

(a) Test for significance of regression.

(b) Compute t-statistics for each of the regression coefficients and provide an interpretation. Specifically, does the model require an interaction term?

(c) Use the partial F-test procedure to determine whether the model requires an interaction term. Is this procedure equivalent to the t-test computed in part (b)?

2.3 Suppose that we wish to use the models from Exercises 2.1 and 2.2 to estimate the mean bearing wear when $x_1 = 25$ and $x_2 = 1000$.

(a) Compute point estimates of the mean wear using both models.

(b) Find a 95% confidence interval on the mean response at the point $x_1 = 25$ and $x_2 = 1000$ for both models. Which interval is narrower? Does this provide any insight about which model is preferable?

2.4 The pull strength of a wire bond is an important characteristic. The table below gives information on pull strength (y), die height (x_1), post height (x_2), loop height (x_3), wire length (x_4), bond width on the die (x_5), and bond width on the post (x_6).

(a) Fit a multiple linear regression model using x_2, x_3, x_4, and x_5 as the regressors.

(b) Test for significance of regression using the analysis of variance with $\alpha = 0.05$. What are your conclusions?

(c) Use the model from part (a) to predict pull strength when $x_2 = 20$, $x_3 = 30$, $x_4 = 90$, and $x_5 = 2.0$.

y	x_1	x_2	x_3	x_4	x_5	x_6
8.0	5.2	19.6	29.6	94.9	2.1	2.3
8.3	5.2	19.8	32.4	89.7	2.1	1.8
8.5	5.8	19.6	31.0	96.2	2.0	2.0
8.8	6.4	19.4	32.4	95.6	2.2	2.1
9.0	5.8	18.6	28.6	86.5	2.0	1.8
9.3	5.2	18.8	30.6	84.5	2.1	2.1
9.3	5.6	20.4	32.4	88.8	2.2	1.9
9.5	6.0	19.0	32.6	85.7	2.1	1.9
9.8	5.2	20.8	32.2	93.6	2.3	2.1
10.0	5.8	19.9	31.8	86.0	2.1	1.8
10.3	6.4	18.0	32.6	87.1	2.0	1.6
10.5	6.0	20.6	33.4	93.1	2.1	2.1
10.8	6.2	20.2	31.8	83.4	2.2	2.1
11.0	6.2	20.2	32.4	94.5	2.1	1.9
11.3	6.2	19.2	31.4	83.4	1.9	1.8

y	x_1	x_2	x_3	x_4	x_5	x_6
11.5	5.6	17.0	33.2	85.2	2.1	2.1
11.8	6.0	19.8	35.4	84.1	2.0	1.8
12.3	5.8	18.8	34.0	86.9	2.1	1.8
12.5	5.6	18.6	34.2	83.0	1.9	2.0

2.5 Consider the wire bond pull strength data in Exercise 2.4.

(a) Estimate σ^2 for this model.

(b) Find the standard errors for each of the regression coefficients.

(c) Calculate the t-test statistic for each regression coefficient. Using $\sigma = 0.05$, what conclusions can you draw? Do all variables contribute to the model?

2.6 For the regression model for the wire bond pull strength data in Exercise 2.4:

(a) Plot the residuals versus $\hat{y}$ and versus the regressors used in the model. What information is provided by these plots?

(b) Construct a normal probability plot of the residuals. Are there reasons to doubt the normality assumption for this model?

(c) Are there any indications of influential observations in the data?

2.7 Consider the wire bond pull strength data in Exercise 2.4.

(a) Find 95% confidence intervals on the regression coefficients.

(b) Find a 95% confidence interval on mean pull strength when $x_2 = 20$, $x_3 = 30$, $x_4 = 90$, and $x_5 = 2.0$.

2.8 Consider the wire bond pull strength data in Exercise 2.4. Fit a regressor model using all six regressors.

(a) Is there an indication that this model is superior to the one from Exercise 2.4?

(b) Find a 95% confidence interval on mean pull strength when $x_0 = 6.0$, $x_2 = 20$, $x_3 = 30$, $x_4 = 90$, $x_5 = 2.0$, and $x_6 = 2.1$. Compare the length of this interval to the one from Exercise 2.7, part (b).

2.9 An engineer at a semiconductor company wants to model the relationship between the device gain or $hFE(y)$ and three parameters: emitter-RS (x_1), base-RS (x_2), and emitter-to-base-RS (x_3). The data are shown below:

x_1 Emitter-RS	x_2 Base-RS	x_3 E-B-RS	y hFE-1M-5V
14.620	226.00	7.000	128.40
15.630	220.00	3.375	52.62
14.620	217.40	6.375	113.90
15.000	220.00	6.000	98.01

x_1 Emitter-RS	x_2 Base-RS	x_3 E-B-RS	y hFE-1M-5V
14.500	226.50	7.625	139.90
15.250	224.10	6.000	102.60
16.120	220.50	3.375	48.14
15.130	223.50	6.125	109.60
15.500	217.60	5.000	82.68
15.130	228.50	6.625	112.60
15.500	230.20	5.750	97.52
16.120	226.50	3.750	59.06
15.130	226.60	6.125	111.80
15.630	225.60	5.375	89.09
15.380	234.00	8.875	171.90
15.500	230.00	4.000	66.80
14.250	224.30	8.000	157.10
14.500	240.50	10.870	208.40
14.620	223.70	7.375	133.40

(a) Fit a multiple linear regression model to the data.

(b) Predict hFE when $x_1 = 14.5$, $x_2 = 220$, and $x_3 = 5.0$.

(c) Test for significance of regression using the analysis of variance with $\alpha = 0.05$. What conclusions can you draw?

(d) Estimate σ^2 for the model you have fit to the data.

(e) Find the standard errors of the regression coefficients.

(f) Calculate the t-test statistic for each regression coefficient. Using $\alpha = 0.05$, what conclusions can you draw?

(g) Find 99% confidence intervals on the regression coefficients.

(h) Find a 99% prediction interval on hFE when $x_1 = 14.5$, $x_2 = 220$, and $x_3 = 5.0$.

(i) Find a 99% confidence interval on mean hFE when $x_1, = 14.5$, $x_2 = 220$, and $x_3 = 5.0$.

2.10 Consider the semiconductor hFE data in Exercise 2.9.

(a) Plot the residuals from this model versus $\hat{y}$. Comment on the information in this plot.

(b) What is the value of R^2 for this model?

(c) Refit the model using In hFE as the response variable.

(d) Plot the residuals versus predicted In hFE for the model in part (c). Does this give any information about which model is preferable?

(e) Plot the residuals from the model in part (d), versus the regressor x_3. Comment on this plot.

(f) Refit the model to In hFE using x_1, x_2, and $1/x_3$ as the regressors. Comment on the effect of this change in the model.

2.11 Heat treating is often used to carburize metal parts, such as gears. The thickness of the carburized layer is considered an important feature of the gear, and it contributes to the overall reliability of the part. Because of the critical nature of this feature, two different lab tests are performed on each furnace load. One test is run on a sample pin that accompanies each load. The other test is a destructive test, where an actual part is cross-sectioned. This test involved running a carbon analysis on the surface of both the gear pitch (top of the gear tooth) and the gear root (between the gear teeth). The data below are the results of the pitch carbon analysis test for 32 parts.

Temp	Soaktime	Soakpct	Difftime	Diffpct	Pitch
1650	0.58	1.10	0.25	0.90	0.013
1650	0.66	1.10	0.33	0.90	0.016
1650	0.66	1.10	0.33	0.90	0.015
1650	0.66	1.10	0.33	0.95	0.016
1600	0.66	1.15	0.33	1.00	0.015
1600	0.66	1.15	0.33	1.00	0.016
1650	1.00	1.10	0.50	0.80	0.014
1650	1.17	1.10	0.58	0.80	0.021
1650	1.17	1.10	0.58	0.80	0.018
1650	1.17	1.10	0.58	0.80	0.019
1650	1.17	1.10	0.58	0.90	0.021
1650	1.17	1.10	0.58	0.90	0.019
1650	1.17	1.15	0.58	0.90	0.021
1650	1.20	1.15	1.10	0.80	0.025
1650	2.00	1.15	1.00	0.80	0.025
1650	2.00	1.10	1.10	0.80	0.026
1650	2.20	1.10	1.10	0.80	0.024
1650	2.20	1.10	1.10	0.80	0.025
1650	2.20	1.15	1.10	0.80	0.024
1650	2.20	1.10	1.10	0.90	0.025
1650	2.20	1.10	1.10	0.90	0.027
1650	2.20	1.10	1.50	0.90	0.026
1650	3.00	1.15	1.50	0.80	0.029
1650	3.00	1.10	1.50	0.70	0.030
1650	3.00	1.10	1.50	0.75	0.028
1650	3.00	1.15	1.66	0.85	0.032
1650	3.33	1.10	1.50	0.80	0.033
1700	4.00	1.10	1.50	0.70	0.039
1650	4.00	1.10	1.50	0.70	0.040
1650	4.00	1.15	1.50	0.85	0.035
1700	12.50	1.00	1.50	0.70	0.056
1700	18.50	1.00	1.50	0.70	0.068

(a) Fit a linear regression model relating the results of the pitch carbon analysis test (PITCH) to the five regressor variables.

(b) Test for significance of regression. Use $\alpha = 0.05$.

(c) Estimate σ^2 for the model.

(d) Find the standard errors of the regression coefficients.

(e) Evaluate the contribution of each regressor to the model using the t-test with $\sigma = 0.05$.

(f) Plot residuals and comment on model adequacy.

(g) Find 95% confidence intervals on the regression coefficients.

(h) Find a 95% confidence interval on mean PITCH for TEMP = 1650, SOAK-TIME = 1.00, SOAKPCT = 1.10, DIFFTIME = 1.00, and DIFFPCT = 0.80.

2.12 Reconsider the heat treating data from Exercise 2.11.

(a) Fit a new model to the response PITCH using new regressors x_1 = SOAKTIME × SOAKPCT and x_2 = DIFFTIME × DIFFPCT.

(b) Test the model in part (a) for significance of regression using $\alpha = 0.05$. Also calculate the t-test for each regressor, and draw conclusions.

(c) Estimate σ^2 for the model from part (a), and compare this to the estimate of σ^2 obtained in Exercise 2.11, part (c). Which estimate is smaller? Does this offer any insight regarding which model might be preferable?

2.13 Reconsider the heat treating data in Exercises 2.11 and 2.12, where we fit a model to PITCH using regressors x_1 = SOAKTIME × SOAKPCT and x_2 = DIFFTIME × DIFFPCT.

(a) Using the model with regressors x_1 and x_2, find a 95% confidence interval on mean PITCH when SOAKTIME = 1.00, SOAKPCT = 1.10, DIFFTIME = 1.00, and DIFFPCT = 0.80.

(b) Compare the length of this confidence interval with the length of the confidence interval on mean PITCH at the same point from Exercise 2.11, part (h), where an additive model in SOAKTIME, SOAKPCT, DIFFTIME, and DIFFPCT was used. Which confidence interval is shorter? Does this tell you anything about which model is preferable?

2.14 An article in the *Journal of Pharmaceuticals Sciences* (Vol. 80, 1991, pp. 971–977) presents data on the observed mole fraction solubility of a solute at a constant temperature to the dispersion, dipolar, and hydrogen bonding Hansen partial solubility parameters. The data are as follows:

Observation	y	x_1	x_2	x_3
1	0.22200	7.3	0.0	0.0
2	0.39500	8.7	0.0	0.3
3	0.42200	8.8	0.7	1.0
4	0.43700	8.1	4.0	0.2
5	0.42800	9.0	0.5	1.0
6	0.46700	8.7	1.5	2.8
7	0.44400	9.3	2.1	1.0
8	0.37800	7.6	5.1	3.4
9	0.49400	10.0	0.0	0.3
10	0.45600	8.4	3.7	4.1
11	0.45200	9.3	3.6	2.0
12	0.11200	7.7	2.8	7.1
13	0.43200	9.8	4.2	2.0
14	0.10100	7.3	2.5	6.8
15	0.23200	8.5	2.0	6.6
16	0.30600	9.5	2.5	5.0
17	0.09230	7.4	2.8	7.8
18	0.11600	7.8	2.8	7.7
19	0.07640	7.7	3.0	8.0
20	0.43900	10.3	1.7	4.2
21	0.09440	7.8	3.3	8.5
22	0.11700	7.1	3.9	6.6
23	0.07260	7.7	4.3	9.5
24	0.04120	7.4	6.0	10.9
25	0.25100	7.3	2.0	5.2
26	0.00002	7.6	7.8	20.7

where y is the negative logarithm of the mole fraction solubility, x_1 is the dispersion Hansen partial solubility, x_2 is the dipolar partial solubility, and x_3 is the hydrogen bonding partial solubility.

(a) Fit the model

$$y = \beta_0 + \beta_1 x_1 + \beta_2 x_2 + \beta_3 x_3 + \beta_{12} x_1 x_2 + \beta_{13} x_1 x_3 + \beta_{23} x_2 x_3 + \beta_{11} x_1^2 + \beta_{22} x_2^2 + \beta_{33} x_3^2 + \varepsilon.$$

(b) Test for significance of regression, using $\alpha = 0.05$.

(c) Plot the residuals and comment on model adequacy.

(d) Use the extra sum of squares method to test the contribution of the second-order terms, using $\alpha = 0.05$.

2.15 In an effort to develop a preliminary personnel equation for estimation of worker-hours per month expended in surgical services at Naval hospitals, the U.S. Navy collected data on y (worker-hours per month) and x (surgical cases) from 15 hospitals. The data (taken from the Navy's

Procedures and Analyses for Staffing Standards Development: Data/Regression Analysis Handbook, are shown in the table below.

y (worker-hours per month)	x Surgical Cases
1275	230
1350	235
1650	250
2000	277
3750	522
4222	545
5018	625
6125	713
6200	735
8150	820
9975	992
12,200	1322
12,750	1900
13,014	2022
13,275	2155

Fit the following models to these data.

(a) $y = \beta_0 + \beta_1 x + \varepsilon$.

(b) In $y = \beta_0 + \beta_1(1/x) + \varepsilon$.

(c) $1/y = \beta_0 + \beta_1(1/x) + \varepsilon$.

(d) $y = \beta_0 + \beta_1 x + \beta_2 x^2 + \varepsilon$.

(e) Comment on the adequacy of each of these models.

2.16 In a study in collaboration with the Engineering Sciences and Mechanics Department at Virginia Tech, personnel at the Statistics Consulting Center were called upon to analyze a data set dealing with cathodic debonding of elastomeric metal bonds. A model relating the amount of debonding as a function of time (x_1), voltage (x_2), pH at time of bonding (x_3), and temperature (x_4) was sought. The following data were collected.

y (cm)	x_1 (min)	x_2 (volts)	x_3 (pH)	x_4 (°F)
2.240	24.00	1000	8.47	299
3.581	43.00	1000	8.55	299
5.131	84.00	1000	8.53	299
5.715	159.00	1000	8.60	297
0.889	3.79	1100	7.72	300

y (cm)	x_1 (min)	x_2 (volts)	x_3 (pH)	x_4 (°F)
2.845	17.00	1100	7.83	299
6.147	37.00	1100	4.91	299
7.016	54.20	1100	3.79	299
7.747	58.00	1100	3.69	299
8.286	85.00	1100	3.70	298
9.321	89.00	1100	3.91	297
9.195	95.00	1100	3.57	296
1.895	1.85	1420	4.43	298
3.708	5.88	1420	3.67	298
3.467	7.02	1420	2.89	298
5.049	13.88	1420	2.56	298
4.846	17.19	1420	2.56	298
6.108	19.80	1420	2.84	298
7.137	33.80	1420	3.22	298
8.295	40.80	1420	3.54	298
1.168	0.83	1825	3.73	298
2.865	8.08	1825	3.45	295
3.912	21.08	1825	2.85	298
3.726	21.08	1825	2.85	298
4.521	7.00	1400	8.88	303
5.664	15.00	1400	8.88	303
7.620	27.00	1400	8.88	303
9.766	52.00	1400	8.88	303
12.497	65.00	1400	8.88	303
2.174	1.76	1310	5.46	313
8.153	7.90	1310	2.15	309
10.757	20.90	1310	2.55	312

(a) Fit a standard multiple linear regression model to these data and comment on model adequacy.

(b) Use a Box–Cox power transformation on the response, and find the appropriate transform y^{λ}.

(c) Does a confidence interval estimate on λ indicate whether or not there is a need to work with a transformed response rather than the natural metric y? Explain.

(d) If transformation is needed, fit a new regression using y^{λ} as the new response and indicate further evidence of the improved regression.

2.17 Consider the following data set in which nitrogen dioxide concentrations in parts per million are collected for 26 days in September 1984 at a monitoring facility in the San Francisco Bay area. These data were taken from Chatterjee and Hadi (1988).

x_1 Windspeed (average mph)	x_2 Maximum Temperature (°F)	x_3 Insolation (langleys per day)	NO
11.1	90	382	6
12.1	86	380	5
12.0	80	372	5
17.8	70	352	3
9.5	90	358	7
7.2	100	362	9
11.5	92	302	6
13.4	74	316	2
10.8	87	339	10
13.8	78	328	7
14.6	73	278	3
12.1	85	339	4
8.0	94	241	13
8.8	91	193	10
12.9	84	268	7
12.7	68	113	3
12.1	81	313	6
11.1	78	317	5
11.3	74	324	4
9.0	78	312	9
9.2	84	349	11
8.4	90	290	8
8.0	90	295	9
13.8	80	283	6
17.8	68	259	2

(a) Fit a standard multiple regression model to these data and comment on model adequacy.

(b) Consider a transformation for the response. Use the Box–Cox procedure to determine an appropriate transformation.

(c) Does a model with transformed response seem more reasonable than the model in part (a)?

2.18 The data in the following table come from a factorial experiment conducted to study the effect of reaction time and reaction temperature on the concentration of a chemical product.

	Temperature		
Time (h)	50°C	75°C	100°C
2	4.70, 2.68	5.52, 3.75	3.98, 4.22
4	6.35, 6.10	5.88, 7.69	6.28, 7.12
6	7.85, 9.25	9.00, 9.78	11.43, 9.62

(a) Fit the main effects model $y = \beta_0 + \beta_1 x_1 + \beta_2 x_2 + \varepsilon$ to the data, where the x's are coded variables taking on the values -1, 0, $+1$ corresponding to the low, medium, and high levels of the design factors time and temperature.

(b) Test for significance of regression.

(c) Analyze the residuals and comment on model adequacy.

(d) Expand the model in part (a) to include an interaction term, say, $y = \beta_0 + \beta_1 x_1 + \beta_2 x_2 + \beta_{12} x_1 x_2 + \varepsilon$. Does the interaction term improve the fit of the model to the data?

2.19 Reconsider the data from the factorial experiment in Exercise 2.18. Use the Box–Cox method to determine if a transformation on the response is necessary.

2.20 Consider a multiple regression model with k regressors. Show that the test statistic for significance of regression can be written as

$$F_k = \frac{R^2/k}{(1 - R^2)/(n - k - 1)}$$

Suppose that $n = 20$, $k = 4$, and $R^2 = 0.90$. If $\alpha = 0.05$, what conclusion would you draw about the relationship between y and the four regressors?

2.21 A regression model is used to relate a response y to $k = 4$ regressors. What is the smallest value of R^2 that will result in a significant regression if $\alpha = 0.05$? Use the results of the previous exercise. Are you surprised by how small the value of R^2 is?

2.22 Show that we can express the residuals from a multiple regression model as

$$\mathbf{e} = (\mathbf{I} - \mathbf{H})\mathbf{y}, \quad \text{where } \mathbf{H} = \mathbf{X}(\mathbf{X}'\mathbf{X})^{-1}\mathbf{X}'$$

2.23 Show that the variance of the ith residual e_i in a multiple regression model is $\sigma^2(1-h_{ii})$ and that the covariance between e_i, and e_j is $-h_{ij}$, where the h's are the elements of $\mathbf{H} = \mathbf{X}(\mathbf{X}'\mathbf{X})^{-1}\mathbf{X}'$.

2.24 Consider the model

$$\mathbf{y} = \mathbf{X}\boldsymbol{\beta} + \boldsymbol{\varepsilon}$$

where $E(\boldsymbol{\varepsilon}) = \mathbf{0}$, and $\mathrm{Var}[\boldsymbol{\varepsilon}] = \mathbf{V}$. Show that

$$\hat{\boldsymbol{\beta}} = (\mathbf{X}'\mathbf{V}^{-1}\mathbf{X})^{-1}\mathbf{X}'\mathbf{V}^{-1}\mathbf{y}$$

is BLUE.

2.25 Suppose that you want to fit a first-order model with all of the two-factor interactions in three factors over the ± 1 range using $n = 12$ runs. Find a D-optimal design for this situation.

2.26 Suppose that you want to fit a first-order model in three factors with all of the two-factor interactions in over the ± 1 range using $n = 12$ runs. Find an I-optimal design for this situation.

2.27 Suppose that you want to fit a first-order model in two factors over the ± 1 range using $n = 8$ runs. The design chosen is a full 2^2 design augmented with four center runs ($x_1 = x_2 = 0$) Is this a D-optimal design?

2.28 Suppose that you want to fit a main-effects first-order model in three factors over the ± 1 range using $n = 8$ runs. Find a D-optimal design for this situation.

2.29 Suppose that you want to fit a second-order model in three factors using $n = 15$ runs. Find a D-optimal design for this situation.

2.30 Suppose that you want to fit a second-order model in three factors using $n = 15$ runs. Find an I-optimal design for this situation.

2.31 Suppose that you want to fit a second-order model in three factors using $n = 12$ runs. Find a D-optimal design for this situation.

2.32 Suppose that you want to fit a second-order model in three factors using $n = 12$ runs. Find a I-optimal design for this situation.

CHAPTER 3

Nonlinear Regression Models

The linear regression models of Chapter 2 provide a flexible framework that suits the needs of many model builders and data analysts. However, linear regression models are not appropriate for all situations. There are many situations where the response variable and the predictor variables are related through a known **nonlinear function**. This leads to a **nonlinear regression model**. Parameter estimation and model inference is somewhat more involved for nonlinear regression models than it is in the linear case. For example, when the method of least squares is applied to a nonlinear regression model, the resulting normal equations are nonlinear and often difficult to solve. The usual approach is to directly minimize the residual sum of squares by an iterative procedure. Furthermore, the normal-theory inference used in the linear regression model does not apply exactly to nonlinear regression models. Instead, inference based on **asymptotic** or **large-sample** theory must be employed. In this chapter we present a description of estimating the parameters in a nonlinear regression model, and show how to make appropriate inferences on the model parameters.

3.1 LINEAR AND NONLINEAR REGRESSION MODELS

3.1.1 Linear Regression Models

In Chapter 2 we focused on the linear regression model

$$y = \beta_0 + \beta_1 x_1 + \beta_2 x_2 + \cdots + \beta_k x_k + \varepsilon \tag{3.1}$$

We observed that linear regression models include not only first-order relationships, such as Equation (3.1), but polynomial models and other more complex relationships as well. In fact we could write a linear regression model as

$$y = \beta_0 + \beta_1 z_1 + \beta_2 z_2 + \cdots + \beta_r z_r + \varepsilon \tag{3.2}$$

Generalized Linear Models, Second Edition, by Myers, Montgomery, Vining, and Robinson

where z_i represents any function of the original regressors $x_1, x_2, \ldots, x_k$, including expressions such as $\exp(x_i)$, $\sqrt{x_i}$, and $\sin^{-1}(x_i)$. The reason that these models are called linear regression models is that they are **linear in the unknown parameters** β_j, $j = 1,2,\ldots, k$.

We may write the linear regression model (3.1) in a general form as

$$y = f(\mathbf{x}, \boldsymbol{\beta}) + \varepsilon \tag{3.3}$$

where $f(\mathbf{x}, \boldsymbol{\beta}) = \beta_0 + \beta_1 x_1 + \beta_2 x_2 + \ldots + \beta_k x_k$ and $\mathbf{x}' = [x_1, x_2, \ldots, x_k]$. Since the expected value of the model errors is zero, the expected value of the response variable is

$$\begin{aligned} E(y) &= E[f(\mathbf{x},\boldsymbol{\beta}) + \varepsilon] \\ &= f(\mathbf{x},\boldsymbol{\beta}) \end{aligned}$$

Consequently, $f(\mathbf{x},\boldsymbol{\beta})$ is often called the **expectation function** for the model. Obviously, the expectation function here is just a linear function of the unknown parameters.

3.1.2 Nonlinear Regression Models

There are many situations where a linear regression model is not appropriate. For example, the analyst has direct knowledge of the form of the relationship between the response variable and the regressors, perhaps from the theory underlying the phenomena. The true relationship between the response and the regressors is a differential equation, or the solution to a differential equation. So the model must be of nonlinear form.

Any model that is not linear in the unknown parameters is a nonlinear regression model. For example, the model

$$y = \beta_1 e^{\beta_2 x} + \varepsilon \tag{3.4}$$

is not linear in the unknown parameters β_1, and β_2. In general, we can write a nonlinear regression model as

$$y = f(\mathbf{x},\boldsymbol{\beta}) + \varepsilon \tag{3.5}$$

where $\boldsymbol{\beta}$ is a $p \times 1$ vector of unknown parameters, and $\boldsymbol{\varepsilon}$ is an uncorrelated random error term with $E(\varepsilon) = 0$ and $\mathrm{Var}(\varepsilon) = \sigma^2$. We often assume that the errors are normally distributed, as in linear regression. Since

$$\begin{aligned} E(y) &= E[f(\mathbf{x},\boldsymbol{\beta}) + \varepsilon] \\ &= f(\mathbf{x},\boldsymbol{\beta}) \end{aligned} \tag{3.6}$$

we call $f(\mathbf{x}, \boldsymbol{\beta})$ the **expectation function** for the nonlinear regression model. This is very similar to the linear regression case, except that now the expectation function is a nonlinear function of the parameters.

In a nonlinear regression model, at least one of the partial derivatives of the expectation function with respect to the parameters depends on at least one of the parameters. In linear regression, these derivatives are *not* functions of the unknown parameters. To illustrate, consider the linear regression model in Equation (3.1), for which the expectation function is

$$f(\mathbf{x},\boldsymbol{\beta}) = \beta_0 + \sum_{j=1}^{k} \beta_j x_j$$

The partial derivatives of the expectation function are

$$\frac{\partial f(\mathbf{x},\boldsymbol{\beta})}{\partial \beta_j} = x_j, \quad j = 0, 1, \ldots, k$$

where $x_0 = 1$ is a dummy variable representing the intercept. Notice that the partial derivatives are not functions of the unknown parameters. Now consider the nonlinear regression model

$$\begin{aligned} y &= f(x, \boldsymbol{\beta}) + \varepsilon \\ &= \beta_1 e^{\beta_2 x} + \varepsilon \end{aligned} \tag{3.7}$$

The partial derivatives of the expectation function with respect to β_1 and β_2 are

$$\frac{\partial f(x, \boldsymbol{\beta})}{\partial \beta_1} = e^{\beta_2 x}$$

$$\frac{\partial f(x, \boldsymbol{\beta})}{\partial \beta_2} = \beta_1 x e^{\beta_2 x}$$

Because the partial derivatives are a function of the unknown parameters β_1 and β_2, the model is nonlinear.

3.1.3 Origins of Nonlinear Models

Nonlinear regression models often strike people as being very ad hoc because these models typically involve mathematical functions that are nonintuitive to people outside the specific application area. Too often, people fail to appreciate the scientific theory underlying these nonlinear regression models. The scientific method uses mathematical models to describe physical phenomena. In many cases, the theory describing the physical relationships involves the solution of a set of differential equations, especially whenever rates of change are the basis for the mathematical model. This section outlines how the differential equations that form the heart of the theory describing physical behavior lead to nonlinear models. Our motivating example deals with reaction rates. Our key point is that nonlinear regression models are almost always deeply rooted in the appropriate science.

Example 3.1. A Chemical Kinetics Model. We first consider formally incorporating the effect of temperature into a second-order reaction kinetics model. For example, the hydrolysis of ethyl acetate is well modeled by a second-order kinetics model. Let A_t be the amount of ethyl acetate at time t. The second-order model is

$$\frac{dA_t}{dt} = -kA_t^2$$

where k is the rate constant. Rate constants depend on temperature, which we will incorporate into our model later. Let A_0 be the amount of ethyl acetate at time zero. The solution to the rate equation is

$$\frac{1}{A_t} = \frac{1}{A_0} + kt$$

With some algebra, we obtain

$$A_t = \frac{A_0}{1+A_0 tk}$$

We next consider the impact of temperature on the rate constant. The Arrhenius equation states

$$k = C_1 \exp\left(-\frac{E_a}{RT}\right)$$

where E_a is the activation energy and C_1 is a constant. Substituting the Arrhenius equation into the rate equation yields

$$A_t = \frac{A_0}{1+A_0 t C_1 \exp(-E_a/RT)}$$

Thus an appropriate nonlinear regression model is

$$A_t = \frac{\beta_1}{1+\beta_2 t \exp(-\beta_3/T)} + \varepsilon_t$$

where $\beta_1 = A_0$, $\beta_2 = C_1 A_0$, and $\beta_3 = E_a/R$. □

3.2 TRANSFORMING TO A LINEAR MODEL

Sometimes an analyst will consider transforming the nonlinear regression model into a linear model. Often only the expectation function is considered when selecting the transformation. For example, consider the model

$$\begin{aligned} y &= f(x, \boldsymbol{\beta}) + \varepsilon \\ &= \beta_1 e^{\beta_2 x} + \varepsilon \end{aligned} \tag{3.8}$$

Now, since the expectation function is $E(y) = f(x, \boldsymbol{\beta}) = \beta_1 e^{\beta_2 x}$, we can easily linearize the expectation function just by taking logarithms:

$$\ln E(y) = \ln \beta_1 + \beta_2 x$$

Therefore it is tempting to consider rewriting the regression model as

$$\begin{aligned} \ln(y) &= \ln \beta_1 + \beta_2 x + \varepsilon \\ &= \alpha_0 + \alpha_1 x + \varepsilon \end{aligned} \tag{3.9}$$

and using *linear* regression to estimate the parameters α_0 and α_1 in this new equation.

Considerable care must be exercised in this approach. In general, the linear least square estimates of the parameters in Equation (3.9) are not equivalent to the nonlinear parameter estimates in the original model of Equation (3.8). The reason is that in the original nonlinear model, least squares implies minimization of the sum of squared residuals on y, whereas in the transformed model we are minimizing the sum of squared residuals on the logarithm of y.

Note that in the original nonlinear model of Equation (3.8) the error structure is *additive*, so taking logarithms *cannot* produce the model in Equation (3.9). However, if the error structure is multiplicative, then

$$\begin{aligned} y &= f(x, \boldsymbol{\beta})(1+\varepsilon) \\ &= \beta_1 e^{\beta_2 x} \varepsilon^* \end{aligned} \tag{3.10}$$

and taking logarithms is appropriate because

$$\begin{aligned} \ln(y) &= \ln \beta_1 + \beta_2 x + \ln \varepsilon^* \\ &= \alpha_0 + \alpha_1 x + \varepsilon^{**} \end{aligned} \tag{3.11}$$

Now if the new error term ε^{**} follows a normal distribution with constant variance all of the standard linear regression model properties and inference procedures apply.

A nonlinear regression model that can be transformed into an equivalent linear regression model is said to be **intrinsically linear**. However, the issue often revolves around the error structure. That is, do the standard assumptions apply to the errors in the transformed or linearized model? This is often not easy to determine.

Example 3.2. The Puromycin Data. Bates and Watts (1988) present data on the velocity of an enzymatic reaction, where the substrate has been treated with puromycin at several concentrations. The velocity and concentration data are shown in Table 3.1 and a scatter diagram is given in Figure 3.1. Bates and Watts propose fitting the Michaelis–Menten model for chemical kinetics

$$E(y_i) = f(x_i, \boldsymbol{\beta}) = \frac{\beta_1 x_i}{\beta_2 + x_i} \tag{3.12}$$

to the data. Note that the expectation function of the Michaelis–Menten model can easily be linearized because

$$\begin{aligned}\frac{1}{f(x_i, \boldsymbol{\beta})} &= \frac{\beta_2 + x_i}{\beta_1 x_i} \\ &= \frac{1}{\beta_1} + \frac{\beta_2}{\beta_1}\frac{1}{x_i} \\ &= \alpha_0 + \alpha_1 z_i\end{aligned}$$

where $z_i = 1/x_i$. Therefore we are tempted to fit the *linear* regression model

$$y_i^* = \alpha_0 + \alpha_1 z_i + \varepsilon_i$$

where $y_i^* = 1/y_i$ is the reciprocal of the observed velocity. The least squares fit is

$$\hat{y}_i^* = 0.005107 + 0.0002472 z_i \qquad \square$$

Figure 3.2a presents a scatter diagram of the transformed data with this straight-line fit superimposed. Since there are replicates in the data, it is relatively

Table 3.1 Reaction Velocity and Substrate Concentration for Puromycin Experiment

Substrate Concentration (ppm)	Velocity (counts/min^2)	
0.02	47	76
0.06	97	107
0.11	123	139
0.22	152	159
0.56	191	201
1.10	200	207

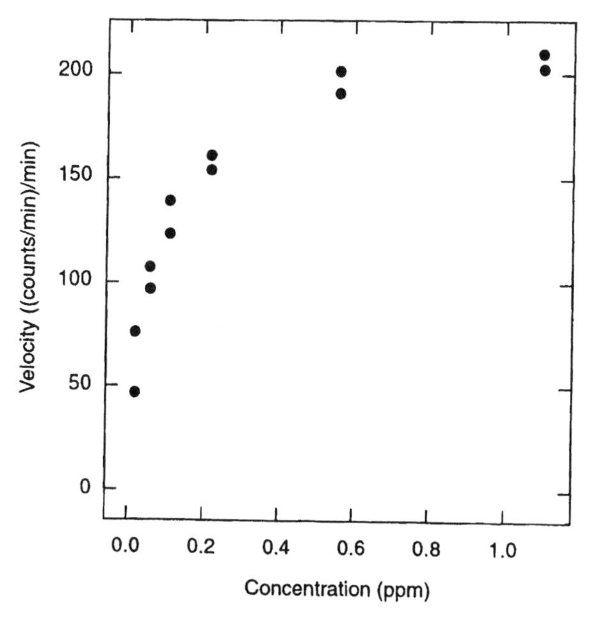

Figure 3.1 Plot of reaction velocity versus substrate concentration for the puromycin experiment. (Adapted from Bates and Watts, 1988, with permission of the publisher.)

easy to see from Figure 3.1 that the variance of the original response data is nearly constant, but Figure 3.2a reveals that the variance of the response in the *transformed* scale is far from constant.

Now the relationship between the parameters is

$$\alpha_0 = \frac{1}{\beta_1}$$

and

$$\alpha_1 = \frac{\beta_2}{\beta_1}$$

Therefore we may set

$$0.005107 = \frac{1}{b_1}$$

and

$$0.0002472 = \frac{b_2}{b_1}$$

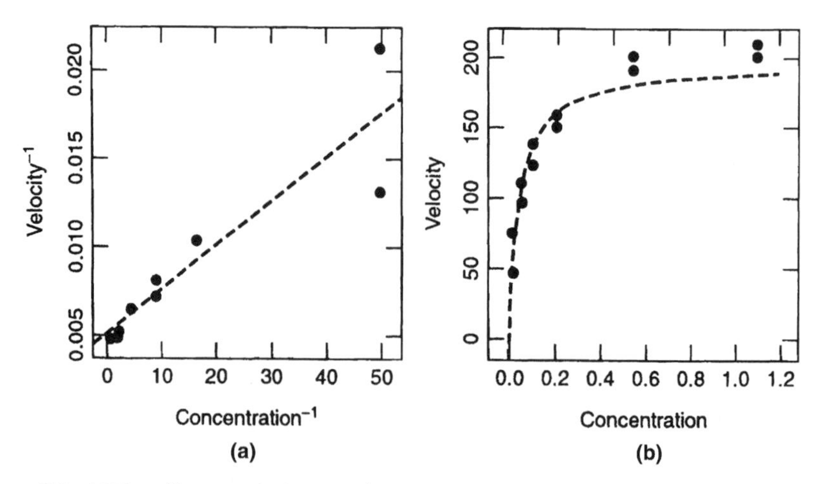

Figure 3.2 (a) Plot of inverse velocity versus inverse concentration for the puromycin data, (b) Fitted curve in the original scale. (Adapted from Bates and Watts, 1988, with permission of the publisher.)

So we can estimate β_1, and β_2 in the *original* nonlinear model as

$$b_1 = 195.81$$

and

$$b_2 = 0.04841$$

Figure 3.2b shows the fitted nonlinear model in the original scale superimposed on the scatter plot of the data. Observe from the plot that the fitted asymptote is too small, and the model does not fit the data well at high concentrations. The variance of the replicated response observations at the low concentrations has been distorted by the transformation, so runs with low concentration (or high reciprocal concentration) have too much influence on the least squares fit.

3.3 PARAMETER ESTIMATION IN A NONLINEAR SYSTEM

3.3.1 Nonlinear Least Squares

Suppose that we have a sample of n observations on the response and the regressors, say, y_i, x_{i1}, $x_{i2}, \ldots,$ x_{ik}, for $i = 1,2,\ldots,$ n. We have observed previously that the method of least squares in *linear* regression involves minimizing the least squares function

$$S(\boldsymbol{\beta}) = \sum_{i=1}^{n} \left[y_i - \left(\beta_0 + \sum_{j=1}^{k} \beta_j x_{ij} \right) \right]^2$$

Because this is a linear regression model, when we differentiate $S(\boldsymbol{\beta})$ with respect to the unknown parameters and equate the derivatives to zero, the resulting normal equations are linear equations, and consequently they are easy to solve.

Now consider the nonlinear regression situation. The general form of the model is

$$y_i = f(\mathbf{x}_i, \boldsymbol{\beta}) + \varepsilon_i, i = 1, 2, \ldots, n$$

where $\mathbf{x}_i' = [x_{i1}, x_{i2}, \ldots, x_{ik}]$ for $i = 1,2,\ldots, n$. The least squares function is

$$S(\boldsymbol{\beta}) = \sum_{i=1}^{n} [y_i - f(\mathbf{x}_i, \boldsymbol{\beta})]^2 \tag{3.13}$$

To find the least squares estimates, we must differentiate (3.13) with respect to $\boldsymbol{\beta}$. This will provide a set of p normal equations for the nonlinear regression situation. The normal equations are

$$\sum_{i=1}^{n} [y_i - f(\mathbf{x}_i, \boldsymbol{\beta})] \left[\frac{\partial f(\mathbf{x}_i, \boldsymbol{\beta})}{\partial \beta_j} \right]_{\boldsymbol{\beta}=\mathbf{b}} = 0 \quad \text{for } j = 1, 2, \ldots, p \tag{3.14}$$

In a nonlinear regression model the derivatives in the large square brackets will be functions of the unknown parameters. Furthermore, the expectation function is also a nonlinear function, so the normal equations can be very difficult to solve.

Example 3.3. Finding the Normal Equations for a Specific Model. Consider the nonlinear regression model in Equation (3.4):

$$y = \beta_1 e^{\beta_2 x} + \varepsilon$$

The least squares normal equations for this model are as follows:

$$\sum_{i=1}^{n} \left[y_i - b_1 e^{b_2 x_i} \right] e^{b_2 x_i} = 0$$

$$\sum_{i=1}^{n} \left[y_i - b_1 e^{b_2 x_i} \right] b_1 x_i e^{b_2 x_i} = 0 \tag{3.15}$$

After simplification the normal equations are

$$\sum_{i=1}^{n} y_i e^{b_2 x_i} - b_1 \sum_{i=1}^{n} e^{2b_2 x_i} = 0$$

$$\sum_{i=1}^{n} y_i x_i e^{b_2 x_i} - b_1 \sum_{i=1}^{n} x_i e^{2b_2 x_i} = 0 \tag{3.16}$$

These equations are not linear in b_1 and b_2, and no simple closed-form solution exists. In general, **iterative methods** must be used to find the values of b_1 and b_2. To further complicate the problem, sometimes there are multiple solutions to the normal equations. That is, there are multiple stationary values for the residual sum of squares function $S(\boldsymbol{\beta})$. □

3.3.2 The Geometry of Linear and Nonlinear Least Squares

Examining the geometry of the least squares problem is helpful in understanding the complexities introduced by a nonlinear model. For a given sample, the residual sum of squares function $S(\boldsymbol{\beta})$ depends only on the model parameters $\boldsymbol{\beta}$. Thus, in the parameter space (the space defined by the $\beta_1, \beta_2, \ldots, \beta_p$), we can represent the function $S(\boldsymbol{\beta})$ with a contour plot, where each contour on the surface is a line of constant residual sum of squares.

Suppose that the regression model is linear. Figure 3.3a shows the contour plot for this situation. If the model is linear in the unknown parameters, the contours are **ellipsoidal** and have a unique global minimum at the least squares estimator **b**.

When the model is nonlinear, the contours often appear as in Figure 3.3b. Notice that these contours are not elliptical, and are in fact quite elongated and irregular in shape. A banana-shaped appearance is very typical. The specific shape and orientation of the residual sum of squares contours depend on the form of the nonlinear model and the sample of data that has been obtained. Often the surface will be very elongated near the optimum, so many solutions for $\boldsymbol{\beta}$ produce a residual sum of squares that is close to the global minimum. This results in a problem that is **illconditioned**, and in such problems it is often difficult to find the global minimum for $\boldsymbol{\beta}$. In some situations the contours may be so irregular that there are several local minimum, and perhaps more than one global minimum. Figure 3.3c shows a situation where there is one local minimum and one global minimum.

3.3.3 Maximum Likelihood Estimation

We have concentrated on least squares in the nonlinear case. If we know the distribution of the error then we can use the method of maximum likelihood to estimate the model parameters. If the errors are normally and independently

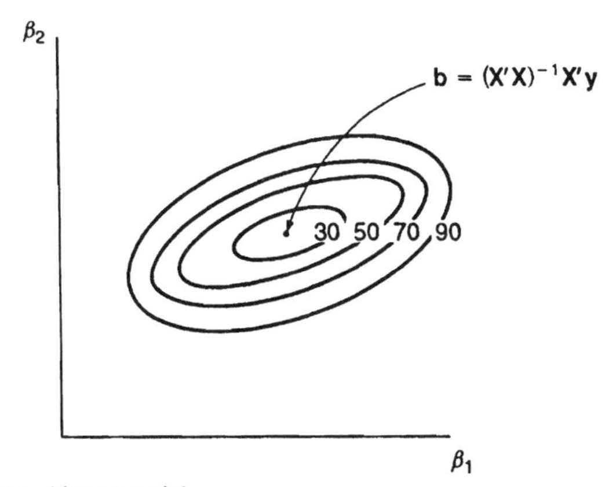

(a) Linear model

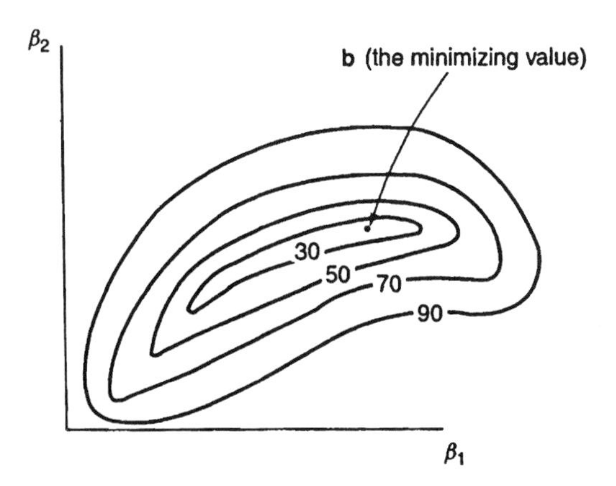

(b) Nonlinear model

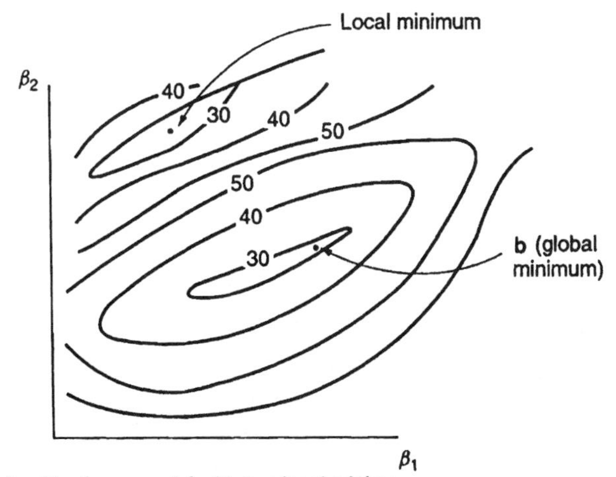

(c) Nonlinear model with two local minima

Figure 3.3 Contours of the residual sum of squares function: (a) linear model, (b) nonlinear model, and (c) nonlinear model with one local and one global minimum.

distributed with constant variance, application of the method of maximum likelihood to the estimation problem will lead to least squares. For example, consider the model

$$y_i = f(\mathbf{x}_i, \boldsymbol{\beta}) + \varepsilon_i, \quad i = 1, 2, \ldots, n \tag{3.17}$$

If the errors are normally and independently distributed with mean zero and variance σ^2, then the likelihood function is

$$\mathscr{L}(\boldsymbol{\beta}, \sigma^2) = \frac{1}{(2\pi\sigma^2)^{n/2}} \exp\left[-\frac{1}{2\sigma^2} \sum_{i=1}^{n} [y_i - f(\mathbf{x}_i, \boldsymbol{\beta})]^2\right] \tag{3.18}$$

Recall that maximizing the likelihood function is equivalent to maximizing the log-likelihood, which is

$$\ln \mathscr{L}(\boldsymbol{\beta}, \sigma^2) = -\frac{n}{2} \ln(2\pi\sigma^2) - \frac{1}{2\sigma^2} \sum_{i=1}^{n} [y_i - f(\mathbf{x}_i, \boldsymbol{\beta})]^2 \tag{3.19}$$

Clearly, choosing the vector of parameters **b** that maximizes the log-likelihood is equivalent to minimizing the residual sum of squares. Therefore, in the normal-theory case, least squares estimates in nonlinear regression are the same as maximum likelihood estimates.

From Equation (3.14) the maximum likelihood estimates must satisfy the **score equations**

$$\frac{1}{\sigma^2} \sum_{i=1}^{n} [y_i - f(\mathbf{x}_i, \boldsymbol{\beta})] \left[\frac{\partial f(\mathbf{x}_i, \boldsymbol{\beta})}{\partial \beta_i}\right]_{\boldsymbol{\beta}=\mathbf{b}} = 0 \text{ for } j = 1, 2, \ldots, p$$

Let $\mu_i = f(\mathbf{x}_i, \boldsymbol{\beta})$ for $i = 1,2,\ldots, n$ and $D_{ij} = \left[\frac{\partial f(\mathbf{x}_i, \boldsymbol{\beta})}{\partial \beta_j}\right]_{\boldsymbol{\beta}=\mathbf{b}}$ for $i = 1,2,\ldots, n$ and $j = 1,2,\ldots, p$. Then the score equations for a nonlinear regression model can be written in matrix notation as

$$\frac{1}{\sigma^2} \mathbf{D}'(\mathbf{y} - \hat{\boldsymbol{\mu}}) = \mathbf{0} \tag{3.20}$$

where $\boldsymbol{\mu}' = [\mu_1, \mu_2, \ldots, \mu_n]$, $\mathbf{D} = [D_{ij}]$, and $\hat{\boldsymbol{\mu}}$ denotes the expectation function with the parameters $\boldsymbol{\beta}$ replaced with their estimators **b**. Note that the score equations are **nonlinear** equations. Furthermore, in *linear* regression we have $\mathbf{D} = \mathbf{X}$ and $\boldsymbol{\mu}, = \mathbf{X}\boldsymbol{\beta}$, so the score equations for nonlinear regression are directly analogous to the score equations for linear regression. Score equations very similar to Equation (3.20) arise in subsequent chapters for generalized linear models.

3.3.4 Linearization and the Gauss–Newton Method

A widely used method for least squares estimation of the parameters in nonlinear regression is **linearization** of the expectation function followed by the Gauss–Newton iteration method. Linearization is accomplished by a Taylor series expansion of $f(\mathbf{x}_i, \boldsymbol{\beta})$ about a point $\mathbf{b}'_0 = [b_{10}, b_{20}, \ldots, b_{p0}]$ with only the linear terms retained. The point $\mathbf{b}_0$ is usually an initial estimate or a set of starting values for the model parameters $\boldsymbol{\beta}$. The Taylor series expansion yields

$$\begin{aligned} y_i &= f(\mathbf{x}_i, \boldsymbol{\beta}) + \varepsilon_i \\ &= f(\mathbf{x}_i, \mathbf{b}_0) + \sum_{j=1}^{p} \left[\frac{\partial f(\mathbf{x}_i, \boldsymbol{\beta})}{\partial \beta_j}\right]_{\boldsymbol{\beta}=\mathbf{b}_0} (\beta_j - b_{j0}) + \varepsilon_i, \quad i = 1, 2, \ldots, n \end{aligned} \tag{3.21}$$

If we set

$$\begin{aligned} f_i^0 &= f(\mathbf{x}_i, \mathbf{b}_0) \\ y_i^0 &= y_i - f_i^0 \\ D_{ij}^0 &= \sum_{j=1}^{p} \left[\frac{\partial f(\mathbf{x}_i, \boldsymbol{\beta})}{\partial \beta_j}\right]_{\boldsymbol{\beta}=\mathbf{b}_0} \\ \theta_j^0 &= (\beta_j - b_{j0}) \end{aligned}$$

then we can write Equation (3.21) as

$$y_i^0 = \sum_{j=1}^{p} \theta_j^0 D_{ij}^0 + \varepsilon_i, \quad i = 1, 2, \ldots, n \tag{3.22}$$

This is a linear regression model with unknown parameters θ_j^0, $j = 1, 2, \ldots, p$.

In matrix notation Equation (3.22) is

$$\mathbf{y}_0 = \mathbf{D}_0 \boldsymbol{\theta}_0 + \boldsymbol{\varepsilon} \tag{3.23}$$

and the least squares estimate of $\boldsymbol{\theta}_0$ is

$$\begin{aligned} \hat{\boldsymbol{\theta}}_0 &= (\mathbf{D}'_0 \mathbf{D}_0)^{-1} \mathbf{D}'_0 \mathbf{y}_0 \\ &= (\mathbf{D}'_0 \mathbf{D}_0)^{-1} \mathbf{D}'_0 (\mathbf{y} - \mathbf{f}_0) \end{aligned} \tag{3.24}$$

Now because $\boldsymbol{\theta}_0 = \boldsymbol{\beta} - \mathbf{b}_0$, we can use

$$\mathbf{b}_1 = \mathbf{b}_0 + \hat{\boldsymbol{\theta}}_0 \tag{3.25}$$

as a **revised** estimate of the unknown parameters $\boldsymbol{\beta}$. We usually call $\hat{\boldsymbol{\theta}}_0$ the **vector of increments**. Now we may use the revised parameter estimates $\mathbf{b}_1$ in Equation (3.21) in the same role originally played by the starting values $\mathbf{b}_0$, obtaining another set of revised estimates, say, $\mathbf{b}_2$. In general, at the k th of these iterations we have

$$\begin{aligned} \mathbf{b}_{k+1} &= \mathbf{b}_k + \hat{\boldsymbol{\theta}}_k \\ &= \mathbf{b}_k + (\mathbf{D}_k'\mathbf{D}_k)^{-1}\mathbf{D}_k'(\mathbf{y} - \mathbf{f}_k) \end{aligned} \tag{3.26}$$

where

$$\begin{aligned} \mathbf{D}_k &= [D_{ij}^k] \\ \mathbf{f}_k &= [f_1^{\,k}, f_2^{\,k}, \ldots, f_n^{\,k}]' \\ \mathbf{b}_k &= [b_{1k}, b_{2k}, \ldots, b_{pk}]' \end{aligned}$$

This iterative process continues until convergence, that is, until there is little meaningful change in the estimates of the parameters. Typically the convergence criteria is based on

$$\left| \frac{b_{j,k+1} - b_{jk}}{b_{jk}} \right| < \delta, \; j = 1, 2, \ldots, p$$

where δ is some small number, say, 10^{-6}. At each iteration the residual sum of squares should also be evaluated to ensure that a reduction in its value has been obtained.

Example 3.4. The Puromycin Data. We reconsider the puromycin data from Example 3.2, and use the Gauss–Newton linearization scheme to fit the Michaelis–Menten model. We use as the starting values for the parameters $b_{10} = 205$ and $b_{20} = 0.08$, based on the chemist's, suggestion. Later we show another approach for obtaining the starting values. The data and other quantities necessary to perform the linearization procedure are given in Table 3.2.

To see how the required quantities are calculated, note that the derivatives of the expectation function for the Michaelis–Menten model are

$$\frac{\partial f(x, \beta_1, \beta_2)}{\partial \beta_1} = \frac{x}{\beta_2 + x}$$

and

$$\frac{\partial f(x, \beta_1, \beta_2)}{\partial \beta_2} = \frac{-\beta_1 x}{(\beta_2 + x)^2}$$

Table 3.2 Data, Derivatives, f_i^0 and $y_i - f_i^0$ for the Puromycin Data at the Starting Values $\mathbf{b}_0' = [205, 0.08]'$

i	x_i	y_i	f_i^0	$y_i - f_i^0$	D_{i1}^0	D_{i2}^0
1	0.02	76	41.00	35.00	0.2000	−410.00
2	0.02	47	41.00	6.00	0.2000	−410.00
3	0.06	97	87.86	9.14	0.4286	−627.55
4	0.06	107	87.86	19.14	0.4286	−627.55
5	0.11	123	118.68	4.32	0.5789	−624.65
6	0.11	139	118.68	20.32	0.5789	−624.65
7	0.22	159	150.33	8.67	0.7333	−501.11
8	0.22	152	150.33	1.67	0.7333	−501.11
9	0.56	191	179.38	11.62	0.8750	−280.27
10	0.56	201	179.38	21.62	0.8750	−280.27
11	1.10	207	191.10	15.90	0.9322	−161.95
12	1.10	200	191.10	8.90	0.9322	−161.95

The first observation on x is $x_1 = 0.02$, so we have

$$D_{11}^0 = \left.\frac{x_1}{\beta_2 + x_1}\right|_{\beta_2=0.08} = \frac{0.02}{0.08 + 0.02} = 0.2000$$

$$D_{12}^0 = \left.\frac{-\beta_1 x_1}{(\beta_2 + x_1)^2}\right|_{\beta_1=205,\,\beta_2=0.08} = \frac{(-205)(0.02)}{(0.08 + 0.02)^2} = -410.00$$

The derivatives D_{ij}^0 are collected into the matrix $\mathbf{D}_0$ and the vector of increments calculated from Equation (3.24) as

$$\hat{\boldsymbol{\theta}}_0 = \begin{bmatrix} 8.03 \\ -0.017 \end{bmatrix}$$

The revised estimate of the parameters $\mathbf{b}_1$, from (3.25) is

$$\begin{aligned} \mathbf{b}_1 &= \mathbf{b}_0 + \hat{\boldsymbol{\theta}}_0 \\ &= \begin{bmatrix} 205.00 \\ 0.08 \end{bmatrix} + \begin{bmatrix} 8.03 \\ -0.017 \end{bmatrix} \\ &= \begin{bmatrix} 213.03 \\ 0.063 \end{bmatrix} \end{aligned}$$

The residual sum of squares at this point is $S(\mathbf{b}_1) = 1206$, which is considerably smaller than $S(\mathbf{b}_0)$. Therefore $\mathbf{b}_1$, is adopted as the revised estimate of $\boldsymbol{\beta}$ and another iteration would be performed.

The Gauss–Newton algorithm converged at the solution $\mathbf{b}' = [212.7, 0.0641]'$ with $S(\mathbf{b}) = 1195$. Therefore the fitted model obtained by linearization is

$$\hat{y} = \frac{b_1 x_i}{b_2 + x_i} = \frac{212.7 x_i}{0.0641 + x_i}$$

Figure 3.4 shows the fitted model. Notice that the nonlinear model provides a much better fit to the data than did the transformation followed by linear regression did in Example 3.3 (compare Figure 3.4 and Figure 3.2b). □

Residuals can be obtained from a fitted nonlinear regression model in the usual way; that is,

$$e_i = y_i - \hat{y}_i, \quad i = 1, 2, \ldots, n$$

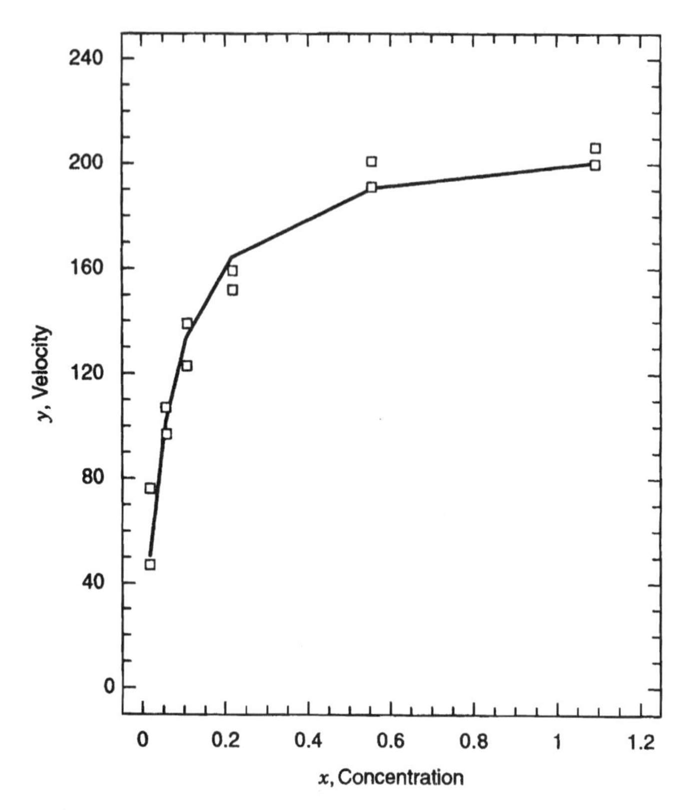

Figure 3.4 Plot of fitted nonlinear regression model, Example 3.4.

In the last example the residuals are computed from

$$e_i = y_i - \frac{b_1 x_i}{p_2 + x_i} = y_i - \frac{212.7x_i}{0.0641 + x_i} \quad i = 1, 2, \ldots, 10$$

The residuals are plotted versus the predicted values in Figure 3.5. A normal probability plot of the residuals is shown in Figure 3.6. There is one moderately large residual; however, the overall fit is satisfactory, and a substantial improvement from that obtained by the transformation approach in Example 3.2.

Estimation of σ^2

When the estimation procedure converges to a final vector of parameter estimates **b**, we can obtain an estimate of the error variance σ^2 from the residual mean square

$$\hat{\sigma}^2 = MS_{\text{res}} = \frac{\sum_{i=1}^{n}(y_i - \hat{y}_i)^2}{n-p} = \frac{\sum_{i=1}^{n}[y_i - f(\mathbf{x}_i, \mathbf{b})]^2}{n-p} = \frac{S(\mathbf{b})}{n-p} \tag{3.27}$$

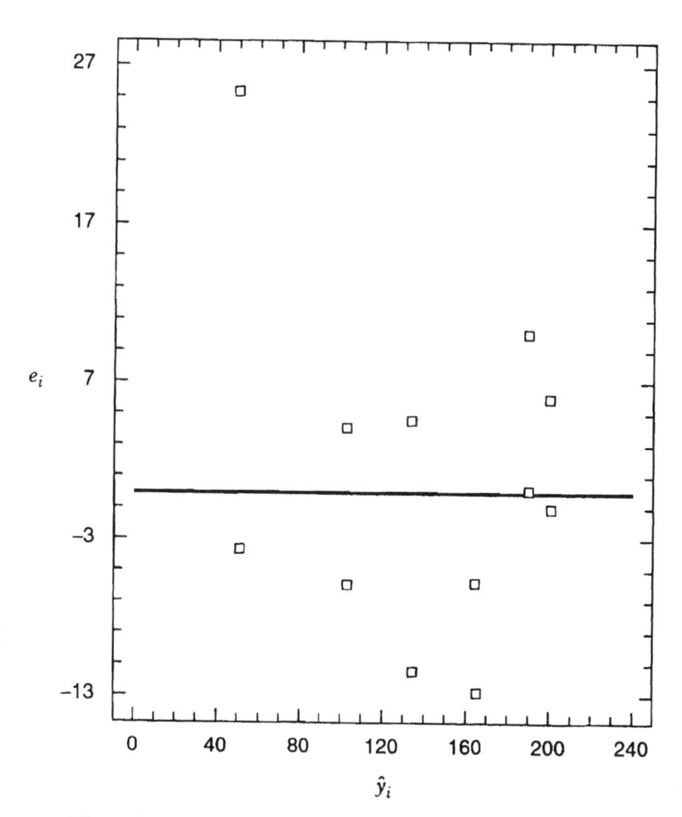

Figure 3.5 Plot of residuals versus predicted. Example 3.4.

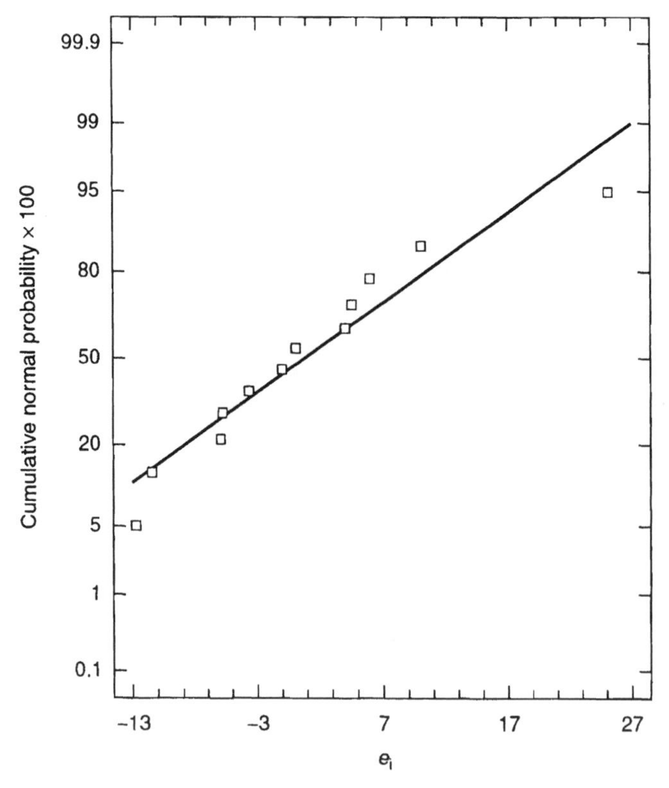

Figure 3.6 Normal probability plot of residuals, Example 3.4.

where p is the number of parameters in the nonlinear regression model. For the puromycin data in Example 3.4, we found that the residual sum of squares at the final iteration was $S(\mathbf{b}) = 1195$, so the estimate of σ^2 is

$$\hat{\sigma}^2 = \frac{S(\mathbf{b})}{n-p} = \frac{1195}{10-2} = 119.5$$

Graduate level mathematical statistics texts, such as Bickel and Doksum (2001), outline the statistical properties of the score function. From these properties the **asymptotic (large-sample) covariance matrix** of the parameter vector **b** can be estimated by

$$\text{Var}(\mathbf{b}) = \hat{\sigma}^2(\mathbf{D}'\mathbf{D})^{-1}$$

where **D** is the matrix of partial derivatives defined previously, evaluated at the final-iteration least squares parameter estimates **b**. This asymptotic covariance is the inverse of the **information matrix**, which can be found from the score equation in (3.20); that is,

$$\mathbf{I}(\mathbf{b}) = \widehat{\text{Var}}\left[\frac{1}{\sigma^2}\mathbf{D}'(\mathbf{y} - \boldsymbol{\mu})\right] = \frac{1}{\hat{\sigma}^2}\mathbf{D}'\mathbf{D}$$

The covariance matrix of the **b** vector for the Michaelis–Menten model in Example 3.4 is

$$\begin{aligned}\widehat{\text{Var}}(\mathbf{b}) &= \hat{\sigma}^2(\mathbf{D}'\mathbf{D})^{-1} \\ &= 119.5\begin{bmatrix} 0.4037 & 36.82 \times 10^{-5} \\ 36.82 \times 10^{-5} & 57.36 \times 10^{-8} \end{bmatrix}\end{aligned}$$

The main diagonal elements of this matrix are approximate variances of the estimates of the regression coefficients. Therefore approximate **standard errors** on the coefficients are

$$\widehat{se}(b_1) = \sqrt{\widehat{\text{Var}}(b_1)} = \sqrt{119.5(0.4037)} = 6.95$$

and

$$\widehat{se}(b_2) = \sqrt{\widehat{\text{Var}}(b_2)} = \sqrt{119.5(57.36 \times 10^{-8})} = 8.28 \times 10^{-3}$$

and the correlation between b_1 and b_2 is approximately

$$\frac{36.82 \times 10^{-5}}{\sqrt{0.4037(57.36 \times 10^{-8})}} = 0.77$$

A Graphical Perspective on Linearization

We have observed that the residual sum of squares function $S(\mathbf{b})$ for a nonlinear regression model is usually an irregular banana-shaped function, as shown in Figure 3.3b,c. On the other hand the residual sum of squares function for linear least squares is very well behaved; in fact, it is elliptical and has the global minimum at the bottom of the bowl. Refer to Figure 3.3a. The linearization technique converts the nonlinear regression problem into a sequence of linear ones, starting at the point $\mathbf{b}_0$.

The first iteration of linearization replaces the irregular contours with a set of elliptical contours. The irregular contours of $S(\boldsymbol{\beta})$ pass exactly through the starting point $\mathbf{b}_0$, as shown in Figure 3.7a. When we solve the linearized problem, we are moving to the global minimum on the set of elliptical contours. This is done by ordinary linear least squares, and it yields the solution $\mathbf{b}_1$. Then

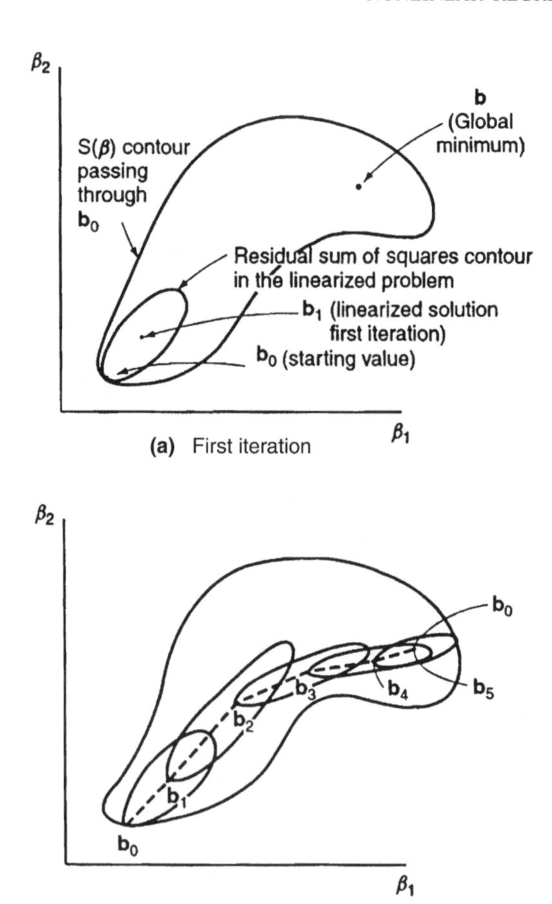

Figure 3.7 A geometric view of linearization: (a) first iteration and (b) evolution of successive linearization iterations.

the next iteration just repeats the process, starting at the new solution $\mathbf{b}_1$. The eventual evolution of linearization is a sequence of linear problems for which the solutions close in on the global minimum of the nonlinear function. This is illustrated in Figure 3.7b. Provided that the nonlinear problem is not too ill conditioned, either because of a poorly specified model or inadequate data, the linearization procedure should converge to a good estimate of the global minimum in a few iterations.

Linearization is facilitated by a good starting value $\mathbf{b}_0$; that is, one that is reasonably close to the true parameter values producing a global minimum on the residual sum of squares surface. When $\mathbf{b}_0$ is close to the true vector of parameter values $\boldsymbol{\beta}$, the actual residual sum of squares contours of the nonlinear problem are usually well approximated by the contours of the linearized problem. We discuss obtaining starting values in Section 3.3.7.

Computer Solution

Many statistics software packages have nonlinear regression modeling capability. For example, the SAS procedure PROC NLIN is a widely used program. It has several procedures for parameter estimation. The program estimates the required derivatives numerically or allows the user to input the derivatives directly. Many PC-based statistics packages also have nonlinear regression model-fitting capability. Table 3.3 gives the appropriate SAS code to analyze the puromycin data. Table 3.4 presents the resulting output from

Table 3.3 SAS Code for Puromycin Data Set

```
data puromycin;
input x y;
cards;
0.02  76
0.02  47
0.06  97
0.06 107
0.11 123
0.11 139
0.22 159
0.22 152
0.56 191
0.56 201
1.10 207
1.10 200
Proc nlin;
  parms t1 = 195.81
        t2 = 0.04841;
  model y = t1*x/ (t2 + x);
  der.t1 = x/ (t2 + x) ;
  der.t2 = -t1*x/ ((t2 + x) * (t2 + x));
  output out = puro2 student = rs p = yp;
run;
goptions device = win hsize = 6 vsize = 6;
  symbol value = star;
proc gplot data = puro2;
  plot rs*yp rs*x;
  plot y*x = ''*''yp*x = ''+''/overlay;
run;
proc capability data = puro2;
  var rs;
  qqplot rs;
run;
```

Table 3.4 Output from SAS PROC NLIN for the Puromycin Data

```
         Nonlinear Least Squares Iterative Phase
Dependent Variable Y               Method: Gauss-Newton
Iter          A                   B                Sum of Squares
 0        205.000000          0.080000            3155.004234
 1        213.028894          0.062892            1205.661845
 2        212.603375          0.063988            1195.477124
 3        212.675434          0.064108            1195.449080
 4        212.682940          0.064120            1195.448817
 5        212.683666          0.064121            1195.448814
NOTE: Convergence criterion met.

    Nonlinear Least Squares      Dependent Variable Y
      Summary Statistics

Source                  DF   Sum of Squares    Mean Square
Regression               2   270213.55119    135106.77559
Residual                10     1195.44881       119.54488
Uncorrected Total       12   271409.00000
(Corrected Total)       11    30858.91667

Parameter  Estimate      Asymptotic          Asymptotic 95%
                         Std. Error        Confidence Interval
                                           Lower          Upper
A         212.6836658 6.9471509268  197.20435150  228.16298000
B           0.0641212 0.0082809366    0.04566997    0.08257235

             Asymptotic Correlation Matrix

           Corr           A                  B
           --------------------------------------
           A      1                     0.7650835687
           B      0.7650835687          1
```

the SAS PROC NLIN. Notice that five iterations were required to produce the least squares estimates of the parameters.

The computer output in Table 3.3 for the puromycin data closely resembles the output from a standard linear regression computer program. The table presents the parameter estimates, where the labels A and B represent b_1, and b_2 respectively. We used the same starting values in these calculations as we did in Example 3.4, $b_{10} = 205$ and $b_{20} = 0.08$. The solution found is virtually identical to the linearization and Gauss–Newton iteration solution from Example 3.4. The output also contains the standard errors of the parameter estimates. These are very similar to the results given earlier.

Notice that one could construct a t-like ratio for each coefficient that is computed as

$$\frac{b}{\widehat{se}(b)} = \frac{\text{coefficient estimate}}{\text{standard error of the coefficient}}$$

In linear regression we would recognize this as a t-statistic for testing the null hypothesis H_0: $\beta = 0$. Table 3.3 also presents an analysis of variance partitioning of the total variability in the response y. We could construct a ratio of the regression mean square to the error or residual mean square. In linear regression, this would be the familiar F-statistic for testing significance of regression, or $H_0 : \beta_1, = \beta_2 = 0$. In nonlinear regression these ratios are not exact t- or F-statistics, but they can usually be interpreted as **approximate** t- or F-statistics. Statistical inference on the model in nonlinear regression rests on **asymptotic** or **large-sample** results. In Section 3.4 we discuss some useful aspects of inference for nonlinear regression models.

3.3.5 Using R to Perform Nonlinear Regression Analysis

In this section we outline the appropriate R code to analyze the puromycin data. This analysis assumes that the data are in a file named "puromycin.txt". The R code to read the data into the package is

```
puro <- read.table ("puromycin.txt", header=TRUE, sep=" ")
```

The object puro is the R data set. The commands

```
puro.model<- nls (y~t1*x/ (t2+x),start=list(t1=205,t2=.08), data=
puro))
summary (puro.model)
```

tell R to estimate the model and to print the estimated coefficients and their tests. The commands

```
yhat <- fitted (puro.model)
e <- residuals (trans.model)
qqnorm (e)
plot (yhat, e)
plot (puro$x, t)
```

set up and then create the appropriate residual plots. The commands

```
puro2 <- cbind (puro, yhat, e)
write.table (puro2, "puromycin_output.txt")
```

create a file "puromycin_output.txt" which the user can then import into his/her favorite package for doing graphics.

3.3.6 Other Parameter Estimation Methods

The basic linearization method for parameter estimation in nonlinear regression described in Section 3.3.4 may converge very slowly in some problems. In other problems it may generate a move in the wrong direction, with the residual sum of squares function $S(\mathbf{b}_k)$ actually *increasing* at the kth iteration, or in extreme situations linearization and Gauss–Newton may fail to converge at all. Consequently, several other techniques for solving the nonlinear regression problem have been developed. Some of them are modifications and refinements of the linearization scheme. In this section we give a brief description of some of these procedures.

The Method of Steepest Descent
The method of steepest descent attempts to find the global minimum on the residual sum of squares function by direct minimization. The objective is to move from an initial starting point $\mathbf{b}_0$ in the vector direction with components given by the derivatives of the residual sum of squares function with respect to the elements of the parameter vector $\boldsymbol{\beta}$. Usually these derivatives are estimated by fitting a first-order or planar approximation around the point $\mathbf{b}_0$. The regression coefficients in the first-order model are taken as approximations to the first derivatives.

The method of steepest descent is widely used in response surface methodology to move from an initial estimate of the optimum conditions for a process to a region more likely to contain the optimum. The major disadvantage of this method in solving the nonlinear regression problem is that it may converge very slowly. Steepest descent usually works best when the starting point is a long way from the optimum. However, as the current solution gets closer to the optimum, the procedure produces shorter and shorter moves and zig-zag behavior. This is the convergence problem mentioned previously.

Fractional Increments
A standard modification to the linearization technique is the use of **fractional increments**. To describe this method, we let $\hat{\boldsymbol{\theta}}_k$ be the standard increment vector in Equation (3.26) at the kth iteration, but continue to the next iteration only if $S(\mathbf{b}_{k+1}) < S(\mathbf{b}_k)$. If $S(\mathbf{b}_{k+1}) > S(\mathbf{b}_k)$, we use $\hat{\boldsymbol{\theta}}_k/2$ as the vector of increments. This halving could be used several times during an iteration, if necessary. If after a specified number of trials a reduction in $S(\mathbf{b}_k)$ is not obtained, the procedure is terminated. The general idea behind this method is to keep the linearization procedure from taking a step at any iteration that is too big. The fractional increments technique is helpful when convergence problems are encountered in the basic linearization procedure.

Marquardt's Compromise

Another popular modification to the basic linearization algorithm was developed by Marquardt (1963). He proposed computing the vector of increments at the kth iteration from

$$(\mathbf{D}_k'\mathbf{D}_k + \lambda\mathbf{I}_p)\hat{\boldsymbol{\theta}}_k = \mathbf{D}_k'(\mathbf{y} - \mathbf{f}_k) \tag{3.28}$$

where $\lambda > 0$. Note the similarity to ridge regression (see Myers, 1990, and Montgomery, Peck, and Vining, 2006). Since the regressor variables are derivatives of the same function, the linearized function invites multicollinearity. Thus the ridgelike procedure in (3.28) is intuitively reasonable. Marquardt (1963) used a search procedure to find a value of λ that would reduce the residual sum of squares at each stage.

Different computer programs select λ in different ways. For example, PROC NLIN in SAS begins with $\lambda = 10^{-8}$. A series of trial-and-error computations are done at each iteration with λ repeatedly multiplied by 10 until

$$S(\mathbf{b}_{k+1}) < S(\mathbf{b}_k) \tag{3.29}$$

The procedure also involves reducing λ by a factor of 10 at each iteration as long as (3.29) is satisfied. The strategy is to keep λ as small as possible while ensuring that the residual sum of squares is reduced at each iteration. This general procedure is often called **Marquardt's compromise** because the resulting vector of increments produced by his method usually lies between the Gauss–Newton vector in the linearization procedure and the direction of steepest descent.

3.3.7 Starting Values

Fitting a nonlinear regression model requires starting values $\mathbf{b}_0$ of the model parameters. Good starting values, that is, values of $\mathbf{b}_0$ that are close to the true parameter values, minimize convergence difficulties. Modifications to the linearization procedure such as Marquardt's compromise make the procedure less sensitive to the choice of starting values, but it is always a good idea to select $\mathbf{b}_0$ carefully. A poor choice could cause convergence to a local minimum on the function, and we might be completely unaware that a suboptimal solution has been obtained.

In nonlinear regression models the parameters often have some physical meaning, and this can be very helpful in obtaining starting values. It may also be helpful to plot the expectation function for several values of the parameters to become familiar with the behavior of the model and how changes in the parameter values affect this behavior.

For example, in the Michaelis–Menten function used for the puromycin data, the parameter b_1 is the asymptotic velocity of the reaction, that is, the maximum value of f as $x \rightarrow \infty$. Similarly b_2 represents the half-concentration or the value of x such that when the concentration reaches that value, the velocity is one-half the maximum value.

The chemist in Example 3.4 took this basic approach. She examined the scatter diagram in Figure 3.1 and concluded that $b_{10} = 205$ and $b_{20} = 0.08$.

In some cases we may transform the expectation function to obtain starting values. For example, the Michaelis–Menten model can be linearized by taking the reciprocal of the expectation function. Linear least squares can be used on the reciprocal data, as we did in Example 3.2, resulting in estimates of the linear parameters. These estimates can then be used to obtain the necessary starting values $\mathbf{b}_0$. Graphical transformation can also be very effective. A nice example of this is given in Bates and Watts (1988, p. 47).

3.4 STATISTICAL INFERENCE IN NONLINEAR REGRESSION

In a linear regression model, when the errors are normally and independently distributed, **exact** statistical tests and confidence intervals based on the t and F distributions are available, and the least squares (or equivalentiy the maximum likelihood) parameter estimates have useful and attractive statistical properties. However, this is not the case in nonlinear regression, even when the errors are normally and independently distributed. That is, in nonlinear regression the least squares (or maximum likelihood) estimates of the model parameters do not enjoy any of the attractive properties that their counterparts do in linear regression, such as unbiasedness, minimum variance, or normal sampling distributions. Statistical inference in nonlinear regression depends on **large-sample** or **asymptotic** results. The large-sample theory generally applies both for normally and nonnormally distributed response variables.

The key asymptotic results may briefly be summarized as follows. In general, when the sample size n is large, the expected value of $\mathbf{b}$ is approximately equal to $\boldsymbol{\beta}$, the true vector of parameter values. Furthermore, the sampling distribution of $\mathbf{b}$ is approximately normal. The asymptotic covariance of $\mathbf{b}$ is the inverse of the information matrix. Recall from our discussion of maximum likelihood estimation in linear regression (Chapter 2) that the information matrix is the negative of the Hessian matrix, which is just the matrix of second derivatives of the log-likelihood function. For a normal response distribution, the Hessian is $-(\mathbf{D}'\mathbf{D}) / \sigma^2$, where $\mathbf{D}$ is the matrix of partial derivatives of the model expectation function evaluated at the final-iteration least squares estimate $\mathbf{b}$. Therefore the asymptotic covariance matrix of $\mathbf{b}$ is

$$\widehat{\text{Var}}(\mathbf{b}) = \sigma^2(\mathbf{D}'\mathbf{D})^{-1} \tag{3.30}$$

As we indicated earlier, we can arrive at this result by finding the information matrix as the variance of the score, and the covariance matrix of $\mathbf{b}$ is the inverse of this information matrix. Consequently, statistical inference for nonlinear regression when the sample size is large is carried out much as it is for linear regression. For example, to test the hypothesis that an individual regression coefficient is equal to zero, or

$$\begin{aligned} H_0 &: \beta = 0 \\ H_1 &: \beta \neq 0 \end{aligned} \tag{3.31}$$

we use a ratio similar to a t-statistic given by

$$t_0 = \frac{b}{\widehat{se}(b)} \tag{3.32}$$

where $\widehat{se}(b)$ is the estimated standard error of b, which we would obtain as a diagonal element of $\widehat{\text{Var}}(\mathbf{b}) = \sigma^2(\mathbf{D}'\mathbf{D})^{-1}$. The asymptotic distribution of t_0 is N(0,1) when the null hypothesis is true. This large-sample approach to statistical inference is usually called **Wald inference**. An approximate $100(1 - \alpha)\%$ Wald confidence interval on the parameter β is

$$b - z_{\alpha/2}\,\widehat{se}(b) \leq \beta \leq b + z_{\alpha/2}\,\widehat{se}(b) \tag{3.33}$$

In many applications the model builder is interested in a function of the parameters in a nonlinear model, and the function is nonlinear in the elements of $\boldsymbol{\beta}$. For example, one may be interested in a confidence interval on the mean response or a prediction interval on a future observed response. In general, suppose that we are interested in a nonlinear function of the elements of $\boldsymbol{\beta}$, say, $g(\boldsymbol{\beta})$. The **point estimate** of $g(\boldsymbol{\beta})$ is $g(\mathbf{b})$, and $g(\mathbf{b})$ is approximately (asymptotically) normally distributed. The mean and variance of $g(\mathbf{b})$ are found by expanding $g(\mathbf{b})$ in a first-order Taylor series around $\boldsymbol{\beta}$ and applying the expected value and variance operators to the resulting expression (this is also sometimes called the **delta method**). The first-order Taylor series expansion is

$$g(\mathbf{b}) = g(\boldsymbol{\beta}) + \mathbf{d}'(\mathbf{b} - \boldsymbol{\beta}) + R \tag{3.34}$$

where R is the remainder and $\mathbf{d}' = [\partial g(\boldsymbol{\beta})/\partial\beta_1, \partial g(\boldsymbol{\beta})/\partial\beta_2, \ldots, \partial g(\boldsymbol{\beta})/\partial\beta_p]$ evaluated at $\boldsymbol{\beta} = \mathbf{b}$. Now ignoring the remainder, the expected value of $g(\mathbf{b})$ is

$$E[g(\mathbf{b})] \cong g(\boldsymbol{\beta}) \tag{3.35}$$

and the covariance matrix of $g(\mathbf{b})$ is

$$\widehat{\text{Var}}[g(\mathbf{b})] \cong \hat{\sigma}^2\mathbf{d}'(\mathbf{D}'\mathbf{D})^{-1}\mathbf{d} \tag{3.36}$$

The applications of these results to obtaining confidence intervals on the mean response and prediction intervals in nonlinear regression are straightforward. Suppose that $\mathbf{x}_0$ is the location for which one wishes to obtain the point estimate or prediction of the response and the corresponding interval. In these applications the vector of derivatives $\mathbf{d}$ is just

$$\mathbf{d}_0' = [\partial f(\mathbf{x}_0, \boldsymbol{\beta})/\beta_1, \partial f(\mathbf{x}_0, \boldsymbol{\beta})/\beta_2, \ldots, \partial f(\mathbf{x}_0, \boldsymbol{\beta})/\beta_p]$$

Consequently, an estimate of the mean response at the point of interest $\mathbf{x}_0$ is

$$\hat{y}(\mathbf{x}_0) = f(\mathbf{x}_0, \mathbf{b}) \tag{3.37}$$

and the approximate standard error of the estimate of the mean response at $\mathbf{x}_0$ is

$$\widehat{se}[\hat{y}(\mathbf{x}_0)] = \sqrt{\hat{\sigma}^2 \mathbf{d}_0'(\mathbf{D}'\mathbf{D})^{-1}\mathbf{d}_0} \tag{3.38}$$

Therefore the approximate 100(1 − α)% confidence interval on the mean response at the point $\mathbf{x}_0$, say, $\mu_{y|\mathbf{x}_0}$, is

$$\hat{y}(\mathbf{x}_0) - z_{\alpha/2}\sqrt{\hat{\sigma}^2 \mathbf{d}_0'(\mathbf{D}'\mathbf{D})^{-1}\mathbf{d}_0} \leq \mu_{y|\mathbf{x}_0} \leq \hat{y}(\mathbf{x}_0) + z_{\alpha/2}\sqrt{\hat{\sigma}^2 \mathbf{d}_0'(\mathbf{D}'\mathbf{D})^{-1}\mathbf{d}_0} \tag{3.39}$$

By analogy with the results from linear regression, the 100(1 − α) prediction interval on the future observation at the point $\mathbf{x}_0$ is

$$\begin{aligned}\hat{y}(\mathbf{x}_0) &- z_{\alpha/2}\sqrt{\hat{\sigma}^2(1 + \mathbf{d}_0'(\mathbf{D}'\mathbf{D})^{-1}\mathbf{d}_0)} \\ &\leq y_0 \leq \hat{y}(\mathbf{x}_0) + z_{\alpha/2}\sqrt{\hat{\sigma}^2(1 + \mathbf{d}_0'(\mathbf{D}'\mathbf{D})^{-1}\mathbf{d}_0)}\end{aligned} \tag{3.40}$$

These are Wald tests and confidence intervals, which are only approximate procedures.

Example 3.5. The Puromycin Data. Reconsider the Michaelis–Menten model for the puromycin data from Example 3.2. The computer output for the model is shown in Table 3.3. To test for significance of regression (i.e., $H_0: \beta_1 = \beta_2 = 0$), we could use the ratio

$$F_0 = \frac{MS_{\text{Model}}}{MS_{\text{res}}} = \frac{135,106,77}{119.54} = 1130.22$$

and compute an approximate P-value from the $F_{2,\,10}$ distribution. This P-value is considerably less than 0.0001 (1.66E − 12), so we are safe in rejecting the null hypothesis and concluding that at least one of the model parameters is nonzero. To test hypotheses on the individual model parameters, H_0: $\beta_1 = 0$ and H_0: $\beta_2 = 0$, we could compute the Wald test statistics from Equation (3.32) as

$$t_0 = \frac{b_1}{\widehat{se}(b_1)} = \frac{212.6836658}{6.9471509268} = 30.61$$

and

$$t_0 = \frac{b_2}{\widehat{se}(b_2)} = \frac{0.0641212}{0.0082809366} = 7.74$$

The approximate P-values for these two test statistics are obtained from the N(0, 1) distribution and are obviously small. Therefore we would conclude that both model parameters are nonzero.

Approximate 95% confidence intervals on β_1 and β_2 are found from Equation (3.33) as follows:

$$b_1 - z_{0.025}\,\widehat{se}(b_1) \leq \beta_1 \leq b_1 + z_{0.025}\widehat{se}(b_1)$$

$$212.6836658 - 1.96(6.9471509268) \leq \beta_1 \leq 212.6836658 + 1.96(6.9471509268)$$

$$199.0672 \leq \beta_1 \leq 226.3001$$

and

$$b_2 - z_{0.025}\widehat{se}(b_2) \leq \beta_2 \leq z_{0.025}\widehat{se}(b_2)$$

$$0.0641212 - 1.96(0.0082809366) \leq \beta_2 \leq 0.0641212 + 1.96(0.0082809366)$$

$$0.0479 \leq \beta_2 \leq 0.0804$$

respectively. In comparing these results to the results from the computer output in Table 3.3, we note that the confidence intervals computed by SAS are slightly different. This is because SAS used the t-distribution with 10 degrees of freedom to construct the confidence intervals. □

Validity of Approximate Inference

Since the test procedures and confidence intervals in nonlinear regression are based on large-sample theory, and typically the sample size in a nonlinear regression problem may not be all that large, it is logical to inquire about the validity of the procedures. It would be desirable to have a simple guideline, or rule of thumb, that would tell us when the sample size is large enough so that the asymptotic results are valid. Unfortunately, no such general guideline is available. However, there are some **indicators** that the results may be valid in a particular application.

1. If the nonlinear regression estimation algorithm converges in only a few iterations, then this indicates that the linear approximation used in solving the problem was very satisfactory, and it is likely that the asymptotic results will apply nicely. If convergence requires many iterations, this can be a symptom that the asymptotic results may not apply, and other adequacy checks should be considered.
2. There are several measures of model curvature and nonlinearity that have been developed. Bates and Watts (1988) discuss these procedures. These measures describe quantitatively the adequacy of the linear approximation. Once again, an inadequate linear approximation would indicate that the asymptotic inference results are questionable.
3. A resampling technique called the **bootstrap** can be used to study the sampling distribution of estimators, to compute approximate standard errors, and to find approximate confidence intervals. We could compute bootstrap estimates of these quantities and compare them to the approximate standard errors and confidence intervals produced by the asymptotic results. Good agreement with the bootstrap estimates is an indication that the large-sample inference results are valid.

When there is some indication that the asymptotic inference results are not valid, the model builder has only a few choices. One possibility is to consider an alternate form of the model, if one exists, or perhaps a different nonlinear regression model, or perhaps a generalized linear model. Sometimes graphs of the data and graphs of different nonlinear model expectation functions are helpful in selecting another nonlinear model. Alternatively, one could use the inference results from resampling or the bootstrap. However, if the model is wrong, or poorly specified, there is little reason to believe that resampling results will be any more valid than the results from large-sample inference.

3.5 WEIGHTED NONLINEAR REGRESSION

The reader quickly discovers that in the analysis of generalized linear models the standard least squares procedure discussed in this chapter does not apply because the response distributions involved (binomial, Poisson) do not possess the property of constant variance. As a result, the maximum likelihood procedure, which in the normal response case reduces to least squares, has an important connection to weighted least squares. The reader should recall the generalized and weighted least squares estimators in linear regression models given by Equations (2.68) and (2.71) and the variance−covariance matrix of the estimator given by Equation (2.69). Also recall that the linearization procedure that stems from the use of the Taylor series heavily involves the derivative matrix **D**. Now, in a situation where one is confronted with a nonlinear model with $E(y_i) = \mu_i$, and $\text{Var}(y_i) = \text{diag}(\sigma_1^2, \sigma_2^2, \ldots, \sigma_n^2)$, the operative procedure is to apply generalized least squares just as in the linear regression model case, and thus choose $\boldsymbol{\beta}$ so as to minimize

$$L = (\mathbf{y} - \boldsymbol{\mu})'\mathbf{V}^{-1}(\mathbf{y} - \boldsymbol{\mu})$$

with resulting score function

$$\mathbf{D}'\mathbf{V}^{-1}(\mathbf{y} - \boldsymbol{\mu}) = \mathbf{0} \tag{3.41}$$

At this point we assume that the matrix of weights $\mathbf{V}$ is known. In Chapters 4 and 5 the elements of $\mathbf{V}$ invariably are functions of the model parameters. If $\mathbf{V}$ is known, Equation (3.41) can be solved through the Gauss–Newton procedure with the jth iteration producing

$$\mathbf{b}_j = \mathbf{b}_{j-1} + (\mathbf{D}'\mathbf{V}^{-1}\mathbf{D})^{-1}\mathbf{D}'\mathbf{V}^{-1}(\mathbf{y} - \boldsymbol{\mu})$$

Here, $\mathbf{D}$ contains $\boldsymbol{\beta}$ replaced by $\mathbf{b}_{j-1}$. Note the similarity to the situation in which $\mathbf{V} = \mathbf{I}$. The procedure is continued to convergence. As before, the asymptotic variance–covariance matrix is obtained from the information matrix as

$$\mathbf{I}(\mathbf{b}) = \text{Var}(\text{score}) = \mathbf{D}'\mathbf{V}^{-1}\mathbf{V}\mathbf{V}^{-1}\mathbf{D} = \mathbf{D}'\mathbf{V}^{-1}\mathbf{D}$$

Thus asymptotically

$$\text{Var}(\mathbf{b}) = (\mathbf{D}'\mathbf{V}^{-1}\mathbf{D})^{-1}$$

Once again, note how similar this result is to the case of weighted least squares for linear regression where $\mathbf{D} = \mathbf{X}$ and $\text{Var}(\mathbf{b}) = (\mathbf{X}'\mathbf{V}^{-1}\mathbf{X})^{-1}$.

3.6 EXAMPLES OF NONLINEAR REGRESSION MODELS

Ideally, a nonlinear regression model is chosen based on **theoretical considerations** from the subject matter field. That is, specific chemical, physical, or biological knowledge leads to a **mechanistic model** for the expectation function rather than an empirical one. Many nonlinear regression models fall into categories designed for specific situations or environments. In this section we discuss a few of these models.

Perhaps the best known category of nonlinear models is **growth models**. These models are used to describe how something grows with changes in a regressor variable. Often the regressor variable is time. Typical applications are in biology, where plants and organisms grow with time, but there are also many applications in economics and engineering. For example, the reliability growth in a complex system over time may often be described with a nonlinear regression model.

The **logistic** growth model is

$$y = \frac{\beta_1}{1 + \beta_2 \exp(-\beta_3 x)} + \varepsilon \tag{3.42}$$

The parameters in this model have a simple physical interpretation. For $x = 0$, $y = \beta_1/(1 + \beta_2)$ is the level of y at time (or level) zero. The parameter β_1, is the limit to growth as $x \rightarrow \infty$. The values of β_2 and β_3 must be positive. Also the term $-\beta_3 x$ in the denominator exponent of Equation (3.42) could be replaced by a more general structure in several regressors.

The **Gompertz** model given by

$$y = \beta_1 \exp(-\beta_2 e^{-\beta_3 x}) + \varepsilon \tag{3.43}$$

is another widely used growth model. At $x = 0$ we have $y = \beta_1 e^{-\beta_2}$ and β_1, is the limit to growth as $x \rightarrow \infty$.

The **Weibull** growth model is

$$y = \beta_1 - \beta_2 \exp(-\beta_3 x^{\beta_4}) + \varepsilon \tag{3.44}$$

When $x = 0$, we have $y = \beta_1 - \beta_2$, while the limiting growth is β_1, as $x \rightarrow \infty$.

In some applications the expected response is given by the solution to a set of linear differential equations. These models are often called **compartment models** and since chemical reactions can frequently be described by linear systems of first-order differential equations, they have frequent application in chemistry, chemical engineering, and pharmacokinetics. Other situations specify the expectation function as the solution to a nonlinear differential equation or an integral equation that has no analytic solution. There are special techniques for the modeling and solution of these problems. The interested reader is referred to Bates and Watts (1988).

3.7 DESIGNS FOR NONLINEAR REGRESSION MODELS

Section 2.7 outlines the basic issues for designing experiments that we plan to analyze by linear regression analysis. These same basic issues carry over to experimental designs for nonlinear models. However, now the situation is more complex since the variance–covariance matrix for the estimated coefficients depends on the very unknown parameters that we seek to estimate.

The most common optimality criterion used to generate nonlinear designs is D. For the linear regression situation,

$$\mathrm{Var}(\mathbf{b}) = \sigma^2(\mathbf{X}'\mathbf{X})^{-1}$$

The D-optimal design maximizes the determinant of $\mathbf{X}'\mathbf{X}$. In this case, $\mathbf{X}$ does not depend on $\boldsymbol{\beta}$. As a result, finding the D-optimal design is a fairly straightforward exercise in nonlinear programming.

For the nonlinear regression situation, the asymptotic variance of the vector of estimated coefficients is

$$\mathrm{Var}(\mathbf{b}) = \sigma^2(\mathbf{D}'\mathbf{D})^{-1}$$

where $\mathbf{D}$ is the matrix of first derivatives of the nonlinear function with respect to the unknown parameters. The D-optimal design minimizes the generalized variance, which is equivalent to maximizing the determinant of $\mathbf{D}'\mathbf{D}$. However, $\mathbf{D}'\mathbf{D}$ is a function of these unknown coefficients. The issue then becomes how to overcome this problem.

Box and Lucas (1959) treated the starting values to be used in the estimation procedure as the known values for the coefficients. They then assumed that the optimal design consisted of p distinct runs, where p is the number of unknown parameters. They established that the optimal design maximizes the determinant of $\mathbf{D}'\mathbf{D}$.

Consider the a D-optimal design for situation for the puromycin data. The model is $y = \frac{\beta_1 x}{x+\beta_2} + \varepsilon$ The resulting derivatives are

$$\frac{\partial y}{\partial \beta_1} = \frac{x}{\beta_2 + x} \quad \text{and} \quad \frac{\partial y}{\partial \beta_2} = \frac{-\beta_1 x}{(\beta_2 + x)^2}$$

There are two parameters to estimate, which implies that an optimal design may require only two runs. As a result,

$$\mathbf{D} = \begin{bmatrix} \frac{x_1}{\beta_2+x_1} & \frac{-\beta_1 x_1}{(\beta_2+x_1)^2} \\ \frac{x_2}{\beta_2+x_2} & \frac{-\beta_1 x_2}{(\beta_2+x_2)^2} \end{bmatrix}$$

The resulting determinant is

$$\frac{\beta_1 x_1 x_2 (x_2 - x_1)}{(\beta_2 + x_1)^2 (\beta_2 + x_2)^2}$$

which is maximized when $x_2 = x_{\max}$, where $x_{\max}$ is the largest theoretical value for x, and

$$x_1 = \frac{\beta_2}{1 + 2\beta_2/x_{\max}} = \frac{\beta_2 x_{\max}}{x_{\max} + 2\beta_2} \approx \frac{x_{\max}}{2}$$

Table 3.4 A Bayesian D-Optimal Design for the Puromycin Experiment

x
0.043422
0.043422
0.043422
0.1
0.026428
0.026428
0.1
0.1
0.043422
0.1
0.1
0.1

If the researcher believes that the largest theoretical value for x is 0.1, then the approximate D-optimal experiment has half the runs at $x = 0.10$ and half at x = 0.05.

It is important to note that this design is approximately optimal only in a very narrow sense. For example, this optimal design does not allow any test for lack of fit. In addition, this design strategy requires a fairly strong assumption that the proposed starting values are valid. Bates and Watts (1988) find the original puromycin experiment given in Table 3.1 "eminently sensible" when compared to the approximate D-optimal design.

Another approach uses a **Bayesian D-optimal** design. This approach assumes a prior distribution for the coefficients $\boldsymbol{\beta}$. The Bayesian D-optimal design is the one that maximizes the expected value of log $[\mathbf{D}'\mathbf{D}]$ with respect to this prior distribution. Chaloner and Larntz (1984) and Chaloner and Verdinelli (1995) discuss this approach. Gotwalt, Jones, and Steinberg (2009) develop a clever quadrature scheme for evaluating this integral.

JMP provides this feature. By default, JMP assumes a normal distribution for each coefficient over the range of the starting value $\pm$ one-half of the starting value. For the puromycin data, the starting value for β_1 is 205 and for β_2 is 0.08. Thus JMP uses as its prior distribution for β_1 a normal distribution over the range (102.5, 307.5) and as its prior distribution for β_2 a normal distribution over the range (0.4, 0.12). The range of possible values for x given to JMP is 0.0 to 0.10. Table 3.4 gives the resulting 12-run Bayesian D-optimal design. This design uses half the runs at 0.10 like the approximate D-optimal design. Interestingly, it uses four runs at 0.043422, which is near the 0.5 used by the approximate design. It also uses two runs at 0.026428. The use of three levels does give the experimenter the opportunity to test for lack of fit.

EXERCISES

3.1 Consider the Michaelis–Menten model in Equation (3.12). Graph the expected value of the response for $\beta_1 = 100$ and $\beta_2 = 0.04, 0.06, 0.08$, and 0.10. Overlay these plots on the same axes. What is the effect of the parameter β_2 on the expected response?

3.2 Consider the Michaelis–Menten model in Equation (3.12). Graph the expected value of the response for $\beta_1 = 100, 150, 200$, and 250 for $\beta_2 = 0.10$. Overlay these plots on the same axes. What is the effect of the parameter β_1 on the expected response?

3.3 Consider the logistic model in Equation (3.42). Graph the expected value of the response for $\beta_1 = 10$, $\beta_2 = 2$, and $\beta_3 = 0.5, 1, 2$, and 3, respectively. Overlay these plots on the same axes. What is the effect of the parameter β_3 on the expected response?

3.4 Consider the logistic model in Equation (3.42). Graph the expected value of the response for $\beta_1 = 1$, $\beta_3 = 1$, and $\beta_2 = 1, 4$, and 8, respectively. Overlay these plots on the same axes. What is the effect of the parameter β_2 on the expected response?

3.5 Consider the Gompertz model in Equation (3.43). Graph the expected value of the response for $\beta_1 = 1$, $\beta_3 = 1$, and $\beta_2 = 1/8, 1, 8$, and 64, respectively, over the range $0 \le x \le 10$. Overlay these plots on the same axes.

(a) What is the effect of the parameter β_2 on the expected response?

(b) Discuss the behavior of the response from this model as $x \to \infty$.

3.6 For the models shown below, determine whether it is a linear model, an intrinsically linear model, or a nonlinear model. If the model is intrinsically linear, show how it can be linearized by a suitable transformation.

(a) $y = \beta_1 e^{\beta_2 + \beta_3 x} + \varepsilon$.

(b) $y = \beta_0 + \beta_1 x_1 + \beta_2 x_2^{\beta_3} + \varepsilon$.

(c) $y = \beta_1 + (\beta_2/\beta_1)x + \varepsilon$.

(d) $y = \beta_1 x_1^{\beta_2} x_2^{\beta_3} + \varepsilon$.

(e) $y = \beta_1 + \beta_2 e^{\beta_3 x} + \varepsilon$.

3.7 Reconsider the regression models in Exercise 3.6 parts (a) to (e). Suppose that the error terms in these models were multiplicative. Rework the problem under this new assumption regarding the error structure.

3.8 Consider the data shown below. Suppose that we are considering fitting the nonlinear model

$$y = \beta_1 e^{\beta_2 x} + \varepsilon$$

to these data.

x	y	
0.5	0.68	1.58
1	0.45	2.66
2	2.50	2.04
4	6.19	7.85
8	56.1	54.2
9	89.8	90.2
10	147.7	146.3

(a) Discuss how you would find starting values for the parameter estimates.
(b) Fit the model to the data.
(c) Test for significance of regression.
(d) Estimate the error variance σ^2.
(e) Test the hypotheses H_0: $\beta_1 = 0$ and H_0: $\beta_2 = 0$. Are the model parameters different from zero? If not, refit an appropriate model.
(f) Analyze the residuals from this model and comment on model adequacy.

3.9 Reconsider the data from the previous exercise. Suppose that the response data were collected on two different days. Fit a new model to the data, say,

$$y = \beta_1 e^{\beta_2 x} + \beta_3 z + \varepsilon$$

where z is an indicator variable with $z = 0$ if the observation was made on the first day and $z = 1$ if the observation was made on the second day. Is there any indication that there is a difference between the two days?

3.10 The model $y = \beta_1 - \beta_2 e^{-\beta_3 x} + \varepsilon$ is called the Mitcherlich equation, and it is often used in chemical engineering to model the relationship between yield and reaction time.
(a) Is this a nonlinear regression model?
(b) Discuss how you would obtain starting values for the parameter estimates.
(c) Graph the expected response for $\beta_1 = 0.5$, $\beta_2 = -0.1$, and $\beta_3 = 0.1$. Discuss the shape of the curve.
(d) Graph the expected response for $\beta_1 = 0.5$, $\beta_2 = 0.1$, and $\beta_3 = 0.1$. Discuss the shape of the curve. How does it compare to the curve obtained in part (c)?

3.11 The data in the following table represent the fraction of active chlorine in a chemical product at a measured time following manufacturing.

Available Chlorine y	Time x
0.49, 0.49	8
0.48, 0.47, 0.48, 0.77	10
0.46, 0.46, 0.45, 0.43	12
0.45, 0.43, 0.43	14
0.44, 0.43, 0.43	16
0.46, 0.45	18
0.42, 0.42, 0.43	20
0.41, 0.41, 0.40	22
0.42, 0.40, 0.40	24
0.41, 0.40, 0.41	26
0.41, 0.40	28
0.40, 0.40, 0.38	30
0.41, 0.40	32
0.40	34
0.41, 0.38	36
0.40, 0.40	38
0.39	40
0.39	42

(a) Construct a scatter diagram of the data.

(b) Fit the Mitcherlich model to these data. Discuss how you obtained the starting values for the parameter estimates.

(c) Test for significance of regression.

(d) Find 95% confidence intervals on the model parameters. Do these confidence intervals indicate that the model parameters are different from zero?

(e) Find an estimate of σ^2 for this model.

3.12 Reconsider the data from Exercise 2.18. Suppose that we now wish to consider fitting a nonlinear model to these data, say, $y = \beta_1(x_1)^{\beta_2}(x_2)^{\beta_3}+\varepsilon$.

(a) Fit the nonlinear model to the data.

(b) Test for significance of regression.

(c) Does it appear that both of the design variables have important effects?

(d) Analyze the residuals and comment on model adequacy.

(e) Which model do you prefer, the linear model from Exercise 2.18 or the nonlinear model? Explain why.

3.13 Repeat Exercise 3.12 using the model $y = \beta_1(x_1)^{\beta_2}(x_2)^{\beta_3}(x_1x_2)^{\beta_4} + \varepsilon$. Discuss the difficulties that you encounter.

3.14 Reconsider the surgical services data introduced in Exercise 2.15. Fit the nonlinear model

$$y = \frac{x}{\beta_0 + \beta_1 x} + \varepsilon$$

to these data. Investigate fully the adequacy of the fit of this model to the data. How does the nonlinear model compare to the models obtained in Exercise 2.15?

3.15 A major problem associated with many mining projects is subsidence, or sinking of the ground above the excavation. The mining engineer needs to control the amount and distribution of this subsidence. This will ensure that structures on the surface survive the excavation. There are several factors that affect the amount and nature of the subsidence. Among these are the depth of the mine and the width of the excavation. An important variable, which aids in characterizing the condition, is known as the angle of draw, y. It is defined as the angle between the perpendicular at the edge of the excavation and the line that connects the same edge of excavation with the point on the surface for which there is zero subsidence. Engineers generally feel that the angle of draw should relate well to the ratio of the width (w) of the excavation to the depth (d) of the mine. It also is suspected that any relationship involved is nonlinear. The following is a data set collected by the Mining Engineering Department at Virginia Tech for mining excavations in West Virginia.

Observation	$w(ft)$	d(ft)	y (deg)
1	610	550	33.6
2	450	500	22.3
3	450	520	22.0
4	430	740	18.7
5	410	800	20.2
6	500	230	31.0
7	500	235	30.0
8	500	240	32.0
9	450	600	26.6
10	450	650	15.1
11	480	230	30.0
12	475	1400	13.5
13	485	615	26.8
14	474	515	25.0
15	485	700	20.4
16	600	750	15.0

(a) Fit a linear regression model of the form $y = \beta_0 + \beta_1 w + \beta_2 d + \varepsilon$ to these data. Investigate the adequacy of this model.

(b) Fit the nonlinear model

$$y = \beta_1 \left[1 - \exp\left(-\beta_2 \left(\frac{w}{d}\right)\right)\right] + \varepsilon$$

to these data. Investigate the adequacy of this model.

(c) Which model do you prefer? Explain why.

3.16 In the field of ecology, the relationship between the concentration of available dissolved organic substrate and the rate of uptake (velocity) of that substrate by heterotrophic microbial communities has been described by the Michaelis–Menten model. The velocity (y) and concentration (x) data shown in the following table were collected by the Department of Biology at Virginia Tech.

Observation	y	x
1	0.0773895	0.417
2	0.0688714	0.417
3	0.0819351	0.417
4	0.0737034	0.833
5	0.0738753	0.833
6	0.0712396	0.833
7	0.0650420	1.670
8	0.0547667	1.670
9	0.0497128	3.750
10	0.0642727	3.750
11	0.0613005	6.250
12	0.0643576	6.250
13	0.0393892	6.250

Fit the Michaelis–Menten model to the data. Investigate the adequacy of this model.

3.17 In a study to develop the growth behavior for protozoa colonization in a particular lake, an experiment was conducted in which 15 sponges were placed in a lake and 3 sponges at a time were gathered. Then the number of protozoa were counted at 1, 3, 6, 15, and 21 days. In this case the MacArthur–Wilson equation was used to describe the growth mechanism. The model is given by

$$y = \beta_1(1 - e^{-\beta_2 t})$$

where

y = total protozoa on the sponge
β_1 = species equilibrium constant
β_2 = parameter that measures how quickly growth rises
t = time, number of days

The data (collected by the Department of Biology at Virginia Tech) are as follows:

Observation	Day	y (Total Protozoa)
1	1	17
2	1	21
3	1	16
4	3	30
5	3	25
6	3	25
7	6	33
8	6	31
9	6	32
10	15	34
11	15	33
12	15	33
13	21	39
14	21	35
15	21	36

(a) Estimate β_1 and β_2 using nonlinear regression. Supply your own starting values.

(b) Give estimated standard errors of the parameter estimates.

3.18 An investigation was made to study age and growth characteristics of selected freshwater mussel species in southwest Virginia. For a particular type of mussel, age and length were measured for 20 females with the following results:

Observation	Length (in.)	Age (yr)
1	29.2	4
2	28.6	4
3	29.4	5
4	33.0	5
5	28.2	6
6	33.9	6
7	33.1	6

Observation	Length (in.)	Age (yr)
8	33.2	7
9	31.4	7
10	37.8	7
11	36.9	8
12	40.2	8
13	39.2	9
14	40.6	9
15	35.2	10
16	43.3	10
17	42.3	13
18	41.4	13
19	45.2	14
20	47.1	18

(a) Use nonlinear regression to fit the model

$$y = \beta_1 - \exp[-(\beta_2 + \beta_3 x)] + \varepsilon$$

where y is the length and x the age, respectively, for a mussel. Supply your own starting values.

(b) Give an estimate of the variance–covariance matrix of the parameter estimates β_1, β_2, and β_3.

3.19 The following data were collected on specific gravity and spectrophotometer analysis for 26 mixtures of NG (nitroglycerine), TA (triacetin), and 2 NDPA (2-nitrodiphenylamine).

Mixture	x_1 (% NG)	x_2 (% TA)	x_3 (% 2 NDPA)	y (Specific Gravity)
1	79.98	19.85	0	1.4774
2	80.06	18.91	1.00	1.4807
3	80.10	16.87	3.00	1.4829
4	77.61	22.36	0	1.4664
5	77.60	21.38	1.00	1.4677
6	77.63	20.35	2.00	1.4686
7	77.34	19.65	2.99	1.4684
8	75.02	24.96	0	1.4524
9	75.03	23.95	1.00	1.4537
10	74.99	22.99	2.00	1.4549
11	74.98	22.00	3.00	1.4565
12	72.50	27.47	0	1.4410
13	72.50	26.48	1.00	1.4414
14	72.50	25.48	2.00	1.4426
15	72.49	24.49	3.00	1.4438

Mixture	x_1 (% NG)	x_2 (% TA)	x_3 (% 2 NDPA)	y (Specific Gravity)
16	69.98	29.99	0	1.4279
17	69.98	29.00	1.00	1.4287
18	69.99	27.99	2.00	1.4291
19	69.99	26.99	3.00	1.4301
20	67.51	32.47	0	1.4157
21	67.50	31.47	1.00	1.4172
22	67.48	30.50	2.00	1.4183
23	67.49	29.49	3.00	1.4188
24	64.98	34.00	1.00	1.4042
25	64.98	33.00	2.00	1.4060
26	64.99	31.99	3.00	1.4068

Source: Raymond H. Myers, *Technometrics*, vol. 6, no. 4 (November 1964): 343–356.

There is a need to estimate activity coefficients from the model

$$y = \frac{1}{\beta_1 x_1 + \beta_1 x_2 + \beta_3 x_3} + \varepsilon$$

The quantity parameters β_1, β_2, and β_3 are ratios of activity coefficients to the individual specific gravity of the NG, TA, and 2 NDPA, respectively.

(a) Determine starting values for the model parameters.

(b) Use nonlinear regression to fit the model.

(c) Investigate the adequacy of the nonlinear model.

CHAPTER 4

Logistic and Poisson Regression Models

In this chapter we make use of the material in preceding chapters to develop the technical machinery for two very important members of the family of generalized linear models, namely, logistic regression and Poisson regression. Both find extensive application in biological, biomedical, and environmental problems. However, they also are finding increasing usage in industrial statistics. One important application of logistic regression is dose–response curve development, which dates back to the 1950s. Dose–response curves define relationships developed by toxicologists and biologists in which one is interested in modeling the response, for example, fraction of patients in remission, as the result of a particular type of intervention, such as a specific chemotherapy protocol. Examples of dose–response relationships are shown later in this chapter.

4.1 REGRESSION MODELS WHERE THE VARIANCE IS A FUNCTION OF THE MEAN

Both logistic and Poisson regression models have a common property that we find in many models that fall under the generalized linear model (GLM) umbrella, namely, that the mean response, which is the expected response at each data point, and the variance of the response are related. Consider first a regression structure in which the response is binary (0 or 1) as one might experience if the endpoint of an experimental run is *whether or not* a patient responds to a drug or whether or not one item in an industrial process is *defective*. It is reasonable to assume that at the ith data point, the response is a Bernoulli random variable y_i, where

$$E(y_i) = \pi_i = \pi(\mathbf{x}_i)$$

Generalized Linear Models, Second Edition, by Myers, Montgomery, Vining, and Robinson

and

$$\mathrm{Var}(y_i) = \pi_i(1 - \pi_i)$$

for $i = 1, 2, \ldots, n$. Here π_i is a probability in a Bernoulli process and $\mathbf{x}_i$ is a vector of predictor variables. The parameter π_i and consequently the variance, is a function of the regressors, $\mathbf{x}_i$. As a result, the variance is a function of the mean.

Consider a second scenario in which the response values are Poisson counts. They might represent the result of endpoints in a biomedical experiment involving the number of cancer cell colonies, or they may represent the number of defects observed in a microelectronic device. We can write the model for the mean as

$$E(y_i) = \mu(\mathbf{x}_i), \quad i = 1, 2, \ldots, n$$

where μ is the parameter of the Poisson distribution. Note that again the mean changes over the observations in the experiment due to changing values of the regressors. However, it is well known that in the case of a Poisson distribution the variance is equal to the mean, and as a result $\mathrm{Var}(y_i) = E(y_i) = \mu(\mathbf{x}_i)$.

From the preceding examples, it becomes apparent that in these two scenarios the use of ordinary least squares estimation is inappropriate for estimation of the model parameters. As we move into the development of GLMs in Chapter 5, we illustrate other types of regression models used with other response distributions in which the variance is a function of the mean. This suggests that the material on *weighted* least squares discussed in Chapter 2 might be used in some form. We demonstrate that there is an interesting analytic connection between the weighted least squares concept and maximum likelihood estimation for the two models discussed here and other members of the family of generalized linear models.

4.2 LOGISTIC REGRESSION MODELS

4.2.1 Models with a Binary Response Variable

Consider the situation where the response variable in a regression problem takes on only two possible values, 0 and 1. These could be arbitrary assignments resulting from observing a qualitative response. For example, the response could be the outcome of a functional electrical test on a semiconductor device for which the results are either a success, which means the device works properly, or a failure, which could be due to a short, an open circuit, or some other functional problem.

Suppose that the model has the form

$$y_i = \mathbf{x}_i'\boldsymbol{\beta} + \varepsilon_i \tag{4.1}$$

where $\mathbf{x}_i' = [1, x_i, x_{i2}, \ldots, x_{ik}]$, $\boldsymbol{\beta}' = [\beta_0, \beta_1, \beta_2, \ldots, \beta_k]$, and the response variable y_i takes on the value either 0 or 1. We assume that the response variable y_i is a **Bernoulli random variable** with probability distribution as follows:

y_i	Probability
1	$P(y_i = 1) = \pi_i$
0	$P(y_i = 0) = 1 - \pi_i$

Now since $E(\varepsilon_i) = 0$, the expected value of the response variable is

$$E(y_i) = 1(\pi_i) + 0(1 - \pi_i) = \pi_i$$

This implies that

$$E(y_i) = \mathbf{x}_i'\boldsymbol{\beta} = \pi_i$$

This means that the expected response given by the response function $E(y_i) = \mathbf{x}_i'\boldsymbol{\beta}$ is just the probability that the response variable takes on the value 1.

There are some very basic problems with the regression model in Eq. (4.1). First, note that if the response is binary, then the error terms ε_i can only take on two values, namely,

$$\varepsilon_i = 1 - \mathbf{x}_i'\boldsymbol{\beta} \quad \text{when } y_i = 1$$
$$\varepsilon_i = -\mathbf{x}_i'\boldsymbol{\beta} \quad \text{when } y_i = 0$$

Consequently, the errors in this model cannot possibly be normal. Second, the error variance is not constant, since

$$\sigma_{y_i}^2 = E\{y_i - E(y_i)\}^2 = (1 - \pi_i)^2\pi_i + (0 - \pi_i)^2(1 - \pi_i) = \pi_i(1 - \pi_i)$$

Notice that this last expression is just

$$\sigma_{y_i}^2 = E(y_i)\,[1 - E(y_i)]$$

since $E(y_i) = \mathbf{x}_i'\boldsymbol{\beta} = \pi_i$. This indicates that the variance of the observations (which is the same as the variance of the errors because $\varepsilon_i = y_i - \pi_i$, and π_i is a constant) is a function of the mean. Finally, there is a constraint on the response function, because

$$0 \leq E(y_i) = \pi_i \leq 1$$

This restriction can cause serious problems with the choice of a **linear response function**, as we have initially assumed in Equation (4.1). It would be possible to fit a model to the data for which the predicted values of the response lie outside the 0, 1 interval.

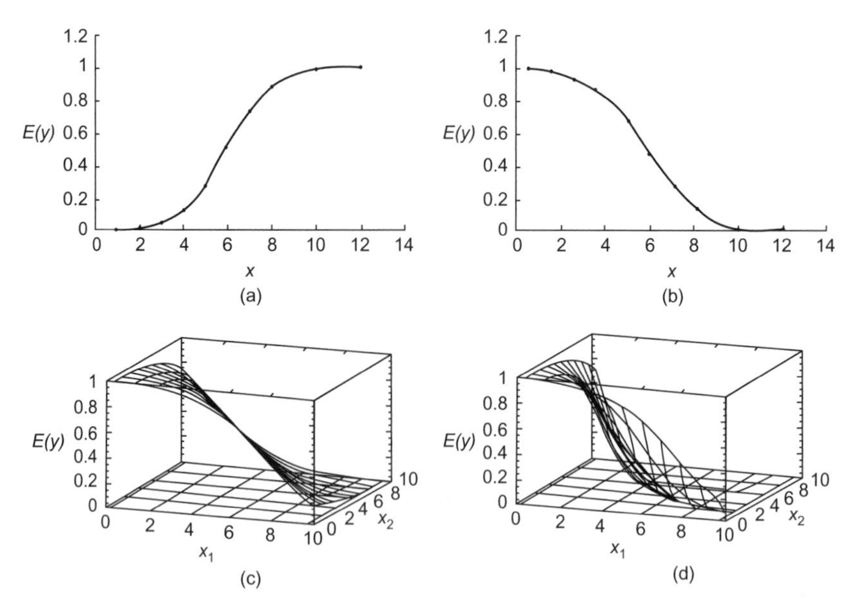

Figure 4.1 Examples of the logistic response function: (a) $E(y) = 1/(1 + e^{-6.0-1.0x})$; (b) $E(y) = 1/(1 + e^{-6.0 + 1.0x})$; (c) $E(y) = 1/(1 + e^{-5.0 + 0.65x_1 + 0.4x_2})$; and (d) $E(y) = 1/(1 + x^{-5.0 + 0.65x_1 + 0.15x_1x_2})$.

Generally, when the response variable is binary, there is considerable empirical evidence indicating that the shape of the response function should be nonlinear. A monotonically increasing (or decreasing) S-shaped (or reverse S-shaped) function, such as shown in Figure 4.1 is usually employed. This function is called the **logistic response function** and has the form

$$E(y) = \frac{\exp(\mathbf{x}'\boldsymbol{\beta})}{1 + \exp(\mathbf{x}'\boldsymbol{\beta})} = \frac{1}{1 + \exp(-\mathbf{x}'\boldsymbol{\beta})} \tag{4.2}$$

The logistic response function can easily be linearized. One approach defines the structural portion of the model in terms of a function of the response function mean. Let

$$\eta = \mathbf{x}'\boldsymbol{\beta} \tag{4.3}$$

be the **linear predictor** where η is defined by the transformation

$$\eta = \ln \frac{\pi}{1 - \pi} \tag{4.4}$$

This transformation is often called the **logit transformation** of the probability π. The logit transformation is a very popular approach for modeling Bernoulli or binomial data. It maps π, which is bounded between 0 and 1, to the real number line. The ratio $\pi/(1 - \pi)$ in the transformation is called the **odds**. Sometimes the logit transformation is called the log-odds.

4.2.2 Estimating the Parameters in a Logistic Regression Model

The general form of the logistic regression model is

$$y_i \sim \text{Bernouilli}(\pi_i) \tag{4.5}$$

where the y_i are assumed to be independent of one another and each with expected value

$$E(y_i) = \pi_i = \frac{\exp(\mathbf{x}_i'\boldsymbol{\beta})}{1 + \exp(\mathbf{x}_i'\boldsymbol{\beta})} \tag{4.6}$$

Note for the logistic model, we do not write the model as

$$y_i = E(y_i) + \varepsilon_i$$

as was done for the linear models discussed in Chapter 2. Similar to the linear model, however, we do express the mean as a function of the regressors in the logistic model. So, for the logistic model, the individual responses (i.e. the y_i) are *modeled* when we state that their distribution is Bernouilli whose mean is a function of the regressors. As was done in the linear model, normal error setting, we use the method of **maximum likelihood** to estimate the parameters in the linear predictor $\mathbf{x}_i'\boldsymbol{\beta}$.

Since each sample observation is assumed to follow a Bernouilli distribution, the probability distribution for the ith observation is given by

$$f(y_i) = \pi_i^{y_i}(1 - \pi_i)^{1-y_i}, \; i = 1, 2, \ldots, n$$

and each observation y_i takes on the value of either 0 or 1. Since the observations are independent, the likelihood function is

$$\mathscr{L}(\boldsymbol{\beta}; y_1, y_2, \ldots, y_n) = \prod_{i=1}^{n} f(y_i) = \prod_{i=1}^{n} \pi_i^{y_i}(1 - \pi_i)^{1-y_i}$$

It is more convenient to work with the log-likelihood

$$\begin{aligned}
\ln \mathscr{L}(\boldsymbol{\beta}; y_1, y_2, \ldots, y_n) &= \ln \prod_{i=1}^{n} f(y_i) \\
&= \sum_{i=1}^{n} \left[y_i \ln\left(\frac{\pi_i}{1 - \pi_i}\right)\right] + \sum_{i=1}^{n} \ln(1 - \pi_i)
\end{aligned}$$

Now since $1 - \pi_i = [1 + \exp(\mathbf{x}_i'\boldsymbol{\beta})]^{-1}$ and $\eta_i = \ln[\pi_i/(1 - \pi_i)] = \mathbf{x}_i'\boldsymbol{\beta}$, the log-likelihood can be written as

$$\begin{aligned}
\ln \mathscr{L}(\boldsymbol{\beta}; \mathbf{y}) &= \sum_{i=1}^{n} y_i \mathbf{x}_i'\boldsymbol{\beta} - \sum_{i=1}^{n} \ln[1 + \exp(\mathbf{x}_i'\boldsymbol{\beta})] \\
&= \boldsymbol{\beta}'\mathbf{X}'\mathbf{y} - \sum_{i=1}^{n} \ln[1 + \exp(\mathbf{x}_i'\boldsymbol{\beta})]
\end{aligned} \tag{4.7}$$

Typically, in logistic regression we have repeated observations or trials at each level of the predictor (**x**) variables. This happens often in designed experiments. Let y_i be the number of 1's observed at the ith observation and n_i be the number of trials at each level of the predictor, with $n = n_1 + n_2 + \cdots + n_m$. Then the kernel of the log-likelihood is

$$\ln \mathscr{L}(\boldsymbol{\beta}; \mathbf{y}) = \boldsymbol{\beta}'\mathbf{X}\mathbf{y} - \sum_{i=1}^{m} n_i \ln[1 + \exp(\mathbf{x}_i'\boldsymbol{\beta})] \tag{4.8}$$

where **X** is the traditional model matrix in linear regression discussed in Chapter 2 and **y** is the response vector. We must now differentiate (4.8) with respect to $\boldsymbol{\beta}$:

$$\frac{\partial \ln \mathscr{L}(\boldsymbol{\beta}; \mathbf{y})}{\partial \boldsymbol{\beta}} = \mathbf{X}'\mathbf{y} - \sum_{i=1}^{m} \left[\frac{n_i}{1 + e^{\mathbf{x}_i'\boldsymbol{\beta}}}\right] e^{\mathbf{x}_i'\boldsymbol{\beta}} \mathbf{x}_i$$

Now, since $e^{\mathbf{x}_i'\boldsymbol{\beta}}/(1 + e^{\mathbf{x}_i'\boldsymbol{\beta}}) = 1/(1 + e^{-\mathbf{x}_i'\boldsymbol{\beta}}) = \pi_i$, we have

$$\frac{\partial \ln \mathscr{L}(\boldsymbol{\beta}; \mathbf{y})}{\partial \boldsymbol{\beta}} = \mathbf{X}'\mathbf{y} - \sum_{i=1}^{m} n_i \pi_i \mathbf{x}_i$$

Since $n_i\pi_i$ represents the mean of the binomial random variable, we can express the right-hand side above in matrix notation as $\mathbf{X}'(\mathbf{y} - \boldsymbol{\mu})$, where

$$\boldsymbol{\mu} = \begin{bmatrix} \mu_1 \\ \mu_2 \\ \vdots \\ \mu_m \end{bmatrix}$$

and $\mu_i = n_i\pi_i$. The $\boldsymbol{\mu}$ notation is motivated by the fact that at the ith data point the mean of the binomial distribution is given by n_i,π_i. As a result the maximum likelihood estimator (MLE) is the solution to the *score equation*

$$\mathbf{X}'(\mathbf{y} - \boldsymbol{\mu}) = \mathbf{0} \tag{4.9}$$

Now, it may appear that Equation (4.9) is trivial, but one must keep in mind that the β's appear in $\boldsymbol{\mu}$ and the elements of $\boldsymbol{\mu}$ are nonlinear according to the model

$$\mu_i = \frac{n_i}{1 + e^{-\mathbf{x}_i'\boldsymbol{\beta}}}, \quad i = 1, 2 \ldots m$$

Consequently, a reasonable procedure to solve these equations uses an iteratively reweighted least squares much like that described in Chapter 3. Such an iterative procedure produces $b_0, b_1, b_2, \ldots, b_k$, the estimators of $\beta_0, \beta_1, \beta_2, \ldots, \beta_k$, the $p = k + 1$ parameters in the logistic regression model.

We can easily tie the maximum likelihood estimation for logistic regression to the development in Section 3.6, since the logistic model is nonlinear in the parameters. In addition, the binomial variance of the random variable y_i is $n_i\pi_i(1-\pi_i)$, and thus the variance is a function of the mean. The distinction between this situation and the one described in Chapter 3 lies in the fact that distribution of the response at each data point here is binomial.

Relationship to Weighted Least Squares

Consider the score function in Equation (4.9) for logistic regression. It is interesting that a relationship exists between weighted least squares and Equation (4.9), which was developed completely from maximum likelihood. In this section we give an informal outline of the relationship. Appendix A.4 gives a more formal presentation.

Take, for example, the use of the weighted residual sum of squares given by

$$S = \sum_{i=1}^{m} \left[\frac{(y_i - \mu_i)^2}{\sigma_i^2} \right] \tag{4.10}$$

where $\mu_i = n_i\pi_i$ and σ_i^2 is the binomial variance at the ith data point with

$$\sigma_i^2 = n_i\pi_i \, [1 - \pi_i] = n_i \frac{e^{-\mathbf{x}'_i\boldsymbol{\beta}}}{(1 + e^{-\mathbf{x}'_i\boldsymbol{\beta}})^2}$$

Recall that iteratively reweighted least squares in Chapter 3 was based on the notion of fixed weights, or weights with fixed $\boldsymbol{\beta}$, or the current $\boldsymbol{\beta}$ in an iterative procedure. In the same spirit we have

$$\min S = \min_{\boldsymbol{\beta}} \sum_{i=1}^{m} \left[\frac{(y_i - \mu_i)^2}{\sigma_i^2} \right]$$

with fixed variance σ_i^2. As a result, we differentiate only the numerator of S and obtain

$$-2 \left[\frac{\sum_{i=1}^{m} (y_i - \mu_i)}{\sigma_i^2} \right] \left(\frac{\partial \mu_i}{\partial \boldsymbol{\beta}} \right)$$

Now $\partial\mu_i/\partial\boldsymbol{\beta} = n_i\pi_i[1 - \pi_i]\mathbf{x}_i = \sigma_i^2\mathbf{x}_i$. Thus the solution given by minimization of the weighted residual sum of squares with fixed weights σ_i^2 is given by

$$\sum_{i=1}^{m} (y_i - \mu_i)\mathbf{x}_i = 0$$

which is identical to $\mathbf{X}'(\mathbf{y} - \boldsymbol{\mu}) = \mathbf{0}$ given by the score equation in Equation (4.9). As a result an iteratively reweighted least squares (IRLS) procedure can be used here to produce the solution to the score equations and thus obtain numerical values $b_0, b_1, \ldots, b_k$ of the MLE.

There are several computer programs that can be used to fit the logistic regression model. SAS PROC LOGISTIC and SAS PROC GENMOD have excellent capability, as do the desktop packages JMP and MINITAB. It is also possible to use R to fit logistic regression models.

Let **b** be the final estimate of the model parameters that the IRLS algorithm produces. If the model assumptions are correct, then we can show that asymptotically

$$E(\mathbf{b}) = \boldsymbol{\beta} \text{ and } \mathrm{Var}(\mathbf{b}) = (\mathbf{X}'\hat{\mathbf{V}}\mathbf{X})^{-1} \tag{4.11}$$

where the matrix $\hat{\mathbf{V}}$ is an $n \times n$ diagonal matrix containing the estimated variance of each observation on the main diagonal; that is, the ith diagonal element of $\hat{\mathbf{V}}$ is

$$V_{ii} = n_i \hat{\pi}_i (1 - \hat{\pi}_i)$$

The estimated value of the linear predictor is $\hat{\eta}_i = \mathbf{x}_i'\mathbf{b}$ and the fitted value of the logistic regression model is written as

$$\hat{y}_i = \hat{\pi}_i = \frac{\exp(\hat{\eta}_i)}{1 + \exp(\hat{\eta}_i)} = \frac{\exp(\mathbf{x}_i'\mathbf{b})}{1 + \exp(\mathbf{x}_i'\mathbf{b})} = \frac{1}{1 + \exp(-\mathbf{x}_i'\mathbf{b})} \tag{4.12}$$

Example 4.1. The* Challenger *Data. We fit the logistic regression model to the *Challenger* O-ring failure data first described in Chapter 1. For convenience, we have given the data again in Table 4.1. Figure 4.2, gives a scatter plot of the data.

As we mentioned above, there are several excellent software packages for fitting logistic regression models. Some of the output from MINITAB is shown in Table 4.2. From the logistic regression table portion of the output, we find that the estimates of the parameters in the linear predictor are $b_0 = 10.8753$ and $b_1 = -0.171321$, so the fitted logistic regression model is

Table 4.1 Temperature and O-Ring Failure Data from the *Challenger* Accident

Temperature at Launch (°F)	At Least One O-Ring Failure	Temperature at Launch (°F)	At Least One O-Ring Failure
53	1	70	1
56	1	70	1
57	1	72	0
63	0	73	0
66	0	75	0
67	0	75	1
67	0	76	0
67	0	76	0
68	0	78	0
69	0	79	0
70	0	80	0
70	1	81	0

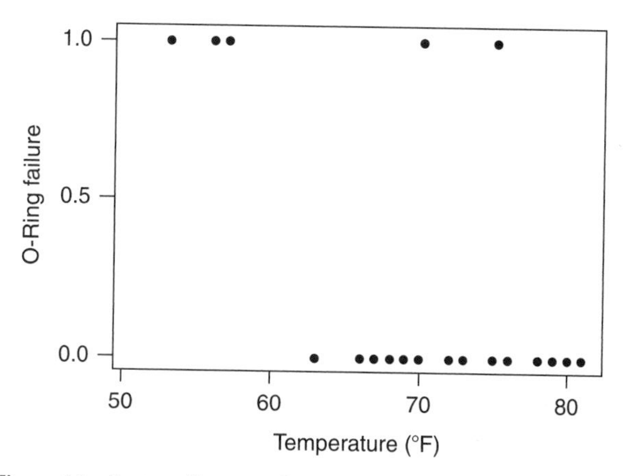

Figure 4.2 Scatter diagram of O-ring failures versus temperature.

$$\hat{y} = \frac{\exp(10.8753 - 0.171321x)}{1 + \exp(10.8753 - 0.171321x)}$$
$$= \frac{1}{\exp(-10.8753 + 0.171321x)}$$

Table 4.3 presents the corresponding output from JMP. To obtain the fitted model from JMP it is necessary to utilize the value ordering feature in the data table so that the program recognizes the value 1 as failure (the default is that

Table 4.2 MINITAB Logistic Regression Output for the Challenger Data

Binary Logistic Regression: Failure versus Temperature

```
Link Function: Logit

Response information

Variable     Value     Count
Failure        1         7     (Event)
               0        17
             Total      24

Logistic Regression Table
                                                          Odds    95%      CI
Predictor       Coef       SE Coef       Z       P       Ratio  Lower   Upper
Constant     10.8753     5.70313       1.91   0.057
Temperature -0.171321  0.0834420    -2.05   0.040      0.84    0.72    0.99

Log-Likelihood = -11.515
Test that all slopes are zero: G = 5.944, DF = 1, P-Value = 0.015
```

failure = 0). Failure to do this will produce a curve that is opposite to the one given here; that is, it descends toward zero as the temperature decreases. The JMP output also contains a plot of the fitted logistic regression model. This plot contains a set of plotted points between zero and one corresponding to each observed level of temperature. The visualization is to imagine these points exerting a pressure that determines the parameters of the logistic curve, just as the pressure of a gas contained in the plot region would determine where a barrier (the fitted curve) would go. The probability measured from the base to the curve is for Failure = 0, and the probability measured from the curve to the top is Failure = 1, as is labeled on the right of the plot.

Both the MINITAB and JMP computer outputs contain other information about the fitted logistic regression model These quantities will be explained in subsequent sections. □

MINITAB and JMP also calculate and display the covariance matrix of the estimated model parameters. For the model of the *Challenger* data, the covariance matrix is

$$\text{Var}(\mathbf{b}) = \begin{bmatrix} 37.5257 & -0.473918 \\ -0.473918 & 0.006963 \end{bmatrix}$$

The estimated standard errors of the model parameter estimates reported in Tables 4.2 and 4.3 are the square roots of the main diagonal elements of this matrix.

4.2.3 Interpertation of the Parameters in a Logistic Regression Model

It is relatively easy to interpret the parameters in a logistic regression model. Consider first the case where the linear predictor has only a single regressor, so that the fitted value of the linear predictor at a particular value of x, say, x_i, is

$$\hat{\eta}(x_i) = b_0 + b_1 x_i$$

The fitted value at $x_i + 1$ is

$$\hat{\eta}(x_i + 1) = b_0 + b_1(x_i + 1)$$

and the difference in the two predicted values is

$$\hat{\eta}(x_i + 1) - \hat{\eta}(x_i) = b_1$$

Now $\hat{\eta}(x_i)$ is just the log-odds when the regressor variable is equal to x_i, and $\hat{\eta}(x_i + 1)$ is just the log-odds when the regressor is equal to $x_i + 1$. Therefore the difference in the two fitted values is

$$\begin{aligned} \hat{\eta}(x_i + 1) - \hat{\eta}(x_i) &= \ln(\text{odds}_{x_i+1}) - \ln(\text{odds}_{x_i}) \\ &= \ln\left(\frac{\text{odds}_{x_i+1}}{\text{odds}_{x_i}}\right) = b_1 \end{aligned}$$

Table 4.3 JMP Logistic Regression Output for the Challenger Data

Ordinal Logistic Fit for Failure Logistic Plot

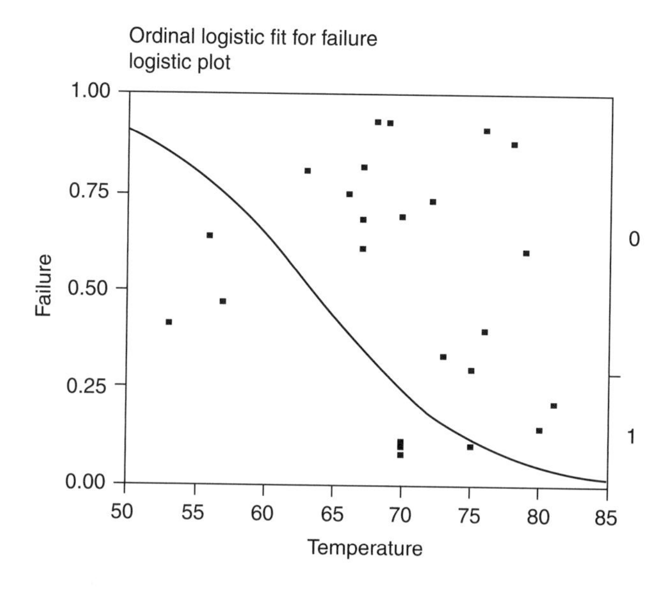

Whole Model Test

Model	-LogLikelihood	DF	ChiSquare	Prob>ChiSq
Difference	2,972069	1	5.944137	0.0148
Full	11.515225			
Reduced	14.487294			

RSquare (U)	0.2052
Observations (or Sum Wgts)	24

Lack Of Fit

Source	DF	-LogLikelihood	ChiSquare
Lack Of Fit	15	7.879590	15.75918
Saturated	16	3.635635	Prob>ChiSq
Fitted	1	11.515225	0.3982

Parameter Estimates

Term	Estimate	Std Error	Chi-Square	Prob>ChiSq	Lower 95%	Upper 95%
Intercept[1]	10.8753	5.7031342	3.64	0.0565		
Temperature	−0.1713205	0.083442	4.22	0.0401	0.3779007	0.03055

Effect Wald Tests

Source	Nparm	DF	Wald ChiSquare	Prob>ChiSq
Temperature	1	1	4.21550393	0.0401

Effect Likelihood Ratio Tests

Source	Npam	DF	L-R ChiSquare	Prob>ChiSq
Temperature	1	1	5.94413724	0.0148

If we take antilogs, we obtain the estimated **odds ratio**

$$\hat{O}_R = \frac{\text{odds}_{x_i+1}}{\text{odds}_{x_i}} = e^{b_1} \tag{4.13}$$

The estimated odds ratio can be interpreted as the estimated increase in the odds of success associated with a one-unit change in the value of the predictor variable. In general, the estimated increase in the odds ratio associated with a change of d units in the predictor variable is $\exp(db_1)$.

Example 4.2. The* Challenger *Data. In Example 4.1 we fit the logistic regression model

$$\hat{y} = \frac{1}{1 + e^{10.8753 - 0.17132x}}$$

to the *Challenger* O-ring failure data of Table 4.1. Since the linear predictor contains only one regressor variable and $b_1 = -0.171321$, we can compute the odds ratio from Equation (4.13) as

$$\hat{O}_R = e^{b_1} = e^{-0.17134} = 0.84 \tag{4.14}$$

This implies that every additional degree of temperature reduces the odds of failure by 16%. If the temperature increases by $d = 5$ degrees then the odds ratio becomes $\exp(db_1) = \exp[5(-0.171321)] = 0.42$. This indicates that the odds of failure are reduced by over 50% with a 5-degree increase in temperature. □

There is a close connection between the odds ratio in logistic regression and the 2×2 contingency table that is widely used in the analysis of categorical data. Consider Table 4.4, which presents a 2×2 contingency table where the categorical response variable represents the outcome (infected, not infected) for a group of patients treated with either an active drug or a placebo. The n_{ij} are the numbers of patients in each cell. The odds ratio in the 2×2 contingency table is defined as

$$\frac{\text{odds infected} \mid \text{placebo drug}}{\text{odds infected} \mid \text{active}} = \frac{n_{11}/n_{01}}{n_{10}/n_{00}} = \frac{n_{11} \cdot n_{00}}{n_{10} \cdot n_{01}}$$

Consider a logistic regression model for these data. The linear predictor is

$$\ln\left(\frac{\pi}{1-\pi}\right) = \beta_0 + \beta_0 x_1$$

When $x_1 = 0$, we have

$$\beta_0 = \ln \frac{P(y = 1 \mid x_1 = 0)}{P(y = 0 \mid x_1 = 0)}$$

Table 4.4 A 2×2 Contingency Table

Response	$x_1 = 0$, Active Drug	$x_1 = 1$, Placebo
$y = 0$, not infected	n_{00}	n_{01}
$y = 1$, infected	n_{10}	n_{11}

Now let $x_1 = 1$:

$$\ln\left(\frac{\pi}{1-\pi}\right) = \beta_0 + \beta_1 x_1$$

$$\ln \frac{P(y=1 \mid x_1 = 1)}{P(y=0 \mid x_1 = 1)} = \ln \frac{P(y=1 \mid x_1 = 0)}{P(y=0 \mid x_1 = 0)} + \beta_1$$

Solving for β_1 yields

$$\beta_1 = \ln \frac{P(y=1 \mid x_1 = 1) \cdot P(y=0 \mid x_1 = 0)}{P(y=0 \mid x_1 = 1) \cdot P(y=1 \mid x_1 = 0)} = \ln \frac{n_{11} \cdot n_{00}}{n_{01} \cdot n_{10}}$$

so $\exp(\beta_1)$ is equivalent to the odds ratio in the 2×2 contingency table. However, the odds ratio from logistic regression is much more general than the traditional 2×2 contingency table odds ratio. Logistic regression can incorporate other predictor variables, and the presence of these variables can impact the odds ratio. For example, suppose that another variable, x_2 = age, is available for each patient in the drug study depicted in Table 4.4. Now the linear predictor for the logistic regression model for the data would be

$$\ln\left(\frac{\pi}{1-\pi}\right) = \beta_0 + \beta_1 x_1 + \beta_2 x_2$$

This model allows the predictor variable age to impact the estimate of the odds ratio for the drug variable. The drug odds ratio is still $\exp(\beta_1)$, but the estimate of β_1 is potentially affected by the inclusion of x_2 = age in the model. It would also be possible to include an interaction term between drug and age in the model, say,

$$\ln\left(\frac{\pi}{1-\pi}\right) = \beta_0 + \beta_1 x_1 + \beta_2 x_2 + \beta_{12} x_1 x_2$$

In this model the odds ratio for drug depends on the level of age and would be computed as $\exp(\beta_1 + \beta_{12} x_2)$. This is an example of an **adjusted odds ratio** where the odds ratio associated with one regressor depends upon the specific values of other regressors.

The interpretation of the regression coefficients in the multiple logistic regression model is similar to that for the case where the linear predictor contains only one regressor. That is, the quantity $\exp(\beta_j)$ is the odds ratio for regressor x_j, assuming that all other predictor variables are constant.

4.2.4 Statistical Inference on Model Parameters

Statistical inference in logistic regression is based on certain properties of maximum likelihood estimators and on likelihood ratio tests. These are large-sample or **asymptotic** results. This section discusses and illustrates these procedures using the logistic regression model fit to the *Challenger* data from Example 4.1.

Likelihood Ratio Tests

A likelihood ratio test can be used to compare a *full* model with a *reduced* model that is of interest. This is analogous to the *extra-sum-of-squares* technique that we have used previously to compare full and reduced models. The likelihood ratio test procedure compares twice the logarithm of the value of the likelihood function for the full model (*FM*) to twice the logarithm of the value of the likelihood function of the reduced model (*RM*) to obtain a test statistic, say,

$$LR = 2\ln\frac{\mathscr{L}(FM)}{\mathscr{L}(RM)} = 2[\ln\mathscr{L}(FM) - \ln\mathscr{L}(RM)] \tag{4.15}$$

For large samples, when the reduced model is correct, the test statistic LR follows a chi-square distribution with degrees of freedom equal to the difference in the number of parameters between the full and reduced models. Therefore, if the test statistic LR exceeds the upper α percentage point of this chi-square distribution, we would reject the claim that the reduced model is appropriate.

The likelihood ratio approach can be used to provide a test for significance of regression in logistic regression. This test uses the current model that has been fit to the data as the full model and compares it to a reduced model that has constant probability of success. This constant-probability-of-success model is

$$E(y) = \pi = \frac{e^{\beta_0}}{1 + e^{\beta_0}}$$

that is, a logistic regression model with no regressor variables. The maximum likelihood estimate of the constant probability of success is just y/n, where y is the total number of successes that have been observed and n is the number of observations. Substituting this into the log-likelihood function in Equation (4.8) gives the maximum value of the log-likelihood function for the reduced model as

$$\ln\mathscr{L}(RM) = y\ln(y) + (n-y)\ln(n-y) - n\ln(n)$$

Therefore the likelihood ratio test statistic for testing significance of regression is

$$LR = 2\left\{\sum_{i=1}^{n} y_i \ln\hat{\pi}_i + \sum_{i=1}^{n}(n_i - y_i)\ln(1-\hat{\pi}_i) - [y\ln(y) + (n-y)\ln(n-y) - n\ln(n)]\right\} \tag{4.16}$$

A large value of this test statistic would indicate that at least one of the regressor variables in the logistic regression model is important because it has a nonzero regression coefficient.

MINITAB computes the likelihood ratio test for significance of regression in logistic regression. In the MINITAB output in Table 4.2 the test statistic in Equation (4.16) is reported as $G = 5.944$ with one degree of freedom (because the full model has only one predictor). The reported P-value is 0.015, so the overall model is significant.

JMP also computes this statistic. It is the chi-square value reported in the Whole Model Test section of the output in Table 4.3. The P-value reported is 0.0148. JMP also reports the negative of the log-likelihood for the full and reduced models as 11.515225 and 14.487294, respectively. The LR in Equation (4.16) is just twice the difference in these two quantities, or

$$\begin{aligned} LR &= 2[-11.515225 - (-14.487294)] \\ &= 2(2.972069) \\ &= 5.944138 \end{aligned}$$

which is (apart from rounding) the value reported by MINITAB.

Testing Goodness of Fit with Deviance

The goodness of fit of the logistic regression model can also be assessed using a likelihood ratio test procedure. This test compares the current model to a **saturated model**, where each unique covariate pattern is allowed to have its own parameter (i.e., a success probability). These parameters or success probabilities are y_i/n_i, where y_i is the number of successes and n_i is the number of observations associated with covariate pattern i. The **deviance** is defined as twice the difference in log-likelihoods between this saturated model and the full model (which is the current model) that has been fit to the data with estimated success probability $\hat{\pi}_i = \exp(\mathbf{x}_i'\mathbf{b})/[1 + \exp(\mathbf{x}_i'\mathbf{b})]$. The deviance is defined as

$$D = 2\ln\frac{\mathcal{L}(\text{saturated model})}{\mathcal{L}(FM)} = 2\sum_{i=1}^{n}\left[y_i \ln\left(\frac{y_i}{n_i\hat{\pi}_i}\right) + (n_i - y_i)\ln\left(\frac{n_i - y_i}{n_i(1-\hat{\pi}_i)}\right)\right] \tag{4.17}$$

In calculating the deviance, note that $y\ln(y/n\hat{\pi}) = 0$ if $y = 0$, and if $y = n$ we have $(n-y)\ln[(n-y)/n(1-\hat{\pi})] = 0$. When the logistic regression model is an adequate fit to the data and the sample size is large, the deviance has a chi-square distribution with $n - p$ degrees of freedom, where p is the number of parameters in the model. Small values of the deviance (or a large P-value imply that the model provides a satisfactory fit to the data, while large values of the deviance imply that the current model is not adequate. A good rule of thumb for logistic regression is to divide the deviance by its number of degrees

Table 4.5 Prison Sentencing Data Ignoring Gender

	Prison	No Prison	Ni
Business crime/No prior arrest	17	75	92
Business crime/Prior arrest	42	109	151
Home crime/No prior arrest	54	359	413
Home crime/Prior arrest	33	175	208

of freedom. If the ratio $D/(n - p)$ is much greater than unity, the current model is not an adequate fit to the data.

The deviance goodness-of-fit statistic is only appropriate when the regressors are categorical, or when we have grouped data as predictors. We illustrate the calculation of the deviance statistic using the data on prison sentencing from Stokes, Davis, and Koch (1995) shown in Table 4.5. These data report on whether or not an offender received a prison sentence as a function of whether the crime involved a business or a home and whether or not the offender had a prior arrest record (later on we use another version of this data that includes the gender of the offender). In the SAS code provided in Table 4.12, ibus = 1 if person committed a business crime and ibus = 0 if person committed a home crime. For iprior, we have that iprior = 1 if person had some prior arrest history and iprior = 0 if no prior arrest history.

The deviance statistic compares the fitted model to that of the saturated model. It is a likelihood ratio test of all higher-order interactions. For the prison data when we fit a model with predictor variables ibus and iprior, the saturated model contains the variables ibus and iprior plus the interaction term ibus x iprior. The SAS code for fitting the model with the variables ibus and iprior using PROC LOGISTIC is shown at the bottom of Table 4.12 and Table 4.6 contains the output. Table 4.7 shows the computation of the deviance for the prison data. This test is a special case of the likelihood ratio test presented earlier but this time the full model is the saturated model, and the reduced model is the fitted model. The test statistic is also sometimes called Wilks's statistic.

To physically compute the deviance statistic for goodness-of-fit we need the log likelihood associated with the saturated model where the saturated model has the variables ibus, iprior, ibus x iprior, and the intercept. This is done using the code in the display below.

```
proc logistic data=sentence descending;
title3'Example For Pearson Chi-Square and Deviance';
model prison=ibus iprior ibus*iprior / aggregate scale=N ;
weight count;
output out=home.good p=phat;
run;
```

Table 4.6 SAS PROC LOGISTIC Output for the Prison Data

```
                        The LOGISTIC Procedure
Data Set: WORK.SENTENCE
Response Variable: PRISON   Prison Sentence?
Response Levels: 2
Number of Observations: 16
Weight Variable: COUNT
Sum of Weights: 864
Link Function: Logit
                              Response Profile
              Ordered                          Total
               Value    PRISON    Count        Weight
                 1        Y         8      146.00000
                 2        N         8      718.00000
          Deviance and Pearson Goodness-of-Fit Statistics
                                                        Pr >
      Criterion    DF      Value      Value/DF     Chi-Square
      Deviance      1     0.5779       0.5779       0.4471
      Pearson       1     0.5717       0.5717       0.4496
              Number of unique profiles: 4
 Model Fitting Information and Testing Global Null Hypothesis BETA=0
                                       Intercept
                 Intercept               and
Criterion          Only               ovariates       Chi-Square for Covariates
AIC               786.974              775.478                  .
SC                787.747              777.796                  .
2 LOG L           784.974              769.478         15.496 with 2 DF (p=0.0004)
Score                .                    .            16.200 with 2 DF (p=0.0003)
```

Table 4.7 Construction of Deviance Chi-Square

Type of Crime	Prior Arrest?	Prison Sentence?	Model-based Probability of Prison	Observed count	Ni	Pi	Expected Count	Chi-Square Contribution i
business	None	N	0.20553	75	92	0.79447	73.091	3.86675
business	None	Y	0.20553	17	92	0.20553	18.909	−3.61784
business	Some	N	0.26551	109	151	0.73449	110.909	−3.78430
business	Some	Y	0.26551	42	151	0.26551	40.091	3.90680
home	None	N	0.12613	359	413	0.87387	360.909	−3.80723
home	None	Y	0.12613	54	413	0.12613	52.091	3.88644
home	Some	N	0.16783	175	208	0.83217	173.091	3.83831
home	Some	Y	0.16783	33	208	0.16783	34.909	−3.71103
								0.5779

Table 4.8 SAS Output for the Saturated Model for the Prison Data

Response Profile

Ordered Value	PRISON	Total Count	Weight
1	Y	8	146.00000
2	N	8	718.00000

Deviance and Pearson Goodness-of-Fit Statistics

Criterion	DF	Value	Value/DF	Pr > Chi-Square
Deviance	0	0	.	.
Pearson	0	1.56E-22	.	.

Number of unique profiles: 4

Model Fitting Information and Testing Global Null Hypothesis BETA=0

Criterion	Intercept Only	Intercept and Covariates	Chi-Square for Covariates
AIC	786.974	776.900	.
SC	787.747	779.991	.
-2 LOG L	784.974	768.900	16.074 with 3 DF (p=0.0011)
Score	.	.	17.434 with 3 DF (p=0.0006)

The SAS output from the saturated model is in Table 4.8. The deviance statistic can be computed as

$$G^2 = -2\log(\mathscr{L}(\boldsymbol{\beta}_R)) - 2\log(\mathscr{L}(\boldsymbol{\beta}_F))$$
$$G^2 = 769.478 - 768.900 = 0.578$$

Sample size guidelines for the deviance goodness-of-fit test are

- Marginal sample sizes at least 10
- 80% of predicted counts at least 5
- All other expected counts are greater than 2

Testing Hypotheses on Subsects of Parameters Using Deviance

We can also use the deviance to test hypotheses on subsets of the model parameters, just as we used the difference in regression (or error) sums of squares to test similar hypotheses in the normal-error linear regression model case. Recall that the linear predictor can be written as

$$\boldsymbol{\eta} = \mathbf{X}\boldsymbol{\beta} = \mathbf{X}_1\boldsymbol{\beta}_1 + \mathbf{X}_2\boldsymbol{\beta}_2 \tag{4.18}$$

where the **full model** has p parameters, $\boldsymbol{\beta}_1$ contains $p - r$ of these parameters, $\boldsymbol{\beta}_2$ contains r of these parameters, and the columns of the matrices $\mathbf{X}_1$ and $\mathbf{X}_2$ contain the variables associated with these parameters.

The deviance of the full model is denoted by $D(\boldsymbol{\beta})$. Suppose that we wish to test the hypotheses

$$H_0: \boldsymbol{\beta}_2 = \mathbf{0}, \quad H_1: \boldsymbol{\beta}_2 \neq \mathbf{0} \tag{4.19}$$

Therefore the **reduced model linear predictor** is

$$\boldsymbol{\eta} = \mathbf{X}_1 \boldsymbol{\beta}_1 \tag{4.20}$$

Assume that the reduced model is fit, and let $D(\boldsymbol{\beta}_1)$ be the deviance for the reduced model. The deviance for the reduced model is never smaller than the deviance for the full model, because the reduced model contains fewer parameters. However, if the deviance for the reduced model is not much larger than the deviance for the full model, then the reduced model is about as good a fit as the full model, so it is likely that the parameters in $\boldsymbol{\beta}_2$ are equal to zero. That is, we cannot reject the null hypothesis above. However, if the difference in deviance is large, at least one of the parameters in $\boldsymbol{\beta}_2$ is likely not zero, and we should reject the null hypothesis. Formally, the difference in deviance is

$$D(\boldsymbol{\beta}_2 | \boldsymbol{\beta}_1) = D(\boldsymbol{\beta}_1) - D(\boldsymbol{\beta}) \tag{4.21}$$

and this quantity has $n - (p - r) - (n - p) = r$ degrees of freedom. If the null hypothesis is true and if n is large, the difference in deviance in Equation. (4.21) has a chi-square distribution with r degrees of freedom. Therefore the test statistic and decision criteria are

$$\begin{aligned} &\text{if } D(\boldsymbol{\beta}_2 | \boldsymbol{\beta}_1) \geq \chi^2_{\alpha,r} \quad \text{reject the null hypothesis} \\ &\text{if } D(\boldsymbol{\beta}_2 | \boldsymbol{\beta}_1) < \chi^2_{\alpha,r} \quad \text{do not reject the null hypothesis} \end{aligned} \tag{4.22}$$

Sometimes the difference in deviance $D(\boldsymbol{\beta}_2 / \boldsymbol{\beta}_1)$ is called the partial deviance.

Example 4.3. The* Challenger *Data. Once again, reconsider the Challenger O-ring failure data of Table 4.1. The model we initially fit to the data is

$$\hat{y} = \hat{\pi} = \frac{1}{1 + e^{0.8753 - 0.17132x}}$$

Suppose that we wish to determine whether adding a quadratic term for temperature in the linear predictor would improve the model. Therefore we consider the full model to be

$$y = \frac{1}{1 + e^{-(\beta_0 + \beta_1 x + \beta_{11} x^2)}}$$

Table 4.9 contains the output from JMP for this model. Notice that JMP automatically *centers* the quadratic temperature term so that the linear

Table 4.9 JMP Output of the Challenger Data with a Quadratic Term in Temperature

```
Ordinal Logistic Fit for Failure
 Whole Model Test
 Model          -LogLikelihood    DF    ChiSquare          Prob ChiSq
 Difference           3.192790     2      6.38558            0.0411*
 Full                11.294504
 Reduced             14.487294
 RSquare (U)                      0.2204
 Observations (or Sum Wgts)      24
 Converged by Gradient

 Lack Of Fit
 Source          DF      -LogLiKelihood        ChiSquare
 Lack Of Fit     14          7.658869           15.31774
 Saturated       16          3.635635          Prob ChiSq
 Fitted           2         11.294504             0.3568

 Parameter Estimates
  Term                Estimate    Std Error   ChiSquare   Prob ChiSq
 Intercept[1]         8.369397   6.5171295      1.65        0.1991
 Temperature        -0.1384684   0.09171        2.28        0.131
 (Temperature 69.9167)*
 (Temperature-69.9167)
 Effect Wald Tests  0.00629863 0.00962         0.43        0.5130
                                             Wald
 Source             Nparm        DF       ChiSquare       Prob ChiSq
 Temperature          1          1       2.27962158        0.1311
 Temperature*         1          1       0.4279422         0.5130
 Temperature
 Effect Likelihood Ratio Tests

                                             L-R
 Source             Nparm        DF       ChiSquare       Prob ChiSq
 Temperature          1          1       2.95011644        0.0859
 Temperature*         1          1       0.44144276        0.5064
 Temperature
```

predictor in the fitted model is $\beta_0 + \beta_1 \times \beta_{11}(x - 69.9167)^2$, where 69.9167 is the average observed temperature. Now the linear predictor for the full model can be written as

$$
\begin{aligned}
\eta &= \mathbf{X}\boldsymbol{\beta} \\
&= \mathbf{X}_1\boldsymbol{\beta}_1 + \mathbf{X}_2\boldsymbol{\beta}_2 \\
&= \beta_0 + \beta_1 x + \beta_{11}(x - 69.9167)^2
\end{aligned}
$$

From Table 4.8, we find that the deviance for the full model is

$$D(\boldsymbol{\beta}) = 15.31774$$

with $n-p = 17-3 = 14$ degrees of freedom. The reduced model linear predictor has $\mathbf{X}_1\boldsymbol{\beta}_1 = \beta_0 + \beta_1 x$, so $\mathbf{X}_2\boldsymbol{\beta}_2 = \beta_{11}(x-69.9167)^2$. Table 4.3 shows the deviance for the reduced model to be

$$D(\boldsymbol{\beta}_1) = 15.75918$$

with $p - r = 3 - 1 = 2$ degrees of freedom. Therefore the difference in deviance between the full and reduced model using Equation. (4.21) is

$$\begin{aligned} D(\boldsymbol{\beta}_2|\boldsymbol{\beta}_1) &= D(\boldsymbol{\beta}_1) - D(\boldsymbol{\beta}) \\ &= 15.75918-15.31774 \\ &= 0.44144 \end{aligned}$$

which should be referred to a chi-square distribution with $r = 1$ degree of freedom. The P-value associated with the difference in deviance is 0.5061, so we conclude that there is no value in including the quadratic term in the regressor variable x = temperature. □

Tests on Individual Model Coefficients

Tests on individual model coefficients, such as

$$H_0: \beta_j = 0, \quad H_1: \beta_j \neq 0 \tag{4.23}$$

can be conducted by using the difference-in-deviance method as illustrated in Example 4.3. There is another approach, also based on the theory of maximum likelihood estimators. For large samples, the distribution of a maximum likelihood estimator is approximately normal. The estimator has little or no bias. Furthermore, the variances and covariances of a set of maximum likelihood estimators can be found from the second partial derivatives of the log-likelihood function with respect to the model parameters, evaluated at the maximum-likelihood estimates. A t-like statistic called the **Wald statistic** can be constructed to test the above hypotheses.

Let $\mathbf{G}$ denote the $p \times p$ matrix of second partial derivatives of the log-likelihood function, that is,

$$G_{ij} = \frac{\partial^2 \mathscr{L}(\boldsymbol{\beta})}{\partial \beta_i \partial \beta_j}, \ i,j = 0, 1, \ldots, k$$

G is called the **Hessian matrix**. If the elements of the Hessian are evaluated at the maximum likelihood estimators $\boldsymbol{\beta} = \mathbf{b}$, the large-sample approximate covariance matrix of the regression coefficients is

$$\hat{\text{V}}\text{ar}(\mathbf{b}) = -\mathbf{G}(\mathbf{b})^{-1} = (\mathbf{X}'\hat{\mathbf{V}}\mathbf{X})^{-1} \tag{4.24}$$

Notice that this is just the estimated covariance matrix of **b** given earlier. The square roots of the diagonal elements of this covariance matrix are the large-sample estimated standard errors of the regression coefficients, so the test statistic for the null hypothesis in

$$H_0 : \beta_j = 0, \quad H_1 : \beta_j \neq 0$$

$$Z_0 = \frac{\hat{\beta}_j}{\hat{se}(\hat{\beta}_j)} \tag{4.25}$$

The reference distribution for this statistic is the standard normal distribution. Some computer packages square the Z_0 statistic and compare it to a chi-square distribution with one degree of freedom.

Example 4.4. The* Challenger *Data. Table 4.9 contains output from JMP for the data, originally given in Table 4.1. The fitted model is

$$\hat{y} = \frac{1}{1 + e^{8.369597 - 0.1384684x + 0.00629863(x - 69.9167)^2}}$$

The JMP output gives the standard errors of each model coefficient. Using Equation (4.25), The Z_0 test statistic for the squared term in temperature is

$$\begin{aligned} Z_0 &= \frac{0.00629863}{0.0096284} \\ &= 0.65417 \end{aligned}$$

JMP reports the square of this statistic:

$$\begin{aligned} Z_0^2 &= (0.65417)^2 \\ &= 0.42794 \\ &\simeq 0.43 \end{aligned}$$

This is compared to a chi-square distribution with one degree of freedom, resulting in a F-value of 0.5130. Thus, the squared term in temperature is not needed. □

Recall from the previous example that when we tested for the significance of β_{11} using the partial deviance method we obtained a different P-value. In linear regression, the t-test on a single regressor is equivalent to the partial F-test on a single variable (recall that the square of the t-statistic is equal to the partial

F statistic). However, this equivalence is only true for **linear models**, and the GLM is a **nonlinear model**.

Confidence Intervals

It is straightforward to use Wald statistics to construct confidence intervals in logistic regression. Consider first finding confidence intervals on individual regression coefficients in the linear prediction. An approximate $100(1 - \alpha)$ percent confidence interval on the jth model coefficient is

$$b_j - Z_{\alpha/2}\hat{se}(b_j) \leq \beta_j \leq b_j + Z_{\alpha/2}\hat{se}(b_j) \qquad (4.26)$$

Example 4.5. The Challenger Data. Using the JMP output in Table 4.9, we can find an approximate 95% confidence interval on β_{11} from Equation (4.26) as follows:

$$b_{11} - Z_{0.025}\hat{se}(b_{11}) \leq \beta_{11} \leq b_{11} + Z_{0.025}\hat{se}(b_{11})$$
$$0.00629863 - 1.96(0.0096284) \leq \beta_{11} \leq 0.00629863 + 1.96(0.0096284)$$
$$-0.01257 \leq \beta_{11} \leq 0.02517$$

Notice that the confidence interval includes zero, so at the 5% significance level, we would not reject the hypothesis that this model coefficient is zero. □

The regression coefficient β_j is also the logarithm of the odds ratio. Because we know how to find a confidence interval (CI) for β_j, it is easy to find a CI for the odds ratio. The point estimate of the odds ratio is $\hat{O}_R = \exp(b_j)$ and the $100(1 - \alpha)$ percent CI for the odds ratio is

$$\exp\left[b_j - Z_{\alpha/2}se(b_j)\right] \leq O_R \leq \exp\left[b_j + Z_{\alpha/2}se(b_j)\right] \qquad (4.27)$$

The CI for the odds ratio is generally not symmetric around the point estimate. Furthermore, the point estimate $\hat{O}_R = \exp(b_j)$ actually estimates the median of the sampling distribution of $\hat{O}_R$.

Example 4.6. The* Challenger *Data. Reconsider the original logistic regression model that we fit to the *Challenger* O-ring data in Example 4.1. JMP does not repeat the odds ratio but MINITAB does. From the MINITAB output for this data shown in Table 4.2 we find that the estimate of β_1 is $b_1 = -0.171321$ and the odds ratio $\hat{O}_R = \exp(b_1) = 0.84$. Because the estimated standard error of b_1 is $\hat{se}(b_1) = 0.0834420$, we can find a 95% CI on the odds ratio as follows:

$$\exp[-0.171321 - 1.96(0.0834420)] \leq O_R \leq \exp[-0.171321 + 1.96(0.0834420)]$$
$$\exp(-0.33487) \leq O_R \leq \exp(-0.00777)$$
$$0.72 \leq O_R \leq 0.99$$

This agrees with the 95% CI reported by MINITAB in Table 4.2. □

It is possible to find a CI on the linear predictor at any set of values of the predictor variables that is of interest. Let $\mathbf{x}_0' = [1, \mathbf{x}_{01}, \mathbf{x}_{02}, \ldots, \mathbf{x}_{0k}]$ be the values of the regressor variables that are of interest. The linear predictor evaluated at $\mathbf{x}_0$ is $\mathbf{x}_0'\mathbf{b}$. The estimated variance of the linear predictor at this point is

$$\widehat{\text{Var}}(\mathbf{x}_0'\mathbf{b}) = \mathbf{x}_0'\widehat{\text{Var}}(\mathbf{b})\mathbf{x}_0 = \mathbf{x}_0'(\mathbf{X}'\hat{\mathbf{V}}\mathbf{X})^{-1}\mathbf{x}_0$$

so the 100(1–α) percent CI on the linear predictor is

$$\mathbf{x}_0'\mathbf{b} - Z_{\alpha/2}\sqrt{\mathbf{x}_0'(\mathbf{X}'\hat{\mathbf{V}}\mathbf{X})^{-1}\mathbf{x}_0} \le \mathbf{x}_0'\boldsymbol{\beta} \le \mathrm{x}'_0\mathbf{b} + Z_{\alpha/2}\sqrt{\mathbf{x}_0'(\mathbf{X}'\hat{\mathbf{V}}\mathbf{X})^{-1}\mathbf{x}_0} \tag{4.28}$$

The CI on the linear predictor given in Equation (4.28) enables us to find a CI on the estimated probability of success π_0 at the point of interest $\mathbf{x}_0' = [1, x_{01}, x_{02}, \ldots, x_{0k}]$. Let

$$L(\mathbf{x}_0) = \mathbf{x}_0'\mathbf{b} - Z_{\alpha/2}\sqrt{\mathbf{x}_0'(\mathbf{X}'\hat{\mathbf{V}}\mathbf{X})^{-1}\mathbf{x}_0}$$

and

$$U(\mathbf{x}_0) = \mathbf{x}_0'\mathbf{b} - Z_{\alpha/2}\sqrt{\mathbf{x}_0'(\mathbf{X}'\hat{\mathbf{V}}\mathbf{X})^{-1}\mathbf{x}_0}$$

be the lower and upper 100(1–α) percent confidence bounds on the linear predictor at the point $\mathbf{x}_0$ from Equation (4.28). Then the point estimate of the probability of success at this point is $\hat{\pi}_0 = \exp(\mathbf{x}_0'\mathbf{b})/[1 + \exp(\mathbf{x}_0'\mathbf{b})]$ and the 100(1–α) percent CI on the probability of success at $\mathbf{x}_0$ is

$$\frac{\exp[L(\mathbf{x}_0)]}{1 + \exp[L(\mathbf{x}_0)]} \le \pi_0 \le \frac{\exp[U(\mathbf{x}_0)]}{1 + \exp[U(\mathbf{x}_0)]} \tag{4.29}$$

Example 4.7. The* Challenger *Data. Suppose that we want to find a 95% CI on the probability of O-ring failure when the temperature is $x = 60$ degrees. From the fitted logistic regression model in Example (4.1), we calculate a point estimate of the probability at 60 degrees as

$$\hat{\pi}_0 = \frac{e^{10.8753-0.171321\,(60)}}{1 + e^{10.8753-0.1713211(60)}} = \frac{e^{0.59604}}{1 + e^{0.59604}} = 0.6448$$

To find the CI, we need to calculate the estimated variance of the linear predictor at this point. The estimated variance is

$$\begin{aligned}\text{Var}\left(\mathbf{x}_0'\mathbf{b}\right) &= \mathbf{x}_0'\left(\mathbf{X}'\hat{\mathbf{V}}\mathbf{X}\right)^{-1}\mathbf{x}_0 \\ &= [1 \quad 60]\begin{bmatrix} 32.5257 & -0.473918 \\ -0.473918 & 0.006963 \end{bmatrix}\begin{bmatrix} 1 \\ 60 \end{bmatrix} = 0.72234\end{aligned}$$

Now

$$L(\mathbf{x}_0) = 0.59604 - 1.96\sqrt{0.72234} = -1.06978$$

and

$$U(\mathbf{x}_0) = 0.59604 \times 1.96\sqrt{0.72234} = -2.26186$$

Therefore the 95% CI on the estimated probability of O-ring failure when the temperature is $X = 60$ degrees is

$$\frac{\exp[L(\mathbf{x}_0)]}{1+\exp[L(\mathbf{x}_0)]} \leq \pi_0 \leq \frac{\exp[U(\mathbf{x}_0)]}{1+\exp[U(\mathbf{x}_0)]}$$

$$\frac{\exp(-1.06978)}{1+\exp(-1.06978)} \leq \pi_0 \leq \frac{\exp(-2.26186)}{1+\exp(-2.26186)}$$

$$0.2555 \leq \pi_0 \leq 0.9057$$

□

4.2.5 Lack-of-Fit Tests in Logistic Regression

An important question in any regression analysis is whether the proposed model adequately fits the data, which leads naturally to the notion of a formal test for lack-of-fit. Put simply, lack-of-fit represents what we could have fit to the data but chose not to fit. The deviance test outlined in the previous section is such a test since it formally tests the current logistic regression model against the saturated model. This approach is most valuable for comparing nested models with grouped data.

A second statistic offered in most statistical software packages is the Pearson chi-square statistic. To provide some intuitive background, consider the simple linear regression model

$$y_i = \beta_0 + \beta_i x_{1i} + \varepsilon_i$$

and the scatterplot in Figure 4.3 displaying the relationship between y and x.

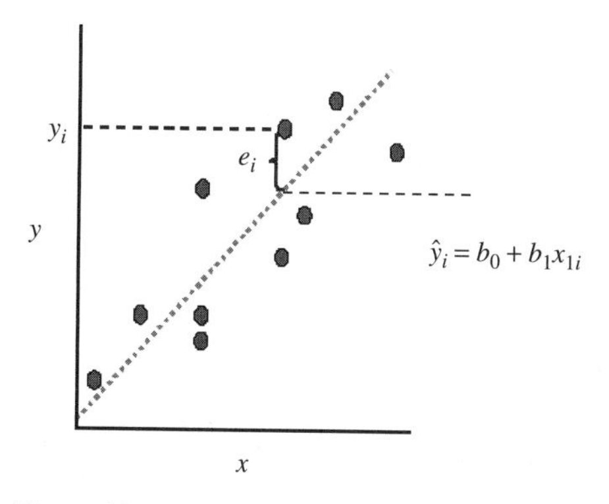

Figure 4.3 A Scatterplot for Linear Regression.

Table 4.10 Data Classification of Observed Counts and Model Predicted Cell Counts for a Logistic Model with Categorical Regressors

Covariate Pattern	Response		Row Marginals
	$Y = 1$	$Y = 0$	
$\mathbf{x}_1$	O_{11} (E_{11})	O_{12} (E_{12})	N_1
$\mathbf{x}_2$	O_{21} (E_{21})	O_{22} (E_{22})	N_2
$\vdots$	$\vdots$	$\vdots$	$\vdots$
$\mathbf{x}_m$	O_{m1} (E_{m1})	O_{m2} (E_{m2})	N_m

To assess fit in the simple linear regression model, we compare the fitted values, the $\hat{y}_i$'s, to the observed responses, the y_i's. A basic summary of the quality of fit is the sum of squared errors given by $\sum_{i=1}^{n} (y_i - \hat{y}_i)^2$. We note that the smaller the sum of squared errors, the better the fit. This notion of comparing the observed values to the model predicted values to assess quality of fit extends to logistic regression. The appropriate summary of model fit depends largely on the structure of the regressors. When the regressors are all categorical or when the covariate patterns of the regressors can be described easily using a contingency table, the Pearson chi-square or deviance chi-square is appropriate.

A contingency table, such as in Table 4.10, is a powerful way to summarize the data and the model predicted counts when we have a binary response and categorical regressors. The m rows correspond to the m possible covariate patterns, and the columns correspond to the two response categories. The O_{ij} denote the number of observed responses that occur in covariate pattern i and response category j. The model predicted counts for the first response category, the E_{i1}, are calculated as the sum of the N_i fitted probabilities for subjects in covariate pattern i, and $E_{i2} = N_i - E_{i1}$. The Pearson chi-square goodness-of-fit, which compares the observed to expected counts, is

$$\chi^2 = \sum_{i=1}^{m} \sum_{j=1}^{2} \frac{(O_{ij} - E_{ij})^2}{E_{ij}} \tag{4.30}$$

This test statistic has an approximate null chi-square distribution with $m - k - 1$, where k is the number of regressors in the model. When the chi-square statistic in Equation (4.30) is larger than $\chi^2_{1-\alpha,\mathrm{df}}$, one rejects the hypothesis of adequate model fit. It is helpful to illustrate the procedure with an example.

Example 4.8. The Prison Sentence Data. Recall the Stokes, Davis, and Koch (1995) data from the last section which involved a prison study whose goal was to model the relationship between receiving a prison sentence (y) and three regressors: type of crime committed (Type), history of prior arrest (Prior), and gender (Gender). The full data are provided in Table 4.11 in a contingency

Table 4.11 Prison Data

			Prison Sentence	
Gender	Type	Prior	Yes	No
Male	Business	Some	28	59
		None	7	35
	Home	Some	19	75
		None	27	200
Female	Business	Some	14	50
		None	10	40
	Home	Some	14	100
		None	27	159

Source: Stokes, Davis, and Roch (1995).

table format. Note for this example that each of the regressors has a qualitative scale.

Consider the logistic model

$$\ln\left(\frac{\pi_i}{1-\pi_i}\right) = \beta_0 + \beta_1 x_{i1} + \beta_2 x_{i2} \tag{4.31}$$

where π_i denotes the probability that the ith person will receive a prison sentence, $x_{i1} = 1$ if the ith person commits a business crime and $x_{i1} = 0$ if the ith person commits a home crime, and $x_{i2} = 1$ if the ith person has some prior arrest history and $x_{i2} = 0$ if the ith person has no prior arrest history. Table 4.12 gives the SAS code to analyze these data by PROC LOGISTIC. The code uses the Descending option in order to ensure appropriate modeling of reception of prison sentence. The Aggregate option in the Model Statement requests the Pearson chi-square and deviance chi-square goodness-of-fit tests. Table 4.13 summarizes the output.

The Pearson chi-square and deviance chi-square goodness-of-fit statistics are produced by SAS in the section *Deviance and Pearson Goodness-of-Fit Statistics*. Calculation of the Pearson chi-square is easily seen upon observing the information in Table 4.14. In Table 4.14, the covariate patterns are listed in the first two columns and then the response categories for each covariate pattern are given in column 3. The model-based probability of receiving a prison sentence (based upon the fit of model 4.31) are provided in the fourth column. The column denoted by "Pi" provides the probabilities of receiving a prison sentence as well as the probability of not receiving a prison sentence and doing element wise multiplication with the row totals in the column denoted by 'Ni', the expected counts in column 8 are generated. The last column, "Chi-Square Contribution" contains the value of each of the $(O_{ij} - E_{ij})^2/E_{ij}$ and the sum of these entries yields

Table 4.12 SAS Code for Analyzing the Prison Data

```
data sentence;
attrib type label='Type of crime'
       prior label='Prior Arrest?'
       prison label='Prison Sentence?' ;
input type $ prior $ prison $ gender $ count @@;
cards;
business Some Y F 14 business Some N F 50
business None Y F 10 business None N F 40
home Some Y F 14 home Some N F 100
home None Y F 27 home None N F 159
business Some Y M 28 business Some N M 59
business None Y M 7 business None N M 35
home Some Y M 19 home Some N M 75
home None Y M 27 home None N M 200
;
run;

*Creating indicator variables for type of crime and prior arrest
history

data sentence; set sentence;
ibus=(type='business');
iprior=(prior='Some');
imale=(gender='M');
run;
*Running Proc Logist for Model 4.31
proc logistic data=sentence descending;
title3'Example For Pearson Chi-Square and Deviance';
model prison=ibus iprior / aggregate scale=N ;
weight count;
output out=home.good p=phat;
run;
```

the Pearson chi-square statistic that is found in the *Deviance* and *Pearson Goodness-of-Fit Statistics* part of the output in Table 4.13. □

Note that when the Pearson chi-square concludes significant lack-of-fit, one can look at the contribution of each of the covariate patterns to the overall test statistic value to diagnose where the model is fitting poorly. Since the Pearson chi-square statistic (as well as the deviance chi-square discussed earlier) only follows a chi-square distribution asymptotically, it is important to keep in mind sample size guidelines when using this goodness-of-fit statistic. Specifically, the marginal sample sizes (the N_i's) should all be at least 10, 80% of the predicted counts (the E_{ij}'s) should be $\geq$ 5, and all other expected counts should be greater than 2.

Table 4.13 Output from PROC LOGISTIC for the Prison Data

```
                    The LOGISTIC Procedure
Data Set: WORK.SENTENCE
Response Variable: PRISON      Prison Sentence?
Response Levels: 2
Number of Observations: 16
Weight Variable: COUNT
Sum of Weights: 864
Link Function: Logit
                        Response Profile
               Ordered                        Total
           Value        PRISON         Count        Weight
               1           Y             8       146.00000
               2           N             8       718.00000
        Deviance and Pearson Goodness-of-Fit Statistics
                                                  Pr >
Criterion            DF        Value        Value/DF        Chi-Square
Deviance              1       0.5779         0.5779            0.4471
Pearson               1       0.5717         0.5717            0.4496
                    Number of unique profiles: 4
 Model Fitting Information and Testing Global Null
 Hypothesis
                                  BETA=0
                            Intercept
              Intercept     and
Criterion     Only          Covariates     Chi-Square for Covariates
AIC           786.974       775.478                      .
SC            787.747       777.796                      .
-2 LOG L      784.974       769.478         15.496 with 2 DF (p=0.0004)
Score             .             .           16.200 with 2 DF (p=0.0003)
```

Assessing Model Fit When Continuous Regressors are Present

While the Pearson and deviance chi-squares are reasonable tests for lack-of-fit when all of the regressors in the logistic model are categorical, they are not appropriate when continuous regressors are present since the number of distinct covariate patterns approaches the sample size, n, thus producing small expected cell counts. To address this shortcoming, Hosmer and Lemeshow (1989) proposed an alternative chi-square test, commonly referred to as the Hosmer–Lemeshow test, for assessing fit in logistic regression models. The Hosmer–Lemeshow test first orders all responses according to their model fitted probabilities and then classifies them into 10 (or possibly less) deciles of risk groups. This grouping by fitted probabilities provides a convenient substitute for the grouping of responses according to distinct covariate patterns that the Pearson and deviance statistics use. A general form of the contingency table constructed by the Hosmer-Lemeshow test is provided in Table 4.15.

Table 4.14 Construction of the Pearson Chi-Square Statistic for the Logistic Analysis of the Model in Equation (4.30)

Type of Crime	Prior Arrest?	Prison Sentence?	Model-based Probability of Prison	Observed count	Ni	Pi	Expected Count	Chi-Square Contribution i
business	None	N	0.20553	75	92	0.79447	73.091	0.04984
business	None	Y	0.20553	17	92	0.20553	18.909	0.19266
business	Some	N	0.26551	109	151	0.73449	110.909	0.03285
business	Some	Y	0.26551	42	151	0.26551	40.091	0.09087
home	None	N	0.12613	359	413	0.87387	360.909	0.01009
home	None	Y	0.12613	54	413	0.12613	52.091	0.6994
home	Some	N	0.16783	175	208	0.83217	173.091	0.02105
home	Some	Y	0.16783	33	208	0.16783	34.909	0.10436
								0.57166

Table 4.15 Contingency Table Constructed in the Hosmer–Lemeshow Procedure

	Response		
Deciles of Risk	$Y = 1$	$Y = 0$	Row Marginals
Decile 1	O_{11} (E_{11})	O_{12} (E_{12})	N_1
Decile 2	O_{21} (E_{21})	O_{22} (E_{22})	N_2
⋮	⋮	⋮	⋮
Decile 10	$O_{10,1}$ $(E_{10,1})$	$O_{10,2}$ $(E_{10,2})$	N_m

Model-based expected counts (the E_{ij}'s) are computed in the same manner as they were constructed for the deviance and Pearson statistics and the test statistic is given by

$$\chi^2_{H-L} = \sum_{i=1}^{g}\sum_{j=1}^{2}\frac{(O_{ij} - E_{ij})^2}{E_{ij}},$$

where g denotes the number of decile groups. The degrees of freedom associated with χ^2_{H-L} is $g-2$. The Hosmer–Lemeshow statistic is widely available in most statistical software packages that accommodate generalized linear models. While the Hosmer–Lemeshow statistic is appropriate for assessing fit from logistic regression models involving continuous regressor variables, it has been shown to have low power when there are both continuous and categorical variables in the logistic regression model. Also, when lack-of-fit is diagnosed with the Hosmer–Lemeshow statistic, it is often difficult to ascertain where the model fits poorly.

Assessing Model Fit When Both Categorical and Continuous Regressors are Present

Pulkstenis and Robinson (2002) propose two statistics that combine the strengths of the Pearson chi-square and deviance tests with a modification for continuous covariates similar to that provided by the Hosmer–Lemeshow procedure. The Pulkstenis–Robinson procedures utilize a two-level sub grouping within each categorical covariate pattern based on fitted probabilities within the specific covariate patterns. Table 4.16 helps to illustrate this procedure. For each covariate pattern in this table, model-based fitted probabilities are sorted, and two sub categories are produced based on the median of fitted probabilities within each of the m rows. Table 4.16 gives the resulting contingency table, where O_{ihj} and E_{ihj} denote the observed and model-based predicted counts, respectively, for the ith covariate pattern, the hth sub division, and the jth response category. Note that Table 4.16 is similar to the Pearson chi-square in Table 4.9. Table 4.16, however, has twice as many rows due to the grouping by the median of fitted probabilities within each of the m categorical regressor covariate patterns.

Table 4.16 Contingency Table of Observed and Model-Based Expected Cell Counts Utilized in the Pulkstenis–Robinson Procedures

	Response		
Covariate Pattern	Y = 1	Y = 0	Row Marginals
$\mathbf{x}_{11}$	O_{111} (E_{111})	O_{112} (E_{112})	N_{11}
$\mathbf{x}_{12}$	O_{121} (E_{121})	O_{122} (E_{122})	N_{12}
$\mathbf{x}_{21}$	O_{211} (E_{211})	O_{212} (E_{212})	N_{21}
$\mathbf{x}_{22}$	O_{221} (E_{221})	O_{222} (E_{222})	N_{22}
$\vdots$	$\vdots$	$\vdots$	$\vdots$
$\mathbf{x}_{m1}$	O_{m11} (E_{m11})	O_{m12} (E_{m12})	N_{m1}
$\mathbf{x}_{m2}$	O_{m21} (E_{m21})	O_{m22} (E_{m22})	N_{m2}

Model-based expected counts are computed exactly as described for the other methods. The Pulkstenis–Robinson statistics are

$$\chi^2_{\text{P-R}} = \sum_{i=1}^{m} \sum_{h=1}^{2} \sum_{j=1}^{2} \frac{(O_{ihj} - E_{ihj})^2}{E_{ihj}} \tag{4.32}$$

and

$$D^2_{\text{P-R}} = 2 \sum_{i=1}^{m} \sum_{h=1}^{2} \sum_{j=1}^{2} O_{ihj} \ln \frac{O_{ihj}}{E_{ihj}} \tag{4.33}$$

The degrees of freedom for $\chi^2_{\text{P-R}}$ and $D^2_{\text{P-R}}$ are $2m - k - 2$, where $2m$ refers to the number of rows of the contingency table and k is the number of *categorical* regressors in the model. These degrees of freedom are analogous to the $g - 2$ degrees of freedom for the Hosmer–Lemeshow statistic; however, the Pulkstenis–Robinson statistics must subtract k for the number of categorical variables in the model. The Pulkstenis–Robinson statistics provide some ability to diagnose which covariate patterns do not fit well when the null hypothesis is rejected because these statistics incorporate the design structure of the categorical regressors in the contingency table formulation. Pulkstenis and Robinson show that these statistics are more powerful than the Hosmer–Lemeshow test in many settings. Certainly an important consideration with the Pulkstenis–Robinson tests involves sample size since these statistics double the number of categorical covariate patterns used in the Pearson chi-square. As a sample size guideline, the authors suggest that the majority (80%) of the model-based expected cell counts exceed five. The authors do not recommend the Pulkstenis–Robinson tests statistics when there are only categorical covariates or only continuous covariates present.

Example 4.9. Treatment of Aneurysms. Pulkstenis and Robinson (2002) consider a clinical trial designed to compare an endovascular approach to a standard surgical procedure for the treatment of abdominal aneurysms. In the study, 102 surgical control subjects were compared to 266 treated subjects with respect to the incidence of moderate and severe bleeding complications within 30 days of surgery (comp30 = 1 if moderate to severe and 0 if not). Covariates of interest included treatment group (control = 1 if subject had standard procedure and 0 if endovascular approach), gender (male = 1 if subject is male and 0 if female), early investigator experience (early = 1 if investigator had limited experience and 0 if investigator had more experience), and baseline aneurysm diameter (diameter). Table 4.17 gives the appropriate PROC LOGISTIC code for the main effects logistic model. Table 4.18 summarizes the results for the fitted model. Data can be found in aneurysms.xls on the accompanying ftp site.

The results indicate that the control group (those that undergo the standard surgery) tend to have more bleeding complications than those who undergo the endovascular approach. Recall that the odds ratio value corresponding to the jth model term is $\exp(\beta_j)$, where β_j is the corresponding coefficient. For these data the *adjusted* odds of having bleeding complications when one undergoes the standard surgery (the control group) versus when one has the endovascular approach is

$$\begin{aligned}\frac{\Pr(y=1)}{\Pr(y=0)} &= \frac{\pi}{1-\pi}\\ &= \frac{\exp(\beta_0+\beta_1\text{control}=1+\beta_2\text{male}+\beta_3\text{early}+\beta_4\text{diameter})}{\exp(\beta_0+\beta_1\text{control}=0+\beta_2\text{male}+\beta_3\text{early}+\beta_4\text{diameter})}\\ &= \exp(\beta_1)\end{aligned}$$

Since the slope coefficient is positive from Table 4.14, the odds of having bleeding complications when one undergoes the standard procedure is greater than for those who undergo the less invasive endovascular approach. More specifically, the estimated odds of having bleeding complications if one has the standard procedure are $\exp(1.69) = 5.42$ greater than the estimated odds of having bleeding complications if you undergo the endovascular approach. Note that this is an *adjusted* odds ratio as the values of the coefficients have been adjusted for the presence of the other variables in the model. The estimated

Table 4.17 PROC LOGISTIC Code for the Aneurysm Data Set

```
proc logistic data=aneurysm descending;
model comp30 = control male early diameter / lackfit;
output out=phats p=phats;
run;
```

Table 4.18 PROC LOGISTIC Output for the Main-Effects Model of the Aneurysm Data

Model Information

Data Set	WORK.ANEURYSM	
Response Variable	comp30	comp30
Number of Response Levels	2	
Model	binary logit	
Optimization Technique	Fisher's scoring	

Number of Observations Road	379
Number of Observations Used	368

Response Profile

Ordered Value	comp30	Total Frequency
1	1	83
2	0	285

Probability modeled is comp30=1.

NOTE: 11 observations were deleted due to missing values for the response or explanatory variables.

Model Convergence Status

Convergence criterion (GCONV=1E-8) satisfied.

Model Fit Statistics

Criterion	Intercept only	Intercept and Covariates
AIC	394.903	355.808
SC	398.811	375.348
2 Log L	392.903	345.808

Testing Global Null Hypothesis: BETA=0

Test	Chi-Square	DF	Pr > ChiSq
Likelihood Ratio	47.0952	4	<0.0001
Score	46.5724	4	<0.0001
Wald	39.5528	4	<0.0001

The LOGISTIC Procedure

Analysis of Maximum Likelihood Estimates

Parameter	DF	Estimate	Standard Error	Wald Chi-Square	Pr > ChiSq
Intercept	1	2.6099	0.8414	9.6212	0.0019
Control	1	1.6911	0.3219	27.5943	<0.0001
male	1	1.0539	0.3446	9.3548	0.0022
early	1	1.2430	0.3516	12.4953	0.0004
diameter	1	0.0242	0.0135	3.2088	0.0732

Odds Ratio Estimates

Effect	Point Estimate	95% Wald Confidence	Limits
Control	5.425	2.887	10.196
male	0.349	0.177	0.685
early	3.466	1.740	6.905
diameter	1.024	0.998	1.052

(Continued)

Table 4.18 *Continued*

Association of Predicted Probabilities and Observed Responses

Percent concordant	73.4	Somers' D	0.473
Percent Discordant	26.1	Gamma	0.475
Percent Tied	0.5	Tau-a	0.166
Pairs	23655	c	0.736

Partition for the Hosmer and Lemeshow Test

		comp30 = 1		comp30 = 0	
Group	Total	Observed	Expected	Observed	Expected
1	38	1	2.59	37	35.41
2	40	3	3.19	37	36.81
3	34	1	3.00	33	31.00
4	37	5	3.77	32	33.23
5	37	9	6.62	28	30.38
6	37	9	8.62	28	28.38
7	37	10	10.40	27	26.60
8	37	17	11.95	20	25.05
9	37	13	14.00	24	23.00
10	34	15	18.86	19	15.14

Hosmer and Lemeshow Goodness-of-Fit Test

Chi-Square	DF	Pr > ChiSq
9.0885	8	0.3349

odds ratios that correspond to the other factors can be interpreted in a similar fashion. In general, it appears that males are less likely to experience bleeding (p = 0.0022), and that subjects with larger aneurysm diameters are marginally more likely to experience bleeding ($p = 0.0732$). Finally, a learning curve appears to exist in terms of those performing the procedure since bleeding complications decrease for experienced physicians as compared to those with little experience ($p = 0.0004$).

Note that the model of interest involves both categorical predictors (control, male, and early) and a continuous predictor (diameter). The Hosmer–Lemeshow goodness-of-fit statistic is requested using the *lackfit* option the PROC LOGISTIC Model statement. The contingency table produced in the Hosmer–Lemeshow procedure appears in the *Partition for the Hosmer and Lemeshow Test* section of the output. The Hosmer–Lemeshow test results appear in the *Hosmer and Lemeshow Goodness-of-Fit Test* section of the output. We note that the test suggests no significant lack of fit in the main-effects model. Using the Pulkstenis–Robinson statistics, the $\chi^2_{\mathrm{P-R}}$ and $D^2_{\mathrm{P-R}}$ statistics result in $p = 0.0083$ and $p = 0.0118$, respectively, suggesting significant lack of fit. Table 4.19 Summarizes the contingency table produced by the Pulkstenis–Robinson procedures.

Upon first inspection, we note that 33% (8/24) of the model-based expected cell counts in Table 4.19 are less than 5; so we should proceed with caution regarding the use of the Pulkstenis–Robinson procedures. Direct comparison of the observed and expected counts as well as the contributions to the $D^2_{\mathrm{P-R}}$

Table 4.19 Contingency Table Produced by the Pulkstenis–Robinson Procedure

Group	Investigator Experience	Gender	Response $Y = 1$	Response $Y = 0$	$D^2_{\text{P-R}}$ Contribution ($Y = 1$)	$D^2_{\text{P-R}}$ Contribution ($Y = 0$)
Treatment	Late	Female	5 (2.0)	5 (8.0)	9.14	−4.70
Treatment	Late	Female	3 (2.6)	7 (7.4)	0.79	−0.72
Treatment	Late	Male	4 (6.0)	77 (75.0)	−3.28	4.11
Treatment	Late	Male	7 (8.3)	74 (72.7)	−2.45	2.70
Treatment	Early	Female	1 (1.8)	3 (2.2)	−1.16	1.83
Treatment	Early	Female	3 (2.1)	1 (1.9)	2.25	−1.32
Treatment	Early	Male	10 (8.3)	28 (29.7)	3.77	−1.33
Treatment	Early	Male	9 (10.9)	29 (27.1)	−3.39	3.86
Control	**Late**	**Female**	**2 (6.1)**	**9 (4.9)**	**−4.43**	**10.81**
Control	Late	Female	7 (6.5)	3 (3.5)	1.14	−1.01
Control	**Late**	**Male**	**19 (12.7)**	**22 (28.3)**	**15.28**	**11.06**
Control	Late	Male	13 (15.8)	27 (24.2)	5.03	5.86

Table 4.20 The PROC LOGISTIC Output for the Main-Effects Plus the Gender by Treatment Interaction Model

The LOGISTIC Procedure

Analysis of Maximum Likelihood Estimates

Parameter	DF	Estimate	Standard Error	Wald Chi-Square	Pr > ChiSq
Intercept	1	1.9405	0.8858	4.7990	0.0285
Control	1	0.4410	0.6068	0.5283	0.4673
male	1	1.8007	0.4564	15.5665	<.0001
early	1	1.3350	0.3674	13.2021	0.0003
diameter	1	0.0223	0.0137	2.6389	0.1043
control*male	1	1.6238	0.6769	5.7548	0.0164

statistic on a cell-by-cell basis reveals that the largest contribution to $D^2_{\text{P-R}}$ is 15.28, which occurs with male patients who undergo the standard surgery with an experienced physician. The next highest contribution to the $D^2_{\text{P-R}}$ value is 10.81, which involves female patients who undergo the standard surgery with an experienced physician. The indication is that, for female controls, too many events are being predicted (since the contribution to $D^2_{\text{P-R}}$ when $Y = 1$ is a relatively large negative value (-4.43) while for male controls, too few are being predicted (since the contribution to $D^2_{\text{P-R}}$ when $Y = 1$ is large positive (15.28). Since the effect of the treatment seems to depend on the gender, one would be inclined to model a treatment by gender interaction term. Table 4.20 provides the relevant PROC LOGISTIC output for the main-effects plus the gender by treatment interaction model.

As suspected, the gender by treatment interaction is statistically significant ($p = 0.0164$). Once this term is added, the $\chi^2_{\text{P-R}}$ and $D^2_{\text{P-R}}$ statistics result in $p = 0.0578$ and $p = 0.0566$, respectively. These values are nonsignificant and one may reasonably assume that the model is now adequate. This example illustrates the general fact that when both categorical and continuous regressors are present in the logistic model, the Hosmer–Lemeshow statistic often lacks the power to detect important factor interactions and the Pulkstenis–Robinson statistics offer a reasonable alternative. SAS macros for performing the Pulkstenis–Robinson procedure can be obtained by downloading GOF.zip at http://lib.stat.cmu.edu/general/. Pulkstenis and Robinson (2004) extended their procedures for the ordinal response case and the corresponding macros for this situation can be obtained by downloading GOF_ordinal.zip at http://lib.stat.cmu.edu/general/. □

4.2.6 Diagnostic Checking in Logistic Regression

Residuals can be used for diagnostic checking and investigating model adequacy in logistic regression. The ordinary residuals are defined as usual,

$$e_i = y_i - \hat{y}_i = y_i - n_i\hat{\pi}_i, \quad i = 1, 2, \ldots, n$$

In linear regression the ordinary residuals are components of the residual sum of squares; that is, if the residuals are squared and summed, the residual sum of squares results. In logistic regression, the quantity analogous to the residual sum of squares is the deviance. This leads to a **deviance residual**, defined as

$$d_i = \pm\left\{2\left[y_i \ln\left(\frac{y_i}{n_i\hat{\pi}_i}\right) + (n_i - y_i)\ln\left(\frac{n_i - y_i}{n_i(1-\hat{\pi}_i)}\right)\right]\right\}^{1/2}, \quad i = 1,2,\ldots,n \tag{4.34}$$

The sign of the deviance residual is the same as the sign of the corresponding ordinary residual; also, when $y_i = 0,\ d_i = -\sqrt{-2n\ln(1-\hat{\pi}_i)}$, and when $y_i = n_i,\ d_i = \sqrt{-2n_i \ln \hat{\pi}_i}$. Similarly, we can define a **Pearson residual**

$$r_i = \frac{y_i - n_i\hat{\pi}_i}{\sqrt{n_i\hat{\pi}_i(1-\hat{\pi}_i)}}, \quad i = 1,2,\ldots,n \tag{4.35}$$

It is also possible to define a hat matrix analog for logistic regression,

$$\mathbf{H} = \hat{\mathbf{V}}^{1/2}\mathbf{X}(\mathbf{X}'\hat{\mathbf{V}}\mathbf{X})^{-1}\mathbf{X}'\hat{\mathbf{V}}^{1/2} \tag{4.36}$$

where $\hat{\mathbf{V}}$ is the diagonal matrix defined earlier that has the variances of each observation on the main diagonal, $\hat{V}_{ii} = n_i\hat{\pi}_i(1-\hat{\pi}_i)$, and these estimated variances are calculated using the estimated probabilities that result from the fitted logistic regression model. The diagonal elements of $\mathbf{H}$, h_{ii}, can be thought of as **leverage** values, and can be used to calculate a **standardized Pearson residual**

$$r_{si} = \frac{r_i}{\sqrt{1-h_{ii}}} = \frac{y_i - n_i\hat{\pi}_i}{\sqrt{(1-h_{ii})n_i\hat{\pi}_i(1-\hat{\pi}_i)}}, \quad i = 1,2,\ldots,n \tag{4.37}$$

The deviance and Pearson residuals are the most appropriate for conducting model adequacy checks. Plots of these residuals versus the estimated probability and a normal probability plot of the deviance residuals are useful in checking the fit of the model at individual data points and in checking for possible outliers.

Table 4.21 displays the deviance residuals, Pearson residuals, hat matrix diagonals, leverage values, and the standardized Pearson residuals for the *Challenger* data. To illustrate the calculations, consider the deviance residual for the first observation (53 degrees). From Equation (4.34)

$$\begin{aligned} d_1 &= +\left\{2\left[y_1 \ln\left(\frac{y_1}{n_1\hat{\pi}_1}\right) + (n_1 - y_1)\ln\left(\frac{n_1 - y_1}{n_1(1-\hat{\pi}_1)}\right)\right]\right\}^{1/2} \\ &= +\left\{2\left[1\ \ln\left(\frac{1}{1\,(0.857583489)}\right) + (1-1)\ln\left(\frac{1-1}{1\,(1-0.857583489)}\right)\right]\right\}^{1/2} \\ &= 0.554322544 \end{aligned} \tag{4.38}$$

Table 4.21 Diagnostic Quantities for the Challenger Logistic Regression Model

Temperature (°F)	Failure	Predicted Failure Probability	Pearson Residuals	Standardized Pearson Residuals	Deviance Residuals	Delta Chi-Square	Delta Deviance	Delta Beta	Leverage
53	1	0.857583489	0.407513467	0.463124821	0.554322544	0.2144846	0.355690857	0.048417374	0.225738231
56	1	0.782688179	0.52692316	0.595849145	0.700029857	0.355036203	0.567429988	0.077388187	0.217972663
57	1	0.752144136	0.574049128	0.645398893	0.7547546666	0.416539731	0.656661935	0.08700733	0.208881226
63	0	0.520527851	−1.041934199	−1.105340324	−1.212492862	1.221777232	1.606289296	0.136150356	0.111436318
66	0	0.393695655	−0.805814301	−0.836046251	−1.00037313	0.698973335	1.050383046	0.049636647	0.071013649
67	0	0.353629202	−1.281131837	−1.422642575	−1.618113621	2.023911897	3.000904805	0.382613114	0.189046329
67	0								
67	0								
68	0	0.315518366	−0.678940041	−0.699468274	−0.870739302	0.489255866	0.786483219	0.028296287	0.057835356
69	0	0.279737227	−0.623203162	−0.641140954	−0.81011008	0.411061722	0.678957883	0.022679542	0.055173081
70	0	0.246552215	2.336164369	2.642316474	2.113906067	6.981836347	5.992771248	1.524172387	0.218305373
70	1								
70	1								
70	1								
72	0	0.188509095	−0.481974793	−0.496443489	−0.64634681	0.246456138	0.431920635	0.014156437	0.057439984
73	0	0.163686925	−0.442407572	−0.456272987	−0.59791679	0.208185038	0.369965067	0.012460579	0.059853382
75	0	0.121993265	1.633419311	1.750266891	1.301995466	3.063434189	2.090567738	0.395375544	0.129062849
75	1								
76	0	0.104798542	−0.483873933	−0.519503781	−0.665451705	0.269884178	0.478576166	0.035750195	0.132464954
76	0								
78	0	0.076728513	−0.288279447	−0.298552964	−0.399579661	0.089133872	0.165692738	0.006028833	0.067637953
79	0	0.065438269	−0.264613445	−0.273990377	−0.367906502	0.075070727	0.140405646	0.005050452	0.067275913
80	0	0.055709091	−0.242890278	−0.251358575	−0.338588228	0.063181133	0.118827434	0.004185446	0.066245188
81	0	0.047353127	−0.222950452	−0.230522722	−0.311483503	0.053140725	0.100455794	0.003433821	0.064617503

which closely matches the value reported in Table 4.21. The sign of the deviance residual d_1 is positive because the ordinary residual $e_1 = y_1 - n_1\hat{\pi}_1 = 1 - 0.857583489 = 0.142416511$ is positive.

Figure 4.4 is the normal probability plot of the deviance residuals and Figure 4.5 plots the deviance residuals versus the estimated probability of failure. These plots indicate that there are no obvious problems with the model fit.

In linear regression we often find it useful to assess the effect that deleting individual observations has on the estimates of the model parameters (e.g., see Section 2.4.4). It is possible to obtain similar measures for logistic regression. For example, the influence that the ith individual observation has on the vector of model parameters to a linear approximation, is given by

$$\begin{aligned}\Delta\mathbf{b} &= (\mathbf{b} - \mathbf{b}_{(-i)})'(\mathbf{X}'\mathbf{V}\mathbf{X})(\mathbf{b} - \mathbf{b}_{(-i)}) \\ &= \frac{r_i^2 h_{ii}}{(1-h_{ii})^2} \\ &= \frac{r_{si}^2}{1-h_{ii}} \end{aligned} \tag{4.39}$$

Hosmer and Lemeshow (2002) point out that by using similar linear approximations, the decrease in the value of the Pearson chi-square statistic due to deletion of the ith individual observation is

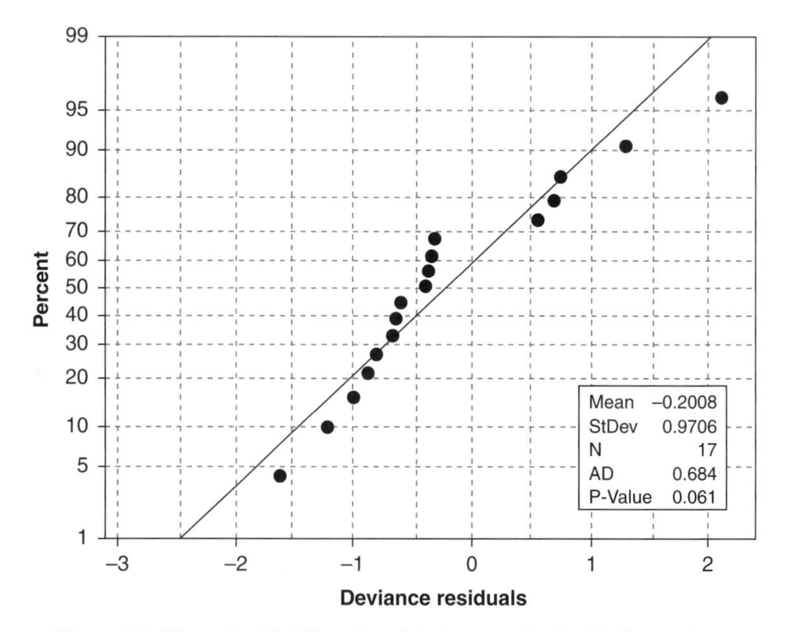

Figure 4.4 Normal probability plot of deviance residuals, *Challenger* data.

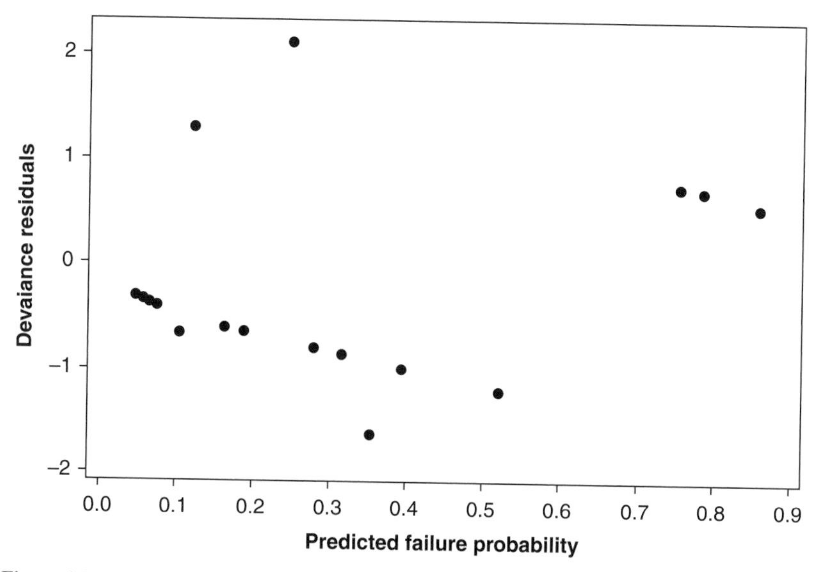

Figure 4.5 Plot of deviance residuals versus predicted failure probability, *Challenger* data.

$$\Delta\chi_i^2 = \frac{r_i^2}{1 - h_{ii}} = r_{si}^2 \tag{4.40}$$

The change in deviance is

$$\Delta D_i = d_i^2 + \frac{r_i^2 h_{ii}}{1 - h_{ii}} \tag{4.41}$$

If we replace r_i^2 by d_i^2 we have

$$\Delta D_i = \frac{d_i^2}{1 - h_{ii}} \tag{4.42}$$

These diagnostics can be very useful in identifying observations that have an unusual impact on the estimates of the model parameters or that fit poorly in terms of the deviance and Pearson chi-square statistics. Generally, values of these quantities larger than 4 indicate influential observations.

It is usually a good idea to assess these statistics graphically. Plotting $\Delta\mathbf{b}_i$, $\Delta\chi_i^2$, and ΔD_i versus the estimated probabilities from the logistic regression model $\hat{\pi}_i$ and against the leverage values h_{ii} is usually recommended. Figures 4.6 through 4.11 present plots of these quantities for the *Challenger* O-ring failure data. The values of $\Delta\chi_i^2$ and ΔD_i for the observation where temperature is 70 degrees are 6.98 and 5.99, respectively, indicating that this

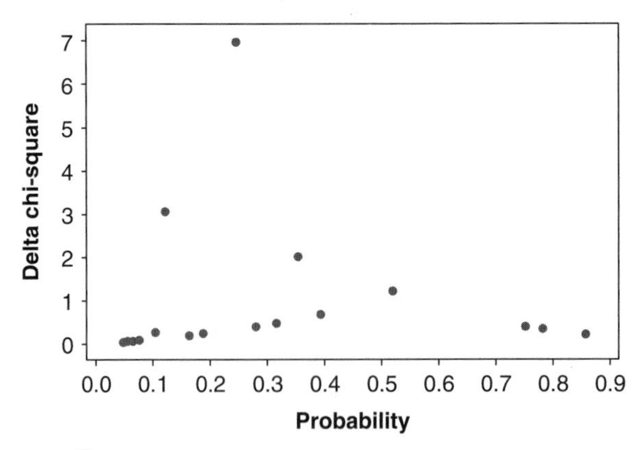

Figure 4.6 Plot of $\Delta\chi_i^2$ versus the estimated probabilities for the *Challenger* O-ring failure data.

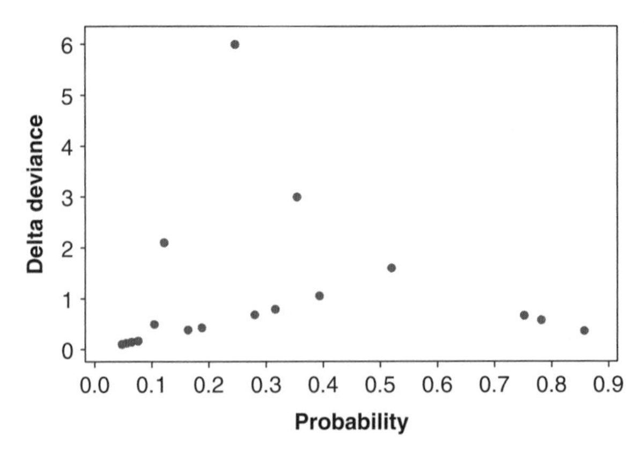

Figure 4.7 Plot of ΔD_i versus the estimated probabilities for the *Challenger* O-ring failure data.

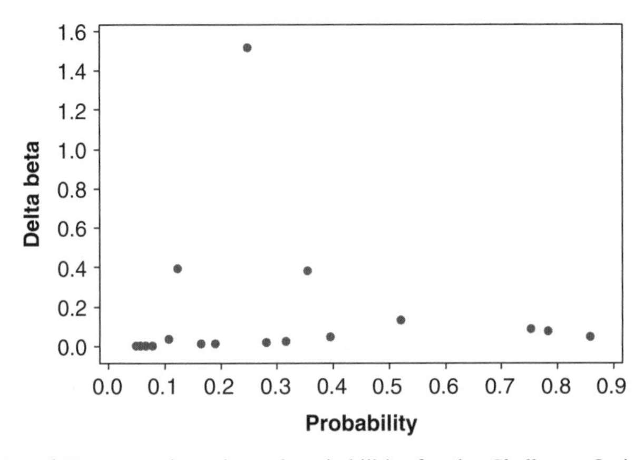

Figure 4.8 Plot of $\Delta \mathbf{b}_i$ versus the estimated probabilities for the *Challenger* O-ring failure data.

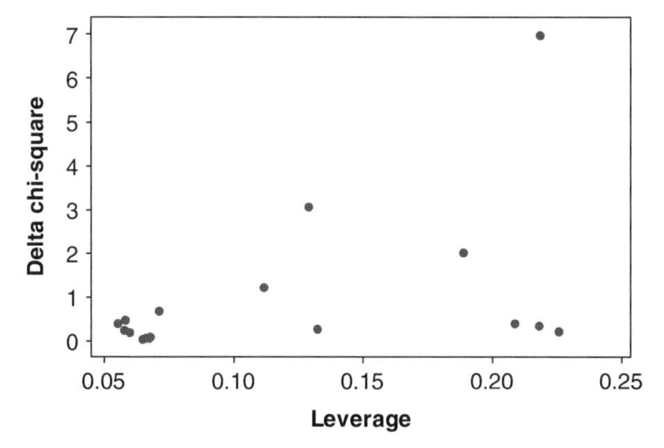

Figure 4.9 Plot of $\Delta\chi_i^2$ versus leverage for the *Challenger* O-ring failure data.

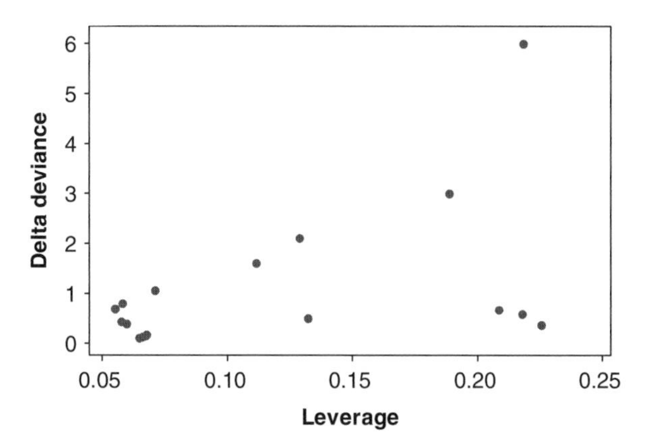

Figure 4.10 Plot of ΔD_i versus leverage for the *Challenger* O-ring failure data.

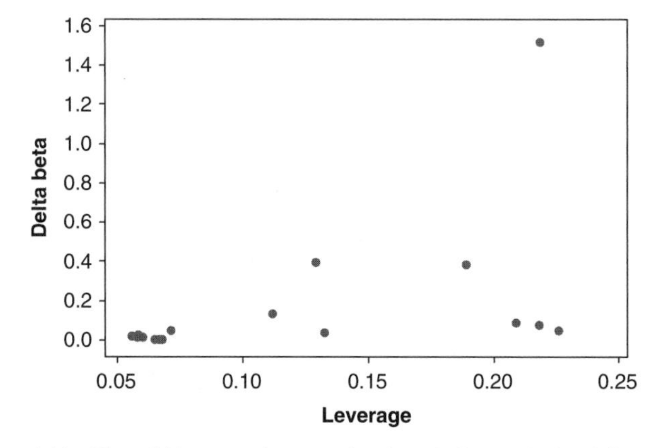

Figure 4.11 Plot of $\Delta \mathbf{b}_i$ versus leverage for the *Challenger* O-ring failure data.

level of temperature is not fit very well by the model. This result is not too surprising because the estimated probability of failure at 70 degrees is quite low (0.24655), yet three of the four observations at 70 degrees resulted in failure. This results in large Pearson and deviance residuals, and the large values of $\Delta\chi_i^2$ and ΔD_i.

4.2.7 Classification and the Receiver Operating Characteristic Curve

A simple and intuitive way to summarize the results of fitting a logistic regression model is through a classification table. This table is produced by cross-classifying the response variable with predicted responses that are derived from the fitted model. To obtain the predictions, we need to define a cut point c, and compare each estimated probability to it. If the estimated probability exceeds c, then the predicted response is unity. If the estimated probability is less than c, the predicted response is zero. The usual value for the cut point is $c = 0.5$; however, $c = 0.5$ is not always the optimal value.

Table 4.22 is the classification table for the *Challenger* O-ring failure data using $c = 0.5$. The overall rate of correct classification is

$$100 \times \frac{16 + 3}{24} = 79.2\%$$

with

$$100 \times \frac{16}{17} = 94.12\%$$

of the successful launches being correctly classified and only

$$100 \times \frac{3}{7} = 42.86\%$$

of the failures being correctly classified. Classification is sensitive to the size of the two groups, and the larger group always has a higher probability of correct classification, regardless of how well the model fits the data.

Table 4.22 Classification Table for the Challenger O-Ring Failure Data

	Observed		
Predicted	Failure = 0	Failure = 1	Total
Failure = 0	16	4	20
Failure = 1	1	3	4
Total	17	7	24

The probability of detecting the true signal (which in the Challenger case is an O-ring failure) is called **sensitivity**. For the *Challenger* data this probability is just the probability of a failed launch being correctly classified, which is 0.4286. The probability of detecting the null signal is called **specificity**. For the *Challengers* data, this is the probability of correctly detecting a successful launch, or 0.9412.

The sensitivity and specificity depend on the cut point c that is chosen. One could construct classification tables for a range of different cut points to see how the logistic regression model works as a classifier. A better way to do this is given by the area under the **receiver operating characteristic (ROC)** curve. This graph plots the probability of detecting the signal of interest (sensitivity) against the probability of getting a false signal (1 −specificity) for a range of possible cut points. The area under the ROC curve lies between zero and unity and measures the ability of the model to discriminate between observations that will lead to the response of interest and those that will not. JMP can display the ROC curve and provide the area under the ROC curve. Figure 4.12 shows this plot for the logistic regression model for the *Challenger* data. The area under the ROC curve is reported by JMP as 0.7227.

Hosmer and Lemeshow (2002) suggest that the area under the ROC curve is a general guide to how well the model discriminates, with the following guidelines:

- ROC = 0.5, no discrimination — you could do as well by tossing a coin.
- $0.7 \leq$ ROC < 0.8, acceptable discrimination.

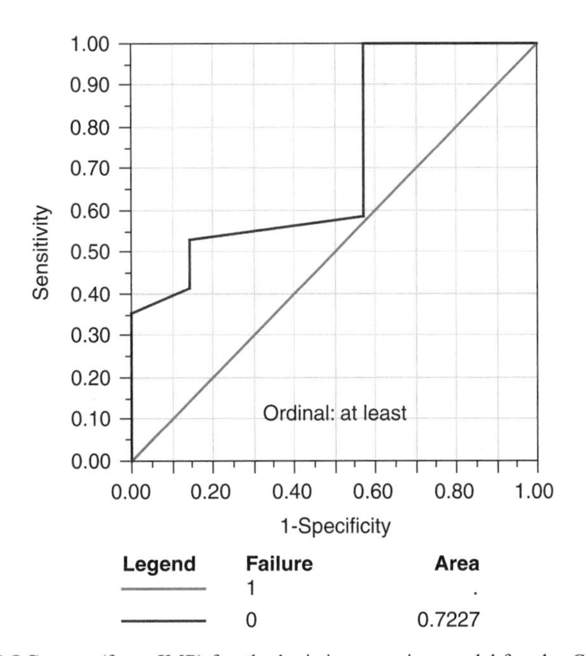

Figure 4.12 ROC curve (from JMP) for the logistic regression model for the *Challenger* data.

- $0.8 \leq \text{ROC} < 0.9$, excellent discrimination.
- $\text{ROC} \geq 0.9$, outstanding discrimination.

Therefore we conclude that the logistic regression model or the *Challenger* O-ring failure data exhibits acceptable discrimination ability.

4.2.8 A Biological Example of Logistic Regression

This example illustrates the use of logistic regression to analyze a single agent quantal bioassay of a toxicity experiment. The results show the effect of different doses of nicotine on the common fruit fly. Table 4.23 gives the data.

The purpose of the study was to use logistic regression to arrive at an appropriate model and to estimate two so-called **effective doses** (EDs), that is, values of the nicotine dose that result in a centain probability, π_i of fly death. These quantities are often used to characterize the results of an assay type experiment. For example, an ED_{50} might be used. In the experiment it was of interest to estimate both the ED_{50} and the ED_{90}, where ED_π is **the value of** x that produces a probability $\pi \times 100$ of an event, in this case the probability that an exposed fruit fly is killed.

Table 4.24 presents the MINITAB logistic regression model output. The fitted logistic regression model is

$$\hat{y} = \frac{e^{-1.73611+6.29539x}}{1 + e^{-1.73611+6.29539x}}$$

The test for significance of regression indicates that the concentration variable is significant, and the Homer-Lemeshow test does not indicate any problems with model adequacy. Table 4.25 contains the diagnostic statistics (Pearson residuals, standardized Pearson residuals, deviance residuals, $\Delta\chi^2$, ΔD, $\Delta\mathbf{b}$, leverage values) and the estimated probabilities from the fitted model. None of the values of $\Delta\chi^2$ or ΔD exceeds 4 and none of the values of $\Delta\mathbf{b}$ exceeds unity. A normal probability plot of the deviance residuals is shown in Figure 4.13. This plot is satisfactory. Figure 4.14 is a scatterplot of the observed proportion

Table 4.23 Toxicity Experiment Data

x Concentration (g/100 cc)	n Number of Insects	y Number Killed	Percentage Killed
0.10	47	8	17.0
0.15	53	14	26.4
0.20	55	24	43.6
0.30	52	32	61.5
0.50	46	38	82.6
0.70	54	50	92.6
0.95	52	50	96.2

Table 4.24 MINITAB Output for the Toxicity Data

```
Binary Logistic Regression: No. Killed, No. of Insects versus
Concentration
Link Function: Logit
Response Information
Variable          Value         Count
No. Killed        Event           216
                  Non-event       143
No. of Insects    Total           359
Logistic Regression Table
                                                Odds      95%     CI
Predictor          Coef    SE Coef   Z      P   ratio   Lower   Upper
Constant       -1.73611  0.242042 -7.17 0.000
Concentration 6.29539  0.742229  8.48 0.000 542.07 126.55 2321.94
Log-Likelihood  171.305
Test that all slopes are zero:  G 140.122,  DF 1,  P-Value  0.000
Goodness-of-Fit Tests
Method            Chi-Square    DF      P
Pearson           6.44510       5       0.265
Deviance          5.89952       5       0.316
Hosmer-lemeshow   6.44510       5       0.265
Table of Observed and Expected Frequencies:
  (See Hosmer-Lemeshow Test for the Pearson Chi-Square Statistic)
                          Group
Value          1     2     3     4     5     6     7     Total
Event
 Obs           8    14    24    32    38    50    50      216
 Exp        11.7  16.5  21.1  28.0  37.0  50.5  51.3
Non-event
 Obs          39    39    31    20     8     4     2      143
 Exp        35.3  36.5  33.9  24.0   9.0   3.5   0.7
Total         47    53    55    52    46    54    52      359
```

Table 4.25 Diagnostic Quantities and Estimated Probabilities for the Toxicity Data

Pearson Residuals	Standardized Pearson Residuals	Deviance Residuals	$\Delta\chi^2$	ΔD	$\Delta\mathbf{b}$	Leverage	Estimated Probability
−1.24213	−1.48284	−1.29453	2.19881	2.33172	0.655918	0.298305	0.248511
−0.74859	−0.89068	−0.76012	0.79332	0.81071	0.232927	0.293612	0.311783
0.81483	0.95057	0.80858	0.90357	0.89343	0.239619	0.265190	0.382954
1.11853	1.27518	1.12530	1.62607	1.64128	0.374973	0.230600	0.538054
0.37710	0.46130	0.38276	0.21280	0.21710	0.070600	0.331762	0.804016
−0.27918	−0.35599	−0.27328	0.12673	0.12347	0.048790	0.384995	0.935273
−1.48547	−1.65619	−1.22664	2.74296	2.04099	0.536345	0.195535	0.985860

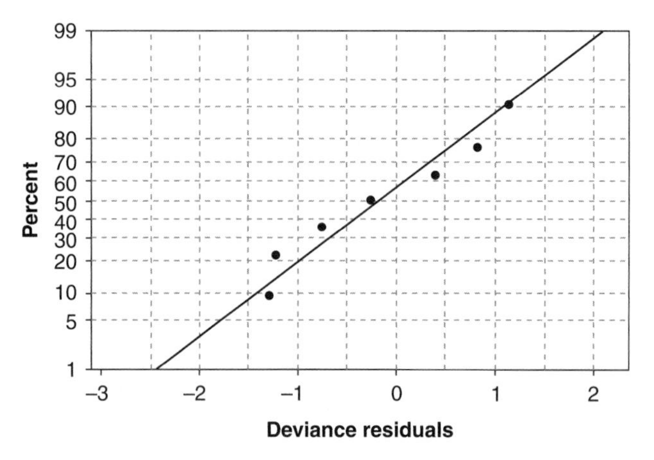

Figure 4.13 Normal probability plot of the deviance residuals for the toxicity data.

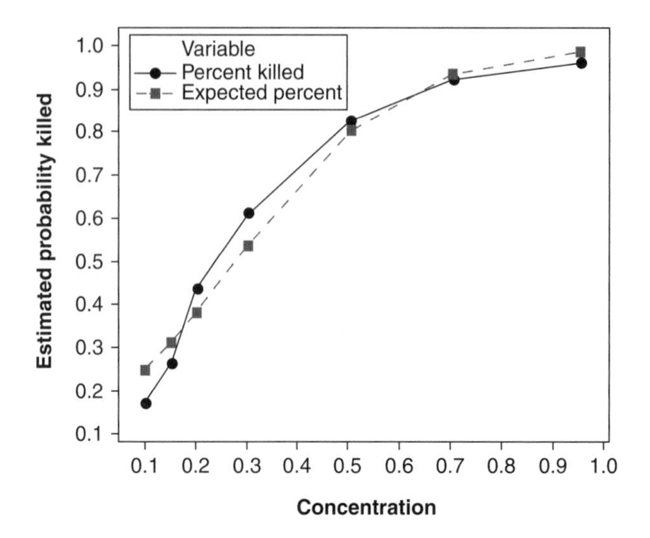

Figure 4.14 Scatterplot of the observed and estimated proportion of insects killed.

of insects that are killed and the estimated probability from the logistic regression models versus concentration. There is reasonably close agreement between the observed and estimated proportions of insects killed. We conclude that there are no obvious problems with the logistic regression model fit.

In quantal assay experiments it is fairly common practice to use the logarithm of concentration as the predictor variable (see Finney, 1950). This can be particularly effective in cases where the range of the predictor variable is large. Table 4.26 is the MINITAB output the for the log model. The summary statistics and goodness of fit tests do not indicate any problems with this model.

Table 4.26 MINITAB Output for the Logistic Regression Model for the Toxicity Data using Log Concentration

```
Binary Logistic Regression: Killed, Number versus Log Concentration
Link Function: Logit
Response Information
Variable      Value      Count
Killed        Success    216
              Failure    143
Number        Total      359
Logistic Regression Table
                                                          Odds     95%     CI
Predictor      Coef       SE Cpef       Z       P        Ratio   Lower   Upper
Constant      3.12361   0.334921   9.33   0.000
Log            2.12785   0.221408   9.61    0.000    8.40    5.44   12.96
Concentration
Log-Likelihood=- 168.722
Test that all slopes are zero: G = 145.288, DF = 1, P-Value = 0.000
```

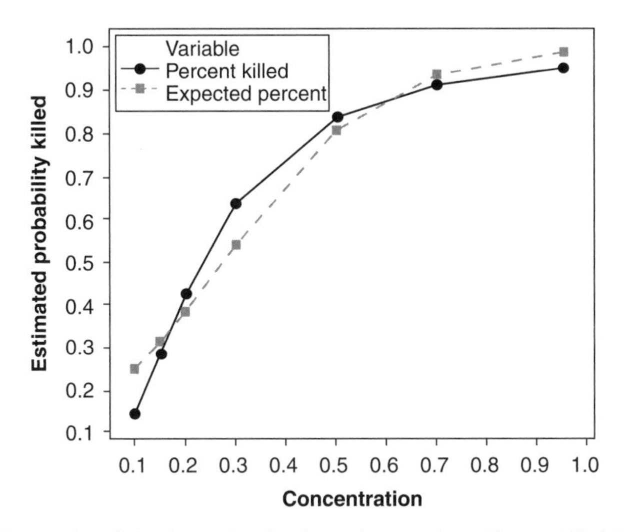

Figure 4.15 Scatterplot of the observed and estimated proportion of insects killed from the model using log concentration.

Figure 4.15 is a scatterplot of the observed proportions of insects killed and the estimated proportion of insects killed from the logistic regression model using log concentration as the predictor variable. There is little difference in the agreement between the observed and predicted responses between the two models.

Recall that one of the primary goals of this study is to determine the concentration in g/100cc that lead to 50% and 90% kill rates, i.e. the ED_{50} and ED_{90} levels. The ED_{50} is the value for concentration (x) such that $\pi(x) = 0.5$.

Similarly, the ED_{90} is the value for X such that $\pi(x) = 0.9$. From the simple logistic model, we have

$$\ln\left(\frac{\pi}{1-\pi}\right) = \beta_0 + \beta_1 x$$

and $\ln\left(\frac{0.5}{1-0.5}\right) = 0$. Thus, to estimate the ED_{50} we set $b_0 + b_1 x$ equal to 0 and solve for x as follows

$$\begin{aligned} -1.73611 + 6.29539x &= 0 \\ x &= 1.73611/6.29539 \\ &= 0.276 \end{aligned}$$

Thus the estimated Ed_{50} is a concentration of 0.2762g/100cc. Note from Figure 4.14 that indeed this concentration level corresponds to an estimated 50% kill rate.

4.2.9 Other Models for Binary Response Data

So far, our discussion of logistic regression uses the logit, defined as $\ln[\pi/(1-\pi)]$, to force the estimated probabilities to lie between zero and unity, which leads to the logistic regression model

$$\pi = \frac{\exp(\mathbf{x}'\boldsymbol{\beta})}{1 + \exp(\mathbf{x}'\boldsymbol{\beta})}$$

Another possibility for modeling a binary response uses the cumulative normal distribution, say, $\Phi^{-1}(\pi)$. The function $\Phi^{-1}(\pi)$ is called the **probit**. A linear predictor can be related to the probit, $\mathbf{x}'\boldsymbol{\beta} = \Phi^{-1}(\pi)$, resulting in a regression model

$$\pi = \Phi(\mathbf{x}'\boldsymbol{\beta})$$

Another possible model is provided by the **complementary log-log** relationship $\log[-\log(1-\pi)] = \mathbf{x}'\boldsymbol{\beta}$, which leads to the regression model

$$\pi = 1 - \exp[-\exp(\mathbf{x}'\boldsymbol{\beta})]$$

Figure 4.16 gives a comparison of all three possible models for the linear predictor $\mathbf{x}'\boldsymbol{\beta} = 1 + 5x$. The logit and probit functions are very similar, except when the estimated probabilities are very close to either 0 or 1. Both of these functions have estimated probability $\pi = \frac{1}{2}$ when $x = -\beta_0/\beta_1$ and exhibit symmetric behavior around this value. The complementary log-log function is not symmetric. In general, it is very difficult to see meaningful differences between these three models when sample sizes are small. A number of software packages including MINITAB will fit the logistic regression model using all three of these functions.

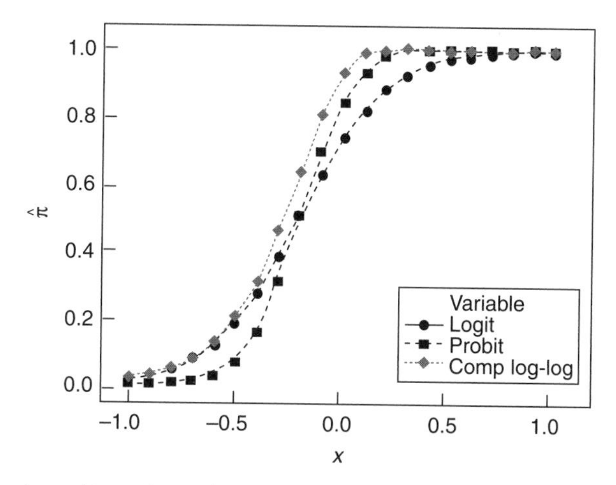

Figure 4.16 Logit, probit, and complementary log–log functions for the linear predictor $\mathbf{x}'\boldsymbol{\beta} = 1 + 5x$.

4.2.10 More than Two Categorical Outcomes

Logistic regression considers the situation where the response variable is categorical, with only two outcomes. We can extend the classical logistic regression model to cases involving more than two categorical outcomes. First consider a case where there are $m + 1$ possible categorical outcomes, but the outcomes are **nominal**. By this we mean that there is no natural ordering of the response categories. Let the outcomes be represented by 0,1,2,..., m. The probabilities that the responses on observation i take on one of the $m + 1$ possible outcomes can be modeled as

$$
\begin{aligned}
P(y_i = 0) &= \frac{1}{1 + \sum_{j=i}^{m} \exp\left[\mathbf{x}_i'\boldsymbol{\beta}^{(j)}\right]} \\
P(y_i = 1) &= \frac{\exp\left[\mathbf{x}'_i\boldsymbol{\beta}^{(1)}\right]}{1 + \sum_{j=i}^{m} \exp\left[\mathbf{x}_i'\boldsymbol{\beta}^{(j)}\right]} \\
&\vdots \qquad\qquad \vdots \\
P(y_i = m) &= \frac{\exp\left[\mathbf{x}'_i\boldsymbol{\beta}^{(m)}\right]}{1 + \sum_{j=i}^{m} \exp\left[\mathbf{x}_i'\boldsymbol{\beta}^{(j)}\right]}
\end{aligned}
\tag{4.43}
$$

Notice that there are m parameter vectors. Comparing each response category to a baseline categroy produces logits

$$\begin{aligned}
\ln\frac{p(y_i = 1)}{p(y_i = 0)} &= \mathbf{x}'_i\boldsymbol{\beta}^{(1)} \\
\ln\frac{p(y_i = 2)}{p(y_i = 0)} &= \mathbf{x}'_i\boldsymbol{\beta}^{(2)} \\
&\vdots \\
\ln\frac{p(y_i = m)}{p(y_i = 0)} &= \mathbf{x}'_i\boldsymbol{\beta}^{(m)}
\end{aligned} \tag{4.44}$$

where our choice of zero as the baseline category is arbitrary. Maximum likelihood estimation of the parameters in these models is fairly straightforward.

Example 4.10. Customer Invoicing Preference. Table 4.27 presents the results of a customer survey conducted by a company to solicit information about their preference for invoicing. There were 202 customers surveyed and each customer was asked to express his/her preference for invoicing and payment. The invoicing choices are traditional mail, electronic invoice, and through a web portal. Customers were classified as to whether they were repeat customers and the size of their accounts. Notice that the three response categories cannot be ordered, so this is a nominal response. Since the web portal is a relatively new procedure, it will be selected as the baseline category for modeling purposes.

Table 4.28 presents the MINITAB output for the nominal logistic regression model. Based on the p-values associated with *Repeat* (i.e., 0.362 for mail vs. web and 0.420 for electronic vs. web), we conclude that whether the customer is a repeat customer or not is not important in determining the invoicing preference. We refit the model with only account size as a predictor, and obtain the results in Table 4.29.

The first logistic regression model compares the preference of a mailed invoice to the use of a web portal, resulting in the following maximum likelihood prediction equation:

$$\ln\left(\frac{\hat{\pi}_{i,mail}}{\hat{\pi}_{i,web}}\right) = -1.48160 + 0.879429x$$

Table 4.27 Customer Inovice Preference Survey

Repeat Customer	Size of Account	Invoice Mail	Invoice Electronically	Invoice by Web Portal	Total
Yes	Small	12	17	26	55
Yes	Large	7	12	36	55
No	Small	11	15	16	42
No	Large	8	12	30	50
Totals		38	56	108	202

Table 4.28 Initial MINITAB Output for the Invoice Preference Data in Example 4.10

```
Nominal Logistic Regression: Preference versus Repeat, Size
Response Information
Variable        Value       Count
Preference      Web Portal   108    (Reference Event)
                Mail          38
                Electronic    56
                Total        202
Frequency: Frequency
Logistic Regression Table
                                                              odds
Predictor        Coef          SE Coef         Z         P    Ratio
Logit 1: (Mail/Web Portal)
Constant     -1.30344       0.341447        3.82      0.000
Repeat
  Yes         0.350678      0.384588        0.91      0.362    0.70
Size
  Small       0.899940      0.387862        2.32      0.020    2.46
Logit 2: (Electronic/Web Portal)
Constant     -0.871234      0.292536        2.98      0.003
Repeat
  Yes         0.270969      0.335846        0.81      0.420    0.76
Size
  Small       0.755478      0.335801        2.25      0.024    2.13
                                           95% CI
Predictor                           Lower         Upper
Logit 1: (Mail/Web Portal)
Constant
Repeat
  Yes                               0.33          1.50
Size
  Small                             1.15          5.26
Logit 2: (Electronic/Web Portal)
Constant
Repeat
  Yes                               0.39          1.47
Size
  Small                             1.10          4.11

Log-Likelihood                    198.435
Test that all slopes are zero: G  9.034, DF-4-P-Value -0.060

Goodness-of-Fit Tests

Method     Chi-Square    DF     P
Pearson    0.0704557      2   0.965
Deviance   0.0704649      2   0.965
```

Table 4.29 Final MINITAB Output for the Invoice Preference Data in Example 4.10

```
Nominal Logistic Regression: Preference versus Size
Response Information
Variable          Value          Count
Preference      Web Portal        108          (Reference Event)
                   Mail            38
                Electronic         56
                   Total          202
Frequency: Frequency
Logistic Regression Table
                                                                 Odds
Predictor          Coef          SE Coef          Z          P    Ratio
Logit I: (Mail/Web Portal)
Constant      -1.48160       0.286039        5.18      0.000
Size
  Small        0.879429      0.386142        2.28      0.023     2.41
Logit 2: (Electronic/Web Portal)
Constant      -1.01160       0.238366        4.24      0.000
Size
  Small        0.739667      0.334481        2.21      0.027     2.10
                                           95% CI
Predictor                            Lower          Upper
Logit 1: (Mail/Web Portal)
Constant
Size
  Small                              1.13            5.14
Logit 2: (Electronic/Web Portal)
Constant
Size
  Small                              1.09            4.04
Log-Likelihood                     199.000
Test that all slopes are zero: G 7.905 DE 2, P-Value 0.019
```

where x is the account size indicator variable, with $x = 1$ indicating a small account and $x = 0$ indicating a large account. Using the equation above, the estimated coefficients can be interpreted just like coefficients in a binary logit model. Specifically, exponentiating the coefficient for x (i.e. *Size*), we get $e^{0.879429} = 2.409$, implying that the odds of a small account customer preferring mailed invoices to invoices via the web are 2.4 times greater than the odds of large account customers preferring mailed invoices over web invoices.

The second logistic regression model compares the preference of electronic invoicing with web invoicing. The resulting prediction model is

$$\ln\left(\frac{\hat{\pi}_{i,electronic}}{\hat{\pi}_{i,web}}\right) = -1.01160 + 0.739667x$$

Using this equation and exponentiating the coefficient for x (i.e. *Size*), we get $e^{0.739667} = 2.095$, implying that the odds of small account customers preferring electronic invoicing to web invoicing are 2.1 times greater than the odds of large account customers preferring electronic invoicing to web invoicing. □

A second case involving a multilevel categorical response is an **ordinal** response. For example, customer satisfaction may be measured on a scale as not satisfied, indifferent, somewhat satisfied, and very satisfied. These outcomes would be coded as 1, 2, 3, and 4 respectively. The usual approach for modeling this type of response data is to use logits of cumulative probabilities:

$$\ln \frac{P(y_i \leq k)}{1 - P(y_i \leq k)} = \alpha_k + \mathbf{x}_i'\boldsymbol{\beta}, \ k = 1, \ldots, m-1 \tag{4.45}$$

The cumulative probabilities are then

$$P(y_i \leq k) = \frac{\exp(\alpha_k + \mathbf{x}_i'\boldsymbol{\beta})}{1 + \exp(\alpha_k + \mathbf{x}_i'\boldsymbol{\beta})}, \ k = 1, \ldots, m-1 \tag{4.46}$$

The cumulative logit models presented above are also known as *proportional odds* models since the only difference in the models from one response category to the next is the intercepts. The effects of the regresssors in $\mathbf{x}_i$ are assumed to be the same for all response categories.

Example 4.11. Modeling Customer Satisfaction. Table 4.30 presents the results of a customer satisfaction survey conducted by a automobile parts wholesaler. There were 210 customers surveyed, and each customer was asked to respond on a five-point scale as to whether he/she was very dissatisfied, dissatisfied, neutral, satisfied, or very satisfied with the company service. Customers were classified as to whether they were repeat customers and the size of their order. Since there was only one customer in the very dissatisfied or dissatisfied categories, it was decided to analyze only the responses in the top

Table 4.30 Ordinal Response Data from Customer Surveys

Repeat Customer	Size of Order	Very Dissatisfied (1)	Dissatisfied (2)	Neutral (3)	Satisfied (4)	Very Satisfied (5)	Total
Yes	Small	0	0	7	10	30	37
Yes	Large	0	0	5	13	25	43
No	Small	0	1	15	33	15	63
No	Large	0	0	16	28	12	56
Total		0	1	43	84	82	210

Table 4.31 Initial MINITAB Output for the Customer Satisfaction Data in Example 4.10

```
Ordinal Logistic Regression: Response versus Repeat, Order
Link Function: Logit
Response Information
Variable          Value          Count
Response            1              43
                    2              84
                    3              82
                  Total           209
Frequency: Satisfaction
Logistic Regression Table
                                                   Odds       95% CI
Predictor  Coef        SE Coef    Z        P      Ratio  Lower  Upper
Const (1) -0.818167  0.237466  3.45   0.001
Const (2)  1.16322   0.246020  4.73   0.000
Repeat
  Yes     -1.44354   0.281491  5.13   0.000   0.24    0.14   0.41
Order
  Small   -0.169757  0.265338  0.64   0.522   0.84    0.50   1.42
Log-Likelihood = -207.309
Test that all slopes are zero: G = 27.934, DF = 2, P-Value = 0.000
Goodness-of-Fit Tests
Method          Chi-Square      DF          P
Pearson           6.06526        4        0.194
Deviance          5.67720        4        0.225
```

three categories, which we renumber as Neutral = 1, Satisfied = 2, and Very Satisfied = 3.

Table 4.31 presents the MINITAB output for the main effects ordinal logistic regression model of the Customer Satisfaction example (Example 4.11). Based on the p-value for *Order*, we conclude that the size of the customer order is not important in determining customer satisfaction. As a result, we refit the model with only repeat order information (i.e. *Repeat*) as a predictor and obtain the results in Table 4.32. Note that with 3 response categories, the model has two intercepts (i.e. the α_k's). Generally the intercepts are only of interest when estimating response probabilities. The estimated effect of being a repeat customer is $b = -1.43736$. Interpreting this coefficient, the estimated odds of a repeat customer being in a less satisfied status direction as compared to a new customer is $e^{-1.43736} = 0.2376$ times the estimated odds for new customers. Stated differently, the odds of new customers being less satisfied is $1/0.2376 = 4.21$ times greater than repeat customers.

If one were interested in computing the estimated probabilities of repeat customers and new customers falling into each of the three response categories, we could do so by first computing the estimated cumulative probabilities

Table 4.32 Final MINITAB Output for the Customer Satisfaction Data in Example 4.10

```
Link Function: Logit
Response Information
Variable      Value   Count
Response        1       43
                2       84
                3       82
              Total    209
Frequency: Satisfaction
Logistic Regression Table
                                                  Odds      95% CI
Predictor  Coef        SE Coef    Z       P      Ratio  Lower  Upper
Const (1) -0.907220  0.193595  4.69  0.000
Const (2)  1.07114   0.199297  5.37  0.000
Repeat
  Yes     -1.43736   0.281192  5.11  0.000   0.24   0.14   0.41
Log-Likelihood = -207.515
Test that all slopes are zero: G = 27.524, DF = 1, P-Value = 0.000
Goodness-of-Fit Tests
Method        Chi-Square      DF         P
Pearson         4.93210        1       0.026
Deviance        4.71868        1       0.030
```

associated with each response category. For repeat customers, the estimated cumulative probabilites are

$$\hat{P}(y_i \leq 1|x = 1) = \frac{\exp[-0.90722 - 1.43736(1)]}{1 + \exp[-0.90722 - 1.43736(1)]} = 0.087$$

and

$$\hat{P}(y_i \leq 2|x = 1) = \frac{\exp[1.07114 - 1.43736(1)]}{1 + \exp[1.07114 - 1.43736(1)]} = 0.409$$

and, trivially, $\hat{P}(y_i \leq 3|x = 1) = 1.0$. The individual category probabilities are then just the differences of cumulative probabilities. So, $\hat{P}(y_i = 1|x = 1) = 0.087$, $\hat{P}(y_i = 2|x = 1) = \hat{P}(y_i \leq 2|x = 1) - \hat{P}(y_i = 1|x = 1) = 0.409 - 0.087 = 0.322$ and finally, $\hat{P}(y_i = 3|x = 1) = 1.0 - \hat{P}(y_i \leq 2) = 1.0 - 0.409 = 0.591$. Going through this same process for new customers, the estimated cumulative probabilities are

$$\hat{P}(y_i \leq 1|x = 0) = \frac{\exp[-0.90722]}{1 + \exp[-0.90722]} = 0.287$$

and

$$\hat{P}(y_i \leq 2|x = 0) = \frac{\exp[1.07114]}{1 + \exp[1.07114]} = 0.7448$$

and, $\hat{P}(y_i \leq 3|x = 0) = 1.0$. The individual category probabilities are then just the differences of cumulative probabilities. So, $\hat{P}(y_i = 1|x = 0) = 0.287$, $\hat{P}(y_i = 2|x = 0) = 0.7448 - 0.287 = 0.4578$ and finally, $\hat{P}(y_i = 3|x = 0) = 1 - 0.7448 = 0.2552$.

Note the p-values associated with test of H_0: adequate model fit given in Table 4.32 under the section *Goodness-of-Fit Tests*. The Pearson chi-square and Deviance chi-square goodness-of-fit tests are general tests of fit and both reject the notion of model fit in this example. While these tests reject the notion of adequate model fit here, the expected counts for repeat customers in the neutral category and the satisfied category are quite small. Given the fact that two of the six cells in the cross-classification of the response categories with repeat/new are small, we have reason to doubt the reliability of the results from these statistics. A general rule of thumb is that 80% of the cross-classified expected cell counts should exceed 5. As discussed for binary logistic regression, when continuous covariates are the model, the Pearson and Deviance chi-squares are inappropriate and an ordinal version of the Hosmer-Lemeshow test is best for models with only continuous covariates and the ordinal version of the Pulkstenis-Robinson test is best when both categorical and continuous covariates are present [see Pulkstenis and Robinson (2004)]. □

4.3 POISSON REGRESSION

We now consider another regression modeling scenario where the response variable of interest is not normally distributed. In this situation, the response variable represents a count of some relatively rare event, such as defects in a unit of manufactured product, errors or bugs in software, or a count of particulate matter or other pollutants in the environment. The analyst is interested, in modeling the relationship between the observed counts and potentially useful regressor or predictor variables. For example, an engineer could be interested in modeling the relationship between the observed number of defects in a unit of product and production conditions when the unit was actually manufactured.

We assume that the response variable y_i is a count, such that the observation $y_i = 0,1,2,\ldots$. A reasonable probability model for count data is often the Poisson distribution

$$f(y_i) = \frac{e^{-\mu_i}\mu_i^{y_i}}{y_i!}, \quad y_i = 0, 1, 2, \ldots \tag{4.47}$$

where the parameter $\mu_i > 0$. The Poisson is another example of a probability distribution where the mean and variance are related. In fact, its is easy to show that for a Poisson random variable,

$$E(y) = \mu \text{ and } Var(y) = \mu \tag{4.48}$$

Note that the mean **and** variance of the Poisson distribution are equal to the parameter μ.

As was the case in the logistic regression model, we do not write the model as

$$y_i = E(y_i) + \varepsilon_i$$

but rather, we write the model in terms of the mean of the response. In particular, we assume that there exists a function, g, that relates the mean of the response to a linear predictor, say

$$g(\mu_i) = \eta_i = \beta_0 + \beta_1 x_1 + \ldots + \beta_k x_k = \mathbf{x}_i'\boldsymbol{\beta} \tag{4.49}$$

The function g is usually called the **link function**. The relationship between the mean and the linear predictor is

$$\mu_i = g^{-1}(\eta_i) = g^{-1}(\mathbf{x}_i'\boldsymbol{\beta}) \tag{4.50}$$

There are several link functions that are commonly used with the Poisson distribution. One of these is the **identity link**

$$g(\mu_i) = \mu_i = \mathbf{x}_i'\boldsymbol{\beta} \tag{4.51}$$

When this link is used, $E(y_i) = \mu_i = \mathbf{x}_i'\boldsymbol{\beta}$ since $\mu_i = g^{-1}(\mathbf{x}_i'\boldsymbol{\beta}) = \mathbf{x}_i'\boldsymbol{\beta}$. Another popular link function for the Poisson distribution is the **log link**

$$g(\mu_i) = \ln(\mu_i) = \mathbf{x}_i'\boldsymbol{\beta} \tag{4.52}$$

For the log link in Equation (4.52), the relationship between the mean of the response variable and the linear predictor is

$$\mu_i = g^{-1}(\mathbf{x}_i'\boldsymbol{\beta}) = e^{\mathbf{x}_i'\boldsymbol{\beta}} \tag{4.53}$$

The log link is particularly attractive for Poisson regression because it ensures that all of the predicted values for the response variable will be nonnegative.

The method of maximum likelihood is used to estimate the parameters in Poisson regression. The development follows closely the approach used for logistic regression. If we have a random sample of n observations on the response y and the predictors x, then the likelihood function is

$$\mathscr{L}(\boldsymbol{\beta}; \mathbf{y}) = \prod_{i=1}^{n} f_i(y_i) = \prod_{i=1}^{n} \frac{e^{-\mu_i}\mu_i^{y_i}}{y_i!} = \frac{\prod_{i=1}^{n} \mu_i^{y_i} \exp\left(-\sum_{i=1}^{n} \mu_i\right)}{\prod_{i=1}^{n} y_i!} \tag{4.54}$$

where $\mu_i = g^{-1}(\mathbf{x}'_i \boldsymbol{\beta})$. Once the link function is selected, we maximize the log-likelihood

$$\ln \mathscr{L}(\boldsymbol{\beta}; \mathbf{y}) = \sum_{i=1}^{n} y_i \ln(\mu_i) - \sum_{i=1}^{n} \mu_i - \sum_{i=1}^{n} \ln(y_i!) \tag{4.55}$$

Iteratively reweighted least squares can be used to find the maximum likelihood estimates of the parameters in Poisson regression, following an approach similar to that used for logistic regression. Once the parameter estimates **b** are obtained, the fitted Poisson regression model is

$$\hat{y}_i = g^{-1}(\mathbf{x}_i'\mathbf{b}) \tag{4.56}$$

For example, if the identity link is used, the prediction equation becomes

$$\hat{y}_i = g^{-1}(\mathbf{x}_i'\mathbf{b}) = \mathbf{x}_i'\mathbf{b}$$

and if the log link is selected, then

$$\hat{y}_i = g^{-1}(\mathbf{x}_i'\mathbf{b}) = \exp(\mathbf{x}_i'\mathbf{b})$$

Inference on the model and its parameters follows exactly the same approach as used for logistic regression. That is, model deviance and the Pearson chi-square statistic are overall measures of goodness of fit, and tests on subsets of model parameters can be performed using the difference in deviance between the full and reduced models. These are likelihood ratio tests. Wald inference, based on large-sample properties of maximum likelihood estimators, can be used to test hypotheses and construct confidence intervals on individual model parameters.

Example 4.12. The Aircraft Damage Data. During the Vietnam War, the United States Navy operated several types of attack (a **bomber** in USN parlance) aircraft, often for low-altitude strike missions against bridges, roads, and other transportation facilities. Two of these included the McDonnell Douglas A-4 Skyhawk and the Grumman A-6 Intruder. The A-4 is a single-engine, single-place light-attack aircraft used mainly in daylight. It was also flown by the Blue Angels, the Navy's flight demonstration team, for many years. The A-6 is a twin-engine, dual-place, all-weather medium-attack aircraft with excellent day/night capabilities. However, the Intruder could not be operated from the smaller Essex-class aircraft carriers, many of which were still in service during the conflict.

Considerable resources were deployed against the A-4 and A-6, including small arms, AAA or antiaircraft artillery, and surface-to-air missiles. Table 4.33 contains data from 30 strike missions involving these two types of aircraft. The regressor x_1 is an indicator variable (A-4 = 0 and A-6 = 1), and the other regressors x_2 and x_3 are bomb load (in tons) and total months of aircrew experience. The response variable is the number of locations where damage was inflicted on the aircraft.

Table 4.33 Aircraft Damage Data

Observation	y	x_1	x_2	x_3
1	0	0	4	91.5
2	1	0	4	84.0
3	0	0	4	76.5
4	0	0	5	69.0
5	0	0	5	61.5
6	0	0	5	80.0
7	1	0	6	72.5
8	0	0	6	65.0
9	0	0	6	57.5
10	2	0	7	50.0
11	1	0	7	103.0
12	1	0	7	95.5
13	1	0	8	88.0
14	1	0	8	80.5
15	2	0	8	73.0
16	3	1	7	116.1
17	1	1	7	100.6
18	1	1	7	85.0
19	1	1	10	69.4
20	2	1	10	53.9
21	0	1	10	112.3
22	1	1	12	96.7
23	1	1	12	81.1
24	2	1	12	65.6
25	5	1	8	50.0
26	1	1	8	120.0
27	1	1	8	104.4
28	5	1	14	88.9
29	5	1	14	73.7
30	7	1	14	57.8

We model the damage response as a function of the three regressors. Since the response is a count, we use a Poisson regression model with the log link. Table 4.34 presents some of the output from JMP for this model.

The model adequacy checks based on deviance and the Pearson chi-square statistics are satisfactory, but we notice that x_3 = crew experience is not significant, using the Wald test (notice that the Wald statistic reported is $[b/se(b)]^2$, which is referred to a chi-square distribution with a single degree of freedom). This is a reasonable indication that x_3 can be removed from the model. The aircraft type also has a relatively large P-value indicating the this factor may have little predictive power. When x_3 is removed, it turns out that now x_1 = type of aircraft is no longer significant (you can easily verify that wald statistic for x_1 in this model has a P-value of 0.2598). A moment of reflection on the data in Table 4.33 reveals that there is a lot of multicollinearity

Table 4.34 JMP Poisson Regression Output for the Aircraft Damage Data

```
Generalized Linear Model Fit
Response: y
Distribution: Poisson
Link: Log
Estimation Method: Maximum Likelihood
Observations (or Sum Wgts) = 28
Whole Model Test
Model          -LogLikelihood        L-R ChiSquare   DF    Prob>ChiSq
Difference       8.36410065             16.7282       3      0.0008*
Full            36.6397945
Reduced         45.0038951
Goodness Of Fit      ChiSquare     DF        Prob>ChiSq
Statistic
Pearson               21.0158      24          0.6378
Deviance              23.3911      24          0.4968
Effect Tests
Source         DF   L-R ChiSquare        Prob>ChiSq
x1              1     2.4058919           0.1209
x2              1     2.8959626           0.0888
x3              1     1.6075308           0.2048
Parameter Estimates
Term          Estimates     Std Error    L-R ChiSquare    Prob>ChiSq
Intercept    -0.414083     0.8827815      0.2238464          0.6361
x1            0.7934384    0.5167237      2.4058919          0.1209
x2            0.1228425    0.0717472      2.8959626          0.0888
x3           -0.010958     0.0086652      1.6075308          0.2048
```

in the data. Essentially, the A-6 is a larger aircraft so it can carry a heavier bomb load, and because it has a two-person crew, it may tend to have more total months of crew experience. Therefore, as x_1 increases, there is a tendency for both of the other regressors to also increase.

To investigate the potential usefulness of various subset models, we fit all three two-variable models and all three one-variable models to the data in Table 4.33. A brief summary of the results obtained is as follows:

Model	Deviance	Difference in Deviance Compared to Full Model	P-Value
$x_1x_2x_3$	23.3911		
x_1x_2	26.9914	3.6003	0.0578
x_1x_3	26.2871	2.8960	0.0888
x_2x_3	26.7280	3.3369	0.0677
x_1	35.2362	11.8451	0.0006
x_2	29.2059	5.8679	0.0154
x_3	42.0783	18.6872	< 0.0001

Table 4.35 Mine Fracture Data

y	x_1	x_2	x_3	x_4	y	x_1	x_2	x_3	x_4
2	50	70	52	1.00	3	65	75	68	5.0
1	230	65	42	6.0	3	470	90	90	9.0
0	125	70	45	1.0	2	300	80	165	9.0
4	75	65	68	0.5	2	275	90	40	4.0
1	70	65	53	0.5	0	420	50	44	17.0
2	65	70	46	3.0	1	65	80	48	15.0
0	65	60	62	1.0	5	40	75	51	15.0
0	350	60	54	0.5	2	900	90	48	35.0
4	350	90	54	0.5	3	95	88	36	20.0
4	160	80	38	0.0	3	40	85	57	10.0
1	145	65	38	10.0	3	140	90	38	7.0
4	145	85	38	0.0	0	150	50	44	5.0
1	180	70	42	2.0	0	80	60	96	5.0
5	43	80	40	0.0	2	80	85	96	5.0
2	42	85	51	12.0	0	145	65	72	9.0
5	42	85	51	0.0	0	100	65	72	9.0
5	45	85	42	0.0	3	150	80	48	3.0
5	83	85	48	10.0	2	150	80	48	0.0
0	300	65	68	10.0	3	210	75	42	2.0
5	190	90	84	6.0	5	11	75	42	0.0
1	145	90	54	12.0	0	100	65	60	25.0
1	510	80	57	10.0	3	50	88	60	20.0

From examining the difference in deviances between each of the subset models and the full model, we notice that all of the subset models are significantly worse than the full model. This leaves us with the full model as the final choice, even though it has one nonsignificant variable and one borderline significant factor. When there is multicollinearity in the regressors this is not an unusual result.

Example 4.13. Mine Fractures. Counts were observed on the number of injuries or fractures that occur in the upper seam of mines in the coal fields of the Appalachian region in western Virginia. A total of 44 observations were collected on mines in this area. Four variables were measured. They all were functions of the material in the land and the mining area.

The data are shown in Table 4.35. The variables are as follows:

x_1: Inner burden thickness in feet (INB)

x_2: Percent extraction of the lower previously mined seam (EXTRP)

x_3: Lower seam height (feet)

x_4: Time that the mine has been opened (years)

Consider a Poisson regression model with log link fit to the data using all of the variables. The *full model* is

$$E(y) = \mu = \exp\left[\beta_0 + \beta_1 x_1 + \beta_2 x_2 + \beta_3 x_3 + \beta_4 x_4\right]$$

where μ is the mean number of fractures. In order to illustrate a model selection process, the deviance was found for all possible subset models, and the results are as follows (the * indicates the minimum deviance model of each subset size):

Model	Deviance
x_1	71.84
x_2	48.62*
x_3	73.84
x_4	72.12
x_1, x_2	42.09
x_1, x_3	71.07
x_1, x_4	70.43
x_2, x_3	47.59
x_2, x_4	41.63*
x_3, x_4	71.28
x_1, x_2, x_3	41.75
x_1, x_2, x_4	38.03*
x_1, x_3, x_4	69.81
x_2, x_3, x_4	41.02
x_1, x_2, x_3, x_4	37.86

One should keep in mind that models with small deviance are models with large log-likelihood. Also, as in the case of the error sum of squares in a linear least squares model, the addition of a new term to the model must lower (at least cannot increase) the deviance. For example, consider the model with x_1 and x_4 with a deviance of 70.43. Relatively speaking, this is not an attractive model, but it gives a smaller deviance than either x_1 or x_4, which individually are not attractive models.

The analyst can arrive at reasonable models through the use of deviance tests involving hierarchical subsets. For instance, if we compare a model with only x_2 to a model with x_2 and x_4, we might question whether x_4 is needed in the presence of x_2. So, to test the significance of x_4, we have the *reduction in deviance* by including x_4 as

$$\begin{aligned} D(x_4|x_2) &= D(x_2) - D(x_2, x_4) \\ &= 48.62 - 41.63 \\ &= 6.99 \end{aligned}$$

with 1 df, which is a significant χ^2 value at less than the 0.01 level. Thus we need x_4 in the presence of x_2. Now focus on the model (x_1, x_2, x_4). Is there a sufficient reduction in deviance by including x_1 in the presence of (x_2, x_4)? We have

$$\begin{aligned} D(x_1|x_2x_4) &= D(x_2, x_4) - D(x_1, x_2, x_4) \\ &= 41.63 - 38.03 \\ &= 3.60 \end{aligned}$$

This χ^2 statistic is significant at the 0.057 level. Thus x_1 is certainly worthy of consideration. What about the full model?

$$D(x_3|x_1x_2x_4) = 38.03 - 37.86 = 0.17$$

which, of course, is not significant, indicating that the (x_1, x_2, x_4) subset model is equivalent to the full model. Table 4.36 gives some summary information about the model (x_1, x_2, x_4). Note that the lack-of-fit information gives quite pleasing results. The deviance divided by degrees of freedom gives a value close to 1.0. Consider the Wald inference labeled *Analysis of Parameter Estimates*. Note that *P*-values are not identical to *P*-values computed earlier in the likelihood inference accomplished through **difference in deviance**. The final fitted model is given by

$$\hat{\mu} = \exp[-3.7207 - 0.0015x_1 + 0.0627x_2 - 0.0317x_4]$$

As in the case of binary regression, an important aspect of the analysis of Poisson regression is the interpretation of coefficients. The nature of the interpretation depends a great deal on the model structure. *Effects* of individual factors can be computed. For example, consider the role of variable x_4, the time that the mine has been opened. Since the coefficient of x_4 in the linear predictor is negative, the aging of the mine reduces the mean number of fractures. For every 10 years of age, the mean number of fractures is reduced by $100(1 - e^{-0.317}) = 27.2\%$, assuming fixed settings for x_1 and x_2. Similar interpretations can be offered for each variable. These effects play the same role as odds ratios play in the logistic regression case. □

Table 4.36 Model Summary Information for Mining Data with Variables x_1, x_2, and x_4

Analysis of Parameter Estimates

Parameter	df	Estimate	Standard Error	Chi Square	Pr > χ
Intercept	1	−3.721	0.9786	14.4471	0.0014
x_1	1	−0.001479	0.0008242	3.2197	0.072
x_2	1	0.06270	0.01227	26.1082	0.0001
x_4	1	−0.03165	0.01631	3.7662	0.0523

$$\text{deviance/df} = \frac{38.03}{40} = 0.9508$$

4.4 OVERDISPERSION IN LOGISTIC AND POISSON REGRESSION

Overdispersion is an important concept that can impact both logistic and Poisson regression models. We initially focus on the logistic regression case. Up to this point a poor fit of the logistic regression model would seem to arise from one of the following sources:

1. The binomial assumption is incorrect.
2. The choice of the logit model is improper (perhaps probit or complementary log–log is more appropriate).
3. The structure used in the linear predictor is incorrect. Perhaps there are additional terms like interactions or other higher-order terms that were ignored, or perhaps the log of a regressor variable should be used rather than a linear term.
4. There are outliers in the data.

The practitioner has at his/her disposal the use of lack-of-fit information. The Hosmer–Lemeshow, Pearson chi-square, or Pulkstenis–Robinson tests may be used, or one may simply look at the rule of thumb that the *mean deviance*, that is, deviance/df should be close to unity. The latter is particularly effective when the data are grouped and there is a reasonable sample size in each group.

The choice of distribution and model may indeed be appropriate, and the data set may be void of outliers, yet the mean deviance may signal a problem. The problem often encountered is called **extra binomial variation** or **overdispersion**. Overdispersion results when the variability accounted for by the binomial assumption, $n\pi(1 - \pi)$, is not sufficient. In other words, we say that the model is overdispersed. Consequently, there is an extra parameter involved, a scale parameter, say, $\sigma^2 > 1$, so the variance of an individual observation becomes $n\pi(1 - \pi)\sigma^2$ rather than $n\pi(1 - \pi)$. If $\sigma^2 < 1$, we call the phenomenon underdispersion. However, this problem does not occur in practice as often as overdispersion. There are reasonable explanations of overdispersion, and we discuss these subsequently. Analysts should not leap to the conclusion of overdispersion until all effort has been put forth to find the correct model. The symptoms of a poorly specified model is the same as the symptom of overdispersion. Thus we have a situation very similar to what is encountered in standard linear regression analysis. A large mean squared error, s^2, in linear regression can come from one of three sources. The experimental error variance may be large, there may be outliers in the data, or the model may be grossly underspecified. The symptom is the same, a large error mean square. In logistic regression the mean deviance replaces the error mean square. We now turn our attention to what causes overdispersion and what is its impact on the logistic regression results. This is important because researchers can often anticipate the existence of overdispersion based on the application.

Almost any cause of overdispersion results when experimental units are not homogeneous. The reader should certainly recognize the analogy with linear regression or simple analysis of variance where nonhomogeneity in experimental units may lead to incorrect F-tests and an improper estimate of residual variance. For the case of logistic regression suppose that the study involves experimental units that are animals, say, rabbits, and a study may involve litters of different rabbits. This type of nonhomogeneity can easily produce binomially distributed responses that result in two different animals in the same group, that is, exposed to the same experimental condition but having different binomial probabilities. One might also describe this condition as one in which two rabbits in the same litter have responses that are positively correlated. The explanation produces the same result; namely, the variance of a response exceeds that which is accounted for by binomial variability.

One fairly simple way of analytically illustrating overdispersion is to assume some type of variability in the binomial parameter π. Let p be the random variable replacing the known binomial parameter, where p has a distribution with mean μ and variance $\phi > 0$. Then, if Y is the binomial random variable,

$$E(Y) = E[E(Y \mid p)] = nE(p) = n\mu$$

but

$$\text{Var}(Y) = \text{Var}[E(Y \mid p)] + E[\text{Var}(Y \mid p)]$$

Now $\text{Var}[E(Y \mid p)] = \text{Var}[np] = n^2\phi$ and $E[\text{Var}(Y \mid p)] = nE[p(1 - p)]$. The term $nE[p(1 - p)] = n[E(p) - E(p^2)] = n[\mu - (\phi + \mu^2)]$. As a result

$$\begin{aligned}\text{Var}(Y) &= n^2\phi + n\mu - \phi n\mu^2 \\ &= n\mu(1 - \mu) + n\phi\,(n - 1) \\ &> n\mu(1 - \mu)\end{aligned}$$

Thus if we visualize that the non homogeneiry of experimental units produces an effect equivalent to randomly varying p, it is clear that the variance of the binomial random variable increases beyond that explained by binomial variance. On the other hand, if $\phi = 0$, this $\text{Var}(Y)$ reduces to the variance of an ordinary binomial random variable.

The effect that overdispersion has on the results of a fitted logistic regression is similar to what one would suspect from what we know about standard linear regression. In standard linear regression if the model residual variability is inflated through faulty modeling that leaves the error mean square large, then standard errors of regression coefficients are underestimated. Using linear regression notation, this is because the variance covariance matrix of estimators is estimated by $(\mathbf{X}'\mathbf{X})^{-1}s^2$, where s^2 is the error mean square and s^2 is inflated. In the case of overdispersion in logistic regression, the scale parameter $\sigma^2 > 1$ enters the variance–covariance matrix in the same fashion:

$$\text{Var}(\mathbf{b}) = (\mathbf{X}'\mathbf{V}\mathbf{X})^{-1}\sigma^2$$

and thus standard errors are underestimated since $\sigma^2 > 1.0$ are ignored. Now, in the case of overdispersion with the proper model, the maximum likelihood estimators of the β's remain asymptotically unbiased.

It is important to note what areas of application are prone to lead to overdispersion. Overdispersion is obviously prominent in biological and biomedical applications where animals are experimental units. In other biological or environmental applications, where laboratory procedures and conditions give rise to clearly independent experimental units, overdispersion is not expected. In *industrial applications* experimental units are not independent at times **by design**, which leads to correlation among observations via a **repeated measures** scenario as in, for example **split plot** designs. We discuss this in more detail in Chapters 6 and 7, which covers **generalized estimating equations and generalized linear mixed models.**

Based on the above discussion it appears as if reasonable adjustments can be made to *correct* standard errors of coefficients. In grouped logistic regression where the sample size in each group is reasonably large, it is reasonable to estimate the scale parameter by the mean deviance, that is, deviance/df which is analogous to the error mean square in standard linear regression. Another estimate that is just as intuitive is the Pearson χ^2 statistic divided *by* $n - p$ degrees of freedom, that is,

$$\frac{1}{n-p}\sum_{i=1}^{m}\left[\frac{(y_i-\hat{y}_i)^2}{n_i\hat{\pi}_i(n_i\hat{\pi}_i(1-\hat{\pi}_i))}\right]=\frac{\chi^2}{n-p}$$

The intuition here should be clear. The division of the squared residual by the binomial variance standardizes for binomial variance. However, an overdispersed situation has the factor σ^2 also imbedded in $\mathrm{Var}(y_i)$. Thus the quantity above estimates 1.0 in a pure binomial situation but estimates $\sigma^2 > 1.0$ in an overdispersed situation when the overdispersion model involves the single scale parameter σ^2 in $\mathrm{Var}(y_i)$. An appropriate correction for the standard error is to multiply the ordinary standard errors by the factor $\sqrt{deviance/(n-p)}$ or $\sqrt{\chi^2/(n-p)}$. Obviously, if these two factors give similar results, there is a level of comfort for the analyst. It is also clear that overdispersion impacts the Wald χ^2 statistics.

Example 4.14. Grain Beetles. The data in Table 4.37 are the results of an experiment designed to investigate the effectiveness of ethylene oxide as a fumigant against the grain beetle. The data are included in Bliss (1940) and further illustrated in Collett (1991). Various concentrations of the fumigant are included with sample sizes and the number of insects affected. Due to the nature of the experimental unit, one may well suspect that overdispersion is present here. Logistic regression is used with the model

$$\pi = \frac{1}{1+\exp[-(\beta_0+\beta_1 \ln x)]}$$

Table 4.37 Number of Grain Beetles Affected by Exposure to Concentrations of Ethylene Oxide

Concentration (mg/L)	Number Exposed	Number of Insects Affected
24.8	30	23
24.6	30	30
23.0	31	29
21.0	30	22
20.6	26	23
18.2	27	7
16.8	31	12
15.8	30	17
14.7	31	10

The lack-of-fit information is

Model	Deviance	df
Intercept	138.00	$10 - 1 = 9$
Intercept + $\ln x$	36.44	$10 - 2 = 8$

One must bear in mind that the mean deviance $36.44/8 = 4.56$ does suggest a problem with the model. A plot of the logit of $\hat{\pi}$ versus $\ln x$ taken from Collett is shown in Figure 4.17. If the model is correct, one would expect to see no systematic variation around a straight line. This plot shows no such variation. As a result the rather large value for the mean deviance does suggest overdispersion. The estimated parameters are $b_0 = -17.87$ and $b_1 = 6.27$ with standard errors 2.272 and 0.779, respectively. These maximum likelihood estimators are appropriate, but the standard errors are subject to doubt. Before we adjust for overdispersion, consider a likelihood inference approach to testing. $H_0 : \beta_1 = 0$. This approach should give the reader a clear illustration of the analog between deviance and error sum of squares. It also introduces what becomes a generalization of analysis of variance—namely, **analysis of deviance**, given in Table 4.38. Just as ratios of mean squares are F–ratios in ordinary normal error linear regression, ratios of mean deviance are approximately F–ratios as ratios of χ^2/df variates. The P-value provides strong evidence of a significant log concentration term in the logistic regression model.

We now adjust the standard error for overdispersion. Recall that the estimated standard errors are taken from the square root of the diagonal elements of $(\mathbf{X}'\mathbf{V}\mathbf{X})^{-1}$ defined earlier in this chapter. The adjustment involves multiplication by $\sqrt{36.44/8} = 2.13$. After the adjustment the standard errors are 4.84 for the intercept term and 1.66 for the log concentration term. Since the multiplication adjustment on $(\mathbf{X}'\hat{\mathbf{V}}\mathbf{X})^{-1}$ involves multiplying by a $\sqrt{\chi_8^2/8}$ variate then

$$\frac{\text{Coefficient}}{\text{Adjusted standard error}}$$

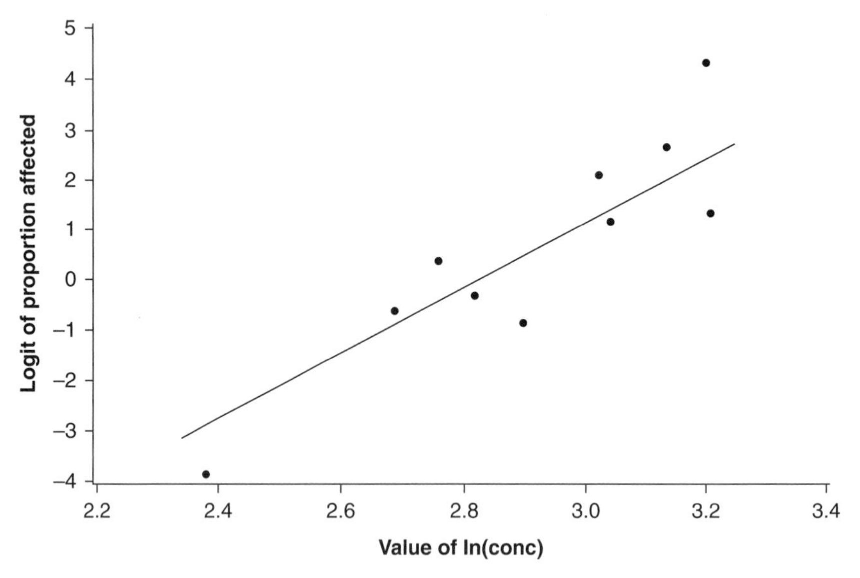

Figure 4.17 Plot of estimated logit against ln x.

Table 4.38 Analysis of Deviance for Data of Table 4.33

Source of Variation	d.f.	Deviance	Mean Deviance	F-Ratio
log (conc.) adjusted for intercept	1	138 −36.44 −101.56	101.56	22.3 ($P \cong 0.002$)
Residual deviance	8	36.44	4.56	
Total (intercept only)	9	138.00		

is approximately a normal random variable divided by $\sqrt{\chi_8^2/8}$ so a t-statistic is reasonable or one can use $t^2 = F_{1,8}$. Thus we have

$$F_{1,8} = \left(\frac{6.27}{1.66}\right)^2 \cong 14.4$$

which is significant at a level less than 0.01. □

Overdispersion also can occur in Poisson regression models. As in logistic regression, we model this with a multiplicative dispersion factor ϕ, so that

$$\text{Var}(y) = \phi\mu$$

The Poisson regression model is fit in the usual manner, and the parameter estimates are not affected by the value of ϕ. We can estimate ϕ by the deviance divided by degrees of freedom. The standard errors of the parameter estimates are multipled by the square root of the deviance divided by the degrees of freedom.

Another common approach to over dispersed data is the use of the negative binomial distribution rather than the Poisson. The negative binomial distribution inherently allows for greater dispersion than the Poisson. Several software packages including SAS PROC GENMOD and the *glm, nb* function in the MASS library of R, can be used for negative binomial regression.

EXERCISES

4.1 Show that the binomial deviance function $D(\boldsymbol{\beta})$ is given by

$$2\sum_{i=1}^{n} y_i \ln\left(\frac{y_i}{\mu}\right) + 2\sum_{i=1}^{n}(n-y_i)\ln\left\{\frac{n-y_i}{n-\mu}\right\}$$

4.2 Show for both Poisson regression and logistic regression that if an intercept is contained in the linear predictor then

$$\sum_{i=1}^{n}(y_i - \hat{\mu}_i) = 0$$

4.3 In Chapter 3 we learned that the Gauss–Newton procedure for computation of coefficients in a nonlinear model with nonhomogeneous variance involves computing

$$\mathbf{b} = \mathbf{b}_0 + (\mathbf{D}'\mathbf{V}^{-1}\mathbf{D})^{-1}\mathbf{D}'\mathbf{V}^{-1}(\mathbf{y} - \boldsymbol{\mu})$$

where

$$\mathbf{V} = \text{diag}\{\sigma_i^2\}$$

and $\mathbf{D}$ is the matrix of derivatives of the type $\partial\boldsymbol{\mu}/\partial\boldsymbol{\beta}$ and $\boldsymbol{\beta}_0$ is a starting value. Now this clearly is a candidate for iteratively reweighted least squares for both logistic and Poisson regression. Show that this reduces to

$$\mathbf{b} = \mathbf{b}_0 + (\mathbf{X}'\mathbf{V}\mathbf{X})^{-1}\mathbf{X}'(\mathbf{y} - \boldsymbol{\mu})$$

for Poisson and logistic regression.

4.4 Let us suppose that a 2×2 factorial with variable coded to $\pm$ 1 is used to fit a logistic regression model in a drug study in which 20 subjects were allocated to each of the 4 treatment combinations. If the response is successful, then $y = 1$, and if the response is unsuccessful, then $y = 0$. The coefficients in the model are $b_0 = 1.4$, $b_1 = 4.2$, and $b_2 = 7.1$. No interaction was found. Do the following:

(a) Compute the odds ratio for each coefficient and draw practical interpretations.

(b) Compute the asymptotic 95% confidence intervals for each odds ratio.

4.5 Anand (1997) discusses an experiment to improve the yield of silica gel in a chemical plant. The factors were (A) sodium silicate density, (B) pH, (C) setting time, and (D) drying temperature. The response was the percent of batches judged to be second grade material out of 1000 batches. The data follow.

(a) Use the logit link and logistic regression to analyze these data.

(b) Repeat this analysis using the probit link.

(c) Discuss the two analyses.

(d) The limiting distribution for the binomial is the Poisson. Repeat part (a) using Poisson regression. Discuss the differences.

A	B	C	D	P
1.125	3.5	40	150	14.0
1.125	3.5	50	120	13.5
1.125	3.5	30	135	18.3
1.125	3.0	40	120	17.4
1.125	3.0	50	135	16.3
1.125	3.0	30	150	13.9
1.125	4.0	40	135	14.1
1.125	4.0	50	150	12.0
1.125	4.0	30	120	12.0
1.121	3.5	40	120	17.0
1.121	3.5	50	135	10.5
1.121	3.5	30	150	15.3
1.121	3.0	40	135	21.0
1.121	3.0	50	150	20.5
1.121	3.0	30	120	19.8
1.121	4.0	40	150	12.3
1.121	4.0	50	120	11.3
1.121	4.0	30	135	10.5
1.123	3.5	40	135	15.2
1.123	3.5	50	150	12.3
1.123	3.5	30	120	12.0
1.123	3.0	40	150	11.1
1.123	3.0	50	120	12.8
1.123	3.0	30	135	13.4
1.123	4.0	40	120	10.3
1.123	4.0	50	135	10.4
1.123	4.0	30	150	11.7

4.6 Nelson (1982, pp. 407–409) discusses an experiment to determine the relationship of time in use to the number of fissures that develop in furbine wheels. The data follow.

(a) Use the logit link and logistic regression to analyze these data.

(b) Repeat this analysis using the probit link.

(c) Discuss the two analyses.

(d) The limiting distribution for the binomial is the Poisson. Repeat part (a) using Poisson regression. Discuss the differences.

Hours	Total Turbines Studied	Number with Fissures
400	39	0
1000	53	4
1400	33	2
1800	73	7
2200	30	5
2600	39	9
3000	42	9
3400	13	6
3800	34	22
4200	40	21
4600	36	21

4.7 A major aircraft manufacturer studied the number of failures of an alloy fastener after each fastener was subjected to a pressure load. The data follow.

(a) Use the logit link and logistic regression to analyze these data.

(b) Repeat this analysis using the probit link.

(c) Discuss the two analyses.

(d) The limiting distribution for the binomial is the Poisson. Repeat part (a) using Poisson regression. Discuss the differences.

Pressure	Total Fasteners Studied	Number Failing
2500	50	10
2700	70	17
2900	100	30
3100	60	21
3300	40	18
3500	85	43
3700	90	54
3900	50	33
4100	80	60
4300	65	51

4.8 Maruthi and Joseph (1999–2000) conducted an experiment to improve the yield of high, dense, inner-layer circuits in a printed circuit board operation. The factors were (A) surface preparation, (B) preheat, (C) lamination speed, (D) lamination pressure, (E) lamination temperature, (F) exposure step, (G) developer speed, and (H) oxidation–reduction potential. They measured two responses. The first response, y_1, was the percentage of shorts out of 400 opportunities. The second response, y_2, was the percent of opens age out of 800 opportunities. The data follow.

(a) For both responses, individually, use the logit link and logistic regression to analyze these data.

(b) 'Repeat this analysis using the probit link.

(c) Discuss the two analyses.

(d) The limiting distribution for the binomial is the Poisson. Repeat part (a) using the Poisson regression. Discuss the differences.

A	*B*	*C*	*D*	*E*	*F*	*G*	*H*	y_1	y_2
−1	−1	−1	−1	−1	−1	−1	−1	26.0	12.2
−1	−1	0	1	0	0	0	0	19.0	16.6
−1	−1	1	−1	1	1	1	1	12.6	16.4
0	−1	−1	1	0	0	1	1	16.4	23.0
0	−1	0	−1	1	1	−1	−1	11.8	18.6
7	−1	1	1	−1	−1	0	0	16.9	7.7
1	−1	−1	−1	−1	1	0	1	12.8	17.8
1	−1	0	1	0	−1	1	−1	19.0	19.4
1	−1	1	−1	1	0	−1	0	17.5	20.5
−1	1	−1	1	1	0	0	−1	11.9	21.2
−1	1	0	−1	−1	1	1	0	9.8	17.2
−1	1	1	1	0	−1	−1	1	13.3	12.5
0	1	−1	−1	1	−1	1	0	16.9	18.6
0	1	0	1	−1	0	−1	1	11.6	13.3
0	1	1	−1	0	1	0	−1	9.2	20.5
1	1	−1	1	0	1	−1	0	7.5	16.9
1	1	0	−1	1	−1	0	1	21.2	20.0
1	1	1	1	−1	0	1	−1	16.4	17.2

4.9 Chowdhury, Gijo, and Raghavan (2000) conducted an experiment to decrease the number of defects on a printed circuit assembly-encoder. The factors were (*A*) bath temperature, (*B*) wave height, (*C*) overhead preheater, (*D*) preheater 2, (*E*) preheater 1, (*F*) air knife, and (*G*) the vibration of the solder wave. The response was the total number of defects on a unit. The data follow. Analyze these data by Poisson regression.

Experiment	(*A*) Bath Tempe-rature (°C)	(*B*) Wave Height	(*C*) Overhead Preheater (°C)	(*D*) Pre-heater 2 (°C)	(*E*) Pre-heater 1(°C)	(*F*) Air Knife	(*G*) *Omega*	1
1	248	4.38	340	340	340	0	0	4
2	248	4.38	360	360	360	3	2	2
3	248	4.38	380	380	380	6	4	1

Experiment	(*A*) Bath Temperature (°C)	(*B*) Wave Height	(*C*) Overhead Preheater (°C)	(*D*) Preheater 2 (°C)	(*E*) Preheater 1(°C)	(*F*) Air Knife	(*G*) *Omega*	1
4	248	4.4	340	340	360	3	4	2
5	248	4.4	360	360	380	6	0	6
6	248	4.4	380	380	340	0	2	15
7	248	4.42	340	360	340	6	2	9
8	248	4.42	360	380	360	0	4	5
9	248	4.42	380	340	380	3	0	8
10	252	4.38	340	380	380	3	2	5
11	252	4.38	360	340	340	6	4	4
12	252	4.38	380	360	360	0	0	11
13	252	4.4	340	360	380	0	4	10
14	252	4.4	360	380	340	3	0	15
15	252	4.4	380	340	360	6	2	4
16	252	4.42	340	380	360	6	0	12
17	252	4.42	360	340	380	0	2	6
18	252	4.42	380	360	340	3	4	7

4.10 A student conducted a project looking at the impact of popping temperature, amount of oil, and the popping time on the number of inedible kernels of popcorn. The data follow. Analyze these data using Poisson regression.

Temperature	Oil	Time	*y*
7	4	90	24
5	3	105	28
7	3	105	40
7	2	90	42
6	4	105	11
6	3	90	16
5	3	75	126
6	2	105	34
5	4	90	32
6	2	75	32
5	2	90	34
7	3	75	17
6	3	90	30
6	3	90	17
6	4	75	50

4.11 Zhang and Zelterman (1999) discuss an experiment where female mice were fed extremely low doses of a known carcinogen, 2-acetylaminofluorene (2-AAF). The following table summarizes the results of the incidences of bladder and liver cancers. Perform an appropriate analysis for each cancer.

Dose	Bladder Cancer		Liver Cancer	
(parts per 10^4 2-AAF)	Mice Exposed	Incidence	Mice Exposed	Incidence
0	101	1	555	6
0.3	443	5	2014	34
0.35	200	0	1102	20
0.45	103	2	550	15
0.6	66	2	441	13
0.75	75	12	382	17
1.0	31	21	213	19
1.5	11	11	211	24

4.12 Slaton, Piegorsch, and Durham (2000) discuss an experiment that examined the in utero damage in laboratory rodents after exposure to boric acid. This particular experiment used four levels of boric acid. The experimenters recorded the number of rodents in the litter and the number of dead embryos. The data are in the table below. Perform an appropriate analysis of them data.

Dose = 0		Dose = 0.1		Dose = 0.2		Dose = 0.3	
Dead	Litter Size	Dead	Litter Size	Dead	Litter Size	Dead	Litter Size
0	15	0	6	1	12	12	12
0	3	1	14	0	12	1	12
1	9	1	12	0	11	0	13
1	12	0	10	0	13	2	8
1	13	2	14	0	12	2	12
2	13	0	12	0	14	4	13
0	16	0	14	4	15	0	13
0	11	3	14	0	14	1	13
1	11	0	10	0	12	0	12
2	8	2	12	1	6	1	9
0	14	3	13	2	13	3	9
0	13	1	11	0	10	0	11
3	14	1	11	1	14	1	14
1	13	0	11	1	12	0	10
0	8	0	13	0	10	3	12
0	13	0	10	0	9	2	21
2	14	1	12	1	12	3	10
3	14	0	11	0	13	3	11
0	11	2	10	1	14	1	11
2	12	2	12	0	13	1	11
0	15	2	15	0	14	8	14
0	15	3	12	1	13	0	15
2	14	1	12	2	12	2	13

Dose = 0		Dose = 0.1		Dose = 0.2		Dose = 0.3	
Dead	Litter Size	Dead	Litter Size	Dead	Litter Size	Dead	Litter Size
1	11	0	12	1	14	8	11
1	16	1	12	0	13	4	12
0	12	1	13	0	12	2	12
0	14	1	15	1	7		

4.13 Bailer and Piegorsch (2000) report on an experiment that examines the effect of a herbicide, nitrofen, on the umber of offspring produced by a particular freshwater invertebrate zooplankton. The data follow. Perform an appropriate analysis of these data.

Dose	Number of Offspring									
Control	27	32	34	33	36	34	33	30	24	31
80	33	33	35	33	36	26	27	31	32	29
160	29	29	23	27	30	31	30	26	29	29
235	23	21	7	12	27	16	13	15	21	17
310	6	6	7	0	15	5	6	4	6	5

4.14 The table below presents the test-firing results for 25 surface-to-air antiaircraft missiles at targets of varying speed. The result of each test is either a hit ($y = 1$) or a miss ($y = 0$).

Test	Target Speed x (knots)	y	Test	Target Speed x (knots)	y
1	400	0	14	330	1
2	220	1	15	280	1
3	490	0	16	210	1
4	210	1	17	300	1
5	500	0	18	470	1
6	270	0	19	230	0
7	200	1	20	430	0
8	470	0	21	460	0
9	480	0	22	220	1
10	310	1	23	250	1
11	240	1	24	200	1
12	490	0	25	390	0
13	420	0			

(a) Fit a logistic regression model to the response variable y. Use a simple linear regression model as the structure for the linear predictor.

(b) Does the model deviance indicate that the logistic regression model from part (a) is adequate?

(c) Provide an interpretation of the parameter β_1 in this model.

(d) Expand the linear predictor to include a quadratic term in target speed. Is there any evidence that this quadratic term is required in the model?

4.15 A study was conducted attempting to relate home ownership to family income. Twenty households were selected and family income was estimated, along with information concerning home ownership ($y = 1$ indicates yes and $y = 0$ indicates no). The data are shown below.

Household	Income	Home Ownership Status	Household	Income	Home Ownership Status
1	38,000	0	11	38,700	1
2	51,200	1	12	40,100	0
3	39,600	0	13	49,500	1
4	43,400	1	14	38,000	0
5	47,700	0	15	42,000	1
6	53,000	0	16	54,000	1
7	41,500	1	17	51,700	1
8	40,800	0	18	39,400	0
9	45,400	1	19	40,900	0
10	52,400	1	20	52,800	1

(a) Fit a logistic regression model to the response variable y. Use a simple linear regression model as the structure for the linear predictor.

(b) Does the model deviance indicate that the logistic regression model from part (a) is adequate?

(c) Provide an interpretation of the parameter β_1 in this model.

(d) Expand the linear predictor to include a quadratic term in income. Is there any evidence that this quadratic term is required in the model?

4.16 The compressive strength of an alloy fastener used in aircraft construction is being studied. Ten loads were selected over the range 2500–4300 psi and a number of fasteners were tested at those loads. The numbers of fasteners failing at each load were recorded. The complete test data are shown below.

Load, x (psi)	Sample Size, n	Number Failing, r
2500	50	10
2700	70	17
2900	100	30
3100	60	21
3300	40	18
3500	85	43
3700	90	54
3900	50	33
4100	80	60
4300	65	51

(a) Fit a logistic regression model to the data. Use a simple linear regression model as the structure for the linear predictor.

(b) Does the model deviance indicate that the logistic regression model from part (a) is adequate?

(c) Expand the linear predictor to include a quadratic term. Is there any evidence that this quadratic term is required in the model?

(d) For the quadratic model in part (c), find Wald statistics for each individual model parameter.

(e) Find approximate 95% confidence intervals on the model parameters for the quadratic model from part (c).

4.17 The market research department of a soft drink manufacturer is investigating the effectiveness of a price discount coupon on the purchase of a two-liter beverage product. A sample of 5500 customers was given coupons for varying price discounts between 5 and 25 cents. The response variable was the number of coupons in each price discount category redeemed after one month. The data are shown below.

Discount, x	Sample Size, n	Number Redeemed, r
5	500	100
7	500	122
9	500	147
11	500	176
13	500	211
15	500	244
17	500	277
19	500	310
21	500	343
23	500	372
25	500	391

(a) Fit a logistic regression model to the data. Use a simple linear regression model as the structure for the linear predictor.

(b) Does the model deviance indicate that the logistic regression model from part (a) is adequate?

(c) Draw a graph of the data and the fitted logistic regression model.

(d) Expand the linear predictor to include a quadratic term. Is there any evidence that this quadratic term is required in the model?

(e) Draw a graph of this new model on the same plot that you prepared in part (c). Does the expanded model visually provide a better fit to the data than the original model from part (a)?

(f) For the quadratic model in part (d), find Wald statistics for each individual model parameter.

(g) Find approximate 95% confidence intervals on the model parameters for the quadratic logistic regression model from part (d).

4.18 A study was performed to investigate new automobile purchases. A sample of 20 families was selected. Each family was surveyed to determine the age of their oldest vehicle and their total family income. A follow-up survey was conducted 6 months later to determine if they had actually purchased a new vehicle during that time period ($y = 1$ indicates yes and $y = 0$ indicates no). The data from this study are shown in the following table.

Income, x_1	Age, x_2	y	Income, x_1	Age, x_2	y
45,000	2	0	37,000	5	1
40,000	4	0	31,000	7	1
60,000	3	1	40,000	4	1
50,000	2	1	75,000	2	0
55,000	2	0	43,000	9	1
50,000	5	1	49,000	2	0
35,000	7	1	37,500	4	1
65,000	2	1	71,000	1	0
53,000	2	0	34,000	5	0
48,000	1	0	27,000	6	0

(a) Fit a logistic regression model to the data.

(b) Does the model deviance indicate that the logistic regression model from part (a) is adequate?

(c) Interpret the model coefficients β_1 and β_2.

(d) What is the estimated probability that a family with an income of $45,000 and a car that is 5 years old will purchase a new vehicle in the next 6 months?

(e) Expand the linear predictor to include an interaction term. Is there any evidence that this term is required in the model?

(f) For the model in part (a), find statistics for each individual model parameter.

(g) Find approximate 95% confidence intervals on the model parameters for the logistic regression model from part (a).

4.19 A chemical manufacturer has maintained records on the number of failures of a particular type of valve used in its processing unit and the length of time (months) since the valve was installed. The data are shown below.

Valve	Number of Failures	Months	Valve	Number of Failures	Months
1	5	18	9	0	7
2	3	15	10	0	12
3	0	11	11	0	3
4	1	14	12	1	7
5	4	23	13	0	2
6	0	10	14	7	30
7	0	5	15	0	9
8	1	8			

(a) Fit a Poisson regression model to the data.

(b) Does the model deviance indicate that the Poisson regression model from part (a) is adequate?

(c) Construct a graph of the fitted model versus months. Also plot the observed number of failures on this graph.

(d) Expand the linear predictor to include a quadratic term. Is there any evidence that this term is required in the model?

(e) For the model in part a, find Wald statistics for each individual model parameter.

(f) Find approximate 95% confidence intervals on the model parameters for the Poisson regression model from part (a).

4.20 The following data come from a dose–response study that investigated a new pharmaceutical product.

Observation	Dose (mg)	Patient Response	Observation	Dose (mg)	Patient Response
1	180	0	16	210	0
2	182	0	17	212	1
3	184	0	18	215	1
4	186	0	19	216	0
5	185	0	20	218	1
6	188	0	21	220	1
7	190	1	22	222	0
8	192	0	23	225	1
9	195	1	24	228	0
10	194	0	25	230	1
11	196	1	26	232	1
12	199	1	27	234	1
13	200	0	28	235	1
14	204	0	29	236	1
15	205	1	30	240	1

(a) Fit a logistic regression model to these data.

(b) Prepare a graph of the fitted function. Does it adequately describe the data?

(c) Investigate the lack of fit of the model. What are your conclusions?

(d) Find 95% confidence intervals on the model parameters.

(e) Estimate the probability of a successful response if the dose is 226 mg.

(f) Find a 96% confidence interval on the probability of success when the dose is 226 mg.

4.21 The following table presents data on the reproduction of *Ceriodaphnia* organisms in a controlled environment in which a varying concentration of a component of jet engine fuel is introduced. We expect that as the concentration increases the mean number of counts of the organisms should decrease. Information is also provided on two different strains of the organism.

Observation	y	x	Strain	Observation	y	x	Strain
1	82	0	1	2	58	0	2
3	106	0	1	4	58	0	2
5	63	0	1	6	62	0	2
7	99	0	1	8	58	0	2

Observation	y	x	Strain	Observation	y	x	Strain
9	101	0	1	10	73	0	2
11	45	0.5	1	12	27	0.5	2
13	34	0.5	1	14	28	0.5	2
15	26	0.5	1	16	31	0.5	2
17	44	0.5	1	18	28	0.5	2
19	42	0.5	1	20	38	0.5	2
21	31	0.75	1	22	19	0.75	2
23	22	0.75	1	24	20	0.75	2
25	16	0.75	1	26	22	0.75	2
27	30	0.75	1	28	20	0.75	2
29	29	0.75	1	30	28	0.75	2
31	22	1	1	32	14	1	2
33	14	1	1	34	14	1	2
35	10	1	1	36	15	1	2
37	21	1	1	38	14	1	2
39	20	1	1	40	21	1	2
41	15	1.25	1	42	9	1.25	2
43	8	1.25	1	44	10	1.25	2
45	6	1.25	1	46	12	1.25	2
47	14	1.25	1	48	10	1.25	2
49	13	1.25	1	50	16	1.25	2
51	10	1.5	1	52	7	1.5	2
53	8	1.5	1	54	3	1.5	2
55	11	1.5	1	56	1	1.5	2
57	10	1.5	1	58	8	1.5	2
59	10	1.5	1	60	7	1.5	2
61	8	1.75	1	62	4	1.75	2
63	8	1.75	1	64	3	1.75	2
65	3	1.75	1	66	2	1.75	2
67	8	1.75	1	68	8	1.75	2
69	1	1.75	1	70	4	1.75	2

(a) Fit a Poisson regression model to these data including both concentration and strain in the linear predictor.

(b) Is there any evidence of lack of fit in the model?

(c) Construct a normal probability plot of the deviance residuals. Is the plot satisfactory?

(d) Is the strain of the organism significant?

CHAPTER 5

The Generalized Linear Model

As we introduce the notion of generalized linear models, two important issues surface: the distribution of the response and the model that relates the mean response to the regression variables. These two issues are not independent of each other because certain types of models are more appropriate for some distributions than for others. For example, the logistic model forces the inequality

$$0 \leq \pi \leq 1$$

where π is the binomial probability. A model that allows π to be either negative or greater than 1.0 is undesirable. In addition the Poisson regression model as discussed in Chapter 4, forces the mean number of counts to be nonnegative. This result, of course, is quite pleasing. In this chapter we use GLM to denote the basic methodology associated with generalized linear models.

5.1 THE EXPONENTIAL FAMILY OF DISTRIBUTIONS

An important unifying concept underlying the GLM is the **exponential family of distributions**. Members of the exponential family of distributions all have probability density functions for an observed response y that can be expressed in the form

$$f(y;\theta,\phi) = \exp\left\{\frac{y\theta - b(\theta)}{a(\phi)} + c(y,\phi)\right\} \tag{5.1}$$

where $a(\cdot)$, $b(\cdot)$, and $c(\cdot)$ are specific functions. The parameter θ is a natural **location parameter**, and ϕ is often called a **dispersion parameter**. The function

Generalized Linear Models, Second Edition, by Myers, Montgomery, Vining, and Robinson

$a(\phi)$ is generally of the form $a(\phi) = \phi \cdot \omega$, where ω is a known constant. The binomial, Poisson, and normal distributions are members of this family. For some common members of the family, $\phi = 1.0$—like the binomial and Poisson—except in situations of overdispersion, in which case the analyst needs to deal with an additional parameter as discussed in the previous chapter.

The most prominent member of the exponential family is the familiar normal distribution. The probability density function for a normal random variable y with parameters μ and σ is given by

$$\begin{aligned} f(y;\, \mu,\sigma) &= \exp\left\{-[y-\mu]^2/2\sigma^2\right\} \cdot \frac{1}{\sqrt{2\pi\sigma}} \\ &= \exp\left\{(y\mu - \mu^2/2)/\sigma^2 - \frac{1}{2}[y^2/\sigma^2 + \ln(2\pi\sigma^2)]\right\} \end{aligned}$$

This density function is of the form given in Equation (5.1) with $\theta = \mu$, $b(\theta) = \mu^2/2$, $a(\phi) = \phi$, $\phi = \sigma^2$, and

$$c(y,\phi) = -\frac{1}{2}\left[\frac{y^2}{\sigma^2} + \ln(2\pi\sigma^2)\right]$$

The location parameter is μ and the natural scale parameter is σ^2, as expected.

Consider now the Poisson distribution. We know that the probability function is given by

$$\begin{aligned} f(y;\mu) &= \frac{e^{-\mu}\mu^y}{y!} \\ &= \exp[y \ln \mu - \mu - \ln(y!)] \end{aligned}$$

As a result, $\theta = \ln \mu$, $b(\theta) = e^\theta$, and $c(y,\, \phi) = -\ln(y!)$. Thus the location parameter is μ, and the scale parameter is $\phi = 1.0$. We do not present all of the details in the case of the binomial distribution. If we assume y is binomial with parameters n and π, we have

$$\begin{aligned} \theta &= \ln\left(\frac{\pi}{1-\pi}\right) \\ b(\theta) &= n \ln(1+e^\theta) \\ \phi &= 1.0 \\ a(\phi) &= 1 \\ c(\cdot) &= \ln\binom{n}{y} \end{aligned}$$

From mathematical statistics (e.g., Bickel and Doksom, 2001), we can show that for members of the exponential family

$$E\left(\frac{\partial \ln \mathscr{L}(\theta)}{\partial \theta}\right) = 0$$

and

$$E\left(\frac{\partial^2 \ln \mathscr{L}}{\partial \theta^2}\right) + E\left(\frac{\partial \ln \mathscr{L}}{\partial \theta}\right)^2 = 0$$

We then apply the results (see Exercise 5.8) to obtain

$$\mu = E(y) = \frac{db(\theta)}{d\theta} = b'(\theta)$$

$$\text{Var}(y) = \frac{d^2b(\theta)}{d\theta^2}a(\phi) = b''(\theta)a(\phi)$$

$$= \frac{d\mu}{d\theta}a(\phi)$$

Let Var_μ be the variance of the response, y, apart from $a(\phi)$; Var_μ denotes the dependence of the variance of the response on its mean. Thus

$$\text{Var}_\mu = \frac{\text{Var}(y)}{a(\phi)} = \frac{d\mu}{d\theta}$$

As a result we have

$$\frac{d\theta}{d\mu} = \frac{1}{\text{Var}_\mu}$$

One can easily check these results for the normal, binomial, or the Poisson case. For example, in the normal case,

$$\theta = \mu$$

$$b(\theta) = \frac{\mu^2}{2}$$

$$a(\phi) = \sigma^2$$

$$E(y) = \frac{db(\theta)}{d\theta} = \mu$$

$$\text{Var}(y) = \frac{d^2b(\theta)}{d\theta^2}a(\phi) = \sigma^2$$

For the Poisson case, we have

$$\theta = \ln(\mu) \text{ and } \mu = \exp(\theta)$$
$$b(\theta) = \mu$$
$$a(\phi) = 1$$
$$c(y, \phi) = -\ln(y!)$$

Thus

$$\mathrm{E}(y) = \frac{db(\theta)}{d\theta} = \frac{db(\theta)}{d\mu}\frac{d\mu}{d\theta}$$

However, since

$$\frac{d\mu}{d\theta} = \exp(\theta) = \mu$$

the mean of the Poisson distribution is

$$E(y) = 1 \cdot \mu = \mu$$

The variance of the Poisson distribution is

$$\mathrm{Var}(y) = \frac{d\mu}{d\theta} = \frac{dE(y)}{d\theta} = \mu$$

which, of course, is the Poisson mean.

Among the other distributions in the exponential family are the **exponential, gamma, inverse Gaussian**, and **geometric** distributions. Some of the distributions are discussed at length later in this chapter. An important distribution that is not a member of the exponential family is the Weilbull.

At this point it is important to discuss the formality of model structure, that is, how the foregoing determines the structure of the model. As this development ensues, we must determine what the characteristics are that define the class of generalized linear models.

5.2 FORMAL STRUCTURE FOR THE CLASS OF GENERALIZED LINEAR MODELS

We now consider the following structure as an extension of the Poisson and binomial development in the previous chapters. The context of this extension is the exponential family defined by the probability distribution given in Equation (5.1). The structure is:

1. We have $y_1, y_2, \ldots, y_n$ independent response observations with means $\mu_1, \mu_2, \ldots, \mu_n$, respectively.
2. The observation y_i has a distribution that is a member of the exponential family.
3. The systematic portion of the model involves regressor or predictor variables $x_1, x_2, \ldots, x_k$.
4. The model is constructed around the **linear predictor** $\eta = \mathbf{x}'\boldsymbol{\beta} = \beta_0 + \sum_{i=1}^{k} \beta_i x_i$. The involvement of this linear predictor suggests the terminology **generalized linear models**.
5. The model is found through the use of a **link function**.

$$\eta_i = g(\mu_i), \quad i = 1, 2, \ldots, n \tag{5.2}$$

   The term link is derived from the fact that the function is the link between the mean and the linear predictor. Note that the expected response is

$$E(y_i) = g^{-1}(\eta_i) = g^{-1}(\mathbf{x}_i'\boldsymbol{\beta})$$

   In fact, in multiple linear regression the model

$$\mu_i = \eta_i = \mathbf{x}_i'\boldsymbol{\beta}, \quad i = 1, 2, \ldots, n$$

   suggests a special case in which $g(\mu_i) = \mu_i$, and thus the link function used is the **identity link**.
6. The link function is a monotonic differentiable function.
7. The variance σ_i^2 $(i = 1, 2, \ldots, n)$ is a function of the mean μ_i.

The reader is referred to McCullagh and Nelder (1987) for further details on the structure of the GLM.

There are many possible choices of the link function. If we choose

$$\eta_i = \theta_i \tag{5.3}$$

then we say that η_i is the **canonical link**. The link function from this formulation results in some interesting theoretical properties that are demonstrated later. Table 5.1 shows the canonical links for the most common choices of distributions employed with the generalized linear model.

There are other link functions that could be used with a generalized linear model, including:

1. The probit link

$$\eta_i = \Phi^{-1}[E(y_i)]$$

   where Φ represents the cumulative standard normal distribution function.

Table 5.1 Canonical Links for the Generalized Linear Model

Distribution	Canonical Link
Normal	$\eta_i = \mu_i$ (identity link)
Binomial	$\eta_i = \ln\left(\frac{\pi_i}{1-\pi_i}\right)$ (logistic link)
Poisson	$\eta_i = \ln(\mu_i)$ (log link)
Exponential	$\eta_i = \frac{1}{\mu_i}$ (reciprocal link)
Gamma	$\eta_i = \frac{1}{\mu_i}$ (reciprocal link)

2. The complimentary log–log link,

$$\eta_i = \ln\{\ln[1 - \mu_i]\}$$

3. The power family link

$$\eta_i = \begin{cases} \mu_i^{\lambda}, & \lambda \neq 0 \\ \ln[\mu_i], & \lambda = 0 \end{cases}$$

A very fundamental idea is that there are two components to a generalized linear model: the response distribution (also called the **error distribution**) and the link function. We can view the selection of the link function in a vein similar to the choice of a transformation on the response. However, it is important to understand that the link function is a **transformation on the population mean**, not the data. Unlike a transformation, the link function takes advantage of the **natural** distribution of the response. Just as not using an appropriate transformation can result in problems with a fitted linear model, improper choices of the link function can also result in significant problems with a generalized linear model.

5.3 LIKELIHOOD EQUATIONS FOR GENERALIZED LINEAR MODELS

The method of maximum likelihood is the theoretical basis for parameter estimation in the GLM. However, the actual implementation of maximum likelihood results in an algorithm based on iteratively reweighted least squares. This is exactly what we saw previously for the special cases of logistic and Poisson regression.

Consider the method of maximum likelihood applied to GLM, and suppose that we use the canonical link. The log-likelihood function is

$$L = \log \mathscr{L}(\boldsymbol{\beta}; \mathbf{y}) = \sum_{\mathbf{i}=1}^{\mathbf{n}} \{[\mathbf{y_i}\theta_\mathbf{i} - \mathbf{b}(\theta_\mathbf{i})]/\mathbf{a}(\phi) + \mathbf{c}(\mathbf{y_i}, \phi)\}$$

For the canonical link we have $\eta_i = g(\mu_i) = \mathbf{x}_i'\boldsymbol{\beta}$; therefore

$$\begin{aligned}\frac{\partial L}{\partial \boldsymbol{\beta}} &= \frac{\partial L}{\partial \theta_i}\frac{\partial \theta_i}{\partial \boldsymbol{\beta}} \\ &= \sum_{i=1}^{n} \frac{1}{a(\phi)}\left[y_i - \frac{\partial b(\theta_i)}{\partial \theta_i}\right]\mathbf{x}_i \\ &= \sum_{i=1}^{n} \frac{1}{a(\phi)}(y_i - \mu_i)\mathbf{x}_i\end{aligned}$$

Use of the canonical link simplifies the mathematics greatly. We can find the maximum likelihood estimates of the parameters by solving the following system of equations for $\boldsymbol{\beta}$:

$$\sum_{i=1}^{n} \frac{1}{a(\phi)}(y_i - \mu_i)\mathbf{x}_i = \mathbf{0}$$

In most cases, $a(\phi)$ is a constant, so these equations become

$$\sum_{i=1}^{n} (y_i - \mu_i)\mathbf{x}_i = \mathbf{0}$$

This is actually a *system* of $p = k + 1$ equations, one for each model parameter. In matrix form these equations are

$$\mathbf{X}'(\mathbf{y} - \boldsymbol{\mu}) = \mathbf{0} \tag{5.4}$$

where $\boldsymbol{\mu}' = [\mu_1, \mu_2, \ldots, \mu_n]$. These are called the **maximum likelihood score equations**, and they are the same equations that we saw previously in the cases of logistic and Poisson regression. Thus the score function, which was operative for multiple linear regression (normal errors), logistic regression, and Poisson regression, is relevant for a broader class of models, namely, for generalized linear models in which the canonical link is used. Appendix A.5 outlines how we can use iteratively reweighted least squares to solve these score equations.

Let $\mathbf{b}$ be the final value that the algorithm above produces as the estimate of $\boldsymbol{\beta}$. If the model assumptions, including the choice of the link function, are correct, then we can show that asymptotically

$$E(\mathbf{b}) = \boldsymbol{\beta}$$

since $\mathbf{b}$ is the solution to the score equations (5.4). The information matrix $\mathbf{I}(\mathbf{b})$ of estimators given by the variance of the score is

$$\mathbf{I}(\mathbf{b}) = \text{Var}\left\{\frac{1}{a(\phi)}[\mathbf{X}'(\mathbf{y} - \boldsymbol{\mu})]\right\} = \frac{\mathbf{X}'\mathbf{V}\mathbf{X}}{[a(\phi)]^2}$$

where $\mathbf{V} = \text{diag}\{\sigma_i^2\}$ and σ_i^2, which is a function of μ_i, depends on the distribution in question. Thus the asymptotic variance–covariance matrix of $\mathbf{b}$ is given by

$$\text{Var}(\mathbf{b}) = \mathbf{I}^{-1}(\mathbf{b}) = [\mathbf{X}'\mathbf{V}\mathbf{X}]^{-1}[a(\phi)]^2 \tag{5.5}$$

Estimated standard errors of coefficients come from square roots of the diagonal elements of the matrix in Equation (5.5) with elements of $\mathbf{V}$ replaced by their appropriate estimators.

Consider the three distributions that we have dealt with until this point. For the normal case all $\sigma_i^2 = \sigma^2, a(\phi) = \sigma$, and hence we have the familiar

$$\text{Var}(\mathbf{b}) = (\mathbf{X}'\mathbf{X})^{-1}\sigma^2$$

For the logistic and Poisson models $a(\phi) = 1.0$, and as we indicated in Chapter 4,

$$\text{Var}(\mathbf{b}) = (\mathbf{X}'\mathbf{V}\mathbf{X})^{-1} \tag{5.6}$$

where for the Poisson case $\sigma_i^2 = e^{\mathbf{x}_i'\boldsymbol{\beta}}$ and for the binomial

$$\sigma_i^2 = n_i\pi_i(1 - \pi_i) = \frac{e^{-\mathbf{x}_i'\boldsymbol{\beta}} \cdot n_i}{(1 + e^{-\mathbf{x}_i'\boldsymbol{\beta}})^2}$$

As we indicated in Chapter 4, the variances on the diagonal elements of Equation (5.6) can be inflated in the case of overdispersion.

Later in this chapter we discuss distributions in which the parameter $a(\phi) \neq 1.0$, and thus ϕ must also be estimated. Maximum likelihood can be used to do this. The most widely used distribution with a scale parameter is the gamma distribution, which has applications in many fields.

It is important to point out that while the canonical link is the most natural link to consider, given the distribution involved, this by no means implies that a noncanonical link function should not be considered. Choosing a link is tantamount to the exercise of model selection in standard multiple linear regression, while one attempts the model that best describes the data mechanism. For example, if the ranges on the regressors are sufficiently narrow that a linear model is sufficient, then an identity link might be considered. We see examples later in this chapter in which the log link is reasonable for an exponential or a gamma response. Thus it is important to review what changes are relevant in the foregoing when one uses a noncanonical link.

If we do not use the canonical link, then $\eta_i \neq \theta_i$, and the appropriate derivative of the log-likelihood is

$$\frac{\partial L}{\partial \boldsymbol{\beta}} = \sum_{i=1}^{n} \frac{dL}{d\theta_i}\frac{d\theta_i}{d\eta_i}\frac{\partial \eta_i}{\partial \boldsymbol{\beta}}$$

We note that

$$\frac{dL}{d\theta_i} = \sum_{i=1}^{n} \frac{1}{a(\phi)}\left[y_i - \frac{db(\theta_i)}{d\theta_i}\right] = \frac{1}{a(\phi)} \sum_{i=1}^{n} (y_i - \mu_i)$$

and

$$\frac{\partial \eta_i}{\partial \boldsymbol{\beta}} = \mathbf{x}_i$$

Putting this all together yields

$$\frac{\partial L}{\partial \boldsymbol{\beta}} = \sum_{i=1}^{n} \frac{y_i - \mu_i}{a(\phi)} \frac{d\theta_i}{d\eta_i} \mathbf{x}_i$$

Typically $a(\phi)$ is a constant. We then can express the score equations as

$$\mathbf{X}'\boldsymbol{\Delta}(\mathbf{y} - \boldsymbol{\mu}) = \mathbf{0} \tag{5.7}$$

where $\boldsymbol{\Delta} = \text{diag}\{d\theta_i/d\eta_i\}$. Appendix A.6 outlines the computational method for finding the solution to the score equations. The information matrix for the resulting MLE, $\mathbf{b}$, of the coefficients, $\boldsymbol{\beta}$, is

$$\mathbf{I}(\mathbf{b}) = \frac{\mathbf{X}'\boldsymbol{\Delta}\mathbf{V}\boldsymbol{\Delta}\mathbf{X}}{[a(\phi)]^2}$$

Thus the asymptotic variance–covariance matrix of $\mathbf{b}$ is

$$\text{Var}(\mathbf{b}) = (\mathbf{X}'\boldsymbol{\Delta}\mathbf{V}\boldsymbol{\Delta}\mathbf{X})^{-1}[a(\phi)]^2$$

While the score equation for the canonical link is simpler than that for the general case in Equation (5.7), it is clear that $\boldsymbol{\mu}$ and $d\theta_i/d\eta_i$ are both nonlinear in general. To gain some sense of the complexity that is involved, consider an example with the Poisson distribution. Even in the case of the canonical link we have

$$\sum_{i=1}^{n} (y_i - e^{\mathbf{x}_i'\boldsymbol{\beta}})\mathbf{x}_i = \mathbf{0}$$

which, of course, cannot be solved in closed form since it is nonlinear in $\boldsymbol{\beta}$. Now suppose that it is decided that an identity link can be justified, thereby resulting in a model that is linear, even though the link is not the canonical link. We have

$\mu = \mathbf{x}'\boldsymbol{\beta}$, and $\partial\theta_i/\partial\eta_i = (1/\mathbf{x}_i'\boldsymbol{\beta})$, and thus using Equation (5.7), we can give the score equation by

$$\sum_{i=1}^{n}\left[\frac{(y_i - \mathbf{x}_i'\boldsymbol{\beta})\mathbf{x}_i}{\mathbf{x}_i'\boldsymbol{\beta}}\right] = 0$$

which must be solved iteratively even though the *model is linear*. Thus iterative procedures must be used.

Some important observations about the GLM are:

1. Typically, when experimenters and data analysts use a transformation, they use ordinary least squares (OLS) to fit the model in the transformed scale.
2. In the GLM we recognize that the variance of the response is not constant, and we use weighted least squares as the basis of parameter estimation.
3. This suggests that the GLM should outperform standard analyses that depend on transformations when a problem remains with constant variance after the transformation. Chapter 8 discusses this issue in more detail.
4. All of the inference we described previously on logistic regression carries over directly to the GLM. That is, model deviance can be used to test for overall model fit, and the difference in deviance between a full and a reduced model can be used to test hypotheses about subsets of parameters in the model. Wald inference can be applied to test hypotheses and construct confidence intervals about individual model parameters.

5.4 QUASI-LIKELIHOOD

In Chapter 4 and in this chapter we emphasize the use of maximum likelihood estimation for the parameter vector $\boldsymbol{\beta}$ in the GLM. The distributions involved are members of the exponential family, and the very important assumption of independence is made. However, these are certainly many modeling problems encountered in situations where (1) responses are independent but do not obey one of the members of the exponential family, even though the variance is a function of the mean; and (2) responses have variances that are a function of the mean but the responses are correlated. It is the latter situation that occurs often in practice in biomedical, industrial process, and ecological studies, as well as many other areas.

The motivation of quasi-likelihood stems from the concept of weighted least squares, or more generally **generalized least squares** for the case where responses are correlated. Wedderburn (1974) developed the notion of quasi-likelihood, which exploits the fact that the score function involves the distribution of the response only though the first two moments. In addition, his work indicates that the use of generalized least squares produces asymptotic properties that are quite similar to those of maximum likelihood estimators. As a result, good efficiency can be obtained even when the likelihood is not known. Also see Carroll and Ruppert (1988) for more information on quasi-likelihood.

We begin with the general case in which the **V** matrix is positive definite but not necessarily diagonal. The method of generalized least squares focuses on the function $(\mathbf{y} - \boldsymbol{\mu})'\mathbf{V}^{-1}(\mathbf{y} - \boldsymbol{\mu})$, which gives as the score function

$$\mathbf{D}'\mathbf{V}^{-1}(\mathbf{y} - \boldsymbol{\mu}) = \mathbf{0} \tag{5.8}$$

where **D** is a matrix of derivatives $d\boldsymbol{\mu}/d\boldsymbol{\beta}$. This **D** matrix is very much like the **D** matrix from Chapter 3 in the nonlinear regression model. Since Chapters 6 and 7 contain the important material on the GLM with correlated data, we postpone discussion of the solution of Equation (5.8) in this general case.

In the special case in which responses are independent, though not necessarily members of the exponential family, $\mathbf{V} = \{\sigma_i^2\}$ and Equation (5.8) reduces to

$$\sum_{i=1}^{n} \frac{(y_i - \mu_i)}{\sigma_i^2} \frac{d\mu_i}{d\boldsymbol{\beta}} = \mathbf{0}$$

where $\sigma_i^2 = a(\phi) \cdot \text{Var}\ \mu_i$. The above is then solved for $\boldsymbol{\beta}$. In the event that the link function and thus the model involve the linear predictor $\mathbf{x}'\boldsymbol{\beta}$, then the score function reduces to

$$\sum_{i=1}^{n} \left(\frac{(y_i - \mu_i)}{\sigma_i^2}\right)\left(\frac{\partial \mu_i}{\partial \eta_i}\right)\mathbf{x}_i = \mathbf{0}$$

or in matrix form

$$\mathbf{X}'\boldsymbol{\Gamma}\mathbf{V}^{-1}(\mathbf{y} - \boldsymbol{\mu}) = \mathbf{0} \tag{5.9}$$

where $\boldsymbol{\Gamma} = \text{diag}[\partial\mu_i/\partial\eta_i]$ and $\mathbf{V} = \text{diag}\{\sigma_i^2\}$.

It is of interest to note that Equation (5.9) does not invoke properties of the exponential family. In particular, it does not use the relationship $\partial\mu_i/\partial\mu_i = \sigma_i^2$ (apart from $a(\phi)$). But if the response is a member of the exponential family, $\partial\mu_i/\partial\eta_i = \partial\mu_i/\partial\theta_i \cdot \delta_i \cdot \mathbf{x}_i$, and thus we obtain $\mathbf{X}'\boldsymbol{\Delta}(\mathbf{y} - \boldsymbol{\mu}) = \mathbf{0}$, which is the most general GLM score function in Equation (5.7).

One must keep in mind that this methodology assumes situations where inferences need to be drawn from experiments in which a **likelihood function may not be constructed**. For example, we may have a model that is reasonable and Var(**y**) is known but one does not have information that suggests the distribution of the response.

5.5 OTHER IMPORTANT DISTRIBUTIONS FOR GENERALIZED LINEAR MODELS

The foregoing material in this chapter deals with generalized linear models in the general framework along with illustrations as to how it applies in the special cases of the binomial and Poisson cases. However, other important distributions occur in practice that deserve attention. Two very important response distributions are the **exponential** and **gamma** distributions. The exponential distribution is a special case of the gamma, but each enjoys important applications. We initially consider the exponential distribution.

The exponential distribution for a response in a regression setting often is found in inter-arrival time problems, such as certain engineering applications, or survival time problems in biomedical applications. In both applications its use is reasonable when the **hazard function**, which measures the instantaneous risk of death or failure, is constant; see Cox (1972). The density function for the exponential distribution is

$$f(y) = \frac{1}{\lambda} e^{-y/\lambda}, \quad y \geq 0; \lambda > 0$$

In the form of the exponential family, we have

$$f(y) = \exp\left\{[-1]\left[y\left(\frac{1}{\lambda}\right) + \ln \lambda\right]\right\}$$

Thus

$$a(\phi) = -1$$

$$\theta = \frac{1}{\lambda}$$

$$b(\theta) = \ln \theta$$

$$c(\cdot) = 0$$

$$\mu = \lambda$$

$$\sigma^2 = \lambda^2$$

The canonical link selects $\eta_i = \theta_i$, which produces $1/\lambda = \mathbf{x}'\boldsymbol{\beta}$; thus the form of the link function is

$$\frac{1}{\mu} = \mathbf{x}'\boldsymbol{\beta}$$

which is termed the **reciprocal link**, giving the model form for the mean

$$\mu = \frac{1}{\mathbf{x}'\boldsymbol{\beta}} \tag{5.10}$$

5.5.1 The Gamma Family

Like the exponential distribution, the gamma distribution also finds applications in inter-arrival time problems. In addition, the gamma distribution has potential applications in regression problems in which the response is continuous and the **variance is not constant** but rather is **proportional to the square of the mean**. Such a condition implies a **constant coefficient of variation**. There are other alternatives to the use of the gamma distribution in this case. One possible option is to use a natural log transformation on the response, which **stabilizes the variance**. In the case of this transformation, all coefficients are unbiased except the intercept. This approach inherently assumes that the distribution of the response is log-normal. The intercept is biased by $(\sigma/\mu)^2/2$, since, from a Taylor series expansion, we know that

$$E[\ln(y)] = \ln \mu - \frac{(\sigma/\mu)^2}{2}$$

A second approach assumes a gamma distribution and appeals to the framework of GLMs. Consider the density function for the gamma distribution, which is

$$f(y) = \frac{1}{\Gamma(r)}\left(\frac{1}{\lambda}\right)^r e^{-y/\lambda} y^{r-1}, \quad y \geq 0, r > 0, \lambda > 0$$

If we put the density into the form of the exponential family, we have (after further simplification; see Exercise 5.6)

$$\begin{aligned}
\theta &= -\frac{1}{\lambda r} = -\frac{1}{\mu} \\
\mu &= r\lambda \\
\text{Var}\, y &= \frac{\mu^2}{r} \Rightarrow \frac{\text{Var}\, y}{\mu^2} = r\lambda^2 \\
a(\phi) &= r^{-1} \\
b(\theta) &= -\ln(-\theta) \\
c(\phi) &= r \ln r - \ln \Gamma(r) + (r-1) \ln y
\end{aligned}$$

The parameter r is a scale parameter. When $r = 1$, the distribution reduces to the exponential distribution. The parameter r is assumed here not to vary but to be constant throughout the regression data set, which implies that the mean is changing through the parameter λ. Though there are many applications involving gamma responses, the distribution arises in a natural way for integer values of r as the time to the rth event in a Poisson process.

For general applications, the parameter r can be estimated by maximum likelihood methods.

5.5.2 Canonical Link Function for the Gamma Distribution

From the foregoing it is clear that the canonical parameter $\theta = \mathbf{x}'\boldsymbol{\beta}$ gives the following model for the mean,

$$\mu^{-1} = \mathbf{x}'\boldsymbol{\beta}$$

which is the reciprocal link just as in the special case of the exponential distribution. The resulting model for the mean is

$$\mu = \frac{1}{\mathbf{x}'\boldsymbol{\beta}}$$

There are certain dangers that occur in using the canonical link in this case. Recall the nonnegative response values of the gamma distribution models. However, there is always the possibility that certain estimates of **b** might lead to negative values of the predicted response. Thus the reciprocal link does not guarantee that the predicted response $\hat{y}$ is positive, which may be problematic.

5.5.3 Log Link for the Gamma Distribution

While technically the log link is not the canonical link, one often may have some success using the log link. Unlike the reciprocal link, the log link does not give rise to negative estimated responses. In addition, the log link conceptually has a close connection to the use of a linear model with $\ln(y)$ as the response. In the one case we are **transforming data**, and in the other we are **transforming the mean**. We should keep in mind that transforming the mean does not alter the error distribution, whereas transforming the observations (i.e., the y's), does. In the case of the log transformation, if we call σ^2 the residual variance, our least squares estimation gives $(\mathbf{X}'\mathbf{X})^{-1}\sigma^2$ as the variance–covariance matrix of regression estimators where σ^2 is the square of the coefficient of variation. Now, if we use a generalized linear model that employs the log link, we obtain $(\mathbf{X}'\mathbf{X})^{-1}\sigma^2$ as the asymptotic variance–covariance matrix. To see this, consider the asymptotic variance–covariance matrix using the noncanonical link:

$$\text{Var}(\mathbf{b}) = (\mathbf{X}'\boldsymbol{\Delta}\mathbf{V}\boldsymbol{\Delta}\mathbf{X})^{-1} \cdot [a(\phi)]^2$$

For the gamma distribution and log link, $\ln \mu = \mathbf{x}'\boldsymbol{\beta}$, and $\theta = -1/\mu = e^{-\mathbf{x}'\boldsymbol{\beta}}$, so

$$\Delta_i = \frac{\partial \theta_i}{\partial(\mathbf{x}_i'\boldsymbol{\beta})} = -e^{-\mathbf{x}_i'\boldsymbol{\beta}}$$

Since $\text{Var}(y) = \mu^2/r = e^{2\mathbf{x}_i'\boldsymbol{\beta}}/r$, we have $\boldsymbol{\Delta}\mathbf{V}\boldsymbol{\Delta} = \text{diag}\{1/r, 1/r, \ldots, 1/r\}$. Thus asymptotically,

$$\begin{aligned}\text{Var}(\mathbf{b}) &= (\mathbf{X}'\boldsymbol{\Delta}\mathbf{V}\boldsymbol{\Delta}\mathbf{X})^{-1} \cdot (a(\phi))^2 \\ &= (\mathbf{X}'\mathbf{X})^{-1}\left(\frac{1}{r}\right)\end{aligned}$$

Since $1/r$ is the square of the coefficient of variation, this result is equivalent to that found if we were to perform a log transformation on the response itself, which should not seem surprising.

5.6 A CLASS OF LINK FUNCTIONS—THE POWER FUNCTION

We have seen that each member of the exponential family of distributions has a *natural* or canonical link developed from the relationship $\theta_i = \eta_i = \mathbf{x}_i'\boldsymbol{\beta}$. However, we have emphasized that there are times when the canonical link may not be appropriate.

For example, the log link, which is the canonical link for the Poisson distribution, may also be very useful for analysis of regression problems with gamma responses. The identity link may very well have application with almost any response distribution when the ranges of the regressor variables are quite narrow and the problem does not require complexity beyond a linear model. Quite often, however, the search for the link and, thus the model, is a time-consuming process. One tool that is often used, though perhaps not often enough, is the class of **power functions**, which includes several other links as special cases. In Section 2.4.2 we outlined the classic Box–Cox (1964) transformations. The power transformations within the class of generalized linear models utilize the same technology, but like all link functions, **the transformation is made on the mean** μ.

The power transformation link is given by

$$\begin{aligned}\mu^{\lambda} &= \mathbf{x}'\boldsymbol{\beta}, \quad \lambda \neq 0 \\ \ln \mu &= \mathbf{x}'\boldsymbol{\beta}, \quad \lambda = 0\end{aligned} \tag{5.11}$$

Formally, in order to have continuity at $\lambda = 0$, we write the link function as

$$\mathbf{x}'\boldsymbol{\beta} = \frac{\mu^{\lambda} - 1}{\lambda}$$

It is easy to see that $\lim_{\lambda \to 0} [(\mu^{\lambda} - 1)/\lambda] = \log \mu$. In the Box–Cox approach in linear regression, the error sum of squares (adjusted for scale) is plotted against λ. In generalized linear models, the value of the deviance can be plotted against λ in order to gain insight regarding what range of λ's give models that best describe the data. In fact, confidence intervals on λ can be found much like they are found in the Box–Cox approach. These bounds can be very helpful in determining what values of λ and thus what power transformations are compatible with the data.

The maximum likelihood estimator of λ can be computed if one needs to have the optimum value of λ. Usually the user will be quite satisfied with *natural* values for λ, say, $\lambda = 0 \Rightarrow$ log link, $\lambda = 1 \Rightarrow$ identity link, $\lambda = \frac{1}{2} \Rightarrow$ square root link, and soon as long as one of these values gives a deviance not significantly different from the minimum value.

The iterative procedure for finding the optimum λ proposed by Pregibon (1980) is based on a Taylor series expansion of μ^{λ} around a preliminary estimate μ^{λ_0},

$$\mu^{\lambda} \cong \mu^{\lambda_0} + (\lambda - \lambda_0)\,\mu^{\lambda_0} \ln \mu \tag{5.12}$$

As a result

$$\mu^{\lambda_0} = \mu^{\lambda} - (\lambda - \lambda_0)\,\mu^{\lambda_0} \ln \mu$$

Since $\mu^{\lambda} = \mathbf{x}'\boldsymbol{\beta}$, we can write

$$\mu^{\lambda_0} = \mathbf{x}'\boldsymbol{\beta} - (\lambda - \lambda_0)\,\mu^{\lambda_0} \ln \mu \tag{5.13}$$

The strategy follows from Equation (5.13). The power function μ^{λ_0} with known λ_0 is used with the $\mathbf{x}$ data, but the linear predictor contains an additional regressor, namely, $\hat{\mu}_0^{\lambda_0} \log \hat{\mu}_0$, where the $\hat{\mu}_0$ are fitted values found in the analysis using the guess λ_0 without the additional regressor. From Equation (5.13) it can be seen that the coefficient of the new regressor is $\widehat{\lambda - \lambda_0}$ and produces the adjustment to λ_0. The procedure continues to convergence. A one-step procedure here might show if a significant improvement can be made on the starting value λ_0 and thus on the initial choice of link. An example of the use of the power link accompanies other examples in a later section. The power link family accounts for the identity, square root, negative square root, reciprocal negative square, and log links.

5.7 INFERENCE AND RESIDUAL ANALYSIS FOR GENERALIZED LINEAR MODELS

Inference

In this section we discuss the use of likelihood and Wald inference that we introduced in Chapter 4 in the context of logistic and Poisson regression. The general goals and applications of both forms of inference extend to the family of generalized linear models. For tests of hypotheses on individual coefficients,

Wald inference makes use of the asymptotic normality of the MLE. Thus under $H_0 : \beta_j = 0$,

$$\left(\frac{b_j}{\widehat{se}(b_j)}\right)^2$$

has a χ_1^2 distribution for large samples. The confidence intervals on mean response and prediction intervals on an individual observation are found just as reported in Chapter 4. The model deviance

$$D(\boldsymbol{\beta}) = -2 \ln\left[\frac{\mathscr{L}(\boldsymbol{\beta})}{\mathscr{L}(\boldsymbol{\mu})}\right]$$

is nominally used as a goodness-of-fit statistic. $\mathscr{L}(\boldsymbol{\beta})$ is the likelihood under the model considered, and $\mathscr{L}(\boldsymbol{\mu})$ is the likelihood of the **saturated model**. Asymptotically, $D(\boldsymbol{\beta})$ has a χ^2 distribution with $n - p$ degrees of freedom. However, as we indicated in Chapter 4, the use of the test may not be appropriate in small-sample problems. On the other hand, the use of deviance or analysis of deviance for nested tests or stepwise procedures is quite appropriate. A good rule of thumb is that lack of fit may be a problem when deviance$/(n - p)$ exceeds 1.0 by a substantial amount. Pearson's χ^2 statistic discussed in Chapter 4 applies in the general case for the GLM. This intuitively appealing statistic is

$$\chi^2 = \sum_{i=1}^{n} \left(\frac{y_i - \hat{\mu}_i}{\sqrt{\widehat{\text{Var}}\, y_i}}\right)^2$$

and is asymptotically distributed as χ_{n-p}^2.

Examples with illustrations of these tools are given later in this chapter. It is of interest for the reader to see expressions for the deviance for various distributions. This display motivates the use of specific types of residuals and residual plots. The deviance expressions are

$$\text{Normal:} \sum_{i=1}^{n} (y - \hat{\mu})^2$$

$$\text{Poisson:}\ 2\sum_{i=1}^{n} [y \ln(y/\hat{\mu}) - (y - \hat{\mu})]$$

$$\text{Binomial:}\ 2\sum_{i=1}^{n} \{y \ln(y/\hat{\mu}) + (m - y) \ln[(m - y)/(m - \hat{\mu})]\}$$

$$\text{Gamma:}\ 2\sum_{i=1}^{n} [-\ln(y/\hat{\mu}) + (y - \hat{\mu})/\hat{\mu}]$$

In Chapter 4 we illustrate the development of deviance for the Poisson case. If the linear predictor contains an intercept, the deviance for the Poisson and gamma cases reduces to

$$D(\boldsymbol{\beta}) = \sum_{i=1}^{n} \left[y_i \ln\left(\frac{y_i}{\hat{\mu}_i}\right) \right] \tag{5.14}$$

See Exercise 5.11. The deviance expression in Equation (5.14) also holds for the exponential distribution.

Use of Residuals

In standard linear regression, the set of regression residuals, the $y_i - \hat{\mu}_i$, is often used to detect such violations of assumptions as nonhomogeneous variance and model specification. In the GLM the ordinary or raw residuals $y_i - \hat{\mu}_i$ are technically not appropriate, since Var (y_i) is not constant. The type of residual that is most intuitive is the **Pearson residual**,

$$r_{\text{P}} = \frac{y_i - \hat{\mu}_i}{\sqrt{\widehat{\text{Var}}\, y_i}}$$

A second type of residual that can be plotted is the **deviance residual**, which we introduced in Section 4.2.6. The deviance value is written in the form $\sum_{i=1}^{n} d_i = D(\boldsymbol{\beta})$. The individual components defined as the deviance residuals are

$$d_{i,r} = [\text{sgn}\,(y_i - \hat{\mu}_i)] \cdot \sqrt{d_i}, \quad i = 1, 2, \ldots, n$$

and thus

$$\sum_{i=1}^{n} d_{i,\,r}^2 = D(\boldsymbol{\beta})$$

The deviance residual has the property that it carries the same sign as $y_i - \hat{\mu}_i$, and the sum of their squares is the deviance.

An interesting question is finding which of these types of residuals is most appropriate for diagnostic plotting purposes. Pierce and Schafer (1986) give a nice discussion of residuals in models based on members of the exponential family. Their work suggests that deviance residuals are very nearly the same as those generated by the best possible normalizing transformation. As a result they recommend that analysts use the deviance residuals for constructing diagnostic plots.

McCullagh and Nelder (1989) recommend plotting the deviance residuals against the fitted values, transformed to the constant variance scale or constant information of the error distribution, and against the regressors. These plots

Table 5.2 Transformations to Constant Information

Distribution	Transformation
Normal	$\hat{\mu}$
Binomial	$2\sin^{-1}\sqrt{\hat{\mu}}$
Poisson	$2\sqrt{\hat{\mu}}$
Exponential	$2\log\hat{\mu}$
Gamma	$2\log\hat{\mu}$

are exactly analogous to the common residual plots used in multiple regression. They have exactly the same interpretation. McCullagh and Nelder also suggest plotting the absolute values of the deviance residuals against the fitted values, again transformed to the constant variance scale. A poorly chosen variance function should produce a trend in this plot. The use of the constant variance scale for both sets of plots, in general, is for aesthetic reasons. For the most part, this transformation spreads the data out across the horizontal axis. Table 5.2 gives the appropriate transformation to a constant information scale for several common error distributions. McCullagh and Nelder also recommend a normal probability plot of the deviance residuals. Again, the interpretation of this plot is directly analogous to the normal probability plot in multiple regression.

5.8 EXAMPLES WITH THE GAMMA DISTRIBUTION

Example 5.1. The Resistivity Data. This example illustrates the use of the log link with an assumed gamma distribution. Myers and Montgomery (1997) discuss the experiment summarized in Table 5.3, which is an unreplicated factorial design that was run at a certain step in a semiconductor manufacturing process. The response variable is the resistivity of the test wafer. Resistivity is well known to have a distribution with a heavy right tail, and thus a gamma distribution may be appropriate. Initially, consider a log transformation on the response, where only three main effects are significant. The model fit is

$$\widehat{\ln(y)} = 2.351 + 0.027x_1 - 0.065x_2 + 0.039x_3$$

As we indicated earlier, a log transformation stabilizes variance in the use of the gamma distribution. As an alternative analysis, consider a generalized linear model with a gamma distribution and log link that was fit using SAS PROC GENMOD, a SAS procedure that can be used to fit the GLM. The initial model chosen is the main-effects plus two-factor interaction model. The PROC GENMOD output is given in Table 5.4. Several things should be noted. First, there is also a scale parameter, r, in the table that is estimated by maximum

Table 5.3 Resistivity Data

Run	x_1	x_2	x_3	x_4	Resistivity (y)
1	−	−	−	−	193.4
2	+	−	−	−	247.6
3	−	+	−	−	168.2
4	+	+	−	−	205.0
5	−	−	+	−	303.4
6	+	−	+	−	339.9
7	−	+	+	−	226.3
8	+	+	+	−	208.3
9	−	−	−	+	220.0
10	+	−	−	+	256.4
11	−	+	−	+	165.7
12	+	+	−	+	203.5
13	−	−	+	+	285.0
14	+	−	+	+	268.0
15	−	+	+	+	169.1
16	+	+	+	+	208.5

likelihood, and the value is 472.3951. The interpretation here is that the quantity is the square of the reciprocal of the constant coefficient of variation (CV), which in this case is 0.046. Next, notice that the standard errors of the coefficient are all equal to 0.0115. Recall in our discussion of the gamma distribution that $\text{Var} = (\mathbf{X}'\mathbf{X})^{-1}/r$. The orthogonality of the design here renders $\mathbf{X}'\mathbf{X}$ and hence $(\mathbf{X}'\mathbf{X})^{-1}$ diagonal, the latter being diag(1/16). As a result the **standard errors** are $(1/4)(1/\sqrt{r}) = (0.25)(0.046) = 0.0115$. This is one of the cases in which the variance–covariance matrix of coefficients is a multiple of $(\mathbf{X}'\mathbf{X})^{-1}$. More discussion of this is given in Chapter 8. The scaled deviance here involves division of the deviance by $1/r = 1/472.3951$. This scaling accounts for $1/a(\phi)$ in the likelihood. In GENMOD, the estimate is $\hat{r}$, not $1/\hat{r}$.

Note also that unlike the case of the transformed response, three interactions as well as the additional main effect (x_4) are significant. All these interactions involve x_3 and thus may well have important engineering significance, which was not discovered in the analysis with the transformation. Table 5.5 gives the PROC GENMOD output for the reduced model which eliminates the x_1x_2, x_1x_4 and x_2x_4 two factor interactions. Figures 5.1–5.8 give the appropriate diagnostic plots of the deviance residuals generated by S-Plus. The fitted model is

$$\hat{y} = e^{a}$$

where $a = 5.414 + 0.0617x_1 - 0.150x_2 + 0.090x_3 - 0.028x_4 - 0.040x_1x_3 - 0.044x_2x_3 - 0.046x_3x_4$. These effects are multiplicative rather than additive. The interpretation of an effect is more tedious because of the existence of interaction. One simple interpretation of, say, the coefficient 0.0613 on x_1 is that

Table 5.4 PROC GENMOD Output for the Resistivity Experiment

The GENMOD Procedure Model Information

Description	Value
Data set	WORK.RESIT
Distribution	GAMMA
Link function	LOG
Dependent variable	RESIST
Observations used	16

Criteria for Assessing Goodness of Fit

Criterion	DF	Value	Value/DF
Deviance	5	0.0339	0.0068
Scaled deviance	5	16.0056	3.2011
Pearson chi-square	5	0.0338	0.0068
Scaled Pearson X2	5	155.9695	3.1939
Log-likelihood		-60.0555	

Analysis of Parameter Estimates

Parameter	DF	Estimate	Std Err	Chi-Square	Pr > Chi
INTERCEPT		5.4141	0.0115	221556.929	0.0000
X1	1	0.0617	0.0115	28.6969	0.0000
X2	1	-0.1496	0.0115	169.0053	0.0000
X1*X2	1	0.0050	0.0115	0.1888	0.6640
X3	1	0.0900	0.0115	61.2434	0.0000
X1*X3	1	-0.0386	0.0115	11.2652	0.0008
X2*X3	1	-0.0441	0.0115	14.6967	0.0001
X4	1	-0.0280	0.0115	5.9088	0.0151
X1*X4	1	0.0020	0.0115	0.0292	0.8644
X2*X4	1	-0.0110	0.0115	0.9097	0.3402
X3*X4	1	-0.0456	0.0115	15.7072	0.0001
Scale	1	472.3951	166.9580		

if x_1 is changed from the low (-1) to high $(+1)$ level while x_3 is held constant at the middle (zero) level, then the resistivity would increase by a factor of $e^{2(0.0613)} - 1 = 0.13$ or 13%. On the other hand, if the same change in x_1 is made with x_3 held at $+1$, the mean resistivity would increase by only $e^{0.1226 - 2(0.0389)} - 1 = 0.0458$, or only 4.58%, which reflects the role of the x_1x_3 interaction. Similar interpretations can be made of other effects. Table 5.5 also provides the deviance, chi-square, and raw residuals discussed in

Table 5.5 Edited Model for the Resistivity Data

The GENMOD Procedure Model Information

Description	Value
Data set	WORK.RESIT
Distribution	GAMMA
Link function	LOG
Dependent variable	RESIST
Observation used	16

Criteria for Assessing Goodness of Fit

Criterion	DF	Value	Value/DF
Deviance	8	0.0363	0.0045
Scaled deviance	8	16.0060	2.0008
Pearson chi-square	8	0.0362	0.0045
Scaled Pearson X2	8	15.9769	1.9971
Log-likelihood		-60.5996	

Analysis of Parameter Estimates

Parameter	DF	Estimate	Std Err	Chi-Square	Pr > Chi
INTERCEPT	1	5.4142	0.0119	207004.927	0.0000
X1	1	0.0613	0.0119	26.5041	0.0000
X2	1	-0.1496	0.0119	157.9262	0.0000
X3	1	0.0899	0.0119	57.1267	0.0000
X4	1	-0.0278	0.0119	5.4555	0.0195
X1*X3	1	-0.0389	0.0119	10.6690	0.0011
X2*X3	1	-0.0441	0.0119	13.7078	0.0002
X3*X4	1	-0.0455	0.0119	14.6046	0.0001
Scale	1	441.3557	155.9839		

Observation Statistics

RESIST	Pred	Xbeta	Std	HessWgt	Lower	Upper
193.4	202.7457	5.3120	0.0339	421.0111	189.6955	216.6937
247.6	247.7051	5.5122	0.0338	441.1684	231.8400	264.6560
168.2	164.1771	5.1009	0.0336	452.1705	153.7250	175.3398
205	200.5838	5.3012	0.0334	451.0729	187.8837	214.1424
303.4	313.7789	5.7487	0.0334	426.7570	293.8678	335.0390
339.9	328.1536	5.7935	0.0339	457.1542	307.0883	350.6639
226.3	213.0273	5.3614	0.0336	468.8546	199.4401	227.5400
208.3	222.7864	5.4062	0.0337	412.6571	208.5402	238.0058

(Continued)

Table 5.5 *Continued*

220	210.0455	5.3473	0.0334	462.2725	196.7463	224.2437
256.4	256.6237	5.5476	0.0336	440.9710	240.2862	274.0721
165.7	170.0882	5.1363	0.0338	429.9688	159.1943	181.7276
203.5	207.8058	5.3366	0.0339	432.2107	194.4299	222.1019
285	270.9985	5.6021	0.0337	464.1590	253.6693	289.5115
268	283.4134	5.6469	0.0336	417.3527	265.3369	302.7213
169.1	183.9833	5.2148	0.0339	405.6524	172.1728	196.6039
208.5	192.4119	5.2596	0.0334	478.2588	180.2022	205.4487

Observation Statistics

Resraw	Reschi	Resdev
-9.3457	-0.0461	-0.0468
-0.1051	-0.000424	-0.000425
4.0229	0.0245	0.0243
4.4162	0.0220	0.0219
-10.3789	-0.0331	-0.0334
11.7464	0.0358	0.0354
13.2727	0.0623	0.0611
-14.4864	-0.0650	-0.0665
9.9545	0.0474	0.0467
-0.2237	-0.000872	-0.000872
-4.3882	-0.0258	-0.0260
-4.3058	-0.0207	-0.0209
14.0015	0.0517	0.0508
-15.4134	-0.0544	-0.0554
-14.8833	-0.0809	-0.0832
16.0881	0.0836	0.0814

previous sections. Also note the lower and upper confidence limits on the mean response; the calculation was made on the normal-theory upper and lower confidence interval on the linear predictor. The residual plots reveal no serious problems. Editing the model substantially reduced the mean scaled deviance, much like the mean squared error in standard linear models. □

Example 5.2. The Worsted Yarn Experiment. Table 5.6 contains data from an experiment conducted to investigate the effects of three factors x_1 = length, x_2 = amplitude, and x_3 = load on the cycles to effects of failure, y, of worsted yarn. The regressor variables are coded, and readers who have familiarity with designed experiments recognize that the experiment used here is a 3^3 factorial design. The data also appear in Box and Draper (1987) and Myers, Montgomery, and Anderson-Cook (2009). These authors use the data to illustrate the utility of variance-stabilizing transformations. Both Box and Draper (1987) and Myers, Montgometry, and Anderson-Cook (2009) show that the log

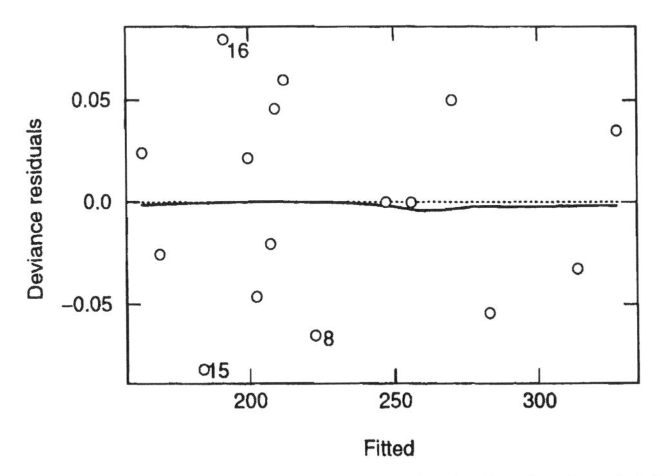

Figure 5.1 Plot of the deviance residuals versus the fitted values for the resistivity data.

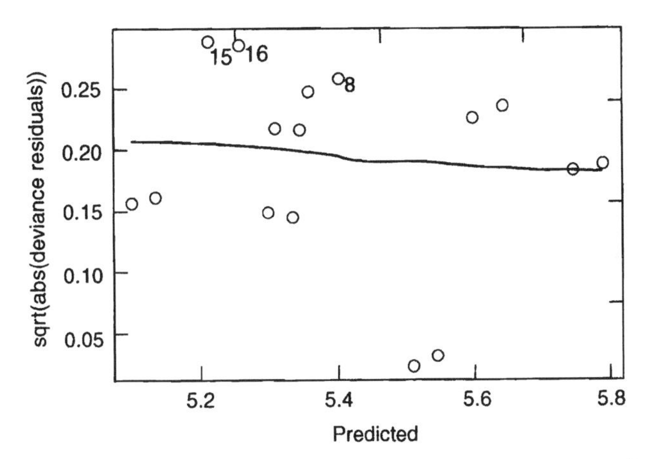

Figure 5.2 Plot of the signed square root of the deviance residuals versus the predicted values of the linear predictors for the resistivity data.

transformation is very effective in stabilizing the variance of the cycles-to-failure response. We also provide this analysis in Section 2.4.1. When a natural log transformation is used for y, the following fitted model is obtained

$$\hat{y} = \exp(6.33+0.82x_1 - 0.63x_2 - 0.38x_3)$$

The response variable in this experiment is an example of a nonnegative response that would be expected to have an asymmetric distribution with a long right tail. Failure data are frequently modeled with exponential, Weibull, lognormal, or gamma distributions both because they possess the anticipated shape and because sometimes there is theoretical or empirical justification for a

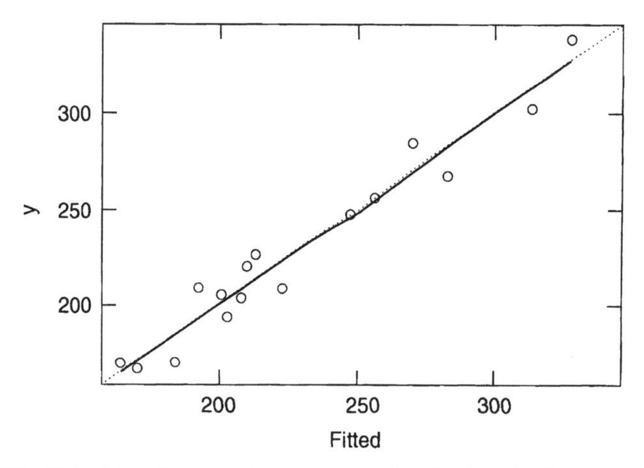

Figure 5.3 Plot of the observed values versus predicted values for the resistivity data.

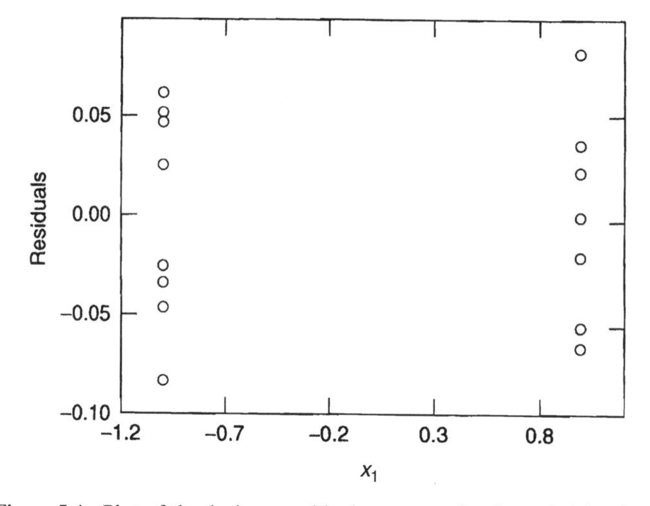

Figure 5.4 Plot of the deviance residuals versus x_1 for the resistivity data.

particular distribution. We model the cycles-to-failure data with a GLM using the gamma distribution and the log link with a first-order predictor.

Table 5.7 presents some summary output information from SAS PROC GENMOD. The appropriate SAS code is

```
proc genmod;
model y = x1 x2 x3 / dist = gamma link = log type1 type3 ;
```

Notice that the fitted model is

$$\hat{y} = \exp(6.35+0.84x_1 - 0.63x_2 - 0.39x_3)$$

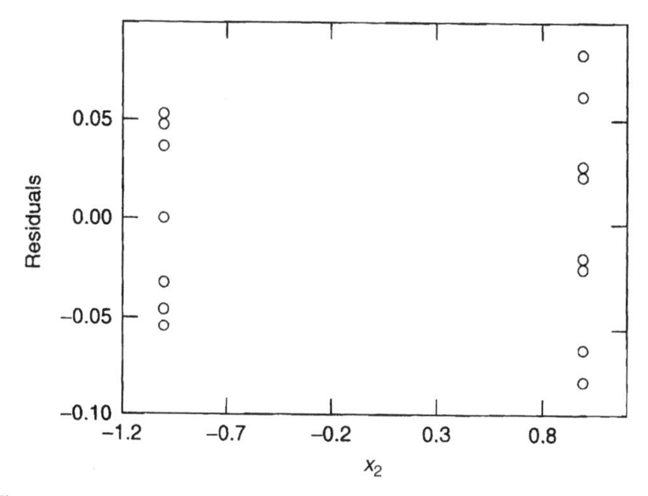

Figure 5.5 Plot of the deviance residuals versus x_2 for the resistivity data.

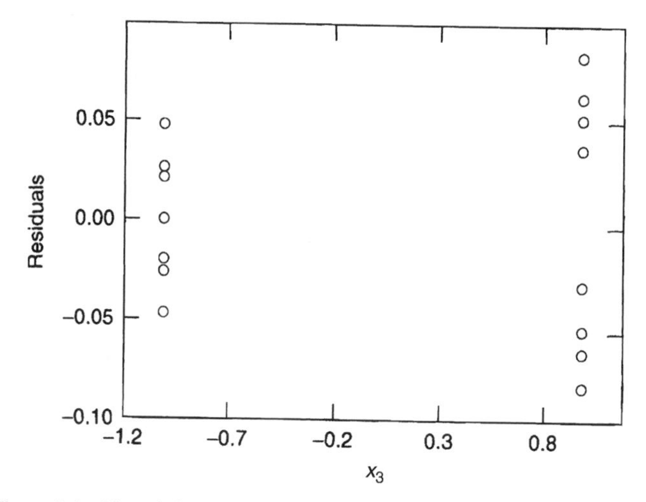

Figure 5.6 Plot of the deviance residuals versus x_3 for the resistivity data.

which is virtually identical to the model obtained via data transformation. Actually, since the log transformation works very well here, it is not too surprising that the GLM produces an almost identical model. Recall that we observed that the GLM is most likely to be an effective alternative to a data transformation when the transformation fails to produce the desired properties of constant variance and approximate normality in the response variable.

For the gamma response case, it is appropriate to use the **scaled deviance** in the SAS output as a measure of the overall fit of the model. This quantity is compared

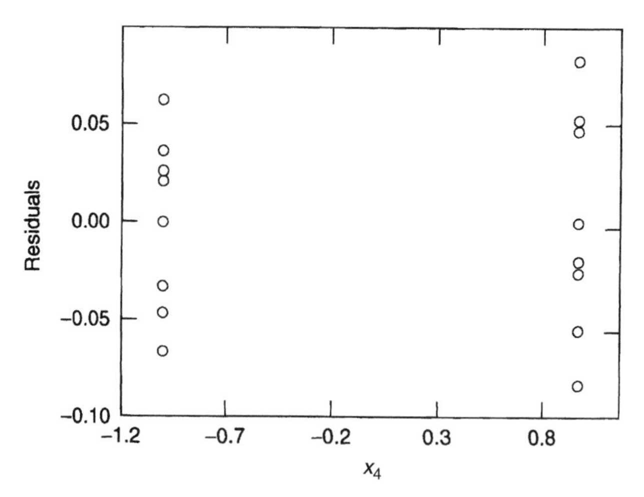

Figure 5.7 Plot of the deviance residuals versus x_4 for the resistivity data.

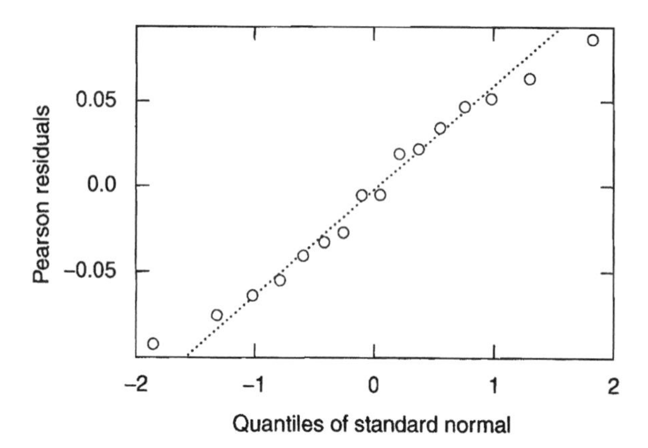

Figure 5.8 Normal probability plot of the Pearson residuals for the resistivity data.

to the chi-square distribution with $n - p$ degrees of freedom, as usual. From Table 5.7 we find that the scaled deviance is 27.1276, and referring this to a chi-square distribution with 23 degrees of freedom gives a P-value of approximately 0.25; so there is no indication of model inadequacy from the deviance criterion. Notice that the scaled deviance divided by its degrees of freedom is also close to unity. Table 5.7 also gives the Wald tests and the partial deviance statistics (both type 1 or *effects added in order* and type 3 or *effects added last* analyses) for each regressor in the model. These test statistics indicate that all three regressors are important predictors and should be included in the model. □

Table 5.6 Data from the Worsted Yarn Experiment

x_1	x_2	x_3	x_4
−1	−1	−1	674
0	−1	−1	1414
1	−1	−1	3636
−1	0	−1	338
0	0	−1	1022
1	0	−1	1568
−1	1	−1	170
0	1	−1	442
1	1	−1	1140
−1	−1	0	370
0	−1	0	1198
1	−1	0	3184
−1	0	0	266
0	0	0	620
1	0	0	1070
−1	1	0	118
0	1	0	332
1	1	0	884
−1	−1	1	292
0	−1	1	634
1	−1	1	2000
−1	0	1	210
0	0	1	438
1	0	1	566
−1	1	1	90
0	1	1	220
1	1	1	360

5.9 USING R TO PERFORM GLM ANALYSIS

The *glm* function within the MASS library is the most popular function for analyzing GLMs in R. The basic form of the statement is

```
glm(formula, family, data)
```

The formula specification is exactly the same as for a standard linear model. For example, the formula for the model $\eta = \beta_0 + \beta_1 x_1 + \beta_2 x_2$ is

```
y~x1+x2
```

The choices for family and the links available are:

- binomial (logit, probit, log, complementary log–log),
- gaussian (identity, log, inverse),

Table 5.7 SAS PROC GENMOD Output for the Worsted Yarn Experiment

The GENMOD Procedure
Model Information

Description	Value
Data Set	WORK.WOOL
Distribution	GAMMA
Link Function	LOG
Dependent Variable	CYCLES
Observations Used	27

Criteria for Assessing Goodness of Fit

Criterion	DF	Value	Value / DF
Deviance	23	0.7694	0.0335
Scaled Deviance	23	27.1276	1.1795
Pearson Chi [Square	23	0.7274	0.0316
Scaled Pearson X2	23	25.6456	1.1150
Log Likelihood		-161.3784	

Analysis of Parameter Estimates

Parameter	DF	Estimate	Std Err	Chi Square	Pr > Chi
INTERCEPT	1	6.3489	0.0324	38373.0419	0.0001
A	1	0.8425	0.0402	438.3606	0.0001
B	1	-0.6313	0.0396	253.7576	0.0001
C	1	-0.3851	0.0402	91.8566	0.0001
SCALE	1	35.2585	9.5511		

Note: The scale parameter was estimated by maximum likelihood.

LR Statistics for Type 1 Analysis

Source	Deviance	DF	Chi Square	Pr > Chi
INTERCEPT	22.8861	0		
A	10.2104	1	23.6755	0.0001
B	3.3459	1	31.2171	0.0001
C	0.7694	1	40.1106	0.0001

LR Statistics for Type 3 Analysis

Source	DF	Chi Square	Pr > Chi
A	1	77.2935	0.0001
B	1	63.4324	0.0001
C	1	40.1106	0.0001

- Gamma (identity, inverse, log)
- inverse.gaussian ($1/\mu^2$, identity, inverse, log)
- poisson (identity, log, square root), and
- quasi (logit, probit, complementary log–log, identity, inverse, log, $1/\mu^2$, square root).

R is case-sensitive, so the family is Gamma, not gamma. By default, R uses the canonical link. To specify the probit link for the binomial family, the appropriate family phrase is binomial(link = probit).

R can produce two different predicted values. The *fit* is the vector of predicted values on the original scale. The *linear.predictor* is the vector of the predicted values for the linear predictor. R can produce the raw, the Pearson, and the deviance residuals. R also can produce the influence.measures, which are the individual observation deleted statistics. The easiest way to put all this together is through examples.

5.9.1 Logistic Regression, Each Response is a Success or Failure

Consider the *Challenger* O-ring data. Suppose these data are in the file "challenger-data.txt". The appropriate R code to analyze these data is

```
o_ring <- read.table ("challenger_data.txt",header=TRUE, sep="")
o_ring.model <- glm(fail~temp, family=binomial, data=o_ring)
summary(o_ring.model)
pred_prob <- o_ring.model$fit
eta_hat <- o_ring.model$linear.predictor
dev_res <- residuals(o_ring.model, c="deviance")
influence.measures(o_ring.model)
df <- dfbetas(o_ring.model)
df_int <- df[,1]
df_temp <- df[,2]
hat <- hatvalues(o_ring.model)
qqnorm(dev_res)
plot(pred_prob,dev_res)
plot (eta_hat,dev_res)
plot(o_ring$temp,dev_res)
plot(hat,dev_res)
plot(pred_prob,df_temp)
plot(hat,df_temp)
o_ring2  <-  cbind(o_ring,pred_prob,eta_hat,dev_res,df_int,df_
  temp,hat)
write.table(o_ring2, "challenger_output.txt")
```

In this code, *pred_prob* is the predicted probability of an O-ring failure. The term *eta_hat* is the vector of predicted value for the linear predictor. The term *dev_res* is the vector of deviance residuals. The terms *df_int*, *df_temp*, and *hat* are the influence measures looking at the change in the estimated intercept, change in the estimated coefficient for temperature, and the hat diagonal, respectively. The *write.table* statement produces a file suitable for plotting the results via another software package, such as EXCEL.

5.9.2 Logistic Regression, Response is the Number of Successes out of *n* Trials

The best way to illustrate the R code is through an example. Ashford (1959) considers the proportion of coal miners who exhibit symptoms of severe pneumoconiosis as a function of the number of years exposure. The data set is small; so we do not need a separate data file. The R code is

```
> years <- c(5.8, 15.0, 21.5, 27.5, 33.5 39.5, 46.0, 51.5)
> cases <- c(0, 1, 3, 8, 9, 8, 10, 5)
> miners <- c(98, 54, 43, 48, 51, 38, 28, 11)
> ymat <- cbind(cases, miners-cases)
> ashford <- data.frame(ymat, years)
> anal <- glm(ymat ~ years, family=binomial, data=ashford)
summary (anal)
pred_prob <- anal$fit
eta_hat <- anal$linear.predictor
dev_res <- residuals(anal, c="deviance")
influence.measures(anal)
df <- dfbetas (anal)
df_int <- df[,1]
df_years <- df[,2]
hat <- hatvalues(anal)
qqnorm(dev_res)
plot(pred_prob,dev_res)
plot(eta_hat,dev_res)
plot (years,dev_res)
plot(hat,dev_res)
plot (pred_prob,df_years)
plot(hat,df_years)
ashford2 <- cbind(ashford,pred_prob,eta_hat,dev_res,df_int,
  df_years,hat)
write.table(ashford2, "ashford_output.txt")
```

5.9.3 Poisson Regression

We next consider the aircraft damage example from Chapter 4. The data are in the file "aircraft_damage_data.txt". The appropriate R code is

```
air <- read.table("aircraft_damage_data.txt" ,header=TRUE, sep="")
air.model <- glm(y~x1+x2+x3, family="poisson", data=air)
summary(air.model)
print(influence.measures(air.model))
yhat <- air.model$fit
dev_res <- residuals(air.model, c="deviance")
```

```
qqnorm(dev_res)
plot(yhat,dev_res)
plot(air$x1,dev_res)
plot(air$x2,dev_res)
plot(air$x3,dev_res)
air2 <- cbind(air,yhat,dev_res)
write.table (air2, "aircraft__damage_output.txt")
```

5.9.4 Using the Gamma Distribution with a Log Link

Now, consider the worsted yarn example from Section 5.8. The data are in the file "worsted_data.txt". The appropriate R code is

```
yarn <- read.table ("worsted_data.txt",header=TRUE, sep="")
yarn.model <- glm(y~x1+x2+x3, family=Gamma(link=log), data=air)
summary(yarn.model)
print(influence.measures(yarn model))
yhat <- air.model$fit
dev_res <- residuals(yarn.model, c="deviance")
qqnorm(dev_res)
plot(yhat,dev_res)
plot(yarn$x1,dev_res)
plot(yarn$x2,dev_res)
plot(yarn$x3,dev_res)
yarn2 <- cbind(yarn,yhat,dev_res)
write.table(yarn2, "yarn_output.txt")
```

5.10 GLM AND DATA TRANSFORMATION

In various parts of Chapter 4 and this chapter we have compared GLMs with data transformations where the response, y, is transformed. In the case of a data transformation, one typically uses ordinary least squares to fit a linear model to the transformed data. A common motivation for the transformation is to achieve an approximately stable variance in the response. In other cases, practitioners use a transformation when the model errors do not appear to follow a normal distribution. When the variance is a function of the mean, the analyst may use an appropriate variance-stabilizing transformation. The reader should not confuse these uses of transformations with the link function in the GLM. The link function serves not as a transformation on the data, y, but as a transformation of the population mean, μ.

In practice, data transformations work reasonably well in many cases. However, when the response is nonnormal, it may be impossible for the same transformation to create normally distributed random errors, to stabilize the variance, and to lead to a linear model. The GLM does not require the assumption that the data follow a normal distribution since the GLM exploits the apparent distribution of the data. Furthermore, constant variance is not an issue in the GLM, which bases its analysis on the natural variance of the data's distribution, as we described in Chapter 4 and parts of this chapter. In addition, the choice of the link function provides the analyst with a great deal of flexibility in the nonlinear models used to fit the data. Finally, with the GLM one loses none of the major elements of ordinary linear models data analysis strategy. The GLM accommodates well the notions of model editing, diagnostic plots, effect plots, and so on. Hamada and Nelder (1997) and Lewis, Montgomery, and Myers (2001a) go further in making comparisons between the GLM and data transformations. The examples that follow illustrate the superiority of the GLM over data transformations.

Example 5.3. The Worsted Yarn Experiment. In Example 5.2 we introduced the worsted yarn experiment, an unreplicated 3^3 factorial design investigating the performance of worsted yarn under cyclic loading (the data are shown in Table 5.6). In Section 2.4.2 we use the Box–Cox method and find that a log transformation leads to a fairly simple model, namely,

$$\hat{y} = e^{6.33+0.83x_1-0.63x_2-0.39x_3}$$

This model has good fit to the data and has satisfactory residual plots. Assuming a gamma distribution and log link, the fitted gamma GLM is

$$\hat{y} = e^{6.35+0.84x_1-0.63x_2-0.39x_3}$$

One measure of performance used for comparing a linear model fir to the log-transformed response to the gamma GLM fit to the data is the lengths of the 95% (ex. can be found using SAS PROC GENMOD) confidence intervals around the mean predicted values. Table 5.8 shows the estimates of the means and the lower and upper 95% confidence intervals for each observation using both the ordinary least squares fit to the log-transformed data and the gamma GLM fit to the raw data. Note that the generalized linear model has consistently shorter confidence intervals than those found from the ordinary least squares fit to the transformed data.

Contour plots showing a two-dimensional graphical representation of the relationship between the response variable and the design factors were also constructed for this experiment. These plots are shown in Figure 5.9. Figure 5.9a shows the contour plots for the ordinary least squares model, and Figure 5.9b shows the contour plots for the GLM. Both sets of contour plots are very similar, indicating that both models produce similar point estimates of cycles to failure. However, Table 5.8 clearly indicates that the GLM confidence intervals are uniformly shorter than their least squares counterparts; consequently, we would expect estimation and response prediction to be more reliable using the GLM.

Table 5.8 Comparison of 95% Confidence Intervals on the Means for Models Generated with Data Transformations and GLM for Worsted Yarn Experiment

	Using Least Squares Methods with Log Data Transformation				Using the Generalized Linear Model		Length of 95% Confidence Interval	
	Transformed		Untransformed					
Observation	Estimate of Mean	95% Confidence Interval	Estimate of Mean	95% Confidence Interval	Estimate of Mean	95% Confidence Interval	Least Squares	GLM
1	2.83	(2.76, 2.91)	682.50	(573.85, 811.52)	680.52	(583.83, 793.22)	237.67	209.39
2	2.66	(2.60, 2.73)	460.26	(397.01, 533.46)	463.00	(407.05, 526.64)	136.45	119.59
3	2.49	(2.42, 2.57)	310.38	(260.98, 369.06)	315.01	(271.49, 365.49)	108.09	94.00
4	2.56	(2.50, 2.62)	363.25	(313.33, 421.11)	361.96	(317.75, 412.33)	107.79	94.58
5	2.39	(2.34, 2.44)	244.96	(217.92, 275.30)	246.26	(222.55, 272.51)	57.37	49.96
6	2.22	(2.15, 2.28)	165.20	(142.50, 191.47)	167.55	(147.67, 190.10)	48.97	42.42
7	2.29	(2.21, 2.36)	193.33	(162.55, 229.93)	192.52	(165.69, 223.70)	67.38	58.01
8	2.12	(2.05, 2.18)	130.38	(112.46, 151.15)	130.98	(115.43, 148.64)	38.69	33.22
9	1.94	(1.87, 2.02)	87.92	(73.93, 104.54)	89.12	(76.87, 103.32)	30.62	26.45
10	3.20	(3.13, 3.26)	1569.28	(1353.94, 1819.28)	1580.00	(1390.00, 1797.00)	465.34	407.00
11	3.02	(2.97, 3.08)	1058.28	(941.67, 1189.60)	1075.00	(972.52, 1189.00)	247.92	216.48
12	2.85	(2.79, 2.92)	713.67	(615.60, 827.37)	731.50	(644.35, 830.44)	211.77	186.09
13	2.92	(2.87, 2.97)	835.41	(743.19, 938.86)	840.54	(759.65, 930.04)	195.67	170.39
14	2.75	(2.72, 2.78)	563.25	(523.24, 606.46)	571.87	(536.67, 609.38)	83.22	72.70

(*Continued*)

Table 5.8 *Continued*

	Using Least Squares Methods with Log Data Transformation							
	Transformed		Untransformed		Using the Generalized Linear Model		Length of 95% Confidence Interval	
Observation	Estimate of Mean	95% Confidence Interval	Estimate of Mean	95% Confidence Interval	Estimate of Mean	95% Confidence Interval	Least Squares	GLM
15	2.58	(2.53, 2.63)	379.84	(337.99, 426.97)	389.08	(351.64, 430.51)	88.99	78.87
16	2.65	(2.58, 2.71)	444.63	(383.53, 515.35)	447.07	(393.81, 507.54)	131.82	113.74
17	2.48	(2.43, 2.53)	299.85	(266.75, 336.98)	304.17	(275.13, 336.28)	70.23	61.15
18	2.31	(2.24, 2.37)	202.16	(174.42, 234.37)	206.95	(182.03, 235.27)	59.95	53.23
19	3.56	(3.48, 3.63)	3609.11	(3034.59, 4292.40)	3670.00	(3165.00, 4254.00)	1257.81	1089.00
20	3.39	(3.32, 3.45)	3433.88	(2099.42, 2821.63)	2497.00	(2200.00, 2833.00)	722.21	633.00
21	3.22	(3.14, 3.29)	1641.35	(1380.07, 1951.64)	1699.00	(1462.00, 1974.00)	571.57	512.00
22	3.28	(3.22, 3.35)	1920.88	(1656.91, 2226.90)	1952.00	(1720.00, 2215.00)	569.98	495.00
23	3.11	(3.06, 3.16)	1295.39	(1152.66, 1455.79)	1328.00	(1200.00, 1470.00)	303.14	270.00
24	2.94	(2.88, 3.01)	873.57	(753.53, 1012.74)	903.51	(793.15, 1029.00)	259.22	235.85
25	3.01	(2.93, 3.08)	1022.35	(859.81, 1215.91)	1038.00	(894.79, 1205.00)	356.10	310.21
26	2.84	(2.77, 2.90)	689.45	(594.70, 799.28)	706.34	(620.99, 803.43)	204.58	182.44
27	2.67	(2.59, 2.74)	464.94	(390.93, 552.97)	480.57	(412.29, 560.15)	162.04	147.86

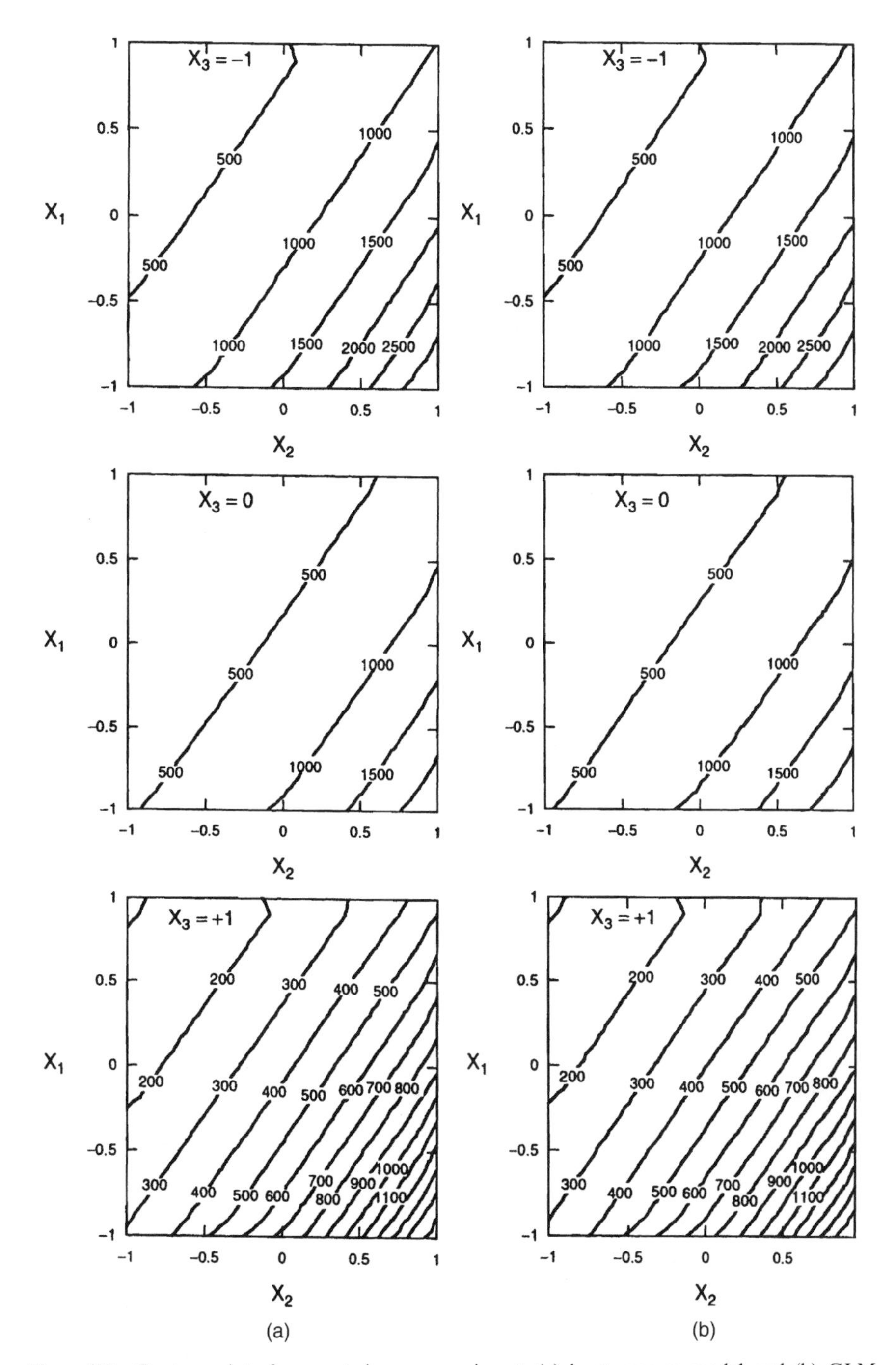

Figure 5.9 Contour plots for worsted yarn experiment: (a) least squares model and (b) GLM model.

Example 5.4. The Semiconductor Manufacturing Experiment. A spherical central composite design (CCD) was used to investigate the count of defects on wafers in a semiconductor manufacturing process. The design matrix and response data are shown in Table 5.9. Defects are nonnegative counts and are usually assumed to follow the Poisson distribution. To obtain the properties necessary to use ordinary least squares methods, the square root transformation would typically be used. This transformation is found to yield an adequate model in the variables, x_1, x_2, x_3, x_1x_2, and x_2^2 on the basis of t-tests on the individual variables, nonsignificance of the lack-of-fit test, and acceptable residual plots. For the GLM model the Poisson distribution with a log link is selected. The best GLM also uses the variables x_1, x_2, x_3, x_1x_2, and x_2^2. The ordinary least squares model fit with the square root transformation is

$$\hat{y} = (2.606 + 0.894x + 1.195x_2 - 0.569x_3 - 0.732x_1x_2 + 0.694x_2^2)^2$$

The GLM fit using a Poisson distribution and log link is

$$\hat{y} = e^{174.690+152.3302x_1+66.017x_2+49.607x_3+22.572x_1x_2+13.820x_2^2}$$

As before, the measure of performance used to compare the ordinary least squares model and the GLM is the length of the 95% confidence intervals around the estimate of the mean response for each observation. Table 5.10 presents the estimates of the means and the 95% confidence intervals for each

Table 5.9 Defects Data Experiment Design Matrix and Response Data

	x_1	x_2	x_3	Observational Value	Square Root of Observation
1	−1	−1	−1	1	1.000
2	1	−1	−1	15	3.873
3	−1	1	−1	24	4.899
4	1	1	−1	35	5.916
5	−1	−1	1	0	0.000
6	1	−1	1	13	3.606
7	−1	1	1	16	4.000
8	1	1	1	13	3.606
9	−1.732	0	0	1	1.000
10	1.732	0	0	17	4.123
11	0	−1.732	0	7	2.646
12	0	1.732	0	43	6.557
13	0	0	−1.732	14	3.742
14	0	0	1.732	3	1.732
15	0	0	0	4	2.000
16	0	0	0	7	2.646
17	0	0	0	8	2.828
18	0	0	0	6	2.450

Table 5.10 Comparison of 95% Confidence Intervals on the Means for Models Generated with Data Transformations and GLM for Defects Data Example

	Using Least Squares Methods with Log Data Transformation				General Linear Model (Gamma Regression with Log Link)		Length of 95% Confidence Interval	
	Transformed		Untransformed					
Observation	$\hat{y}_i$	95% Confidence Interval	$\hat{y}_i$	95% Confidence Interval	$\hat{y}_i$	95% Confidence Interval	OLS	GLM
1	1.05	(0.53, 1.56)	1.10	(0.29, 2.44)	2.00	(1.04, 3.86)	2.15	2.82
2	4.30	(3.79, 4.81)	18.50	(14.34, 23.18)	19.27	(13.67, 27.16)	8.84	13.49
3	4.90	(4.39, 5.42)	24.03	(19.26, 29.33)	23.19	(17.01, 31.60)	10.07	14.59
4	5.23	(4.71, 5.74)	27.31	(22.21, 32.95)	27.73	(20.92, 36.76)	10.74	15.84
5	−0.09	(−0.60, 0.42)	0.01	*	0.99	(0.50, 1.96)	*	1.47
6	3.16	(2.65, 3.68)	10.01	(7.02, 13.52)	9.51	(6.52, 13.86)	6.50	7.33
7	3.76	(3.25, 4.28)	14.17	(10.57, 18.30)	11.44	(8.01, 16.34)	7.74	8.33
8	4.09	(3.58, 4.60)	16.72	(12.78, 21.18)	13.69	(9.94, 18.45)	8.40	8.91
9	1.06	(0.61, 1.51)	1.12	(0.37, 2.29)	2.39	(1.49, 3.84)	1.92	2.35
10	4.15	(3.70, 4.61)	17.25	(13.69, 21.21)	19.79	(14.56, 26.90)	7.52	12.34
11	2.62	(1.99, 3.25)	6.85	(3.96, 10.53)	4.28	(2.18, 8.40)	6.58	6.23
12	6.75	(6.12, 7.38)	45.58	(37.49,54.46)	48.79	(37.66, 63.20)	16.97	25.54
13	3.59	(3.14, 4.04)	12.89	(9.84, 16.35)	12.67	(9.29, 17.29)	6.50	8.00
14	1.62	(1.17, 2.08)	2.63	(1.37, 4.31)	3.74	(2.49, 5.60)	2.94	3.11
15	2.61	(2.35, 2.86)	6.79	(5.54, 8.17)	6.88	(5.51, 8.60)	2.62	3.09
16	2.61	(2.35, 2.86)	6.79	(5.54, 8.17)	6.88	(5.51, 8.60)	2.62	3.09
17	2.61	(2.35, 2.86)	6.79	(5.54, 8.17)	6.88	(5.51, 8.60)	2.62	3.09
18	2.61	(2.35, 2.86)	6.79	(5.54, 8.17)	6.88	(5.51, 8.60)	2.62	3.09

Note: Indicates a negative predicted value. No inverse transformation is made.

observation for both models. The intervals are narrower for the ordinary least squares fit than their GLM counterparts for most portions of the design space. However, for the ordinary least squares fit, the fifth experimental run shows a negative estimate of the mean, which is clearly a nonsensical value. This is a run where variable x_3 is at the $+1$ level, and it is probably a region of the design space where it may be desirable to run the process, since it results in a low number of defects. Therefore the ordinary least squares model actually performs poorly in the region where it is likely to be most useful to the experimenters. Contour plots for this experiment are shown in Figure 5.10. Figure 5.10a shows the contour plots for the ordinary least squares model, and Figure 5.10b shows the contour plots for the GLM. However, the contour plot in the original units for the least squares model is not meaningful when $x_3 = +1$ because negative predicted values result. Only the GLM contours are useful in the region of likely interest. □

5.11 MODELING BOTH A PROCESS MEAN AND PROCESS VARIANCE USING GLM

An important industrial problem involves modeling both the process mean and the process variance. This problem lends itself to the use of generalized linear models rather nicely. This section discusses two different approaches: when there is true replication, and when there is no replication, in which case we use the residuals as the basis for modeling the variance.

5.11.1 The Replicated Case

True replication allows us to generate information on the variance independent of the model structure, since we can generate the sample variances, s_i^2, at the replicated points. The analysis still depends on the distribution of the data. However, if the data follow a normal distribution, then the sample mean and the sample variance at each of the replicated points are independent of one another. In addition

$$\frac{(n_i - 1)s_i^2}{\sigma_i^2} \sim \chi^2_{n_i - 1} \tag{5.15}$$

which is also a gamma distribution with $\lambda = 2$ and $r = \frac{1}{2}, 1, \frac{3}{2}, \ldots$. As a result we may use a GLM to generate a model for the variances. We then may use that model for the variances to generate the appropriate weights to perform generalized least squares to create an appropriate model for the process mean.

An example helps to illustrate this process. Vining and Myers (1990) use a dual response approach for modeling the process mean and process variance on a printing ink study from Box and Draper (1987). The purpose of the experiment is to study the effect of speed, x_1, pressure, x_2, and distance, x_3, on a printing machine's ability to apply coloring inks upon package labels.

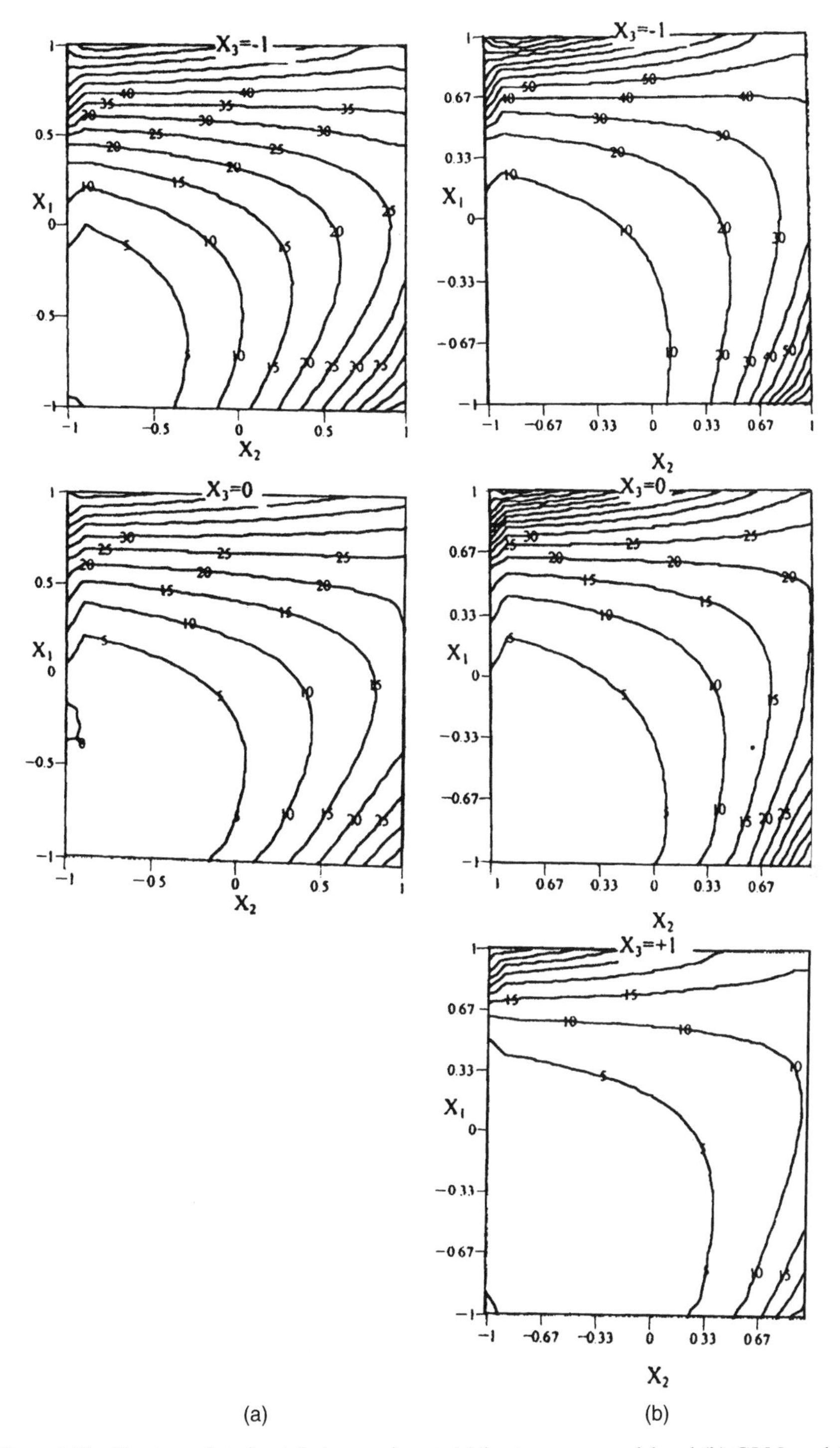

Figure 5.10 Contour plots for defects experiment: (a) least squares model and (b) GLM model.

Vining and Myers use ordinary least squares with a square root transformation on the sample variances (thus they analyze the sample standard deviations). Vining and Myers comment that limitations in the software available for the analysis dictated this approach. The data are in Table 5.11. Note that two of the sample standard deviations are 0. This can lead to complications, particularly with a log link. As a result we add one to each standard deviation when we fit the model.

We fit a quasi-likelihood model with a log link. We assume that the remaining variance is constant. Table 5.12 gives the summary of the fit obtained from S-Plus. The resulting prediction model is

$$\hat{s}^2 = \exp\{3.737 + 0.678x_3\}$$

Figures 5.11 through 5.13 give the residual plots, which appear satisfactory but not great. Historically, analysts have assumed that the process variance is

Table 5.11 The Printing Ink Experiment (3^3 Factorial)

x_1	x_2	x_3	$\bar{y}_i$	s_i
− 1.00	− 1.00	− 1.00	24	12.5
0.00	− 1.00	− 1.00	120.3	8.4
1.00	− 1.00	− 1.00	213.7	42.8
− 1.00	0.00	− 1.00	86	3.7
0.00	0.00	− 1.00	176.7	80.4
1.00	0.00	− 1.00	340.7	16.2
− 1.00	1.00	− 1.00	112.3	27.6
0.00	1.00	− 1.00	256.3	4.6
1.00	1.00	− 1.00	271.7	23.6
− 1.00	− 1.00	0.00	81	0
0.00	− 1.00	0.00	101.7	17.7
1.00	− 1.00	0.00	357	32.9
− 1.00	0.00	0.00	171.3	15
0.00	0.00	0.00	372	0
1.00	0.00	0.00	501.7	92.5
− 1.00	1.00	0.00	264	63.5
0.00	1.00	0.00	427	88.6
1.00	1.00	0.00	730.7	21.1
− 1.00	− 1.00	1.00	220.7	133.8
0.00	− 1.00	1.00	239.7	23.5
1.00	− 1.00	1.00	422	18.5
− 1.00	0.00	1.00	199	29.4
0.00	0.00	1.00	485.3	44.6
1.00	0.00	1.00	673.7	158.2
− 1.00	1.00	1.00	176.7	55.5
0.00	1.00	1.00	501	138.9
1.00	1.00	1.00	1010	142.4

Table 5.12 S-Plus Summary of the Quasi-likelihood Fit for the Process Variance Model

```
*** Generalized Linear Model ***

Call: glm(formula = smod ~ x3, family = quasi (link = log,
variance = "constant"), data = vining.myers,
na.action = na.exclude, control = list (epsilon
= 0.0001, maxit = 50, trace = F) )
Deviance Residuals:
      Min          1Q          Median       3Q         Max
  -63.20802   -26.09204     -8.076064   35.07394   76.49198

Coefficients:
                Value            Std. Error          t value
(Intercept) 3.737100           0.2346486           15.926366
         x3 0.678217           0.2615315            2.593252

(Dispersion Parameter for Quasi-likelihood family taken to be
1723.607)

        Null Deviance: 59790.8 on 26 degrees of freedom

    Residual Deviance: 43090.17 on 25 degrees of freedom

 Number of Fisher Scoring Iterations: 5
```

constant. As a result it is no surprise to see a reasonable but not great fit for the process variance model. There is sufficient evidence to suggest the use of a linear model, but it is not a strong linear relationship.

We next can take the estimated model for the process variance as the basis for generating the weights for a weighted least squares estimate of the model for the

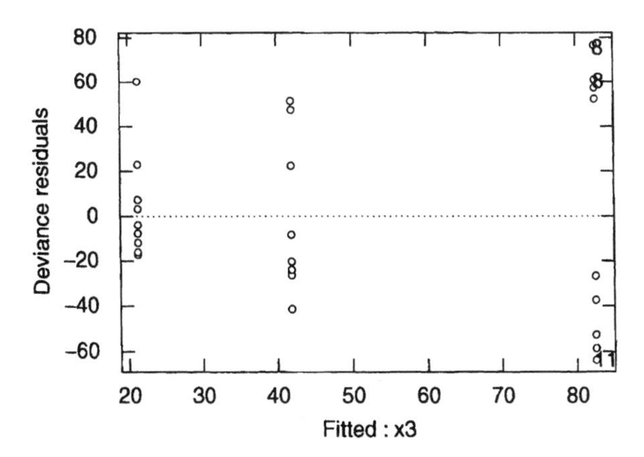

Figure 5.11 Deviance residuals versus fitted values for the process variance model.

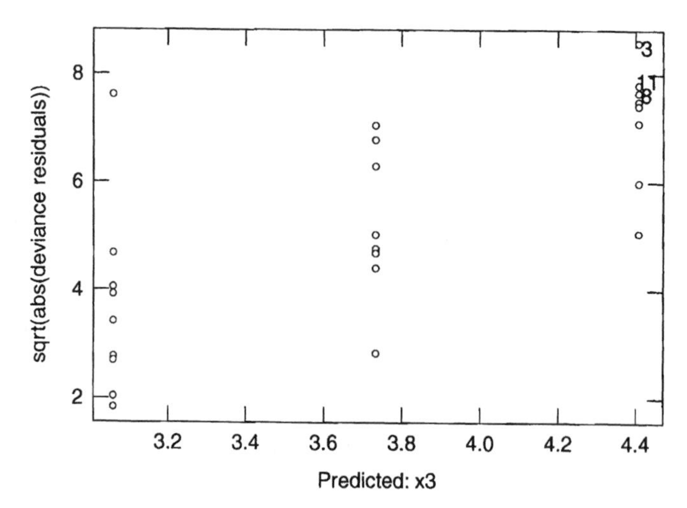

Figure 5.12 Plot of the signed square root of the deviance residuals versus the predicted values of the linear predictors for the process variance.

process mean. Table 5.13 summarizes the resulting output from MINITAB. Interestingly, the final model contains no pure quadratic terms. Instead, it contains all the main effects as well as all interactions, including the three-factor interaction. Figures 5.14 through 5.18 give the appropriate residual plots. Notice that the residual plots do not reveal anything alarming about the chosen model.

5.11.2 The Unreplicated Case

In the case of unreplicated experiments more complications arise in the case of the variance model as was pointed out in earlier discussions in this chapter. In fact, as

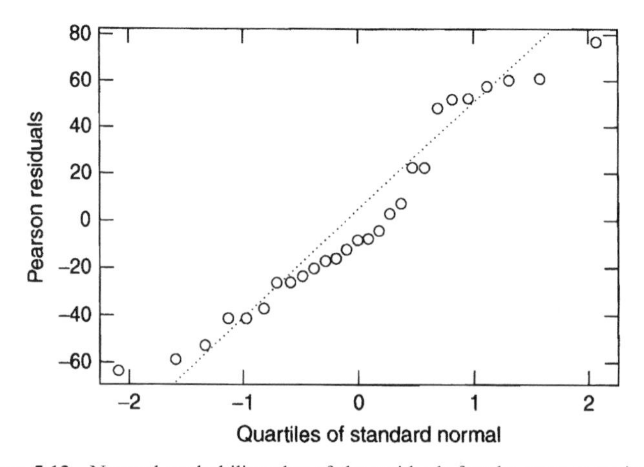

Figure 5.13 Normal probability plot of the residuals for the process variance.

Table 5.13 Weighted Least Squares Regression Output for the Printing Ink Data

```
Weighted analysis using weights in w

The regression equation is
y = 320 + 177 x1 + 119 x2 + 144 x3 + 61.2 x12 + 76.7 x13 + 68.7 x23 + 70.5
   x123
```

Predictor	Coef	StDev	T	P
Constant	319.76	14.50	22.05	0.000
x1	177.50	17.76	9.99	0.000
x2	119.30	17.76	6.72	0.000
x3	144.41	16.16	8.94	0.000
x12	61.21	21.75	2.81	0.011
x13	76.69	19.79	3.88	0.001
x23	68.68	19.79	3.47	0.003
x123	70.53	24.24	2.91	0.009

```
S = 1.453    R-Sq = 93.7%    R-Sq(adj) = 91.4%

Analysis of Variance
```

Source	DF	SS	MS	F	P
Regression	7	599.746	85.678	40.57	0.000
Residual Error	19	40.130	2.112		
Total	26	639.876			

Source	DF	Seq SS
x1	1	266.585
x2	1	87.891
x3	1	168.694
x12	1	1.536
x13	1	31.718
x23	1	25.437
x123	1	17.885

before, it should be noted that the quality of the variance model is very much dependent on a proper choice of the model for the mean since the source of variability is the *squared residual* $e_i = (y_i - \mathbf{x}_i'\mathbf{b})^2$ for the $i = 1, 2, \ldots, n$ design runs.

For the fundamental variance model we still deal with the log linear structure. In order to motivate the potential use of the GLM for this case, consider the unobserved conceptual model error $\varepsilon_i = y_i - \mathbf{x}_i'\boldsymbol{\beta}$. Let $\mathbf{u}_i$ be the ith setting for the regressors used in the variance model. As a starting point, we assume that $\mathbf{x}_i = \mathbf{u}_i$

$$\sigma_i^2 = E(\varepsilon_i^2) = e^{\mathbf{u}_i'\gamma}$$

where $\varepsilon_i^2 \sim \sigma_i{}^2\chi_1^2$ and $\boldsymbol{\gamma}$ is the vector of the variance model coefficients.

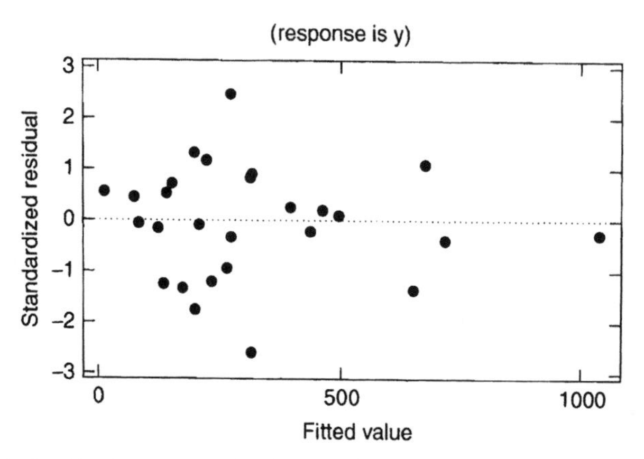

Figure 5.14 Residuals versus the fitted values for the printing ink study (response is y).

Maximum Likelihood Estimation of β and γ Using GLM

In the dual modeling procedure for unreplicated experiments, it is helpful to achieve efficiency of joint estimation of the mean model coefficients $\boldsymbol{\beta}$ and the variance model coefficients γ through maximum likelihood with GLM as the analytical tool. We can write the mean model as $\mathbf{y} = \mathbf{X}\boldsymbol{\beta} + \boldsymbol{\varepsilon}$ with $\text{Var}(\boldsymbol{\varepsilon}) = \mathbf{V}_{n \times n}$ and again $\sigma_i^2 = e^{\mathbf{u}_i'\gamma}$. Now, the MLE for $\boldsymbol{\beta}$ is merely weighted least squares,

$$\mathbf{b} = (\mathbf{X}'\mathbf{V}^{-1}\mathbf{X})^{-1}\mathbf{X}'\mathbf{V}^{-1}\mathbf{y} \tag{5.16}$$

If we use the random vector $\boldsymbol{\varepsilon}' = (\varepsilon_1^2, \varepsilon_2^2, \ldots \varepsilon_n^2)$ we have a set of independent χ_1^2 random variables that follow a gamma distribution with a scale parameter of 2. It

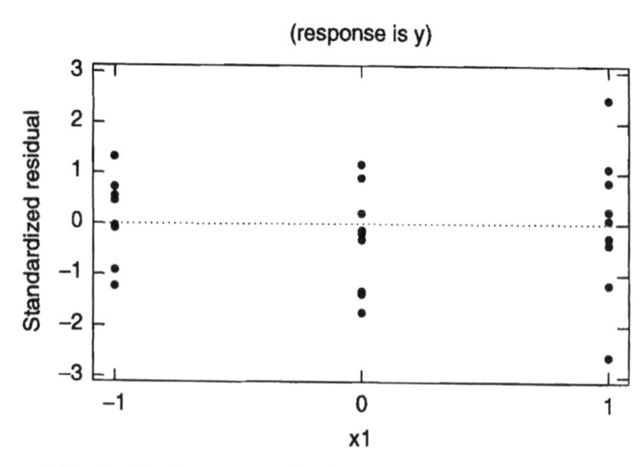

Figure 5.15 Residuals versus x_1 for the printing ink study (response is y).

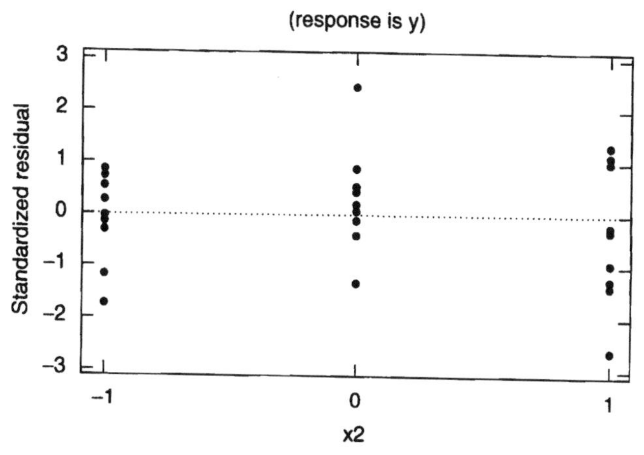

Figure 5.16 Residuals versus x_2 for the printing ink study (response is y).

is interesting, though not surprising, that if we consider the MLE for $\boldsymbol{\beta}$ and γ two separate algorithms emerge. However, the maximum likelihood estimator of $\boldsymbol{\beta}$ involves γ through the $\mathbf{V}$ matrix, and the MLE of γ clearly involves $\boldsymbol{\beta}$ since the *data* in the variance model, namely, the ε_i, involves $\boldsymbol{\beta}$. As a result, an iterative procedure is required. We do not write the likelihood function here but details appear in Aitkin (1987). Aitkin also points out the simplicity provided by the fact that computation of the MLE for $\boldsymbol{\beta}$ and γ can both be done with GLMs with an identity link and normal distribution on the mean model for estimation of $\boldsymbol{\beta}$ and log link with a gamma distribution on the variance model for estimation of γ. Indeed, these two procedures can be put together into an iterative procedure. The scale parameter for the gamma link is again equal to 2. We must use the **squared**

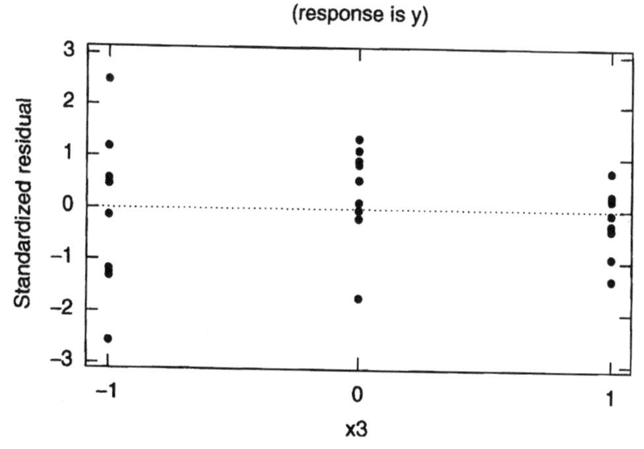

Figure 5.17 Residuals versus x_3 for the printing ink study (response is y).

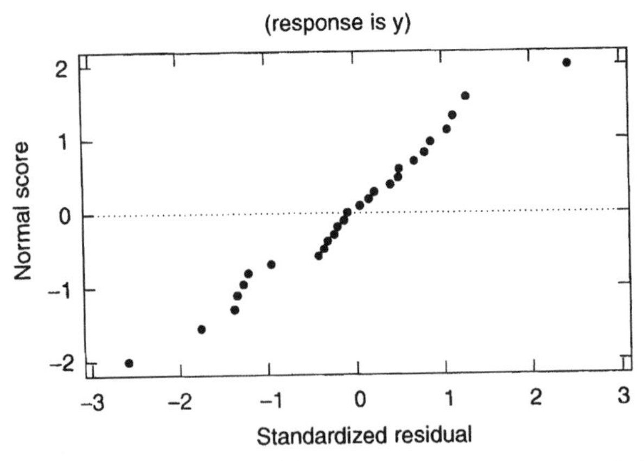

Figure 5.18 Normal probability plot of the residuals in the printing ink study (response is y).

residuals $e_i^2 = (y_i - \mathbf{x}_i'\mathbf{b})^2$ as the *data* for the gamma GLM and we use the y_i values as data for the normal GLM. The method follows:

1. Use OLS to obtain $\mathbf{b}_0$ for the mean model $y_i = \mathbf{x}'_i\boldsymbol{\beta} + \varepsilon_i$.
2. Use $\mathbf{b}_0$ to compute n residuals where $e_i = y_i - \mathbf{x}'_i\boldsymbol{\beta}_0$ for $i = 1, 2, \ldots, n$.
3. Use the e_i^2 as data to fit the variance model with regressors $\mathbf{u}$ and a log link with scale parameter 2. This is done via IRLS with GLM technology.
4. Use the parameter estimates from step 3, namely the $\hat{\gamma}_i$ to form the $\hat{\mathbf{V}}$ matrix.
5. Use $\hat{\mathbf{V}}$ with weighted least squares to update $\mathbf{b}_0$ to say, $\mathbf{b}_1$.
6. Go back to step 2 with $\mathbf{b}_1$ replacing $\mathbf{b}_0$.
7. Continue to convergence.

The complete algorithm for the model is not difficult to construct and is certainly a useful alternative for dual estimation of the mean and variance model. However, one disadvantage is that the estimation of the parameters in the variance model are biased because the MLE of γ does not account for estimation of $\boldsymbol{\beta}$ in formation of the residuals. A somewhat different procedure serves as an alternative. This procedure makes use of **restricted maximum likelihood (REML)**. The REML procedure does adjust nicely for the bias due to estimation of $\boldsymbol{\beta}$ in the squared residuals. In addition, this procedure can also be put into a GLM framework.

Restricted Maximum Likelihood for Dual Modeling

The source of the possible difficulties with ordinary MLE as discussed above is that the response data for the variance model, namely, the e_i^2 are not truly reflecting σ^2 unless the estimator $\mathbf{b}$ is the true parameter $\boldsymbol{\beta}$. As a result, one cannot expect the estimator of γ from the procedure to be unbiased. Of course, the amount of bias may be quite small. We deal with this subsequently. However, the IRLS procedure developed for REML is quite simple and should be considered by the practitioner. Consider now $E(e_i^2)$ for $i = 1,2,\ldots, n$. We know that for a weighted least squares procedure

$$\begin{aligned}
\text{Var}(\mathbf{e}) = \text{Var}\begin{bmatrix} e_1 \\ e_2 \\ \vdots \\ e_n \end{bmatrix} &= \text{Var}[\mathbf{I} - \mathbf{X}(\mathbf{X}'\mathbf{V}^{-1}\mathbf{X})^{-1}\mathbf{X}'\mathbf{V}^{-1}]\mathbf{y} \\
&= [\mathbf{I} - \mathbf{X}(\mathbf{X}'\mathbf{V}^{-1}\mathbf{X})^{-1}\mathbf{X}'\mathbf{V}^{-1}]\mathbf{V}[\mathbf{I} - \mathbf{X}(\mathbf{X}'\mathbf{V}^{-1}\mathbf{X})^{-1}\mathbf{X}'\mathbf{V}^{-1}] \\
&= \mathbf{V} - \mathbf{X}(\mathbf{X}'\mathbf{V}^{-1}\mathbf{X})^{-1}\mathbf{X}' \qquad (5.17)
\end{aligned}$$

Since $E(e_i) = 0$, then $E(e_i^2) = \text{Var}(e_i)$. The matrix $\mathbf{X}(\mathbf{X}'\mathbf{V}^{-1}\mathbf{X})^{-1}\mathbf{X}'$ may seem like a hat matrix as defined in ordinary multiple regression. However, the matrix that is most like a **hat matrix** for weighted regression is

$$\mathbf{H} = \mathbf{V}^{-1/2}\mathbf{X}(\mathbf{X}'\mathbf{V}^{-1}\mathbf{X})^{-1}\mathbf{X}'\mathbf{V}^{-1/2} \qquad (5.18)$$

The matrix $\mathbf{H}$ is idempotent and plays the same role as $\mathbf{X}(\mathbf{X}'\mathbf{X})^{-1}\mathbf{X}'$ does for standard nonweighted regression. Now, $\mathbf{H}$ has the same diagonal elements as $\mathbf{X}(\mathbf{X}'\mathbf{V}^{-1}\mathbf{X})^{-1}\mathbf{X}'\mathbf{V}^{-1}$. Thus the diagonal elements in $\mathbf{X}(\mathbf{X}\mathbf{V}'\mathbf{V}^{-1}\mathbf{X})^{-1}\mathbf{X}'$ are the diagonals of $\mathbf{HV}$. As a result, if we consider only the diagonal element involvement in Var($\mathbf{e}$),

$$E(e_i^2) = \sigma_i^2 - h_{ii}\,\sigma_i^2 \qquad (5.19)$$

where h_{ii} is the ith hat diagonal, that is, the ith diagonal of $\mathbf{H}$. Thus the adjustment on the response is to use $e_i^2 + h_{ii}\hat{\sigma}_i^2$ rather than e_i^2. Here $h_{ii}\,\sigma_i^2$ is a **bias correction**. The estimation of $\boldsymbol{\beta}$ and $\boldsymbol{\gamma}$ are interwoven and thus an iterative scheme is used, once again with the aid of GLM. The iterative procedure is:

1. Calculate $\mathbf{b}_0$ from OLS.
2. Calculate the e_i and a set of responses $z_{0i} = e_i^2 + h_{ii}s^2$, where the h_{ii} are the OLS hat diagonals from $\mathbf{X}(\mathbf{X}'\mathbf{X})^{-1}\mathbf{X}'$, and s^2 is the error mean square.
3. Use the $\mathbf{z}_{0i}$, ($i = 1, 2,\ldots, n$) as responses in a GLM with gamma distribution and log link to calculate the initial $\hat{\boldsymbol{\gamma}}$, say, $\hat{\boldsymbol{\gamma}}_0$.
4. Use the $\hat{\boldsymbol{\gamma}}_0$ to produce weights from $\hat{\sigma}_i^2 = e^{\mathbf{u}_i'\hat{\boldsymbol{\gamma}}_0}$ and calculate the weight matrix $\hat{\mathbf{V}}_0$. Calculate the estimate of $\boldsymbol{\beta}$ as $\mathbf{b}_1 = (\mathbf{X}'\hat{\mathbf{V}}_0^{-1}\mathbf{X})^{-1}\mathbf{X}'\hat{\mathbf{V}}_0^{-1}\mathbf{y}$.
5. Calculate new residuals from $\mathbf{b}_1$, and that new $z_i = e_i^2 + h_{ii}\,\hat{\sigma}_i^2$, where h_{ii} is the ith diagonal from the hat matrix. Here the $\hat{\mathbf{V}}$ matrix is diagonal with the current values $\hat{\sigma}_i^2$ on the main diagonal.

6. Use the new z_i as responses with a gamma GLM and log link to calculate a new $\hat{\gamma}$. The procedure continues the methodology of using variance *fitted values* for weights in the development of the mean model. The mean model continues to provide the residuals from which the z_i are developed. Upon convergence, both mean and variance models emerge.

The reader should note the close resemblance between the iterative procedure here and the MLE in the previous section. The structure is the same except that at each iteration the gamma GLM is conducted with an **adjusted response**, the adjustment coming via $h_{ii} \cdot \hat{\sigma}_i^2$, where $\hat{\sigma}_i{}^2$ is the estimator of $\sigma_i{}^2$ from the previous iteration. The procedure does maximize a **restricted log-likelihood** which is not shown here. For more details the reader should see Engel and Huele (1996). Nelder and Lee (1991), and Lee and Nelder (1998a, b). For more information on the computational details, consult Harvey (1976).

Obviously, the correction for bias introduced through REML may be very helpful. However, for large experiments in which $n \gg p$, where p is the number of parameters, the bias adjustment often is negligible.

5.12 QUALITY OF ASYMPTOTIC RESULTS AND RELATED ISSUES

Many of the inference procedures associated with the GLM make use of **asymptotic results**. We believe that the use of asymptotic results are satisfactory in most real-life situations in which GLMs are used. Obviously, it is not possible to study all distributions, links, sample sizes, and design scenarios, and each of these, even the design, plays a role. However, our experience based simulation studies, most of which focused on the use of two-level factorial designs, indicates that the asymptotic results seem to hold well in practical situations. Some of these results are shown in this section. The studies deal generally with the nature of accuracy of asymptotic variance–covariance matrices of parameter estimates and the accuracy of coverage probabilities on confidence intervals on the mean response.

In terms of accuracy of variances of parameter estimates, our experience suggests that for 8 run designs, the actual or **small-sample variance** may be, on the average, roughly 5–10% higher than the computed asymptotic variance. This, of course, implies that P-value calculations may be slightly underestimated. However, for 16 run designs, the practitioner can feel confident about the use of the computed asymptotic variance. Our simulation results consistently revealed values that are well within the simulation error.

5.12.1 Development of an Alternative Wald Confidence Interval

We first investigate the coverage properties of Wald-type confidence intervals on the mean response in situations involving designed experiments with 8, 16, and 32 runs. We use the Wald inference confidence intervals from SAS PROC

GENMOD. Recall that the confidence interval on the mean response at the point $\mathbf{x}'_0 = [1, x_{10}, x_{20}, \ldots, x_{k0}]$ is

$$g^{-1}\left[\mathbf{x}_0'\mathbf{b} \pm z_{\alpha/2}\sqrt{\mathbf{x}_0^1(\mathbf{D}'\hat{\mathbf{V}}^{-1}\mathbf{D})^{-1}\mathbf{x}_0}\right]$$

where g is the link function.

There is another way to develop a Wald confidence interval on the mean response in the GLM. Recall that in normal-theory linear regression, a confidence interval on $E(y|\mathbf{x} = \mathbf{x}_0)$ for a model with p parameters is

$$\hat{y}(\mathbf{x}_0) \pm t_{\alpha/2,\ n-p}\ s\sqrt{\mathbf{x}_0'(\mathbf{X}'\mathbf{X})^{-1}\mathbf{x}_0} \tag{5.20}$$

Myers and Montgomery (1997) present an analogous expression for generalized linear models. The asymptotic $100(1-\alpha)\%$ confidence interval on the mean response at the point $\mathbf{x}_0$ is

$$\hat{\mu}(\mathbf{x}_0) \pm z_{\alpha/2}\sqrt{\mathbf{d}_0'(\mathbf{D}'\hat{\mathbf{V}}^{-1}\mathbf{D})^{-1}\mathbf{d}_0} \tag{5.21}$$

where $\mathbf{D}$ is the matrix of derivatives induced by the Taylor series expansion used in the GLM, $\mathbf{d}_0$ is the vector of these derivatives evaluated at the point of interest $\mathbf{x}_0, \hat{\mathbf{V}} = \text{diag}\{\widehat{\text{Var}}(y_i)\}$ is a diagonal matrix whose main diagonal elements are the estimated variances of the response variable, and $\hat{\mu}(\mathbf{x}_0)$ is the estimate of the mean response at the point of interest $\mathbf{x}_0$. The variance of the response is a function of the mean through the relationships established for the exponential family. The scale parameter $a(\phi)$ in the exponential family is incorporated in the variance of the response y_i and hence in $\mathbf{V}$.

The development of Equation (5.21) is fairly straightforward. From McCullagh and Nelder (1989, p. 28) the form of the log-likelihood of y for the exponential family is $L(\theta) = \{y\theta - b(\theta)\}/a(\phi) + c(y;\ \phi)$. In addition $\mu = b'(\theta)$ and $\text{Var}(y) = b''(\theta)a(\phi)$. For a link, say, $g(\mu) = \mathbf{x}'\boldsymbol{\beta}$, we have $\mu = g^{-1}(\mathbf{x}'\boldsymbol{\beta})$ and $\hat{\mu} = g^{-1}(\mathbf{x}'\mathbf{b})$, where $\mathbf{b}$ is the maximum likelihood estimator. In general, μ is a nonlinear function of $\boldsymbol{\beta}$. Important results stem from the asymptotic covariance matrix of $\mathbf{b}$. The information matrix $\mathbf{I}(\mathbf{b})$ can be written several ways. One useful form is

$$\mathbf{I}(\mathbf{b}) = \mathbf{D}'\mathbf{V}^{-1}\mathbf{D}$$

where $\mathbf{D}$ is a matrix whose ith row is $(\partial\mu_i/\partial\boldsymbol{\beta})'$, which is the derivative of the mean function with respect to the parameters. The matrix $\mathbf{V} = \text{diag}\{\text{Var}(y_i)\}$. We know from characteristics of the exponential family that

$$\frac{\partial\mu}{\partial\boldsymbol{\beta}} = \frac{\partial\mu}{\partial\theta}\frac{\partial\theta}{\partial\mathbf{x}'\boldsymbol{\beta}}\mathbf{x}$$

which produces $\mathbf{D} = \mathbf{V\Delta X}/a(\phi)$. Thus $\mathbf{D'V}^{-1}\mathbf{D} = \mathbf{X'\Delta V\Delta X}/[a(\phi)]^2 = \mathbf{X'WX}/[a(\phi)]^2$. The matrix $\mathbf{W}$ is the Hessian weight matrix that has been described throughout the text. Here the matrix $\mathbf{\Delta} = \text{diag}\{\partial\theta_i/\partial\mathbf{x}'_i\boldsymbol{\beta}\}$. For the canonical link, $\theta = \mathbf{x}'\boldsymbol{\beta}$ and

$$\mathbf{I}(\mathbf{b}) = \frac{\mathbf{X'VX}}{[a(\phi)]^2} \tag{5.22}$$

We must keep in mind that $a(\phi)$ is also involved in $\mathbf{V}$ through the expression $\text{Var}(y) = b''(\theta)a(\phi)$.

For the confidence interval on $\mu(\mathbf{x}_0)$, we must use the delta method to approximate $\text{Var}[\hat{\mu}(\mathbf{x}_0)]$. The delta method allows for approximation of the variance of a quantity that is a nonlinear function of random variables whose variances are known. In this context, $\hat{\mu}(\mathbf{x}_0)$ is a nonlinear function of parameter estimates in $\mathbf{b}$. The application of the delta method results in

$$\text{Var}[\hat{\mu}(\mathbf{x}_0)] = \mathbf{d}'_0\ [\text{Var}(\mathbf{b})]\ \mathbf{d}_0$$

where $\mathbf{d}_0$ is a vector of derivatives of $\mu(\mathbf{x}_0)$ with respect to $\mathbf{b}$; that is, $\mathbf{d}_0 = \partial\hat{\mu}(\mathbf{x}_0)/\partial\mathbf{b}$. Now Var[(**b**)] is the asymptotic variance–covariance matrix of $\mathbf{b}$, which is given by $\mathbf{I}(\mathbf{b})^{-1} = (\mathbf{D'V}^{-1}\mathbf{D})^{-1}$. As a result $\text{Var}[\hat{\mu}(\mathbf{x}_0)]$ is approximated by $\mathbf{d}'_0(\mathbf{D'V}^{-1}\mathbf{D})^{-1}\mathbf{d}_0$. We then have that the asymptotic distribution of

$$\frac{\hat{\mu}(\mathbf{x}_0) - \mu(\mathbf{x}_0)}{\sqrt{\mathbf{d}'_0(\mathbf{D'V}^{-1}\mathbf{D})^{-1}\mathbf{d}_0}}$$

is $N(0, 1)$, and thus the approximate $100(1 - \alpha)\%$ confidence interval on $\mu(\mathbf{x}_0)$ is

$$\hat{\mu}(\mathbf{x}_0) \pm z_{\alpha/2}\sqrt{\mathbf{d}'_0(\mathbf{D}'\hat{\mathbf{V}}^{-1}\mathbf{D})^{-1}\mathbf{d}_0}$$

This is a useful general form. Let us consider some special cases. Through the chain rule and relationships in the exponential family, $\partial\mu/\partial\boldsymbol{\beta} = (\partial\mu/\partial\theta)[\partial\theta/\partial(\mathbf{x}'\boldsymbol{\beta})][\partial(\mathbf{x}'\boldsymbol{\beta})/\partial\boldsymbol{\beta}] = \text{Var}(y)\mathbf{\Delta x}/a(\phi)$. So for a canonical link, $\delta = 1$ and $\mathbf{D} = [\mathbf{VX}/a(\phi)](\mathbf{D'V}^{-1}\mathbf{D})^{-1} = (\mathbf{X'VX})^{-1}[a(\phi)]^2$, $\mathbf{d}_0 = [\text{Var}(y_0)]\mathbf{x}_0/a(\phi)$, and thus we have as the desired confidence interval

$$\hat{\mu}(\mathbf{x}_0) \pm z_{a/2}\widehat{\text{Var}(y_0)}\sqrt{\mathbf{x}'_0(\mathbf{X'VX})^{-1}\mathbf{x}_0}$$

Note that even though $a(\phi)$ cancels, it still appears in the variance of the response and in the matrix $\mathbf{V}$. In the normal case, since $\mathbf{V} = \sigma^2\mathbf{I}$ and $\text{Var}(y) = \sigma^2$, we have

$$\mu(\mathbf{x}_0) \pm z_{a/2}\sigma\sqrt{\mathbf{x}'_0(\mathbf{X'X})^{-1}\mathbf{x}_0}$$

Table 5.14 Wald Confidence Intervals on the Mean Response for Canonical Link Models for Various Exponential Family Members

Distribution	Link	Model	Confidence Intervals
Normal	Identity	$\mu_i = \mathbf{x}'_i\boldsymbol{\beta}$	$\hat{y}(\mathbf{x}_i) \pm t_{a/2,\,n-p}\, s\sqrt{\mathbf{x}_i'(\mathbf{X}'\mathbf{X})^{-1}\mathbf{x}_i}$
Binomial	Logit	$\mu_i = 1/(1 + \mathrm{e}^{-\mathbf{x}'_i\boldsymbol{\beta}})$	$\left(\frac{1}{1+e^{-\mathbf{x}_i'\mathbf{b}}}\right) \pm z_{a/2}\left(\frac{1}{1+e^{-\mathbf{x}_i'\mathbf{b}}}\right)\left(1-\left(\frac{1}{1+e^{-\mathbf{x}_i'\mathbf{b}}}\right)\right)\sqrt{\mathbf{x}_i'(\mathbf{X}'\hat{\mathbf{V}}\mathbf{X})^{-1}\mathbf{x}_i}$
Gamma	Inverse	$\mu_i = 1/\mathbf{x}'_i\boldsymbol{\beta}$	$\left(\frac{1}{\mathbf{x}_i'\mathbf{b}}\right) \pm z_{a/2}\left(\frac{1}{\mathbf{x}_i'\mathbf{b}}\right)^2 \upsilon\sqrt{\mathbf{x}_i'(\mathbf{X}'\hat{\mathbf{V}}\mathbf{X})^{-1}\mathbf{x}_i}$
Poisson	Log	$\mu_i = e_i^{\mathbf{x}'\boldsymbol{\beta}_i}$	$e^{\mathbf{x}_i'\mathbf{b}} \pm z_{a/2}e^{(\mathbf{x}_i'\mathbf{b})}\sqrt{\mathbf{x}_i'(\mathbf{X}'\hat{\mathbf{V}}\mathbf{X})^{-1}\mathbf{x}_i}$

Note: υ is the gamma scale parameter.

Replacing σ by s and $z_{\alpha/2}$ by $t_{\alpha/2}$ gives the familiar expression shown in Equation (5.20). We also can obtain the same results if we use $\mathbf{I}^{-1}(\mathbf{b})$ and Equation (5.22). Tables 5.14 and 5.15 summarize several of these Wald confidence intervals for binomial, Poisson, and gamma responses for both canonical and noncanonical links.

Lewis, Montgomery, and Myers (2001b) evaluate the coverage and precision (i.e., the length) of confidence intervals on the mean response in the GLM using Monte Carlo simulation. They study two-level factorial designs with $n = 8$, 16, and 32 runs fitted with the GLM. For each experimental design situation, a model is built with known parameters and a known true mean for each experimental run. That is, a model $\mu_i = g^{-1}(\mathbf{x}'\boldsymbol{\beta})$ is generated with known parameters $\boldsymbol{\beta}$, a design matrix $\mathbf{X}$, and therefore a known mean μ_i, for each experimental run. The actual observation is obtained by adding an error drawn at random from a specified distribution to the linear predictor, namely, $y_i = g^{-1}(\mathbf{x}_i'\boldsymbol{\beta}) + \varepsilon_i$. A GLM is fit to the data using the SAS GENMOD procedure and the 95% confidence intervals examined. Each treatment combination is simulated 5000 times. Coverage and precision are calculated over all the observational points. Therefore, for the 8 run designs, a total of 40,000 (i.e., 5000 $\times$ 8) observational points are used in the coverage calculation; 80,000

Table 5.15 Wald Confidence Intervals on the Mean Response for some Noncanonical Link Models for Various Exponential Family Members

Distribution	Link	Model	Confidence Intervals
Binomial	Identity	$\mu = \mathbf{x}'_i\boldsymbol{\beta}$	$(\mathbf{x}_i'\mathbf{b}) \pm z_{a/2}\sqrt{\mathbf{x}_i'(\mathbf{X}'\Delta\hat{\mathbf{V}}\Delta\mathbf{X})^{-1}\mathbf{x}_i}$
Gamma	Log	$\mu = e^{\mathbf{x}'_i\boldsymbol{\beta}}$	$e^{\mathbf{x}_i'\mathbf{b}} \pm z_{a/2}e^{\mathbf{x}_i'\mathbf{b}}\upsilon\sqrt{\mathbf{x}_i'(\mathbf{X}'\mathbf{X})^{-1}\mathbf{x}_i}$
Poisson	Identity	$\mu = \mathbf{x}'_i\boldsymbol{\beta}$	$(\mathbf{x}_i'\mathbf{b}) \pm z_{a/2}\sqrt{\mathbf{x}_i'(\mathbf{X}'\Delta\hat{\mathbf{V}}\Delta\mathbf{X})^{-1}\mathbf{x}_i}$

Note: υ is the gamma scale parameter.

Table 5.16 Scenarios Evaluated in Monte Carlo Simulation

Distribution	Link	Linear Predictor
Binomial	Logit (canonical)	$3 + 2x_1 + x_2$
Binomial	Identity (noncanonical)	$10 + 5x_1 + 3x_2$
Gamma	Inverse (canonical)	$100 + 50x_1 + 30x_2$
Gamma	Log (noncanonical)	$10 + 5x_1 + 3x_2$
Poisson	Log (canonical)	$10 + 5x_1 + 3x_2$
Poisson	Identity (noncanonical)	$100 + 50x_1 + 30x_2$

(i.e., 5000 × 16) observational points for the 16 run designs; and 160,000 (i.e., 5000 × 32) observational points for the 32 run designs. The specific scenarios selected for the analysis of confidence interval coverage and precision studies are shown in Table 5.16

In each case, two design factors, x_1, and x_2, are included. A 2^2 factorial design is used to generate the **X** matrix for each specific scenario. That is, two replicates of the 2^2 factorial are used to generate the 8 experimental run design matrix. Four replicates of the 2^2 give the 16 experimental run design matrix, and eight replicates of the 2^2 give the 32 experimental run design matrix.

Illustration of Binomial Distribution with Logit (Canonical) Link

To illustrate, the true model used in the Monte Carlo simulation for the binomial response distribution with logit link situation is $\mu_i = g^{-1}(3 + 2x_{i1} + x_{i2})$. Therefore the model parameters are $\beta_0 = 3$, $\beta_1 = 2$, *and* $\beta_2 = 1$. A 2^2 factorial is used and the response is modeled as a full factorial with 2 replicates for the 8 run design, 4 replicates for the 16 run design, and 8 replicates for the 32 run design. As discussed previously, the binomial distribution is used to produce an observation y_i for each experimental run. This process is repeated for 5000 iterations. Results are collected and summarized across experimental runs and over all the 5000 iterations.

Tables 5.17, 5.18, and 5.19 give the Monte Carlo results for the binomial distribution with the logit link for $n = 8$, 16, and 32, respectively. Each table shows the results for both the SAS approximate 95% confidence intervals and the 95% Wald confidence intervals. The columns labeled "Coverage" show the number of times the 95% confidence interval contained the true mean out of the 5000 trials. For example, for trial 1, 4752 of the 5000 simulation runs resulted in a confidence interval containing the true mean. The columns labeled "Precision" show the average length of the confidence interval obtained for each of the experimental trials. For example, for trial 1, the precision is 0.0684. As a point of reference for the precision value, the true mean for trial 1 is 0.

Monte Carlo Simulation Confidence Interval Coverage and Precision Results

Table 5.20 and 5.21 give a summary of the Monte Carlo simulations for all of the models in Table 5.17. The coverage for the SAS confidence intervals are in Table 5.20 and the coverage for the Wald confidence intervals are in Table 5.21.

Table 5.17 Confidence Interval Results for Binomial Distribution with Logit (Canonical) Link for 2^2 Factorial with 8 Runs

	SAS PROC GENMOD CI		Wald CI	
Trial	Coverage	Precision	Coverage	Precision
1	4752	0.0684	4742	0.0688
2	4812	0.0203	4770	0.0177
3	4767	0.0447	4724	0.0442
4	4822	0.0033	4662	0.0028
5	4752	0.0684	4742	0.0688
6	4812	0.0203	4770	0.0177
7	4767	0.0447	4724	0.0442
8	4822	0.0033	4662	0.0028
Total	38306		37796	
Coverage %	95.77%		94.49%	

Note: Target confidence level is 95%.

Table 5.18 Confidence Interval Results for Binomial Distribution with Logit (Canonical) Link for 2^2 Factorial with 16 Runs

	SAS PROC GENMOD CI		Wald CI	
Trial	Coverage	Precision	Coverage	Precision
1	4762	0.0154	4761	0.0154
2	4761	0.0039	4764	0.0039
3	4749	0.0099	4753	0.0099
4	4751	0.0006	4731	0.0006
5	4762	0.0154	4761	0.0154
6	4761	0.0039	4764	0.0039
7	4749	0.0099	4753	0.0099
8	4751	0.0006	4731	0.0006
9	4762	0.0154	4761	0.0154
10	4761	0.0039	4764	0.0039
11	4749	0.0099	4753	0.0099
12	4751	0.0006	4731	0.0006
13	4762	0.0154	4761	0.0154
14	4761	0.0039	4764	0.0039
15	4749	0.0099	4753	0.0099
16	4751	0.0006	4731	0.0006
Total	76092		76036	
Coverage %	95.12%		95.05%	

Note: Target confidence level is 95%.

Inspection of Tables 5.20 and 5.21 shows that the SAS approximate confidence intervals calculated using PROC GENMOD closely match the Wald confidence intervals for the binomial and Poisson distributions. As important, the results also show that the coverage observed in the Monte

Table 5.19 Confidence Interval Results for Binomial Distribution with Logit (Canonical) Link for 2^2 Factorial with 32 Runs

	SAS PROC GENMOD CI		Wald CI	
Trial	Coverage	Precision	Coverage	Precision
1	4694	0.0333	4691	0.0334
2	4622	0.0090	4741	0.0087
3	4702	0.0215	4687	0.0215
4	4644	0.0014	4759	0.0014
5	4694	0.0333	4691	0.0334
6	4622	0.0090	4741	0.0087
7	4702	0.0215	4687	0.0215
8	4644	0.0014	4759	0.0014
9	4694	0.0333	4691	0.0334
10	4622	0.0090	4741	0.0087
11	4702	0.0215	4687	0.0215
12	4644	0.0014	4759	0.0014
13	4694	0.0333	4691	0.0334
14	4622	0.0090	4741	0.0087
15	4702	0.0215	4687	0.0215
16	4644	0.0014	4759	0.0014
17	4694	0.0333	4691	0.0334
18	4622	0.0090	4741	0.0087
19	4702	0.0215	4687	0.0215
20	4644	0.0014	4759	0.0014
21	4694	0.0333	4691	0.0334
22	4622	0.0090	4741	0.0087
23	4702	0.0215	4687	0.0215
24	4644	0.0014	4759	0.0014
25	4694	0.0333	4691	0.0334
26	4622	0.0090	4741	0.0087
27	4702	0.0215	4687	0.0215
28	4644	0.0014	4759	0.0014
29	4694	0.0333	4691	0.0334
30	4622	0.0090	4741	0.0087
31	4702	0.0215	4687	0.0215
32	4644	0.0014	4759	0.0014
Total	149296		151024	
Coverage %	93.31%		94.39%	

Note: Target confidence level is 95%.

Carlo simulation for small samples is very close to the asymptotic theoretical claim of 95%. In particular, for the binomial and Poisson distributions, with as few as 8 runs, the coverage is very close to 95%, which is a crucial finding. It implies that confidence interval lengths are an appropriate way to evaluate the predictive performance of models built from designed experiments using the GLM and that they compare well with other approaches.

Table 5.20 Summary of SAS Approximate Confidence Intervals

	Binomial		Poisson		Gamma	
Runs	Canonical Logit Link	Noncanonical Identity Link	Canonical Log Link	Noncanonical Identity Link	Canonical Inverse Link	Noncanonical Log Link
8	95.77	95.48	95.11	94.92	86.39	89.77
16	95.12	93.98	94.93	95.02	93.31	93.16
32	93.31	94.51	95.28	95.20	95.41	95.06

Note: Target confidence level is 95%.

Table 5.21 Summary of Wald Confidence Intervals

Runs	Binomial		Poisson		Gamma	
	Canonical Logit Link	Noncanonical Identity Link	Canonical Log Link	Noncanonical Identity Link	Canonical Inverse Link	Noncanonical Log Link
8	94.49	95.48	95.07	94.92	79.94	85.27
16	95.05	93.98	94.94	95.02	89.65	89.35
32	94.39	94.52	95.25	95.20	92.99	93.73

Note: Target confidence level 95%.

We do not display the precision results; however, in all instances, as the number of experimental runs increases, the precision of estimation improves. That is, the interval tightens around the correct mean value. (See Lewis, 1998, for detailed precision results.)

The coverage results for the gamma distribution are somewhat less impressive for the 8 run designs, but quickly approach 95% as the number of experimental runs increases to 16 and then to 32. One possible explanation for the less impressive results observed with the gamma distribution involves the scale parameter υ. This parameter is unknown for the gamma distribution and must be estimated. There are several methods for estimating υ, and they do affect the coverage results. The next section examines this issue further.

5.12.2 Estimation of Exponential Family Scale Parameter

Myers and Montgomery (1997) and McCullagh and Nelder (1989), among others, suggest several methods for estimating the scale parameter. The scale parameter can be estimated based on the deviance or Pearson's χ^2, or it can be estimated directly using maximum likelihood. To investigate the behavior of the various methods for estimating the exponential family scale parameter υ, Myers and Montgomery (1997) run a Monte Carlo simulation model for the gamma distribution using each of the three estimation methods. They consider both the inverse link (canonical) and the log link (noncanonical). Table 5.22 gives the coverage results. Since precision, as well as coverage, is important, precision is examined through plots showing the average length of the confidence interval over the 5000 simulation runs at the design points. Figure 5.19 is a plot of average confidence interval endpoints for the particular trial in which the true mean is 7.39.

Inspection of Table 5.22 shows that the deviance-based and Pearson's χ^2-based estimation methods give similar results for confidence interval coverage. The maximum likelihood based estimator exhibits a slightly reduced coverage in comparison to the other estimators. However, examination of Figure 5.19 shows that the maximum likelihood estimator yields **shorter** intervals; that is, a more precise estimate is obtained. Similarly, Pearson's χ^2 estimator appears slightly more precise than the deviance-based estimator. Both confidence interval coverage and precision are important in evaluating model performance.

Table 5.22 Coverage Results for Gamma Distribution with Inverse (Canonical) and Identity (Noncanonical) Links for 2^2 Factorial with 8 Runs Using Various Methods of Estimating the Scale Parameter

	Deviance	Pearson's χ^2	Maximum Likelihood
Canonical Link	86.39%	86.40%	82.34%
Noncanonical Link	89.77%	87.60%	82.41%

Note: Target confidence level is 95%.

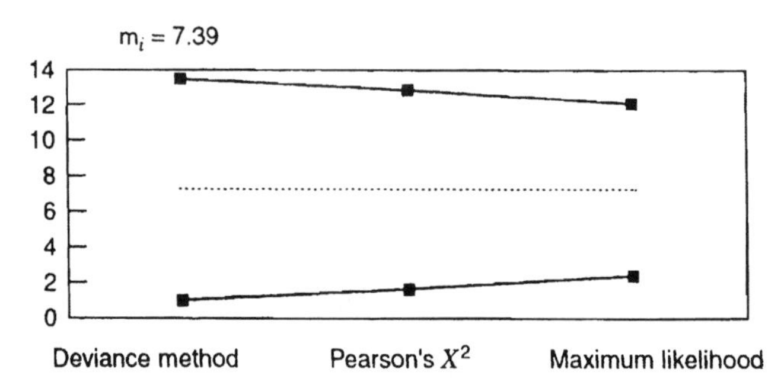

Figure 5.19 Average confidence interval endpoints for various scale parameter estimators for Monte Carlo simulations of gamma distribution with log link.

5.12.3 Impact of Link Misspecification on Confidence Interval Coverage and Precision

As mentioned earlier, one of the three components in any generalized linear model is the choice of link function. In some situations the canonical link may be chosen by default, possibly for ease of interpretation. In other situations the appropriate link may not be obvious. Therefore it is useful to consider the impact of misspecifying the link function on the confidence interval coverage and precision. In a sense, misspecification of the link is tantamount to model misspecification. Three situations involving misspecified links (see Table 5.23) are investigated for the $n = 8$, 16, and 32 run designs.

5.12.4 Illustration of Binomial Distribution with a True Identity Link but with Logit Link Assumed

One common situation is a designed experiment where the response is a binomial variable. Often binomial data are modeled using a logit transformation. Therefore an experimenter can easily elect to use the canonical logit link in the model-building process. However, suppose that the true link is well approximated by an identity link. That is, the binomial parameter is very nearly linear in the design variables, but the experimenter chooses a logit link. To investigate the impact on model performance, the confidence interval coverage and precision are examined.

To simulate this situation, a 2^2 factorial design is selected, and the true model is generated with a set of known parameters, and therefore a known

Table 5.23 Modeling Situations Used to Evaluate Impact of Link Misspecification

Situation	Distribution	True Link	Assumed Link
1	Binomial	Identity (Noncanonical)	Logit (Canonical)
2	Poisson	Identity (Noncanonical)	Log (Canonical)
3	Gamma	Inverse (Canonical)	Log (Noncanonical)

mean for each experimental run. For the true model Lewis, Montgomery, and Myers (2001b) choose $\mu_i = g^{-1}(\mathbf{x}'_i\boldsymbol{\beta}) = g^{-1}(10 + 5x_{i1} + 3x_{i2})$. The true model involves an identity link and a binomial distribution to produce an observation for each experimental run. A GLM is built for the observations using a binomial distribution. However, rather than applying the correct identity link, the model is fit with a logit link. This situation is simulated 5000 times. This process is repeated for the $n = 8$, 16, and 32 run designs.

A summary of confidence interval coverage for this situation is given in Table 5.24. When the incorrect link is assumed, the coverage decreases as the number of experimental runs increases. Again, without displaying the precision results, we report that as the number of experimental runs increases the intervals become shorter. However, in the case of a misspecified link, the intervals tighten around the wrong mean value. Consider the first observation in this example. The average length of the confidence interval changes from 0.0374 for $n = 8$, to 0.0260 for $n = 16$, and to 0.0183 for the $n = 32$ run case. See Lewis (1998) for detailed precision results.

Clearly, link misspecification affects the ability of the model to provide reliable estimates of the mean response. Therefore a complementary issue is whether the model fit with the incorrect link exhibits any behavior that the experimenter can use to identify the problem. For example, consider one of the 8 run designs from the simulation. Since model adequacy is often reflected in the model residuals, a normal probability plot of deviance residuals for this particular design is constructed and shown in Figure 5.20. Examination of the plot does suggest potential inadequacies in the model fit. A model builder could make use of such plots as a potential diagnostic. An unusual pattern on the normal probability plot of the deviance residuals could indicate a link misspecification.

We caution readers that only the deviance residuals should be plotted. The ordinary residuals from a GLM and the Pearson residuals do not have equal variance. The sum of the squared deviance residuals is equal to the model deviance, and in the case of a normal distribution and the identity link the deviance is equal to the residual sum of squares. Furthermore, the asymptotic distribution of the deviance is approximately chi-square. Thus plotting deviance residuals is somewhat analogous to plotting the usual normal-theory residuals. Note also that the deviance residuals are not independent, just as the

Table 5.24 Summary of Impact on Confidence Interval Coverage for Binomial Distribution with True Identity Link but Assumed Log Link

Experimental Runs	True Link Assumed	Incorrect Link Assumed
8	95.48%	81.24%
16	93.98%	69.54%
32	94.51%	53.58%

Note: Target confidence level is 95%.

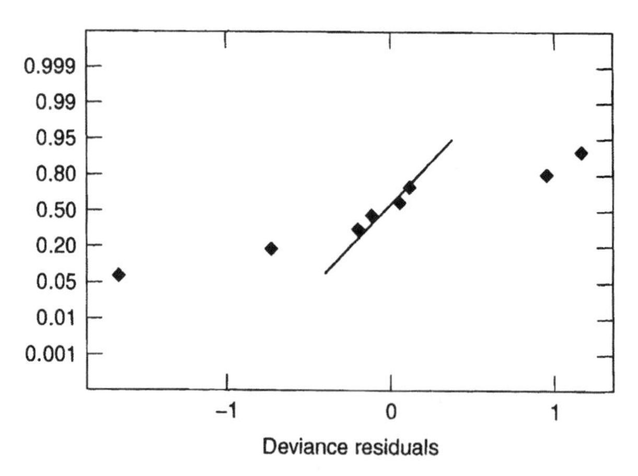

Figure 5.20 Normal probability plot of deviance residuals for binomial response with true identity link but assumed logit link, $n = 8$.

usual normal-theory residuals are not independent, so deviance residual plots based on small sample sizes need to be interpreted cautiously.

5.12.5 Poisson Distribution with a True Identity Link but with Log Link Assumed

Another common situation is a designed experiment where the response is a Poisson variable. Often Poisson data are modeled using a log transformation. Therefore an experimenter is likely to elect to use a log link in the model-building process. However, suppose that a situation arises where the true link is well approximated by an identity link so that the model is approximately linear, but the experimenter chooses the log link. To investigate the impact on potential model performance, the confidence interval on the mean response is examined. As in the binomial illustration, a 2^2 factorial design is used and the true model is generated with a set of known parameters, and therefore a known mean for each experimental run. Lewis, Montgomery, and Myers (2001b) choose the true model as $\mu_i = g^{-1}(100 + 50x_{i1} + 30x_{i2})$. The true model is assumed to involve an identity link and the Poisson distribution is used to obtain an observation for each experimental run. A GLM is built using a Poisson distribution. However, rather than the correct identity link, the model is fit with a log link. This situation is simulated 5000 times.

Table 5.25 gives a summary of the confidence interval coverage for this situation. When the incorrect link is assumed, the coverage declines drastically from the correct link situation. Additionally, as the number of experimental runs increases, the coverage declines even further. In general, the intervals become shorter as the number of experimental runs increases. However, in the case of a misspecified link, they tighten around the wrong mean value.

In this situation, link misspecification severely impacts model performance. Just as in the binomial illustration, deviance residuals can be used to provide an

Table 5.25 Summary of Impact on Confidence Interval Coverage for Poisson Distribution with True Identity Link but Assumed Log Link

Experimental Runs	True Link Assumed	Incorrect Link Assumed
8	94.92%	31.60%
16	95.02%	25.67%
32	95.20%	2.33%

Note: Target confidence level is 95%.

indication of model inadequacy. Using one of the 5000 simulations trials of the 8 run design, a normal probability plot of deviance residuals for this run is constructed and shown in Figure 5.21. The central portion of this plot does not deviate much from linearity, but the tails do, indicating potential inadequacies in the model fit. However, as noted previously, these plots need to be interpreted cautiously.

5.12.6 Gamma Distribution with a True Inverse Link but with Log Link Assumed

The final modeling situation of misspecification considered arises with the gamma-distributed response variable. Often a gamma distributed response variable is analyzed following a log transformation. Therefore an experimenter may likely choose a log link in the model-building process. A 2^2 factorial design is selected and, as before, the true model is generated with a set of known parameters. The true model chosen by Lewis, Montgomery, and Myers (2001b) is $\mu_i = g^{-1}(10 + 50x_{i1} + 3x_{i2})$. A gamma distribution with the inverse link was then used to generate the observations for this experiment. A GLM model is built using a gamma distribution. However, rather than the correct inverse link, the model is fitted with a log link. This situation is simulated 5000 times.

A summary of the confidence interval coverage results obtained over the 5000 runs is provided in Table 5.26. The results for the same model with

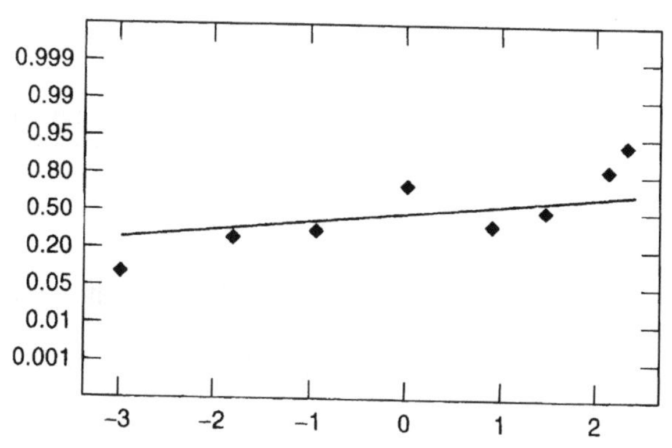

Figure 5.21 Normal probability plot of deviance residuals for Poisson response with true identity link but assumed log link, $n = 8$.

Table 5.26 Summary of Impact on Confidence Interval Coverage for Gamma Distribution with True Identity Link but Assumed Log Link

Experimental Runs	True Link Assumed	Incorrect Link Assumed
8	86.39%	88.65%
16	93.31%	88.72%
32	95.41%	84.25%

Note: Target confidence level is 95%.

the correct link chosen are also shown for comparison purposes. In this situation, link misspecification does not seriously affect confidence interval coverage. This is similar behavior to that experienced when a log transformation or an inverse transformation is made to long right-tailed distributions, such as the gamma distribution. That is, both transformations often produce similar results. This behavior likely explains the similar confidence interval coverage results obtained from using either the log or inverse link functions. Notice also that for the $n = 8$ run design, the coverage is actually slightly worse when the correct link is used than when the incorrect link is used. This is likely due to unstable behavior of the inverse link for small sample sizes. It can be shown that using the incorrect link generally leads to lower precision of estimation.

5.12.7 Summary of Link Misspecification on Confidence Interval Coverage and Precision

The impact of link misspecification is serious for the binomial and Poisson distributions. The severity of the loss in coverage appears to increase as the number of experimental runs increases. The lengths of the confidence intervals also appear to increase from the model analyzed with a correct link to the same model analyzed with an incorrect link. The length of the confidence intervals always decreases as the number of experimental runs increases. However, for the incorrect link, the intervals tighten around the wrong estimate of the response mean.

The impact of link misspecification, for the specific scenario considered, was much less severe for the gamma distribution in terms of confidence interval coverage. This is not completely surprising, since often, in practice, the gamma distribution is well modeled with either the log or the inverse link. That is, good models can often be produced using either link. However, the precision of the intervals is substantially less for the incorrect link models. Therefore considerable care is selecting the most appropriate link necessary to optimize the capability of the model in response variable estimation and prediction.

5.12.8 Impact of Model Misspecification on Confidence Interval Coverage and Precision

Clearly, one of the important tasks involved in analyzing designed experiments is the selection of variables to include in the model. We now briefly examine the

Table 5.27 2^{4-1} Fractional Factorial Design and True Mean Poisson Response

Run	x_1	x_2	x_3	x_4	Response
1	−	−	−	−	41
2	+	−	−	+	51
3	−	+	−	+	49
4	+	+	−	−	119
5	−	−	+	+	89
6	+	−	+	−	99
7	−	+	+	−	141
8	+	+	+	+	211

impact of selecting the wrong subset of variables to include in the model on confidence interval coverage and precision. To perform this investigation, a fractional factorial design is used. This experiment involves a 2^{4-1} design with a Poisson response distribution and identity link. The true model in the coded variables is $\mu_i = 100 + 20x_{i1} + 30x_{i2} + 35x_{i3} + 15x_{i1}x_{i2} + 11x_{i2}x_{i3}$. The design matrix is shown in Table 5.27.

Examination of the resulting SAS generated Wald χ^2 statistics (not provided here) reveals that only the main effects x_1, x_2, x_3, and the x_1x_2 interaction should be retained in the model. To make these decisions, one could utilize other model selection and evaluation statistics, such as PRESS. If one uses a PRESS statistic for variable selection, variable x_2x_3 would have been retained in the model as well. However, suppose that the experimenter relies on only the Wald statistics for variable selection and chooses to leave out the interaction x_2x_3. The confidence interval coverage and precision results comparing the misspecified model results to the correctly specified model are given in Table 5.28.

Inspection of Table 5.28 shows a significant degradation in confidence interval coverage; that is, from 94.9% for the correct model to 68.5% for the incorrect model. It is interesting to note that variable x_2x_3 has the smallest effect in the true model, yet it has a large impact on coverage if omitted from the model. Therefore just as in standard least squares model building, selecting the correct subset of variables in a GLM is critical. This is certainly not surprising. In other words, not only is the choice of link function crucial, but so also is the choice of variables in the linear predictor.

Summary

One major finding in the Lewis, Montgomery, and Myers (2001b) work is that confidence interval coverage for the GLM when applied to data from designed experiments performs closely to theoretical claims even for very small samples. This means that confidence intervals provide an effective method for evaluating the estimates of the mean response given by a particular model and for making comparisons to other models. The binomial, Poisson, and gamma distributions, each with a canonical and noncanonical link, are considered. The only cases

Table 5.28 Confidence Interval Coverage and Precision Results for 2^{4-1} Design for Correct Model Specification and Incorrect Model Specification

	Correct Model Specification Variables x_1, x_2, x_3, x_1x_2, and x_2x_3		Incorrect Model Specification Variables x_1, x_2, x_3, and x_1x_2	
Observation	% Coverage	Precision	% Coverage	Precision
1	4729	23.14	4141	20.54
2	4753	25.29	4204	22.82
3	4732	26.04	3938	27.24
4	4756	37.50	4094	36.78
5	4758	30.55	2978	28.71
6	4723	31.35	3192	29.64
7	4750	39.74	1712	30.04
8	4759	43.88	3042	37.92
Total	37,960		27,301	
Coverage %	94.90		68.25	

Note: Target confidence level is 95%.

showing moderate departures from advertised coverage are for the gamma distribution at $n = 8$ and $n = 16$ runs. For the binomial and Poisson distributions, the SAS PROC GENMOD confidence intervals and the Wald confidence intervals have very similar performance. However, some discrepancy is observed for the gamma distribution. The method of estimating the scale parameter is likely responsible for the discrepancies.

An important consideration in fitting a generalized linear model is the choice of link function. We have shown that there is serious degradation in coverage for a misspecified link when using the binomial and Poisson distributions. Furthermore, as the number of experimental runs increases, the precision of estimation increases as well, but the confidence interval tightens around the wrong predicted value. We also show that misspecifying the link with the gamma distribution (i.e., assuming a log link, when the true link is inverse) exhibits some degradation in coverage, but not nearly as extreme as observed for the binomial and Poisson cases. In practice, GLMs using the gamma distribution commonly employ both the log and inverse link. This suggests that, in practice, often the incorrect link is selected. Yet the results of the analysis often still provide the experimenter with an adequate model. Showing that actual coverage is close to the advertised value whether the correct or incorrect link is used offers an explanation as to why a gamma-distributed response analysis often tolerates the misspecified link.

Lewis, Montgomery, and Myers (2001b) also investigated the effect of misspecifying the model variables in terms of the effect on the confidence intervals on the mean response for the model. They show that serious degradation in confidence interval coverage can occur with the omission of even one variable.

Lewis, Montgomery, and Myers (2001b) recommend the use of confidence intervals to evaluate the potential performance of competing models from designed experiments. They caution, however, that this comparison assumes that (1) the correct link is selected and (2) the appropriate subset of variables is selected. If uncertainty exists with a chosen link, the authors recommend that the experimenter rebuild the model with other link functions and observe the outcome.

EXERCISES

5.1 Consider the binomial distribution with $n = 1$.

(a) Put the probability function in the form of the exponential family and show that

$$b(\theta) = \log(1 + e^{\theta})$$
$$\theta = \log\left(\frac{\pi}{1 - \pi}\right)$$

(b) Use $b(\theta)$ in part (a) to show that

$$\mu = \pi$$
$$\text{Var}\,(y) = \pi(1 - \pi)$$

5.2 For both logistic and Poisson regression, show that elements in $\mathbf{X}'\mathbf{y}$ are sufficient statistics for $\boldsymbol{\beta}$.

5.3 Show formally for a normal distribution and identity link that

$$D(\boldsymbol{\beta}) = \sum_{i=1}^{n} \frac{(y_i - \hat{\mu}_i)^2}{\sigma^2}$$

5.4 Consider a situation in which a normal distribution is assumed with a log link.

(a) Describe the score function for maximum likelihood estimation.

(b) Give the asymptotic covariance matrix for $\mathbf{b}$ and explain all terms.

5.5 Consider an exponential distribution with log link.

(a) Write out the score function in terms of the linear predictor $\mathbf{x}'\boldsymbol{\beta}$.

(b) Show that the information matrix is $\mathbf{X}'\mathbf{X}$. Comment.

5.6 Consider the gamma distribution in the context of the GLM. Put the density function into the form of the exponential family and show that $\theta = 1/\mu$ and $\text{Var}\,(y_i) = \mu^2/r$.

5.7 Fit a generalized linear model in which the gamma distribution is assumed with an identity link. Write out the score function in terms of linear predictor $\mathbf{x}'\boldsymbol{\beta}$. Give an expression for the asymptotic variance–covariance matrix of $\hat{\boldsymbol{\beta}}$.

5.8 It is well known from principles of statistical inference that

$$E\left(\frac{\partial \ln \mathscr{L}}{\partial \theta}\right) = 0$$

$$E\left(\frac{\partial^2 \ln \mathscr{L}}{\partial \theta^2}\right) + E\left(\frac{\partial \ln \mathscr{L}}{\partial \theta}\right)^2 = 0$$

Use these expressions to show that for the generalized linear model

$$\mathrm{Var}(y) = b\,(\theta)\,a\,(\phi)$$

where $b''(\theta) = \partial^2 b(\theta)/\partial\theta^2$.

5.9 Consider an experiment with replications, namely, r_i observations at each of m design points. In addition, suppose that the response is Poisson and overdispersion is expected. Can you suggest an estimator of the overdispersion parameter σ^2 which is free of the deviance $D(\boldsymbol{\beta})$? Explain.

5.10 In linear regression much is made of the so-called Hat matrix and resulting set of Hat diagonals that are used in standard regression diagnostics. If such diagnostics were to be necessary in generalized linear models, what matrix would be used to replace the Hat matrix? Be sure to define all terms.

5.11 Show that if the model contains an intercept term, then the deviance for both the Poisson and gamma cases reduces to Equation (5.14).

5.12 Show that Equations (5.7) and (5.9) are in fact equivalent.

5.13 Derive the score equations for generalized linear models when $a(\phi)$ is not constant. Find the solution to the resulting score equations.

5.14 Chapman (1997–98) conducted an experiment using accelerated life testing to determine the estimated shelf life of a photographic developer. The data follow. Lifetimes often follow an exponential distribution. This company has found that the maximum density is a good indicator of overall developer/film performance; correspondingly using generalized linear models. Perform appropriate residual analysis of your final model.

t (h)	D_{max} (72°C)	t (h)	D_{max} (82°C)	t (h)	D_{max} (92°C)
72	3.55	48	3.52	24	3.46
144	3.27	96	3.35	48	2.91
216	2.89	144	2.50	72	2.27
288	2.55	192	2.10	96	1.49
360	2.34	240	1.90	120	1.20
432	2.14	288	1.47	144	1.04
504	1.77	336	1.19	168	0.65

5.15 Gupta and Das (2000) performed an experiment to improve the resistivity of a urea formaldehyde resin. The factors were amount of sodium hydroxide, A, reflux time, B, solvent distillate, C, phthalic anhydride, D, water collection time, E, and solvent distillate collection time, F. The data follow, where y_1 is the resistivity from the first replicate of the experiment and y_2 is the resistivity from the second replicate. Assume a gamma distribution. Use both the canonical and the log link to analyze these data. Perform appropriate residual analysis of your final models.

A	B	C	D	E	F	y_1	y_2
−1	−1	−1	−1	−1	−1	60	135
−1	−1	−1	1	−1	−1	220	160
0	−1	−1	−1	1	1	85	180
0	−1	−1	1	1	1	330	110
0	1	1	1	−1	−1	95	130
0	1	1	−1	−1	−1	190	175
−1	1	1	1	1	1	145	200
−1	1	1	−1	1	1	300	210
1	−1	1	1	−1	1	110	100
1	−1	1	−1	−1	1	125	130
1	−1	1	1	1	−1	300	170
1	−1	1	−1	1	−1	65	160
1	1	−1	−1	−1	1	170	90
1	1	−1	1	−1	1	70	250
1	1	−1	−1	1	−1	380	80
1	1	−1	1	1	−1	105	200

5.16 Consider the semiconductor data in Exercise 2.9. Assume a gamma distribution. Analyze these data using both the canonical and log links. Perform appropriate residual analyses of your final models. Discuss your results.

5.17 Consider the staffing of naval hospitals data in Exercise 2.15. Analyze these data assuming a Poisson distribution. Use both the canonical and the power links. Perform appropriate residual analyses of your final models. Discuss your results.

5.18 Consider the cathodic bonding of elastomeric metal bond data in Exercise 2.16. Fit a GLM to these data. Perform appropriate residual analyses of your final models. Analyze these data with a power link.

5.19 Consider the nitrogen dioxide data in Exercise 2.17. Analyze these data assuming a Poisson distribution. Use both the canonical and the log links. Perform appropriate residual analyses of your final models. Discuss your results.

5.20 Bast et al. (1983) measured the levels of the antibody CA 125 in blood serums of patients with specific cancers. Antigen levels of 35 and of 65 units per mL were considered significant. The data follow. Fit an appropriate GLM and perform the diagnostic analysis of these data.

Group	Total Tested	> 35 Units/ mL	> 65 Units/ mL
Healthy males	537	4	2
Healthy females	351	5	0
Patients, benign diseases	143	9	3
Patients, pancreatic cancer	29	17	13
Patients, lung cancer	25	8	6
Patients, breast cancer	25	3	2
Patients, colorectal cancer	71	16	12
Patients, gastrointestinal	30	8	6
Patients, nongastrointestinal	20	5	5
Patients, ovarian cancer	101	83	75

5.21 The negative binomial distribution is often used to model the number of trials until the rth success. However, this distribution provides an interesting alternative to the Poisson distribution in the presence of overdispersion. The probability function for the negative binomial distribution is

$$p(y) = \binom{y-1}{r-1} \pi^r (1-\pi)^{y-r}$$

(a) Show that this distribution belongs to the exponential family.

(b) Derive the mean and the variance for this distribution.

(c) Derive the canonical link for this distribution.

(d) Assume that the Poisson mean follows a gamma distribution. Show that the compound distribution becomes the negative binomial.

5.22 Schubert et al. (1992) conducted an experiment using a catapult to determine the effects of hook (x_1), arm length (x_2), start angle (x_3), and stop angle (x_4) on the distance that the catapult throws a ball. They threw the ball three times for each setting of the factors. The data follow.

x_1	x_2	x_3	x_4		y	
−1	−1	−1	−1	28.0	27.1	26.1
−1	−1	1	1	46.3	43.5	46.5
−1	1	−1	1	21.9	21.0	20.1
−1	1	1	−1	52.9	53.7	52.0
1	−1	−1	1	75.0	73.1	74.3
1	−1	1	−1	127.7	126.9	128.7
1	1	−1	−1	86.2	86.5	87.0
1	1	1	1	195.0	195.9	195.7

(a) Use GLMs to fit an appropriate model for the variance.

(b) Use this model as a basis for a weighted least squares estimation of the model for the distance.

(c) Discuss your results.

5.23 Byrne and Taguchi (1987) discuss an experiment to see the effect of interference (x_1), connector wall thickness (x_2), insertion depth (x_3), and amount of adhesive (x_4) on the pull-off force for an elastometric connector to a nylon tube in an automotive application. The data follow.

x_1	x_2	x_3	x_4				y				
−1	−1	−1	−1	15.6	9.5	16.9	19.9	19.6	19.6	20.0	19.1
−1	0	0	0	15.0	16.2	19.4	19.2	19.7	19.8	24.2	21.9
−1	1	1	1	16.3	16.7	19.1	15.6	22.6	18.2	23.3	20.4
0	−1	0	1	18.3	17.4	18.9	18.6	21.0	18.9	23.2	24.7
0	0	1	−1	19.7	18.6	19.4	25.1	25.6	21.4	27.5	25.3
0	1	−1	0	16.2	16.3	20.0	19.8	14.7	19.6	22.5	24.7
1	−1	1	0	16.4	19.1	18.4	23.6	16.8	18.6	24.3	21.6
1	0	−1	1	14.2	15.6	15.1	16.8	17.8	19.6	23.2	24.2
1	1	0	−1	16.1	19.9	19.3	17.3	23.1	22.7	22.6	28.6

(a) Use GLMs to fit an appropriate model for the variance.

(b) Use this model as a basis for a weighted least squares estimation of the model for the distance.

(c) Discuss your results.

CHAPTER 6

Generalized Estimating Equations

While applications of generalized linear models are abundant, there are many situations in which repeated response measurements are made on the *same unit*, and thus this information forms a cluster of correlated observations. The classical example is a longitudinal study where the responses are being measured repeatedly on the same subject across time. The defining advantage with this type of study is that one can distinguish changes over time within individuals from differences at fixed times. In this example the unit that characterizes a cluster is the subject. Observations on the same subject are correlated. The assumption of correlated observations is not confined to biomedical studies. In agriculture studies we may have observations on the same small plot of ground. One may envision industrial studies in which observations taken in the same oven are correlated.

The basic modeling tool considered in this chapter remains generalized linear models. However, the need to account for the correlation among observations in a cluster does provide some complications (see Liang and Zeger, 1986, and Zeger and Liang, 1986). The presence of correlation renders the use of standard maximum likelihood methods to be problematic, at best. As a result we make use of quasi-likelihood methods that were touched on briefly in Chapter 5. In the following section we introduce the data layout for a situation involving longitudinal data.

6.1 DATA LAYOUT FOR LONGITUDINAL STUDIES

One must keep in mind that the clustering of response observations that results in the violation of the independence assumption can come from several different sources, depending on the application. The oven, plot of

Generalized Linear Models, Second Edition, by Myers, Montgomery, Vining, and Robinson

ground, or batch of raw materials that represent the unit describing the structure is referred to rather generically as the subject, and thus observations on the same subject are correlated. As a result the data structure is as follows.

$$\begin{array}{cccccc}
\textbf{Subject} & & & & & \\
1 & \mathbf{x}_{11} & \mathbf{x}_{21} & \cdots & \mathbf{x}_{k1} & \mathbf{y}_1 \\
2 & \mathbf{x}_{12} & \mathbf{x}_{22} & \cdots & \mathbf{x}_{k2} & \mathbf{y}_2 \\
\vdots & \vdots & & & & \vdots \\
s & \mathbf{x}_{1s} & \mathbf{x}_{2s} & \cdots & \mathbf{x}_{ks} & \mathbf{y}_s
\end{array} \tag{6.1}$$

The vector $\mathbf{x}_{ij}$ contains values of the ith variable or covariate on the jth subject. The vector $\mathbf{y}_j$ contains response values on the jth subject. The vectors are t-dimensional. That is, there are t regression observations on each subject. The rationale for the notation t comes from biomedical applications in which t refers to the number of time periods and the data are taken over time for each subject. In these applications the variables $x_1, x_2, \ldots, x_k$ may be dose values of a specific drug or some covariate characteristic associated with the subject, such as heart rate or blood pressure. Typically, the response values obey distributions from the exponential family. Initially, we consider the *normal distribution, identity link case*. As a result the model is linear, and we are dealing with the multivariate normal distribution. Consider the model

$$\mathbf{y} = \begin{bmatrix} \mathbf{y}_1 \\ \mathbf{y}_2 \\ \vdots \\ \mathbf{y}_s \end{bmatrix} = \begin{bmatrix} \mathbf{X}_1 \\ \mathbf{X}_2 \\ \vdots \\ \mathbf{X}_s \end{bmatrix} \boldsymbol{\beta} + \begin{bmatrix} \boldsymbol{\varepsilon}_1 \\ \boldsymbol{\varepsilon}_2 \\ \vdots \\ \boldsymbol{\varepsilon}_s \end{bmatrix} \tag{6.2}$$

Here the matrices $\mathbf{X}_1, \mathbf{X}_2, \ldots, \mathbf{X}_s$ are model matrices that put the regressor information in Equation (6.1) in model form; thus these $\mathbf{X}_j$ define the contributions to the model of the main effects, any interactions, categorical variables, quadratic terms, and so on. As a result, $\mathbf{X}_j$, $j = 1, 2, \ldots, s$, is $t \times p$. If the model is linear in the x's, then $p = k + 1$. For a specific subject, say, subject j, we have the linear model

$$\mathbf{y}_j = \mathbf{X}_j\boldsymbol{\beta} + \boldsymbol{\varepsilon}_j, \quad j = 1, 2, \ldots, s$$

where $\boldsymbol{\beta}$ is $p \times 1$ and $\mathbf{y}_j$ is multivariate normal with mean vector $\mathbf{X}_j\boldsymbol{\beta}$ and variance–covariance matrix $\mathbf{V}_j$. The matrix $\mathbf{V}_j$ is $t \times t$ and accounts for correlation of observations on subject j. Let $N = t \cdot s$ be the total sample size in the experiment.

6.2 IMPACT OF THE CORRELATION MATRIX R

Much of the notation for this case transfers over to the nonnormal case. However, for the latter situation there are further details that raise concerns. In the normal case each subject possesses the same $\mathbf{V}_j$ matrix, since the correlation structure is assumed the same for each subject *and* the variance is homogeneous, described by the scale parameter σ^2. Thus

$$\begin{aligned}\mathrm{Var}(\boldsymbol{\varepsilon}_j) = \mathbf{V}_j &= \begin{bmatrix} \sigma^2 & \sigma_{12} & \cdots & \cdots & \sigma_{1t} \\ & \sigma^2 & \sigma_{23} & \cdots & \sigma_{2t} \\ & & \ddots & & \vdots \\ & & & \ddots & \vdots \\ & & & & \sigma^2 \end{bmatrix} \quad \forall i,j \qquad (6.3) \\ &= \mathbf{R} \cdot \sigma^2\end{aligned}$$

where $\mathbf{R}$ is a $t \times t$ correlation matrix with typical element $\rho_{i,j} = \mathrm{corr}\,(y_{ji}, y_{j,i'})$ for $i \neq i'$ at subject j. The overall variance–covariance matrix for the random errors is

$$\mathrm{Var}(\boldsymbol{\varepsilon}) = \mathrm{V}^* = \begin{bmatrix} \mathbf{V}_1 & & & 0 \\ & \mathbf{V} & & \\ & & \ddots & \\ & 0 & & \mathbf{V}_s \end{bmatrix}$$

which is block diagonal with equal variance–covariance matrices displayed for each subject. It is instructive to consider an iterative procedure here for estimation of $\boldsymbol{\beta}$ since many of the same principles apply in situations discussed later. However, it is important to discuss $\mathbf{R}$ first.

In the methodology that is termed generalized estimating equations, the user may impart a correlation structure that is often called a **working correlation matrix**. One often does not know what the true correlation is, hence, the term working correlation. Common correlation structures include:

1. *Unspecified.* This implies that all correlations are to be independently estimated from the data.
2. *Exchangeable.* All correlations within subjects are equal.
3. *Independent.* All correlations are assumed to be zero.
4. 1-*Dependent.* Only correlations between adjacent observations (adjacent time periods) have nonzero (and equal) correlations.

5. *Autoregressive* (AR). Adjacent correlations are higher in magnitude than nonadjacent ones. Those observations further away in time possess the smallest correlations. The relationship is given by

$$\rho_{ii'} = \rho^{|i-i'|}$$

where ρ is estimated from the data.

The iterative routine for estimation of regression coefficients also provides iterative estimation of correlation with methodology that utilizes and is consistent with the structure that is chosen by the user. Examples are given later. The user must understand that the correlation structure chosen should be one that best describes the scenario involved. However, the generalized estimating equation methodology provides a consistent estimator of $\boldsymbol{\beta}$, even if the correlation structure is incorrect because our procedures use consistent estimates of the variance–covariance structure. We shed more light on the assumption of correlation structure subsequently.

6.3 ITERATIVE PROCEDURE IN THE NORMAL CASE, IDENTITY LINK

Let the matrix $\mathbf{R}^*$ describe the correlation structure through the entire experiment. Thus

$$\mathbf{R}^* = \begin{bmatrix} \mathbf{R} & & & \\ & \mathbf{R} & & \\ & & \ddots & \\ & & & \mathbf{R} \end{bmatrix} \tag{6.4}$$

where $\mathbf{R}^*$ is $ts \times ts$. Since $\mathbf{V}^* = \sigma^2\mathbf{R}$, we can replace $\mathbf{V}^*$ by $\mathbf{R}^*$ in a generalized least squares estimator for $\boldsymbol{\beta}$. As a result an iterative scheme for estimation is as follows.

Step 1. Use OLS for estimation of $\boldsymbol{\beta}$. Call it $\mathbf{b}_0$.

Step 2. Use $\mathbf{b}_0$ to compute residuals and hence correlations in conjunction with the assumed correlation structure.

Step 3. Use estimated correlations and compute

$$\mathbf{b} = (\mathbf{X}'\hat{\mathbf{R}}^{*-1}\mathbf{X})^{-1}\mathbf{X}'\hat{\mathbf{R}}^{*-1}\mathbf{y}$$

where $\hat{\mathbf{R}}^*$ contains the estimated correlations.

Step 4. Go back to step 2 to compute residuals and correlations.

Step 5. Iterate to convergence.

The Scale Parameter, σ^2

The computation of the correlations, as well as the standard errors of coefficients, involves the scale parameter σ^2. We must deal with similar considerations when we leave the normal case. For the normal case the scale parameter estimation is fairly straightforward,

$$\hat{\sigma}^2 = \mathrm{tr}\left[\sum_{j=1}^{s} \frac{(\mathbf{y}_j - \mathbf{X}_j\mathbf{b})(\mathbf{y}_j - \mathbf{X}_j\mathbf{b})'}{N-p}\right] \tag{6.5}$$

where $\mathbf{X}_j\mathbf{b}$ is the vector of estimated mean responses. Note that $(\mathbf{y}_j - \mathbf{X}_j\mathbf{b})(\mathbf{y}_j - \mathbf{X}_j\mathbf{b})'$ is a sum of squares and products matrix of residuals for the jth subject. The $\sum_{j=1}^{s}$ operator essentially performs a pooling across subjects, since it is assumed that the variance–covariance matrix $\mathbf{V}$ is the same for all subjects. The trace operator in Equation (6.5) produces the residual sum of squares over $N = t \cdot s$ data point, while $N-p$ plays the role of error degrees of freedom. It should be understood that the simplicity of the estimate of σ^2 stems from the assumption of homogeneous variance in the case of the normal distribution. When we discuss other distributions in GLM, the residuals are standardized to account for the nonconstant variance.

Standard Errors of Coefficients

The iterative procedure described is tantamount to iteratively reweighted least squares where the weights in **R** are updated at each iteration. There are two schools of thought on the computation of the variance–covariance properties of **b**, the estimator of $\boldsymbol{\beta}$. Some advocate a **model-based** estimate of the variance–covariance matrix of coefficients. This procedure is based on the assumption that the *assumed correlation structure is correct*. One must keep in mind that the notion of a working correlation suggests that while much thought should go into the selection of the structure in **R**, the estimator of β *is a consistent estimator even if* **R** *is not correct*. If **R** is correct, then standard generalized least squares procedures suggest that the model-based variance–covariance of **b** is

$$\mathrm{Var}(\mathbf{b}) = \left(\mathbf{X}'\mathbf{R}^{*-1}\mathbf{X}\right)^{-1}\sigma^2 \tag{6.6}$$

Thus the standard errors come from the diagonal elements of the matrix

$$(\mathbf{X}'\mathbf{R}^{*-1}\mathbf{X})^{-1} \cdot \hat{\sigma}^2 = \left[\sum_{j=1}^{s}\left(\mathbf{X}'_j\hat{\mathbf{R}}^{-1}\mathbf{X}_j\right)\right]^{-1} \cdot \hat{\sigma}^2$$

An alternative estimation procedure for the standard error of coefficients allows for a more **robust** estimator, that is, one that is insensitive to incorrect specification of **R**. The genesis of this robust estimator comes from the use of the variance operator on the generalized least square estimator. Ideally,

$$
\begin{aligned}
\mathrm{Var}(\mathbf{b}) &= \mathrm{Var}\left[\mathbf{X}'\mathbf{R}^{*-1}\mathbf{X}\right]^{-1}\mathbf{X}'\mathbf{R}^{*-1}\mathbf{y} \\
&= \left[\mathbf{X}'\mathbf{R}^{*-1}\mathbf{X}\right]^{-1}\mathbf{X}'\mathbf{R}^{*-1}[\mathrm{Var}(\mathbf{y})]\mathbf{R}^{*-1}\mathbf{X}\left[\mathbf{X}'\mathbf{R}^{*-1}\mathbf{X}\right]^{-1}
\end{aligned} \tag{6.7}
$$

Now $\mathrm{Var}(\mathbf{y}) = \mathbf{V}^*$ is the *true*, but *unknown*, variance–covariance matrix. Note that if the correlation structure **R** is correct, then $\mathbf{V}^* = \mathbf{R}^*\,\sigma^2$, and Equation (6.7) reduces to (6.6). However, the robust approach estimates $\mathbf{V}^*$ in (6.7) empirically from the data. For this reason the estimator that is derived from Equation (6.7) is also called the **empirical estimator**. As a result $\mathbf{V}^*$ in (6.7) is replaced by

$$
\hat{\mathbf{V}}^* = \begin{bmatrix} \hat{\mathbf{V}} & & & \mathbf{0} \\ & \hat{\mathbf{V}} & & \\ & & \ddots & \\ & \mathbf{0} & & \hat{\mathbf{V}} \end{bmatrix}
$$

where $\hat{\mathbf{V}}$ is the matrix in (6.5) from which the trace is taken. Often the empirical or robust estimator and the model-based estimator give similar results, which implies that the correlation structure assumed is a good practical choice. There is more discussion later in this chapter with examples regarding these two types of estimators of standard errors.

6.4 GENERALIZED ESTIMATING EQUATIONS FOR MORE GENERALIZED LINEAR MODELS

The foundation for the use of generalized estimating equations for nonnormal responses is similar to what we have developed for normally distributed responses. Chapters 4 and 5 lay the foundation for the complications brought about by the case of nonnormal responses. Suppose now that the response has a distribution from the exponential family and that within subjects observations exhibit a correlation structure as defined in Equations (6.1), (6.2), and (6.3). The regression data and model are again characterized by

$$
\mathbf{X} = \begin{bmatrix} \mathbf{X}_1 \\ \mathbf{X}_2 \\ \vdots \\ \mathbf{X}_s \end{bmatrix}
$$

where $\mathbf{X}_j$ is $t \times p$, as before. In this situation, however, the variance–covariance matrix $\mathbf{V}^*$ is given by the block diagonal matrix

$$\mathbf{V}^* = \begin{bmatrix} \mathbf{V}_1 & & & \\ & \mathbf{V}_2 & & \\ & & \ddots & \\ & & & \mathbf{V}_s \end{bmatrix}$$

where the $\mathbf{V}$'s are different from subject to subject. These differences result from the fact that from subject to subject the regressor values (or design levels) are different; thus the mean responses differ, and the response variances are functions of the mean. Let us define a mean vector as

$$\boldsymbol{\mu} = \begin{bmatrix} \boldsymbol{\mu}_1 \\ \boldsymbol{\mu}_2 \\ \vdots \\ \boldsymbol{\mu}_s \end{bmatrix}$$

where $\boldsymbol{\mu}'_j$ is $E(\mathbf{y}'_j)$ and is t-dimensional.

6.4.1 Structure of $\mathbf{V}_j$

We assume that a scale parameter $\sigma^2 > 0$ exists such that υ_{ji}, the variance of y_{ji} the ith observation on the jth subject, is given by

$$\upsilon_{ji} = \sigma^2 a_{ji} \tag{6.8}$$

In Equation (6.8), a_{ji} is the component of the variance delivered by the characterization of the distribution. For example, in the case of the binomial distribution, $a_{ji} = n_{ji}\,\mu_{ji}(1-\mu_{ji})$. For the Poisson, $a_{ji} = \mu_{ji}$. As a result we can write

$$\mathbf{V}_j = \mathbf{A}_j^{1/2}\mathbf{R}\mathbf{A}_j^{1/2} \cdot \sigma^2 \tag{6.9}$$

where

$$\mathbf{A}_j = \begin{bmatrix} a_{j1} & & & \phi \\ & a_{j2} & & \\ & & \ddots & \\ & \phi & & a_{jt} \end{bmatrix}$$

and as before, $\mathbf{R}$ is the correlation matrix which is common to all subjects. As a result $\mathbf{V}_j$, the variance–covariance matrix of observations on the jth subject, contains variances $(a_{ji} \cdot \sigma^2)$ on the main diagonals and covariance (derived from $\mathbf{R}$) on the off-diagonals. In what follows we must once again define $\mathbf{R}^*$ as a $ts \times ts$ block diagonal matrix containing $\mathbf{R}$ on all diagonal blocks.

Quasi-likelihood Estimation

As we indicated earlier, the existence of the correlation among observations in the same subject takes the problem to multivariate analysis. However, the operative method of estimation does not make use of the multivariate distribution involved. It uses certain information associated with the marginal distribution rather than the actual likelihoods. The application of generalized least squares in this context is referred to as quasi-likelihood estimation, which we introduced in Section 5.4.

Let us assume that the basic marginal distribution of the response is one of the members of the exponential family, such as binomial, Poisson, or gamma, and that correlations exist within subjects as discussed earlier in this section. Let us also assume that the link $g(\mu) = \mathbf{x}'\boldsymbol{\beta}$, and thus $\mu = g^{-1}(\mathbf{x}'\boldsymbol{\beta})$, is used. Quasi-likelihood estimation involves the use of generalized least squares as we discussed in Chapter 5. Iterative minimization of the quadratic form $(\mathbf{y}-\boldsymbol{\mu})'\mathbf{V}^{-1}(\mathbf{y}-\boldsymbol{\mu})$ with $\mathbf{V}$ fixed but updated at each iteration leads to the score function

$$\mathbf{D}'\mathbf{V}^{-1}(\mathbf{y} - \boldsymbol{\mu}) = \mathbf{0} \tag{6.10}$$

where $\mathbf{D}$ is, once again, the $n \times p$ matrix of partial derivatives with the (i, j) element $\partial\mu_i/\partial\beta_j$, $i = 1, 2, \ldots, N; j = 1, 2, \ldots, p$. The $\mathbf{D}$ matrix can be partitioned like the $\mathbf{X}$ matrix, namely, as

$$\mathbf{D} = \begin{bmatrix} \mathbf{D}_1 \\ \mathbf{D}_2 \\ \vdots \\ \mathbf{D}_s \end{bmatrix}$$

Obviously, the $\mathbf{D}$ matrices differ from subject to subject since the means differ through changing regressor values. Thus an alternative way to write the score function is given by

$$\sum_{j=1}^{s} \mathbf{D}_j'\mathbf{V}_j^{-1}(\mathbf{y}_j - \boldsymbol{\mu}_j) = \mathbf{0} \tag{6.11}$$

As in the discussion of the nonlinear regression material in Chapter 3, the solution of the score equation involves a Gauss–Newton procedure with the adjustment to the initial coefficient estimate $\mathbf{b}_0$ being

$$\mathbf{b}_1 = \mathbf{b}_0 + (\mathbf{D}_0'\mathbf{V}_0^{-1}\mathbf{D}_0)^{-1}\mathbf{D}_0'\mathbf{V}_0^{-1}(\mathbf{y} - \boldsymbol{\mu}_0) \tag{6.12}$$

The rationale for Equation (6.12) is much like that discussed in Chapter 3. The $\mathbf{V}$ matrix above, as well as $\mathbf{D}$ and $\boldsymbol{\mu}$, make use of the initial estimate $\mathbf{b}_0$ through Equation (6.9). The scale parameter σ^2 can be ignored, since it cancels out in (6.12). The vector $\mathbf{b}_0$ is also contained in both $\mathbf{R}$ and $\mathbf{A}_j$, $j = 1, 2, \ldots, s$. The iterative procedure is then as follows:

Step 1. Obtain an initial estimate $\mathbf{b}_0$.

Step 2. Use $\mathbf{b}_0$ to obtain residuals, $\mathbf{R}$, $\mathbf{A}_j$, $\mathbf{V}_0$, $\mathbf{D}_0$, and $\boldsymbol{\mu}_0$.

Step 3. Use Equation (6.12) to obtain the new estimator $\mathbf{b}_1$.

Step 4. Substitute $\mathbf{b}_1$, for $\mathbf{b}_0$ and go back to step 2.

Step 5. Continue iterating to convergence.

Properties of Quasi-likelihood Estimators

We consider here model-based and empirical standard error estimates. Again, we note that the elements described by the iterative procedure are not a maximum likelihood estimator. The properties of the quasi-likelihood estimator of $\boldsymbol{\beta}$, however, resemble the properties of a maximum likelihood estimator. If the correlation structure is correct, the estimator $\mathbf{b}$ is asymptotically normal with mean $\boldsymbol{\beta}$ and variance–covariance matrix

$$\mathrm{Var}(\mathbf{b}) = \left(\mathbf{D}'\mathbf{V}^{-1}\mathbf{D}\right) \tag{6.13}$$

As in the normal response case described earlier in this chapter, the estimator $\mathbf{b}$ is a consistent estimator of $\boldsymbol{\beta}$, even if the user-specified correlation structure is wrong. In this sense we are once again incorporating a working correlation matrix. Obviously, if the correlation matrix is correct, the estimator will be more efficient than the case of an estimator used that is not correct. However, one can certainly be quite successful using generalized estimating equations with a specified structure on $\mathbf{R}$ that is incorrect. We display more evidence on this issue in a subsequent section. Indeed, Liang and Zeger (1986) show that the generalized estimating equations procedure for selecting $\mathbf{b}$ is considerably more robust to the choice of $\mathbf{R}$ than the estimation of standard errors that come from Equation (6.13). Clearly, if one uses (6.13), the standard errors come from the square root of the diagonal elements of $(\hat{\mathbf{D}}\hat{\mathbf{V}}^{-1}\hat{\mathbf{D}})\,\hat{\sigma}^2$, where $\hat{\mathbf{D}}$, $\hat{\mathbf{V}}$, and $\hat{\sigma}^2$ make use of the estimator $\mathbf{b}$. The estimate of the scale parameter σ^2 is computed much like that in Equation (6.5) for the normal case. However, the residuals must be standardized to account for the natural nonhomogeneous variance that stems from the distribution involved. Since $\mathrm{Var}(y_{ji})$ (ith observation for jth subject) $=$ $\sigma^2 a_{ji}$, then the squared residual $(y_{ji} - \hat{\mu}_{ji})^2$ must be scaled by a_{ji}. As a result the scale parameter is estimated by

$$\hat{\sigma}^2 = \mathrm{tr}\left[\sum_{j=1}^{s} \frac{(\mathbf{y}_j - \hat{\boldsymbol{\mu}}_j)\hat{\mathbf{A}}_j^{-1}(\mathbf{y}_j - \hat{\boldsymbol{\mu}}_j)'}{N - p}\right] \tag{6.14}$$

Note that the above displays the sum of the squares of the *Pearson residuals* that we discussed in Chapter 5.

Equation (6.13) produces a *model-based* set of standard errors of the regression coefficients. These model-based standard errors strictly depend on the chosen structure in **R** and thus may not always be appropriate. Once again, as in the normal response case an empirical or robust set of estimates of standard errors can be used. This empirical variance–covariance matrix is derived by incorporating a variance operator on the structure in Equation (6.12) that was based on generalized least squares. Again, much of this development is much like that of the normal case discussed in Section 6.3. For the general case, asymptotically

$$\mathrm{Var}(\mathbf{b}) = \left(\mathbf{D}'\mathbf{V}^{-1}\mathbf{D}\right)^{-1}\mathbf{D}'\mathbf{V}^{-1}[\mathrm{Var}(\mathbf{y})]\mathbf{V}^{-1}\mathbf{D}\left(\mathbf{D}'\mathbf{V}^{-1}\mathbf{D}\right)^{-1} \tag{6.15}$$

The [Var(**y**)] portion of Equation (6.15) is then found without the use of the input **R** but rather from actual residuals in the data; that is, it is found empirically. As we substitute empirical information for [Var(**y**)], it is informative to partition **D** and **V** in the center of the expression in Equation (6.15). Thus the robust (to choice of **R**) or empirical estimator of Var (**b**) is

$$\mathrm{Var}(\mathbf{b}) = \left[\hat{\mathbf{D}}'\mathbf{V}^{-1}\hat{\mathbf{D}}\right]^{-1}\left[\sum_{j=1}^{s}\hat{\mathbf{D}}_j'\hat{\mathbf{V}}_j^{-1}(\mathbf{y}_j-\hat{\boldsymbol{\mu}}_j)(\mathbf{y}_j-\hat{\boldsymbol{\mu}}_j)'\hat{\mathbf{V}}_j^{-1}\hat{\mathbf{D}}_j\right]\left[\hat{\mathbf{D}}'\hat{\mathbf{V}}^{-1}\hat{\mathbf{D}}\right]^{-1} \tag{6.16}$$

Comments on Empirical Estimates

The estimates of the variance–covariance estimator in Equation (6.16) are consistent estimators despite the input structure on **R**, the correlation matrix. In addition, the reader should notice that it does not require the estimator of the scale parameter σ^2 since $\mathbf{V} = \mathbf{R}\sigma^2$ and σ^2 cancels out in (6.16). As a result the $\mathbf{V}_j$ in (6.16) for specific subjects can truly be replaced by $\mathbf{R}_j$. Two very important issues that do reflect some fragility in this robust estimator should, however, be pointed out. First, it is important to use a high-quality estimate of Var(**y**) in Equation (6.16) because it is purely based on residuals and thus depends on the model specification in $\boldsymbol{\mu}_j$ developed through the link. *Sloppy model specification on the mean can lead to poor estimation of standard errors.* Second, if the data set contains few subjects, the empirical estimator in (6.16) can be very poor. This stems from the fact that Var(**y**) is based on so few residuals. For example, one may encounter an industrial situation in which the generalized estimating equations technology is used in a split plot type of experiment where we have few whole plots (e.g., ovens in a manufacturing scenario). The empirical estimator in (6.16) may suffer considerably. For more information on this, one should consult Robinson and Myers (2001). Those cases where the model-based estimator and the empirical estimator give similar results provide support for the choice of the correlation structure. However, if the correlation structure chosen is faulty, one would expect that the model-based standard errors would be too optimistic and hence smaller than those given by the empirical or robust method.

Adjustment for Leverage

It is well known that the use of residuals for estimation of variances and covariances can be hazardous since the fitted values, $\hat{\mu}$, tend to be close to the observed value, y, rendering the residual too small. The term $(\mathbf{y}_j-\hat{\boldsymbol{\mu}}_j)(\mathbf{y}_j-\hat{\boldsymbol{\mu}}_j)'$ is a consistent estimator of the appropriate variance covariance matrix, but in small samples (i.e., for a small number of subjects), the quantity may exhibit bias that is not negligible. The end result is to make standard errors of regression coefficients too small. A simple illustration of this concept in simple linear regression with iid errors brings forth the point. Suppose we have the model

$$y_i = \beta_0 + \beta_1 x_i + \varepsilon_i \quad (i = 1, 2, \ldots, n)$$

with the ε_i being iid $N(0, \sigma)$. We know that the squared residual $e_i^2 = (y_i - b_0 - b_1 x_i)^2$ is not unbiased for σ^2. Indeed, $E(e_i^2) = \sigma^2\ (1-h_{ii})$, where h_{ii} is the hat-diagonal at the ith data point. We know from regression theory that h_{ii} is positive but bounded from above by 1. It can be shown that

$$E\left[\sum_{i=1}^{n} e_i^2\right] = \sigma^2 \sum_{i=1}^{n} (1 - h_{ii})$$
$$= \sigma^2 (n - p)$$

Thus we have the familiar result which involves a correction for leverage.

Much like the simple case above, Mancl and De Rouen (2001) provide an *adjustment for leverage* on $(\mathbf{y}_j - \boldsymbol{\mu}_j)\ (\mathbf{y}_j-\boldsymbol{\mu}_j)'$ in the expression in Equation (6.16). They point out that

$$E(\mathbf{y}_j-\hat{\boldsymbol{\mu}}_j)\,(\mathbf{y}_j-\hat{\boldsymbol{\mu}}_j)' \equiv (\mathbf{I}_j-\mathbf{H}_{jj})\,[\mathrm{Var}\,(\mathbf{y})](\mathbf{I}_j-\mathbf{H}_{jj})'$$

and thus the adjustment in the empirical estimator is to replace $(\mathbf{y}_j-\hat{\boldsymbol{\mu}}_j)\ (\mathbf{y}_j-\hat{\boldsymbol{\mu}}_j)'$ in Equation (6.16) by

$$\left(\mathbf{I}_j - \mathbf{H}_{jj}\right)^{-1}\left(\mathbf{y}_j - \hat{\boldsymbol{\mu}}_j\right)\left(\mathbf{y}_j - \hat{\boldsymbol{\mu}}_j\right)'\left(\mathbf{I}_j - \mathbf{H}'_{jj}\right)^{-1}$$

In the above $\mathbf{H}_{jj} = \mathbf{D}_j\left(\mathbf{D}'\mathbf{V}^{-1}\mathbf{D}\right)^{-1}\mathbf{D}'_j\mathbf{V}_j^{-1}$ and $\mathbf{I}_j$ is an identity matrix of dimension t. Mancl and De Rouen demonstrate the superiority of the adjusted estimator for cases with relatively small data sets.

6.4.2 Iterative Computation of Elements in R

As we indicated in the discussion of the iterative scheme for the computation of the quasi-likelihood estimate, the residuals are used at each iteration to update $\hat{\mathbf{R}}$. This repeated update is done with pooling over the time slots that have common correlation as stipulated by the input correlation structure. For example, in an exchangeable correlation structure only one correlation coefficient is computed, and thus pooling of information is done over all pairs of time points. To gain more insight for details, consider the *unspecified* structure, where there is essentially no fixed structure and different empirical correlations are calculated for each pair of time points. Thus the estimated correlations of a particular iteration are calculated from the off-diagonal elements of the matrix

$$\hat{\mathbf{C}} = \sum_{j=1}^{s} \frac{\mathbf{A}_j^{-1/2}(\mathbf{y}_j - \hat{\boldsymbol{\mu}}_j)(\mathbf{y}_j - \hat{\boldsymbol{\mu}}_j)'\mathbf{A}_j^{-1/2}}{\hat{\sigma}^2(s-p)} \tag{6.17}$$

Notice that the numerator in Equation (6.17) is a matrix containing empirical covariances for each pair of time points divided by the product of the standard deviations, which are in elements of $\mathbf{A}_j^{1/2}\cdot\sigma$.

Consider an example with an exchangeable correlation structure. In this case a single correlation, $\hat{\alpha}$, is computed as

$$\hat{\alpha} = \frac{1}{\hat{\sigma}^2} \sum_{j=1}^{s} \sum_{i<i'} \left[\frac{r_{ji} r_{j,i'}}{[s(t-1)/2]\cdot p} \right]$$

where $r_{ji} = (y_{ji} - \hat{\mu}_{ji})/\sqrt{\hat{a}_{ji}}$, which is the ($ji$)th Pearson residual. As an additional example, consider the 1-dependence case. This structure assumes that only one type of nonzero correlation exists; thus let $\alpha = \text{corr}(y_{j,i}, y_{j,i+1})$ for all j. Here α is the single correlation among adjacent observations on the same subject. Consequently, we have

$$\hat{\alpha} = \frac{1}{\hat{\sigma}^2} \sum_{j=1}^{s} \sum_{i=1}^{t-1} \frac{r_{ji}\cdot r_{j,i+1}}{(t-1)\,s-p}$$

6.5 EXAMPLES

Example 6.1. The Respiratory Example. We begin with a simple example taken from the biomedical literature, which we refer to as the *respiratory* example. The data are from Stokes, Davis, and Koch (1995) and appear in Table 6.1. The data involve the use of two treatments A and P, where A is an active drug and P is a placebo. During treatment the respiratory status was described as $0 =$ good and $1 =$ poor at each of four visits. A baseline measurement was taken

Table 6.1 Data for the Respiratory Example

Patient	Trt	Sex	Age	Baseline	Visit1	Visit2	Visit3	Visit4
1	P	M	46	1	1	1	1	1
2	P	M	28	1	0	0	0	0
3	A	M	23	1	1	0	0	0
4	P	M	44	1	1	1	1	0
5	P	F	13	1	1	1	1	1
6	A	M	34	1	1	0	0	0
7	P	M	43	1	1	0	1	1
8	A	M	28	1	1	0	0	0
9	A	M	31	1	1	1	0	0
10	P	M	37	1	0	1	1	0
11	A	M	30	1	1	1	0	0
12	A	M	14	1	1	1	0	0
13	P	M	23	1	1	0	0	0
14	P	M	30	0	0	0	0	0
15	P	M	20	1	1	1	1	1
16	A	M	22	1	0	0	0	0
17	P	M	25	1	0	0	0	0
18	A	F	47	1	1	1	1	0
19	P	F	31	0	1	1	1	1
20	A	M	20	1	1	1	0	0
21	A	M	26	1	1	0	0	0
22	A	M	46	1	1	1	1	1
23	A	M	32	1	1	1	0	0
24	A	M	48	1	1	1	0	0
25	P	F	35	1	1	1	1	1
26	A	M	26	1	1	0	0	0
27	P	M	23	1	1	0	0	1
28	P	F	36	0	1	1	1	0
29	P	M	19	1	1	1	1	0
30	A	M	28	1	0	0	0	0
31	P	M	37	0	1	1	1	1
32	A	M	23	1	1	0	0	0
33	A	M	30	1	1	1	0	0
34	P	M	15	1	0	1	0	0
35	A	M	26	1	1	0	0	0
36	P	F	45	1	1	1	1	1
37	A	M	31	1	0	0	0	0
38	A	M	50	1	1	1	1	0
39	P	M	28	1	0	0	1	0
40	P	M	26	0	0	0	0	0
41	P	M	14	1	0	0	1	1
42	A	M	31	1	0	0	0	0
43	P	M	13	1	1	1	0	1
44	P	M	27	1	0	0	1	0

(*Continued*)

Table 6.1 *Continued*

Patient	Trt	Sex	Age	Baseline	Visit1	Visit2	Visit3	Visit4
45	P	M	26	1	1	0	1	1
46	P	M	49	1	1	1	1	0
47	P	M	63	1	1	1	1	1
48	A	M	57	1	1	1	1	0
49	P	M	27	0	1	1	1	1
50	A	M	22	1	1	0	0	0
51	A	M	15	1	0	0	0	0
52	P	M	43	1	1	1	1	0
53	A	F	32	0	1	1	1	0
54	A	M	11	1	1	1	0	0
55	P	M	24	1	1	1	1	1
56	A	M	25	1	1	0	0	0
;								
run;								

as well. Covariates include sex, baseline status, and age in years. Tables 6.2 and 6.3 give the SAS PROC GENMOD output for assumed independent and AR(1) correlation structures, respectively. Table 6.4 gives the appropriate SAS PROC GENMOD code. Table 6.5 gives the appropiate R code.

The reader should note that two correlation structures are used for the analysis, independent and AR(1). Also note that the four observations on the same subject create the cluster that necessitates the consideration of correlation. The regressors include the categorical variable treatment (P and A), sex, and baseline observation (0 or 1). The quantitative variable age is also used. Much of the output under "Independent Correlation Structure" is standard output for the independence case. A binomial response is used. The goodness-of-fit information suggests no appreciable lack of fit using the logit link. The material under *Analysis of Initial Parameter Estimates* involves Wald inference using standard GLM technology with the *model-based standard error*. Note that significant effects include treatment, sex, and age, with active treatment being effective as compared to placebo. In addition, older patients tend to have poorer results as expected. Female patients also tend to get poorer results compared to males.

Note that the empirical variance–covariance matrix appears as well as the model-based variance–covariance matrix. For the most part the two matrices are not radically different. The *Analysis of GEE Parameter Estimates* uses empirical standard errors. As expected, the parameter estimates are identical to those from *Analysis of Initial Parameter Estimates,* since both are based on independence, but the standard errors are different because they come from different sources. In other words, despite the assumption of independence in the estimation procedure, standard errors are computed in a way that is robust to the failure of the independence assumption. The slightly different standard

Table 6.2 Output for Analysis of Respiratory Data with an Independent Correlation Structure

The GENMOD Procedure Independent Correlation Structure

Model Information

Description	Value
Data set	WORK.RESP2
Distribution	BINOMIAL
Link function	LOGIT
Dependent variable	OUTCOME
Observations used	224
Number of events	115
Number of trials	224

Class Level Information

Class	Levels	Values
PATIENT	56	1 2 3 4 5 6 7 8 9 10 11 12 13 14 15 16 17 18 19 20 21 22 23 24 25 26 27 28 29 30 31 32 33 34 35 36 37 38 39 40 41 42 43 44 45 46 47 48 49 50 51 52 53 54 55 56
TRT	2	A P
SEX	2	F M
BASELINE	2	0 1

Parameter Information

Parameter	Effect	TRT	SEX	BASELINE
PRM1	INTERCEPT			
PRM2	AGE			
PRM3	TRT	A		
PRM4	TRT	P		
PRM5	SEX		F	
PRM6	SEX		M	
PRM7	BASELINE			0
PRM8	BASELINE			1

Criteria for Assessing Goodness of Fit

Criterion	DF	Value	Value/DF
Deviance	219	263.7478	1.2043
Scaled deviance	219	263.7478	1.2043
Pearson chi-square	219	223.3980	1.0201
Scaled Pearson X2	219	223.3980	1.0201

Analysis of Initial Parameter Estimates

Parameter		DF	Estimate	Std Err	Chi Square	Pr > Chi
INTERCEPT		1	-1.0136	0.4513	5.0455	0.0247
AGE		1	0.0481	0.0138	12.0938	0.0005
TRT	A	1	-1.0702	0.3093	11.9687	0.0005
TRT	P	0	0.0000	0.0000	-	-

(Continued)

Table 6.2 *Continued*

SEX	F	1	2.1780	0.6819	10.2006	0.0014
SEX	M	0	0.0000	0.0000	-	-
BASELINE	0	1	-0.4980	0.5046	0.9740	0.3237
BASELINE	1	0	0.0000	0.0000	-	-
Scale[a]		0	1.0000	0.0000	-	-

GEE Model Information

Description	Value
Correlation structure	Independent
Subject effect PATIENT	(56 levels)
Number of clusters	56
Correlation matrix dimension	4
Maximum cluster size	4
Minimum cluster size	4

Covariance Matrix (Model-Based) Covariances Are Above the Diagonal and Correlations Are Below

Parameter Number	PRM1	PRM2	PRM3	PRM5	PRM7
PRM1	0.20308	-0.005363	-0.03785	-0.000395	-0.03323
PRM2	-0.86224	0.0001905	-0.000442	-0.000255	-0.000354
PRM3	-0.27192	-0.10373	0.09543	-0.01006	0.04098
PRM5	-0.001288	-0.02717	-0.04784	0.46378	-0.11260
PRM7	-0.14631	-0.05091	0.26325	-0.32809	0.25395

Covariance Matrix (Empirical)
Covariances Are Above the
Diagonal and Correlations Are Below

Parameter Number	PRM1	PRM2	PRM3	PRM5	PRM7
PRM1	0.27128	-0.005800	-0.08586	-0.02986	-0.04526
PRM2	-0.85321	0.0001704	0.0004291	0.0003803	0.0002764
PRM3	-0.50127	0.09998	0.10814	0.003484	0.02708
PRM5	-0.08442	0.04290	0.01560	0.46123	-0.39458
PRM7	-0.10213	0.02489	0.09679	-0.68288	0.72387

Working Correlation Matrix

	COL1	COL2	COL3	COL4
ROW1	1.0000	0.0000	0.0000	0.0000
ROW2	0.0000	1.0000	0.0000	0.0000
ROW3	0.0000	0.0000	1.0000	0.0000
ROW4	0.0000	0.0000	0.0000	1.0000

Analysis of GEE Parameter Estimates
Empirical Standard Error Estimates
Empirical 95% Confidence Limits

Parameter	Estimate	Std Err	Lower	Upper	Z	Pr>|Z|
INTERCEPT	-1.0136	0.5208	-2.0345	0.0072	-1.946	0.0516

(Continued)

Table 6.2 *Continued*

AGE		0.0481	0.0131	0.0225	0.0736	3.6826	0.0002
TRT	A	-1.0702	0.3288	-1.7147	-0.4256	-3.254	0.0011
TRT	P	0.0000	0.0000	0.0000	0.0000	0.0000	0.0000
SEX	F	2.1780	0.6791	0.8469	3.5091	3.2070	0.0013
SEX	M	0.0000	0.0000	0.0000	0.0000	0.0000	0.0000
BASELINE	0	-0.4980	0.8508	-2.1656	1.1695	-0.5853	0.5583
BASELINE	1	0.0000	0.0000	0.0000	0.0000	0.0000	0.0000
Scale[b]		0.9987	-	-	-	-	-

[a]The scale parameter is held fixed.
[b]The scale parameter for GEE estimation is computed as the square root of the normalized Pearson's chi-square.

errors have little effect on the conclusion made earlier. Note that, as expected, the working correlation is the identity matrix.

The next set of output in Table 6.3 is based on the AR(1) assumption on the correlation structure. The R code for the GEE analysis with AR(1) structure is given in Table 6.5 and the associated output appears in Table 6.6. It is standard for the GENMOD procedure to begin with analysis of initial parameter estimates despite the assumed structure. However, following this, the model-based and empirical variance–covariance matrices are given. Next are the generalized estimating equations parameter estimates, standard errors, and Wald inference information. The following should be noted:

1. Using model-based estimates, the assumption of independence generally gives smaller variances of parameter estimates than those in which AR(1) is assumed. This supports the general rule that not taking correlation into account results in overoptimism regarding variability in parameter estimates.
2. For this case (though not in general) when one considers parameter estimates and *empirical* standard errors, the results for the independence assumption and the AR(1) assumption are remarkably close, suggesting that the independence assumption would be quite appropriate. This is further supported by the working correlation matrix, which shows correlations that are quite mild.
3. The odds ratio in this situation must be dealt with somewhat differently, since a 0 is good and a 1.0 is poor. The probability of a desirable result is then interpreted to be $1-\pi$. Thus the appropriate odds ratio is

$$\frac{[(1-\hat{\pi})/\hat{\pi}]_{\mathrm{A}}}{[(1-\hat{\pi})/\hat{\pi}]_{\mathrm{P}}} = e^{1.0011} \cong 2.7$$

Consequently the odds of a favorable result increase by a factor of about 2.7 with the use of the drug. One could model the outcome as a success by

Table 6.3 Output for the Analysis of the Respiratory Data with an AR(1) Correlation Structure

The GENMOD Procedure

Model Information

Description	Value
Data set	WORK.RESP2
Distribution	BINOMIAL
Link function	LOGIT
Dependent variable	OUTCOME
Observations used	224
Number of events	115
Number of trials	224

Class Level Information

Class	Levels	Values
PATIENT	56	1 2 3 4 5 6 7 8 9 10 11 12 13 14 15 16 17 18 19 20 21 22 23 24 25 26 27 28 29 30 31 32 33 34 35 36 37 38 39 40 41 42 43 44 45 46 47 48 49 50 51 52 53 54 55 56
TRT	2	A P
SEX	2	F M
BASELINE	2	0 1

Parameter Information

Parameter	Effect	TRT	SEX	BASELINE
PRM1	INTERCEPT			
PRM2	AGE			
PRM3	TRT	A		
PRM4	TRT	P		
PRM5	SEX		F	
PRM6	SEX		M	
PRM7	BASELINE			0
PRM8	BASELINE			1

Criteria for Assessing Goodness of Fit

Criterion	DF	Value	Value/DF
Deviance	219	263.7478	1.2043
Scaled deviance	219	263.7478	1.2043
Pearson chi-square	219	223.3980	1.0201
Scaled Pearson X2	219	223.3980	1.0201
Log-likelihood	-	-131.8739	-

Analysis of Initial Parameter Estimates

Parameter	DF	Estimate	Std Err	Chi Square	Pr>Chi
INTERCEPT	1	-1.0136	0.4513	5.0455	0.0247
AGE	1	0.0481	0.0138	12.0938	0.0005

(Continued)

Table 6.3 *Continued*

TRT	A	1	-1.0702	0.3093	11.9687	0.0005
TRT	P	0	0.0000	0.0000	-	-
SEX	F	1	2.1780	0.6819	10.2006	0.0014
SEX	M	0	0.0000	0.0000	-	-
BASELINE	0	1	-0.4980	0.5046	0.9740	0.3237
BASELINE	1	0	0.0000	0.0000	-	-
Scale[a]		0	1.0000	0.0000	-	-

GEE Model Information

Description	Value
Correlation	AR(1)
Subject effect PATIENT	(56 levels)
Number of clusters	56
Correlation matrix dimension	4
Maximum cluster size	4
Minimum cluster size	4

Covariance Matrix (Model-Based) Covariances Are Above the Diagonal and Correlations Are Below

Parameter Number	PRM1	PRM2	PRM3	PRM5	PRM7
PRM1	0.28149	-0.007354	-0.05568	0.000866	-0.04762
PRM2	-0.86150	0.0002589	-0.0005	-0.000475	-0.000402
PRM3	-0.28914	-0.08569	0.13176	-0.01108	0.05557
PRM5	0.002127	-0.03846	-0.03979	0.58870	-0.15301
PRM7	-0.15129	-0.04216	0.25803	-0.33609	0.35205

Covariance Matrix (Empirical) Covariances Are Above the Diagonal and Correlations Are Below

Parameter Number	PRM1	PRM2	PRM3	PRM5	PRM7
PRM1	0.26575	-0.005558	-0.09011	-0.02891	-0.04658
PRM2	-0.84598	0.0001624	0.0005158	0.0002474	0.0003569
PRM3	-0.53537	0.12394	0.10661	0.007989	0.02589
PRM5	-0.08368	0.02896	0.03651	0.44917	-0.37709
PRM7	-0.10771	0.03338	0.09455	-0.67076	0.70360

Working Correlation Matrix

	COL1	COL2	COL3	COL4
ROW1	1.0000	0.2749	0.0756	0.0208
ROW2	0.2749	1.0000	0.2749	0.0756
ROW3	0.0756	0.2749	1.0000	0.2749
ROW4	0.0208	0.0756	0.2749	1.0000

Analysis of GEE Parameter Estimates Empirical Standard Error Estimates Empirical 95% Confidence Limits

Parameter	Estimate	Std Err	Lower	Upper	z	Pr > |Z|
INTERCEPT	-0.8866	0.5155	-1.8970	0.1237	-1.720	0.0854

(Continued)

Table 6.3 *Continued*

AGE		0.0431	0.0127	0.0181	0.0681	3.3801	0.0007
TRT	A	-1.0011	0.3265	-1.6411	-0.3612	-3.066	0.0022
TRT	P	0.0000	0.0000	0.0000	0.0000	0.0000	0.0000
SEX	F	2.0029	0.6702	0.6893	3.3165	2.9885	0.0028
SEX	M	0.0000	0.0000	0.0000	0.0000	0.0000	0.0000
BASELINE	0	-0.4918	0.8388	-2.1358	1.1522	-0.5863	0.5577
BASELINE	1	0.0000	0.0000	0.0000	0.0000	0.0000	0.0000
Scale[b]		0.9802	-	-	-	-	-

[a]The scale parameter is held fixed.
[b]The scale parameter for GEE estimation is computed as the square root of the normalized Pearson's chi-square.

Table 6.4 SAS PROC GENMOD Code for Output in Tables 6.2 and 6.3

```
*Code for producing output in Table 6.2

proc genmod data=RESP2 desc;
class PATIENT TRT SEX BASELINE;
model y = AGE TRT SEX BASELINE/dist=bin link=logit;
repeated subject=PATIENT/ type=ind corrw covb;
output out=reout pred=predict;
run;

*Code for producing output in Table 6.3

SAS Code:

proc genmod data=RESP2 desc;
class PATIENT TRT SEX BASELINE;
model y = AGE TRT SEX BASELINE/dist=bin link=logit;
repeated subject=PATIENT/ type=ar corrw covb;
output out=reout pred=predict;
run;
```

Table 6.5 R Code for Producing the AR(1) Analysis of Respiratory Data

```
library(gee)

model1 < gee(y~AGE + factor(TRT) + factor(SEX) + factor(BASELINE),
   family=binomial(logit), data=RESP2,id=PATIENT,corstr="AR-M")

summary(model1)
```

omitting the *desc* option in the *proc* statement. In such a case, the signs of the estimated coefficients change and the odds ratio comparing the active drug to the placebo is $exp(b_{\mathrm{TRT}})$. □

Table 6.6 R Output for the AR(1) Analysis of Respiratory Data

```
> model1 <-gee(y~AGE + factor(TRT) + factor(SEX)+factor(BASELINE),
         family=binomial(logit),data=RESP2,id=PATIENT,corstr="AR-M")

running glm to get initial regression estimate

     (Intercept)          AGE       factor(TRT)P     factor(SEX)M
     -0.40380749       0.04806584     1.07015398      -2.17798295
   factor(BASELINE)1
      0.49801320

> summary(model1)

GEE: GENERALIZED LINEAR MODELS FOR DEPENDENT DATA

Model:
Link:                          Logit
Variance to Mean Relation:     Binomial
Correlation Structure:         AR-M, M = 1

Call:
gee(formula = y ~ AGE + factor(TRT) + factor(SEX) + factor(BASELINE),
         id = PATIENT, data = RESP2, family = binomial(logit), corstr = "AR-M")

Summary of Residuals:

        Min              1Q          Median          3Q           Max
    -0.89801467     -0.36534284   0.04500683   0.44631869   0.80437516
```

```
Coefficients:

                       Estimate   Naive S.E.     Naive z  Robust S.E.   Robust z
(Intercept)         -0.37667898   0.97640432  -0.3857818   0.71241177 -0.5287377
Age                  0.04308092   0.01627185   2.6475734   0.01274542  3.3801107
factor(Trt)P         1.00112346   0.36710405   2.7270837   0.32651206  3.0661148
factor(Sex)M        -2.00288235   0.77598136  -2.5810959   0.67020167 -2.9884771
factor (Baseline)1   0.49180418   0.60007660   0.8195690   0.83881069  0.5863113

Estimated Scale Parameter: 0.9828147
Number of Iterations: 3

Working Correlation
                  [,1]        [,2]        [,3]        [,4]
        [1,] 1.00000000   0.2748647   0.0755506  0.02076619
        [2,] 0.27486468   1.0000000   0.2748647  0.07555059
        [3,] 0.07555059   0.2748647   1.0000000  0.27486468
        [4,] 0.02076619   0.0755506   0.2748647  1.00000000
```

In the preceding example the correlations are sufficiently small that the use of the independence correlation structure appears to be satisfactory, which often is the case when all correlations are smaller in magnitude than about 0.3. Some empirical evidence of this appears in Liang and Zeger (1986). □

Example 6.2. Leukemia. This is a biomedical example with 30 subjects (rats) that have a leukemic condition. Three chemotherapy type drugs are used. White and red blood cell counts are collected as covariates and the response is the number of cancer cell colonies. The data are collected on each subject at four different time periods. Poisson responses using a log link are assumed. Recall that the log is the canonical link for this situation. The data appear in Table 6.7. Both independence and AR(1) correlation structures are used for purposes of illustration. The results are in Table 6.8 and 6.9 for the independent and AR(1) analyses, respectively.

Table 6.7 Data on Leukemic Study in Rats

OBS	SUBJECT	DRUG	WBC1	WBC2	WBC3	WBC4	RBC1	RBC2	RBC3	RBC4	Y1	Y2	Y3	Y4
1	1	1	15	18	19	24	2	3	2	5	14	14	12	11
2	2	1	8	11	14	14	2	4	4	5	17	18	18	16
3	3	1	4	5	6	4	7	5	4	4	23	20	19	19
4	4	1	16	14	14	12	3	4	4	2	13	12	12	11
5	5	1	6	4	4	4	7	6	5	2	24	20	20	19
6	6	1	22	20	21	18	4	3	3	2	12	12	10	9
7	7	1	18	17	17	16	5	3	5	2	16	16	14	12
8	8	1	4	7	4	4	8	7	4	4	28	26	26	26
9	9	1	14	12	12	10	3	4	4	5	14	13	12	10
10	10	1	10	10	10	10	3	4	5	2	16	15	15	14
11	11	2	14	14	16	17	6	6	7	6	16	15	15	14
12	12	2	7	7	6	5	4	4	4	2	36	32	30	29
13	13	2	9	8	9	11	8	8	7	4	18	16	17	15
14	14	2	21	20	20	20	3	3	4	3	14	13	13	12
15	15	2	18	17	17	17	4	4	2	2	19	19	18	17
16	16	2	3	6	6	2	10	10	8	7	38	38	37	37
17	17	2	8	9	9	8	3	3	2	2	18	18	17	16
18	18	2	29	30	29	29	6	6	5	4	8	8	7	6
19	19	2	8	8	8	7	9	9	8	8	19	19	18	17
20	20	2	5	4	4	3	8	7	7	7	36	35	30	29
21	21	3	16	17	17	18	2	3	4	2	15	16	17	15
22	22	3	13	11	12	12	6	4	5	4	17	16	16	18
23	23	3	7	8	6	5	3	2	2	3	28	25	27	31
24	24	3	9	8	9	9	4	5	3	3	29	30	32	30
25	25	3	18	19	21	20	3	2	5	4	11	12	12	13
26	26	3	23	25	24	24	5	5	4	4	8	10	9	8
27	27	3	27	28	27	30	7	6	6	4	7	8	8	7
28	28	3	30	32	33	35	6	7	8	7	4	5	5	4
29	29	3	17	19	20	21	4	3	3	2	14	13	13	12
30	30	3	12	12	13	11	3	5	4	5	17	15	16	16

Table 6.8 Output for the Independent Correlation Structure Analysis of the Rat Data

The GENMOD Procedure

Model Information

Description	Value	Label
Data set	WORK.RAT	
Distribution	POISSON	
Link function	LOG	
Dependent variable	Y	Y1
Observations used	120	

Class Level Information

Class	Levels	Values
DRUG	3	1 2 3
SUBJECT	30	1 2 3 4 5 6 7 8 9 10 11 12 13 14 15 16 17 18 19 20 21 22 23 24 25 26 27 28 29 30

Parameter Information

Parameter	Effect	DRUG
PRM1	INTERCEPT	
PRM2	DRUG	1
PRM3	DRUG	2
PRM4	DRUG	3
PRM5	RBC	
PRM6	WBC	

Criteria for Assessing Goodness of Fit

Criterion	DF	Value	Value/DF
Deviance	115	88.2076	0.7670
Scaled deviance	115	88.2076	0.7670
Pearson chi-square	115	88.6115	0.7705
Scaled Pearson X2	115	88.6115	0.7705
Log-likelihood	-	4037.5056	-

Analysis of Initial Parameter Estimates

Parameter	DF	Estimate	Std Err	Chi Square	Pr > Chi
INTERCEPT	1	3.6285	0.0836	1886.1181	0.0001
DRUG 1	1	-0.2403	0.0588	16.7341	0.0001
DRUG 2	1	-0.0068	0.0598	0.0129	0.9096
DRUG 3	0	0.0000	0.0000	-	-
RBC	1	0.0021	0.0113	0.0355	0.8506
WBC	1	-0.0563	0.0035	255.1377	0.0001
Scale[a]	0	1.0000	0.0000	-	-

GEE Model Information

Description	Value
Correlation structure	Independent
Subject effect SUBJECT	(30 levels)
Number of clusters	30

(Continued)

Table 6.8 *Continued*

Correlation matrix dimension	4
Maximum cluster size	4
Minimum cluster size	4

Covariance Matrix (Model-Based) Covariances Are Above the Diagonal and Correlations Are Below

Parameter Number	PRM1	PRM2	PRM3	PRM5	PRM6
PRM1	0.005155	-0.001766	-0.001126	-0.000439	-0.000154
PRM2	-0.48733	0.002549	0.001453	-0.000012	0.000041
PRM3	-0.30517	0.55998	0.002640	-0.000167	0.0000377
PRM5	-0.62774	-0.02501	-0.33263	0.000095	5.1105E-6
PRM6	-0.70832	0.26826	0.24211	0.17308	9.1756E-6

Covariance Matrix (Empirical) Covariances Are Above the Diagonal and Correlations Are Below

Parameter Number	PRM1	PRM2	PRM3	PRM5	PRM6
PRM1	0.01667	-0.006315	-0.003207	-0.001676	-0.000373
PRM2	-0.58762	0.006928	0.004396	-0.000034	0.0001867
PRM3	-0.25078	0.53315	0.009811	-0.000563	0.0001128
PRM5	-0.6445	-0.02004	-0.28224	0.0004056	6.2419E-6
PRM6	-0.69376	0.53844	0.27341	0.07439	0.0000174

Working Correlation Matrix

	COL1	COL2	COL3	COL4
ROW1	1.0000	0.0000	0.0000	0.0000
ROW2	0.0000	1.0000	0.0000	0.0000
ROW3	0.0000	0.0000	1.0000	0.0000
ROW4	0.0000	0.0000	0.0000	1.0000

Analysis of GEE Parameter Estimates
Empirical Standard Error Estimates

Parameter	Estimate	Std Err	Empirical 95% Confidence Limits Lower	Upper	z	Pr > |Z|
INTERCEPT	3.6285	0.1291	3.3755	3.8816	28.103	0.0000
DRUG 1	-0.2403	0.0832	-0.4035	-0.0772	2.887	0.0039
DRUG 2	-0.0068	0.0990	-0.2009	0.1873	-0.0686	0.9453
DRUG 3	0.0000	0.0000	0.0000	0.0000	0.0000	0.0000
RBC -	0.0021	0.0201	-0.0373	0.0416	0.1061	0.9155
WBC	-0.0563	0.0042	-0.0645	-0.0481	13.51	0.0000
Scale[b]	0.8593	-	-	-	-	-

[a]The scale parameter is held fixed.
[b]The scale parameter for GEE estimation is computed as the square root of the normalized Pearson's chi-square.

Table 6.9 Output for the AR(1) Correlation Structure Analysis for the Rat Data

The GENMOD Procedure

Model Information

Description	Value	Label
Data	set WORK.RAT	
Distribution	POISSON	
Link function	LOG	
Dependent variable	Y	Y1
Observations used	120	

Class Level Information

Class	Levels	Values
DRUG	3	1 2 3
SUBJECT	30	1 2 3 4 5 6 7 8 9 10 11 12 13 14 15 16 17 18 19 20 21 22 23 24 25 26 27 28 29 30

Parameter Information

Parameter	Effect	DRUG
PRM1	INTERCEPT	
PRM2	DRUG	1
PRM3	DRUG	2
PRM4	DRUG	3
PRM5	RBC	
PRM6	WBC	

Criteria for Assessing Goodness of Fit

Criterion	DF	Value	Value/DF
Deviance	115	88.2076	0.7670
Scaled deviance	115	88.2076	0.7670
Pearson chi-square	115	88.6115	0.7705
Scaled Pearson X2	115	88.6115	0.7705
Log-likelihood	-	4037.5056	-

Analysis of Initial Parameter Estimates

Parameter	DF	Estimate	Std Err	Chi Square	Pr > Chi
INTERCEPT	1	3.6285	0.0836	1886.1181	0.0001
DRUG 1	1	-0.2403	0.0588	16.7341	0.0001
DRUG 2	1	-0.0068	0.0598	0.0129	0.9096
DRUG 3	0	0.0000	0.0000	-	-
RBC	1	0.0021	0.0113	0.0355	0.8506
WBC	1	-0.0563	0.0035	255.1377	0.0001
Scale[a]	0	1.0000	0.0000	-	-

GEE Model Information

Description	Value
Correlation structure	AR(1)

(Continued)

Table 6.9 *Continued*

Subject effect SUBJECT	(30 levels)
Number of clusters	30
Correlation matrix dimension	4
Maximum cluster size	4
Minimum cluster size	4

Covariance Matrix (Model-Based) Covariances Are Above the Diagonal and Correlations Are Below

Parameter Number	PRM1	PRM2	PRM3	PRM5	PRM6
PRM1	0.01438	-0.009524	-0.009142	-0.000263	-0.000381
PRM2	-0.64682	0.01508	0.008180	8.256E-6	0.0001435
PRM3	-0.64963	0.56767	0.01377	-0.000114	0.0001479
PRM5	-0.26917	-0.008250	-0.11905	0.0000664	-1.345E-6
PRM6	-0.63750	0.23474	0.25324	-0.03314	0.0000248

Covariance Matrix (Empirical) Covariances Are Above the Diagonal and Correlations Are Below

Parameter Number	PRM1	PRM2	PRM3	PRM5	PRM6
PRM1	0.02374	-0.01721	-0.01816	-0.000218	-0.000535
PRM2	-0.86326	0.01674	-0.01418	0.0001546	0.0002418
PRM3	-0.79266	0.73717	-0.02212	0.0000521	0.0003371
PRM5	-0.21805	0.18373	-0.05385	0.0000423	1.766E-6
PRM6	-0.68623	0.36916	0.44773	-0.05363	0.0000256

Working Correlation Matrix

	COL1	COL2	COL3	COL4
ROW1	1.0000	0.9228	0.8516	0.7858
ROW2	0.9228	1.0000	0.9228	0.8516
ROW3	0.8516	0.9228	1.0000	0.9228
ROW4	0.7858	0.8516	0.9228	1.0000

Analysis of GEE Parameter Estimates
Empirical Standard Error Estimates
Empirical 95% Confidence Limits

Parameter	Estimate	Std Err	Lower	Upper	z	Pr > |Z|
INTERCEPT	3.1360	0.1541	2.8341	3.4380	20.355	0.0000
DRUG 1	-0.1242	0.1294	-0.3778	0.1293	-0.9601	0.3370
DRUG 2	0.0953	0.1487	-0.1962	0.3868	0.6408	0.5217
DRUG 3	0.0000	0.0000	0.0000	0.0000	0.0000	0.0000
RBC -	0.0221	0.0065	0.0094	0.0349	3.4024	0.0007
WBC -	0.0305	0.0051	-0.0404	-0.0206	6.028	0.0000
Scale[b]	1.1152	-	-	-	-	-

[a]The scale parameter is held fixed.
[b]The scale parameter for GEE estimation is computed as the square root of the normalized Pearson's chi-square.

Note that once again we have the lack-of-fit material. However, it should be emphasized that this output is only valid when observations are uncorrelated, since the likelihood material makes use of the independence assumption. As before, the *Analysis of Initial Parameter Estimates* presumes independence and uses the model-based standard errors. Note here in this very preliminary analysis that drug 1 is significant and the negative sign suggests a negative (reduction) effect on the mean counts. A clear indication of an erroneous choice of the correlation structure (independence at this point) is that a large discrepancy occurs between the model-based and empirical variance–covariance structures. The analysis using empirical standard errors gives somewhat different results. At this point there is every reason to believe that a nonindependence correlation structure is preferable.

The output under the AR(1) structure in Table 6.9 begins with the same lack-of-fit information as before and the same analysis of initial parameter estimates. The first result that deals with the new correlation structure is the model-based variance–covariance structure followed by the empirical variance–covariance structure. There certainly are some discrepancies here, and the empirical structure is probably more appropriate. The working correlation under the AR(1) structure supports the need to accommodate the correlation within subjects. The final portion of the analysis gives the results of the Wald inference. Note that the parameter estimates are quite different from what they were in the independence case. The comparison is a classic case of the independence analysis producing overly optimistic results. The extremely large correlations results in much larger standard errors associated with the drugs, and indeed, the analysis shows insufficient evidence to support any one of the drugs over the other two. The only significant model terms in the exponential model are the white and red blood cell counts. □

Example 6.3. Substrate Camber. This example is taken from Myers, Montgomery, and Anderson-Cook (2009). It involves a designed experiment in a semiconductor plant. Six factors are employed, and it is of interest to study the curvature or camber of the substrate devices produced in the plant. There is a lamination process, and the camber measurement is made four times on the same device produced. The goal is to model the camber taken in 10^{-4} in./in. as a function of the design variables. Each design variable is taken at two levels and the design is a 2^{6-2} fractional factorial. The camber measurement is known to be nonnormal with a heavy right-tailed distribution. In addition, it is clear that the measurement taken on the same device may well be correlated. As a result we have repeated measures, which represent the source of the correlation. A gamma response is assumed with a log link. Thus

$$\ln \mu = \beta_0 + \beta_1 x_1 + \beta_2 x_2 + \beta_3 x_3 + \beta_4 x_4 + \beta_5 x_5 + \beta_6 x_6$$

Table 6.10 Factor Levels in the Semiconductor Experiment

Run	Lamination Temperature (°C)	Lamination Trime (s)	Lamination Pressure (ton)	Firing Temperature (°C)	Firing Cycle Time (h)	Firing Dew Point (°C)
1	−1	−1	−1	−1	−1	−1
2	+1	−1	−1	−1	+1	+1
3	−1	+1	−1	−1	+1	−1
4	+1	+1	−1	−1	−1	+1
5	−1	−1	+1	−1	+1	+1
6	+1	−1	+1	−1	−1	−1
7	−1	+1	+1	−1	−1	+1
8	+1	+1	+1	−1	+1	−1
9	−1	−1	−1	+1	−1	+1
10	+1	−1	−1	+1	+1	−1
11	−1	+1	−1	+1	+1	+1
12	+1	+1	−1	+1	−1	−1
13	−1	−1	+1	+1	+1	−1
14	+1	−1	+1	+1	−1	+1
15	−1	+1	+1	+1	−1	−1
16	+1	+1	+1	+1	+1	+1

Table 6.11 Response Data in the Semiconductor Experiment

Run	1	2	3	4
1	0.0167	0.0128	0.0149	0.0185
2	0.0062	0.0066	0.0044	0.0020
3	0.0041	0.0043	0.0042	0.0050
4	0.0073	0.0071	0.0039	0.0030
5	0.0047	0.0047	0.0040	0.0089
6	0.0219	0.0258	0.0147	0.0296
7	0.0121	0.0090	0.0092	0.0086
8	0.0255	0.0250	0.0226	0.0169
9	0.0032	0.0023	0.0077	0.0069
10	0.0078	0.0158	0.0060	0.0045
11	0.0043	0.0027	0.0028	0.0028
12	0.0186	0.0137	0.0158	0.0159
13	0.0110	0.0086	0.0101	0.0158
14	0.0065	0.0109	0.0126	0.0071
15	0.0155	0.0158	0.0145	0.0145
16	0.0093	0.0124	0.0110	0.0133

Note: Camber for measurement (in./in.).

Tables 6.10 and 6.11 give data for a ± 1 coding on the design variables. Table 6.12 is an abbreviated printout from SAS PROC GENMOD. An AR(1) correlation structure is assumed. All of the printout shown reflects the AR(1) analysis.

Table 6.12 GENMOD Results for the Semiconductor Experiment

Covariance Matrix (Model-Based)
Covariances Are Above the Diagonal and Correlations Are Below

Parameter Number	PRM1	PRM2	PRM3	PRM4	PRM5	PRM6	PRM7
PRM1	0.003162	1.226E-19	-2.06E-19	-6.02E-20	-6.02E-20	4.9494E-20	-2.36E-20
PRM2	3.876E-17	0.003162	-1.33E-19	-6.02E-20	-6.02E-20	1.294E-20	8.604E-20
PRM3	-6.53E-17	-4.21E-17	0.003162	1.294E-20	8.604E-20	-6.02E-20	1.294E-20
PRM4	-1.9E-17	-1.9E-17	4.091E-18	0.003162	8.604E-20	-1.33E-19	-6.02E-20
PRM5	-1.9E-17	-1.9E-17	2.721E-17	2.721E-17	0.003612	1.294E-20	-6.02E-20
PRM6	1.565E-17	4.091E-18	-1.9E-17	-4.21E-17	4.091E-18	0.003162	8.604E-20
PRM7	-7.47E-18	2.721E-17	4.091E-18	-1.9E-17	-1.9E-17	2.721E-17	0.003162

Analysis of GEE Parameter Estimates
Model-Based Standard Error Estimates
Model-Based 95% Confidence Limits

Parameter	Estimate	Std Err	Lower	Upper	z	Pr > |Z|
INTERCEPT	-4.6789	0.0562	-4.7891	-4.5686	-83.20	0.0000
X1	0.1620	0.0562	0.0518	0.2722	2.8804	0.0040
X2	0.0107	0.0562	-0.0996	0.1209	0.1895	0.8497
X3	0.3039	0.0562	0.1937	0.4141	5.4042	0.0000
X4	-0.0288	0.0562	-0.1390	0.0814	-0.5122	0.6085
X5	-0.1971	0.0562	-0.3074	-0.0869	-3.506	0.0005
X6	-0.3790	0.0562	-0.4892	-0.2688	-6.739	0.0000
Scale	0.3486	-	-	-	-	-

(Continued)

Table 6.12 *Continued*

Covariance Matrix (Empirical)

Covariances Are Above the Diagonal and Correlations Are Below

Parameter Number	PRM1	PRM2	PRM3	PRM4	PRM5	PRM6	PRM7
PRM1	0.003412	-0.000363	0.0003458	-0.000812	-0.000903	-0.000453	-0.001044
PRM2	-0.10638	0.003412	0.001082	0.002248	0.002157	-0.000877	0.002545
PRM3	0.10134	0.31717	0.003412	0.000877	0.0006464	0.002248	0.0008035
PRM4	-0.23796	0.65881	0.25706	0.003412	0.002545	-0.001082	0.002157
PRM5	-0.26479	0.63231	0.18946	0.74584	0.003412	-0.0008035	0.002248
PRM6	-0.13284	0.25706	0.65881	0.31717	0.23550	-0.003412	0.0006464
PRM7	-0.30607	0.74584	0.23550	0.63231	0.65881	-0.18946	0.003412

Working Correlation Matrix

	COL1	COL2	COL3	COL4
ROW1	1.0000	0.3632	0.1319	0.0479
ROW2	-0.3632	1.0000	0.3632	0.1319
ROW3	0.1319	0.3632	1.0000	0.3632
ROW4	0.0479	0.1319	0.3632	1.0000

Analysis of GEE Parameter Estimates

Empirical Standard Error Estimates

Empirical 95% Confidence Limits

Parameter	Estimate	Std Err	Lower	Upper	z	Pr > |Z|
INTERCEPT	-4.6789	0.0584	-4.7933	-4.5644	-80.10	0.0000
X1	0.1620	0.0584	0.0475	0.2765	2.7731	0.0056
X2	0.0107	0.0584	-0.1038	0.1251	0.1824	0.8552
X3	0.3039	0.0584	0.1894	0.4184	5.2029	0.0000
X4	-0.0288	0.0584	-0.1433	0.0857	-0.4932	0.6219
X5	-0.1971	0.0584	-0.3116	-0.0827	3.375	-0.0007
X6	-0.3790	0.0584	-0.4935	-0.2645	6.488	0.0000
Scale	0.3486	-	-	-	-	-

Note: The scale parameter for GEE estimation is computed as the square root of the normalized Pearson's chi square

Note that there is essentially no difference between the analysis with the model-based standard errors and the analysis using the empirical standard errors. Variables x_1, x_3, x_5, and x_6 have a significant effect on the camber. Note that the working correlation matrix reflects fairly small correlations. Two of these variables, x_1, and x_3, have positive effects while x_5 and x_6 have negative effects on curvature. The effect x_6 may be defined as the ratio of mean response at high level to that of low level. In this experiment, this estimated effect is $e^{2b_6} = e^{-0.758} = 0.46$; thus the mean camber at the high level is 46% of that at the low level. The multiplicative effects of the other variables may be computed as well.

One very important experimental design issue should be considered here. Since the log link was used with the gamma distribution, an ordinary GLM analysis with assumption of no correlation yields a diagonal asymptotic variance–covariance matrix of parameter estimates as we discussed in Chapter 5. Note also in this case that the model-based and the empirical variance–covariance matrices of parameter estimates are diagonal with equal values on the main diagonal. This simplification produces equal standard errors in both cases. Chapter 8 provides more discussion on the effect of experimental design. □

Example 6.4. Wafer Resistivity. This study involves five factors in a manufacturing process for an integrated circuit in a 2^{5-1} fractional factorial design in which it was of interest to learn how the factors influence the resistivity of the wafer. The five factors are A = implant dose, B = temperature, C = time, D = oxide thickness, and E = furnace position. Each factor is run at two levels. As in the previous example there is a lack of complete randomization that forces a correlation structure, but in addition there are design levels that define whole plots and subplots in a *split plot* type structure. Table 6.13

Table 6.13 Resistivity Experiment

Run	A	B	C	D	$E = ABCD$	Resistivity
1	−	−	−	−	+	15.1
2	+	−	−	−	−	20.6
3	−	+	−	−	−	68.7
4	+	+	−	−	+	101.0
5	−	−	+	−	−	32.9
6	+	−	+	−	+	46.1
7	−	+	+	−	+	87.5
8	+	+	+	−	−	119.0
9	−	−	−	+	−	11.3
10	+	−	−	+	+	19.6
11	−	+	−	+	+	62.1
12	+	+	−	+	−	103.2
13	−	−	+	+	+	27.1
14	+	−	+	+	−	40.3
15	−	+	+	+	−	87.7
16	+	+	+	+	+	128.3

gives the data. The implementation of the design could not be accomplished in a completely randomized fashion due to cost constraints. The temperature variable was difficult to change; hence all of the low (−1) values are run first followed by the high (+ 1) values. As a result the implementation is that of a split plot design with temperature as the whole plot variable and the other variables as subplot variables. Consequently, the observations within each whole plot are assumed to be correlated with an exchangeable correlation structure. Resistivity is known to approximately obey a gamma distribution. A log link is used, and all main effects are included in addition to the two-factor interaction, *AB*. Included in Table 6.14 is the SAS PROC GENMOD output for generalized estimating equations. Note from the analysis using model-based estimates of the standard errors, that the main effects, *A*, *B*, and *C*, are significant. The only modeled interaction, *AB*, is not statistically significant. Also included in the analysis is the model-based estimate of the variance–covariance matrix of parameter estimates. It is of interest to note that the correlations (below the main diagonal) are quite small. Much of this, of course, is derived from the quality of the experimental design, which is an *orthogonal design* in the case of linear models. Also listed here is the working correlation matrix, which suggests that the correlation among observations within whole plots is quite mild. Thus the analysis shows that correlation is near zero.

It may be instructive to reveal an analysis of the same data set using generalized linear models with a gamma distribution, log link, and independent observations. The following shows the Wald inference for this analysis. The parameter estimates vary little from those using the generalized estimating equations analysis.

Coefficient	Estimate	Standard Error	*P*-Value
Inferent	3.8643	0.0374	0.0001
B	0.6705	0.0374	0.0001
C	0.2592	0.0374	0.0001
A	0.1995	0.0374	0.0001
D	−0.0428	0.0374	0.2523
E	0.0032	0.0374	0.9339
AB	0.0031	0.0374	0.9342
Scale	44.7379	15.7586	

The empirical standard errors are not shown. It is not likely that these estimates are appropriate because there are only two whole plots. These whole plots play the role of subjects in the generalized estimating equations scenario. As a result the pooling over subjects does not allow sufficient

Table 6.14 GENMOD Output for Analysis of Resistivity Data with 2^{5-1} Design in a Split Plot Structure

Analysis of GEE Parameter Estimates
Model-Based Standard Error Estimates
Model-Based 95% Confidence Limits

Parameter	Estimate	Std Err	Lower	Upper	z	Pr > |Z|
INTERCEPT	3.8648	0.0406	3.7853	3.9443	95.264	0.0000
B	0.6707	0.0406	0.5912	0.7502	16.532	0.0000
C	0.2597	0.0373	0.1867	0.3327	6.9713	0.0000
A	0.1992	0.0338	0.1329	0.2656	5.8861	0.0000
D	-0.0440	0.0394	-0.1212	0.0332	-1.116	0.2643
E	0.0095	0.0354	-0.0599	0.0789	0.2691	0.7878
AB	0.0055	0.0338	-0.0608	0.0718	0.1636	0.8700
Scale	0.1483	-	-	-	-	-

Covariance Matrix (Empirical)
Covariances Are Above the Diagonal and Correlations Are Below

Parameter Number	PRM1	PRM2	PRM3	PRM4	PRM5	PRM6	PRM7
PRM1	0.001646	5.873E-22	-4.33E-20	.-4E-20	1.648E-19	-3.79E-20	-9.17E-21
PRM2	3.568E-19	0.001646	-3.4E-21	-7.54E-21	3.317E-20	1.294E-19	-3.98E-20
PRM3	-2.86E-17	-2.25E-18	0.001388	0.0000345	0.0000463	2.015E-20	5.938E-20
PRM4	-2.91E-17	-5.49E-18	0.02735	0.001146	0.0000384	-2.5E-20	-1.68E-20
PRM5	1.031E-16	2.076E-17	0.03156	0.02882	0.001551	3.046E-20	1.914E-20
PRM6	-2.64E-17	9.011E-17	1.528E-I7	-2.08E-17	2.184E-17	0.001254	0.0000302
PRM7	-6.68E-18	-2.9E-17	4.712E-17	-1.47E-17	1.436E-17	0.02517	0.001144

(Continued)

Table 6.14 *Continued*

Working Correlation Matrix

	COL1	COL2	COL3	COL4	COL5	COL6	COL7	COL8
ROW1	1.0000	-0.1329	-0.1329	-0.1329	-0.1329	-0.1329	-0.1329	-0.1329
ROW2	-0.1329	1.0000	-0.1329	-0.1329	-0.1329	-0.1329	-0.1329	-0.1329
ROW3	-0.1329	-0.1329	1.0000	-0.1329	-0.1329	-0.1329	-0.1329	-0.1329
ROW4	-0.1329	-0.1329	-0.1329	1.0000	-0.1329	-0.1329	-0.1329	-0.1329
ROW5	-0.1329	-0.1329	-0.1329	-0.1329	1.0000	-0.1329	-0.1329	-0.1329
ROW6	-0.1329	-0.1329	-0.1329	-0.1329	-0.1329	1.0000	-0.1329	-0.1329
ROW7	-0.1329	-0.1329	-0.1329	-0.1329	-0.1329	-0.1329	1.0000	-0.1329
ROW8	-0.1329	-0.1329	-0.1329	-0.1329	-0.1329	-0.1329	-0.1329	1.0000

information for estimation of the variance–covariance matrix of parameter estimates.

In each analysis temperature appears to be the most significant factor. Increasing the temperature increases the resistivity. The multiplicative effect is to increase the resistivity by a factor of $e^{2(0.6705)} = 3.822$ as one changes the temperature from the -1 level to the $+1$ level.

Example 6.5. Yield of Grits. This example is taken from the SAS PROC MIXED user's guide in which four factors are varied and their effect on yield of grits is sought. The process involves milling of corn in which the corn is milled into a grit product. The factors are A, moisture content of the corn, B, roll gap, C, screen size, and D, roller speed. One factor, moisture content, is very difficult to change. The experiment is displayed in Table 6.15. The response is the number of grams of grits per 1000 grams of corn. Note that all four factors are at three levels and there are 10 batches, indicating batches of corn. Three experiments are run on each batch. A second-order model is fit to the data. A normally distributed response is assumed with an identity link and, of course, constant variances. Observations within the same batch are assumed to be correlated via an exchangeable correlation structure. Table 6.16 gives the GEE analysis and Table 6.17 provides results from using ordinary least squares and thus ignoring the correlation. The estimated standard errors in Table 6.17 come from the matrix $(\mathbf{X}'\mathbf{X})^{-1} \cdot s^2$, which assumes that the analyst ignored the correlation in total. Table 6.18 provides the estimated correlation matrix. Note that the correlations are substantially higher than the correlations that appear in the previous example. As a result the discrepancy among like coefficients between GEE and OLS is marked. Note also that the GEE analysis produced a significant value for the linear and quadratic term in B while the OLS analysis resulted in insignificant values. In addition, the quadratic term in C is marginally significant using GEE's and quite insignificant using OLS. Note that the resulting conclusions using the model–based and empirical standard errors are essentially the same even though the actual standard errors differ a great deal in some cases.

This example provides a distinct lesson. If the correlations within the whole plot are not moderate or small, the GEE analysis can provide completely different conclusions from that of an OLS approach. While there is no clear yardstick, correlations along the order of 0.3 or higher need to be considered seriously and then taken into account. Values of 0.3 or lower may well indicate that independence and their OLS is quite appropriate. In addition, one should be quite careful in the use of the empirical method for computing standard error when the render of whole plots/subplots is small. □

Table 6.15 Design Matrix and Response Data for Grits Example

	Factor				
Batch	A	B	C	D	Yield
1	1	1	1	1	505
1	1	−1	−1	−1	493
1	1	−1	1	−1	491
2	1	1	−1	0	498
2	1	1	−1	−1	504
2	1	−1	1	0	500
3	−1	0	−1	−1	494
3	−1	0	1	0	498
3	−1	−1	0	1	498
4	0	−1	−1	0	496
4	0	0	1	1	503
4	0	−1	0	−1	496
5	−1	−1	1	1	503
5	−1	1	1	−1	495
5	−1	−1	1	−1	494
6	0	0	0	0	486
6	0	1	1	−1	501
6	0	1	−1	1	490
7	−1	1	0	0	494
7	−1	1	1	1	497
7	−1	−1	1	−1	492
8	1	−1	1	1	503
8	1	0	0	−1	499
8	1	0	−1	1	493
9	1	1	1	−1	505
9	1	1	0	1	500
9	1	−1	−1	1	490
10	−1	−1	−1	1	494
10	−1	1	−1	−1	497
10	−1	−1	−1	−1	495

6.6 SUMMARY

The use of generalized estimating equations can be a valuable tool in either biological or industrial applications when response values are clearly correlated and the user is primarily interested in a marginal model. The analyst has available several decisions to make over and above those discussed in previous chapters for analysis. In addition to linear predictor terms, distributions, and link decisions, one is faced with the issue of what type of correlation structure to use. However, independence should always be considered. In many cases if one deviates from independence, the results may be quite insensitive to the choice of correlation statute, a concept that we saw illustrated in our examples.

Table 6.16 GEE Analysis for Corn Data Using Exchangeable Correlation Structure

Analysis of GEE Parameter Estimates Empirical Standard Error Estimates

			Empirical 95% Confidence Limits			
Parameter	Estimate	Std Err	Lower	Upper	Z	Pr > |Z|
A	1.6475	0.4478	0.7698	2.5251	3.6792	0.0002
B	1.6054	0.3137	0.9905	2.2203	5.117	0.0000
C	2.7427	0.3109	2.1333	3.352	8.8217	0.0000
D	-0.0726	0.0943	-0.2574	0.1122	-0.7696	0.4416
AA	0.0524	2.0077	-3.8827	3.9875	0.0261	0.9792
BB	2.7439	1.3892	0.0211	5.4668	1.9752	0.0482
CC	1.429	0.9651	-0.4625	3.3206	1.4807	0.1387
DD	0.4941	0.9771	-1.4211	2.4092	0.5056	0.6131
AB	2.1386	0.2726	1.6043	2.6730	7.8439	0.0000
AC	0.429	0.3912	-0.3378	1.1957	1.0966	0.2728
AD	-0.4261	0.2146	-0.8468	-0.0054	-1.985	0.0471
BC	0.3674	0.2336	-0.0905	0.8253	1.5727	0.1158
BD	-1.599	0.3608	-2.3061	-0.8918	-4.432	0.0000
CD	2.5858	0.3116	1.975	3.1966	8.2976	0.0000
Scale[a]	2.2220	-	-	-	-	-

Analysis of GEE Parameter Estimates

Model-Based Standard Error Estimates

			Model-Based 95% Confidence Limits			
Parameter	Estimate	Std Err	Lower	Upper	Z	Pr > |Z|
INTERCEPT	492.8582	1.5588	489.8030	495.9134	316.18	0.0000
A	1.6475	0.6584	0.3570	2.9379	2.5021	0.0123
B	1.6054	0.3656	0.8889	2.3219	4.3915	0.0000
C	2.7427	0.3930	1.9724	3.5130	6.9785	0.0000
D	-0.0726	0.3377	-0.7344	0.5893	-0.2149	0.8298

(Continued)

Table 6.16 *Continued*

AA	0.0524	1.4784	-2.8453	2.9501	0.0355	0.9717
BB	2.7439	0.9221	0.9366	4.5513	2.9756	0.0029
CC	1.4290	0.8145	-0.1673	3.0254	1.7545	0.0793
DD	0.4941	0.8703	-1.2116	2.1998	0.5677	0.5702
AB	2.1386	0.3892	1.3759	2.9014	5.4952	0.0000
AC	0.4290	0.4589	-0.4705	1.3285	0.9348	0.3499
AD	-0.4261	0.3973	-1.2047	0.3526	-1.072	0.2835
BC	0.3674	0.4087	-0.4335	1.1684	0.8991	0.3686
BD	-1.5990	0.4483	-2.4775	-0.7204	-3.567	0.0004
CD	2.5858	0.4031	1.7959	3.3758	6.4155	0.0000
Scale[a]	2.2220	-	-	-	-	-.

[a] The scale parameter of GEE estimation is computed as the square root the normalized Pearson's chi-square.

Table 6.17 OLS Analysis Results Using Corn Data

	OLS		
Source	b	Standard	*p*-value
A	1.763	0.616	**0.0119**
B	1.023	0.608	0.1133
C	2.626	0.608	**0.0006**
D	−0.067	0.608	0.9134
AA	0.022	1.398	0.9879
BB	1.592	1.385	0.2685
CC	1.060	1.398	0.4598
DD	0.476	1.394	0.7374
AB	2.304	0.671	**0.0037**
AC	0.388	0.678	0.5754
AD	−0.156	0.690	0.8236
BC	0.311	0.669	0.6486
BD	−2.127	0.684	**0.0072**
CD	3.006	0.687	**0.0005**

Table 6.18 Estimated Correlation Matrix Using Corn Data

	Working Correlation Matrix		
	COL1	COL2	COL3
ROW1	1.0000	0.5237	0.5237
ROW2	0.5237	1.0000	0.5237
ROW3	0.5237	0.5237	1.0000

Nevertheless, a sense of comfort is realized if the analyst performs the analysis with the use of more than one type of structure. In the next chapter we discuss models for correlated data which offer both subject-specific and marginal interpretations.

EXERCISES

6.1 An engineer studied the effect of pulp preparation and processing temperature on the strength of paper. The nature of the process dictated that pulp preparation was hard-to-change. As a result one large batch of pulp was split into smaller batches, which then receive the temperature treatment. Thus the experimental structure is a split plot. The data follow. Use generalized estimating equations to analyze these data.

(a) Assume an exchangeable structure.

(b) The engineers actually conducted the experiment by ramping up the temperature, which means that the first batch from a large pulp batch always receives a temperature of 200, the next receives a temperature of 225, and so on. In such a case, an AR(1) structure may make sense. Perform the analysis accordingly.

(c) Discuss these analyses.

	Replicate (or Block) 1			Replicate (or Block) 2			Replicate (or Block) 3		
Pulp Preparation Method	1	2	3	1	2	3	1	2	3
Temperature (°F)									
200	30	34	29	28	31	31	31	35	32
225	35	41	26	32	36	30	37	40	34
250	37	38	33	40	42	32	41	39	39
275	36	42	36	41	40	40	40	44	45

6.2 Steel is normalized by heating above the critical temperature, soaking, and then air cooling. This process increases the strength of the steel, refines the grain, and homogenizes the structure. An experiment is performed to determine the effect of temperature and heat treatment time on the strength of normalized steel. Two temperatures and three times are selected. The experiment is performed by heating the oven to a randomly selected temperature and inserting three specimens. After 10 minutes one specimen is removed, after 20 minutes the second is removed, and after 30 minutes the final specimen is removed. Then the temperature is changed to the other level and the process is repeated. Four shifts are required to collect the data, as are shown below.

(a) Analyze assuming an exchangeable structure.

(b) By the actual conduct of the experiment an AR(1) structure is reasonable. Reanalyze accordingly.

(c) Discuss these analyses.

		Temperature (°F)	
Shift	Time (min)	1500	1600
	10	63	89
1	20	54	91
	30	61	62
	10	50	80
2	20	52	72
	30	59	69
	10	48	73
3	20	74	81
	30	71	69
	10	54	88
4	20	48	92
	30	59	64

6.3 An experiment is designed to study pigment dispersion in paint. Four different mixes of a particular pigment are studied. The procedure consists of preparing a particular mix and then applying that mix to a panel by three application methods (brushing, spraying, and rolling). The response measured is the percentage reflectance of pigment. Three days are required to run the experiment, and the data obtained follow. Analyze these data, assuming an exchangeable structure.

Day	Application Method	Mix			
		1	2	3	4
1	1	64.5	66.3	74.1	66.5
	2	68.3	69.5	73.8	70.0
	3	70.3	73.1	78.0	72.3
2	1	65.2	65.0	73.8	64.8
	2	69.2	70.3	74.5	68.3
	3	71.2	72.8	79.1	71.5
3	1	66.2	66.5	72.3	67.7
	2	69.0	69.0	75.4	68.6
	3	70.8	74.2	80.1	72.4

6.4 Reiczigel (1999) analyzed an experiment that studies the effect of cholagogues on changes in gallbladder volume (GBV) of dogs. The experiment used two different cholagogues (A and B) and tap water (W) as a control. Six healthy dogs were assigned to each treatment. The GBVs were measured by ultrasound every ten minutes for two hours. The data follow.

(a) It is probably reasonable to assume an AR(1); therefore perform a thorough analysis of these data using this assumption.

(b) Repeat part (a) using the robust approach.

(c) Discuss any differences, if any, in these two analyses.

6.5 Stiger, Barnhart, and Williamson (1999) analyzed a double-blind clinical trail that compared a hypnotic drug to a placebo in subjects with insomnia. The researchers surveyed each patient about the amount of time it took them to go to sleep prior to treatment and again after 2 weeks of treatment. The data follow. Perform a thorough analysis of these data using the robust approach.

Treatment	Dog	Minutes after treatment												
		0	10	20	30	40	50	60	70	80	90	100	110	120
Cholecystokinin	1	17.70	10.35	10.78	11.44	11.20	12.38	12.68	12.30	14.00	14.64	14.96	14.18	16.78
	2	17.22	11.30	11.30	13.28	14.08	13.98	14.74	15.63	17.60	17.34	17.38	17.36	17.64
	3	14.24	9.20	9.40	9.62	10.10	10.08	9.60	9.70	11.23	11.20	11.96	12.20	13.98
	4	39.58	26.88	26.20	29.80	31.50	32.75	34.45	35.64	36.62	38.65	38.56	39.20	39.36
	5	13.33	7.15	7.82	7.94	8.40	8.94	9.28	9.95	10.40	10.95	11.70	12.10	12.35
	6	16.16	8.36	9.53	9.80	9.64	9.84	10.70	11.26	12.12	12.60	13.98	14.52	14.78
8CIanobutin	1	16.35	13.65	13.10	13.58	14.03	15.45	15.58	15.56	15.62	16.10	16.28	16.74	16.25
	2	15.65	13.08	12.35	12.76	13.78	13.76	13.54	14.18	14.40	15.16	15.20	15.18	13.40
	3	12.68	9.68	10.70	10.98	11.12	11.78	12.02	11.95	12.16	12.25	12.40	12.55	12.54
	4	21.88	15.92	15.18	17.04	18.6	18.98	19.26	20.38	21.32	21.03	21.8	21.08	22.65
	5	12.78	9.03	9.28	9.54	9.38	9.88	9.94	10.14	10.34	11.5	11.83	11.78	12.08
	6	15.58	11.5	11.88	12.06	12.58	12.98	13.00	13.00	13.04	13.18	13.88	13.43	13.85
Control	1	20.75	19.83	19.98	18.84	19.10	19.50	19.75	19.64	20.00	19.13	20.15	19.45	19.43
	2	13.88	13.60	13.73	13.16	13.44	13.62	13.86	13.58	14.28	14.10	13.12	13.53	13.42
	3	11.92	11.74	11.84	10.90	11.75	11.45	11.98	12.38	11.70	11.48	11.80	11.20	12.03
	4	26.38	26.90	27.73	27.73	27.56	28.43	27.54	26.50	27.94	27.58	27.56	27.64	28.83
	5	13.30	13.18	13.52	13.43	13.4	13.25	13.28	13.24	13.44	12.98	12.60	13.48	13.08
	6	13.80	13.86	13.06	13.76	13.82	13.80	13.86	13.84	13.76	13.82	13.50	13.72	13.70

Treatment	Initial	Time to Falling asleep (min) Follow-up < 20	20-30	30-60	> 60	Total	Proportion
Active	< 20	7	4	1	0	12	0 101
	20-30	11	5	2	2	20	0 168
	30-60	13	23	3	1	40	0 336
	> 60	9	17	13	8	47	0 395
Total		40	49	19	11	119	
Proposition		0.336	0.412	0.16	0.092		
Placebo	< 20	7	4	2	1	14	0·117
	20–30	14	5	1	0	20	0·167
	30-60	6	9	18	2	35	0·292
	> 60	4	11	14	22	51	0·425
Total		31	29	35	25	120	
Proportion		0.258	0.242	0.292	0.208		

6.6 Potthoff and Roy (1964) analyzed a longitudinal study of dental growth in children. Specifically, they measured the distance from the center of the pituitary gland to the maxillary fissure for children at ages 8, 10, 12, and 14. The data follow.

Individual Girl	Age (in years) 8	10	12	14	Individual Boy	Age (in years) 8	10	12	14
1	21	20	21.5	23	1	26	25	29	31
2	21	21.5	24	25.5	2	21.5	22.5	23	26.5
3	20.5	24	24.5	26	3	23	22.5	24	27.5
4	23.5	24.5	25	26.5	4	25.5	27.5	26.5	27
5	21.5	23	22.5	23.5	5	20	23.5	22.5	26
6	20	21	21	22.5	6	24.5	25.5	27	28.5
7	21.5	22.5	23	25	7	22	22	24.5	26.5
8	23	23	23.5	24	8	24	21.5	24.5	25.5
9	20	21	22	21.5	9	23	20.5	31	26
10	16.5	19	19	19.5	10	27.5	28	31	31.5
11	24.5	25	28	28	11	23	23	23.5	25
					12	21.5	23.5	24	28
					13	17	24.5	26	29.5
					14	22.5	25.5	25.5	26
					15	23	24.5	26	30
					16	22	21.5	23.5	25

(a) Perform a thorough analysis of these data assuming an exchangeable structure.

(b) Repeat part (a) assuming an AR(1) structure.

(c) Repeat (a) part using the robust approach.

(d) Compare and contrast the three analyses. Which seems most appropriate?

6.7 Thall and Vail (1990) compared a new treatment medication for epilepsy to a placebo. They recorded each patient's age and 8-week baseline seizure counts. They then monitored the 2-week seizure counts for each patient during four successive but nonoverlapping periods. The data follow. Perform a thorough analysis of these data. Is there a problem with overdispersion? Discuss the effect of the assumed correlation structure on your results.

Y_1	Y_2	Y_3	Y_4	Trt	Base	Age
5	3	3	3	0	11	31
3	5	3	3	0	11	30
2	4	0	5	0	6	25
4	4	1	4	0	8	36
7	18	9	21	0	66	22
5	2	8	7	0	27	29
6	4	0	2	0	12	31
40	20	23	12	0	52	42
5	6	6	5	0	23	37
14	13	6	0	0	10	28
26	12	6	22	0	52	36
12	6	8	4	0	33	24
4	4	6	2	0	18	23
7	9	12	14	0	42	36
16	24	10	9	0	87	26
11	0	0	5	0	50	26
0	0	3	3	0	18	28
37	29	28	29	0	111	31
3	5	2	5	0	18	32
3	0	6	7	0	20	21
3	4	3	4	0	12	29
3	4	3	4	0	9	21
2	3	3	5	0	17	32
8	12	2	8	0	28	25
18	24	76	25	0	55	30
2	1	2	1	0	9	40
3	1	4	2	0	10	19
13	15	13	12	0	47	22
11	14	9	8	1	76	18
8	7	9	4	1	38	32
0	4	3	0	1	19	20
3	6	1	3	1	10	30

Y_1	Y_2	Y_3	Y_4	Trt	Base	Age
2	6	7	4	1	19	18
4	3	1	3	1	24	24
22	17	19	16	1	31	30
5	4	7	4	1	14	35
2	4	0	4	1	11	27
3	7	7	7	1	67	20
4	18	2	5	1	41	22
2	1	1	0	1	7	28
0	2	4	0	1	22	23
5	4	0	3	1	13	40
11	14	25	15	1	46	33
10	5	3	8	1	36	21
19	7	6	7	1	38	35
1	1	2	3	1	7	25
6	10	8	8	1	36	26
2	1	0	0	1	11	25
102	65	72	63	1	151	22
4	3	2	4	1	22	32
8	6	5	7	1	41	25
1	3	1	5	1	32	35
18	11	28	13	1	56	21
6	3	4	0	1	24	41
3	5	4	3	1	16	32
1	23	19	8	1	22	26
2	3	0	1	1	25	21
0	0	0	0	1	13	36
1	4	3	2	1	12	37

6.8 Lipsitz, Kim, and Zhao (1994) analyzed data from an arthritis clinical trail. Patients were randomly assigned to receive either auranofin or a placebo. They surveyed each patient at baseline, 1 month, 2 months, and 3 months on how they felt. A value of 3 indicated "good," a value of 2 indicated "fair," and a value of 1 indicated "poor." The data follow. They report data on 23 patients; however, we have dropped the three patients that have missing values for at least one survey. Perform a thorough analysis of these data.

				Self-assessment		
Sex	Age	Treatment	Baseline	1 month	2 months	3 months
M	49	P	3	3	3	3
M	59	A	3	2	1	1
M	65	P	2	3	2	3
M	60	P	3	2	2	3

				Self-assessment		
Sex	Age	Treatment	Baseline	1 month	2 months	3 months
M	56	P	2	2	1	2
M	34	A	1	3	3	3
M	53	A	2	1	1	2
M	36	A	3	3	3	3
F	31	A	2	2	2	2
M	66	P	1	2	2	2
M	55	A	2	3	3	3
F	60	A	1	2	2	2
M	28	A	2	1	3	1
F	47	P	1	3	1	2
F	44	P	1	3	3	2
F	55	P	2	2	1	2
F	31	P	1	1	2	2
F	27	A	2	2	3	3
F	63	A	2	2	2	3
M	55	P	1	1	2	2

CHAPTER 7

Random Effects in Generalized Linear Models

To this point, our statistical models involved only **fixed effects**. For all of our models up to now, the levels of the regressors used in the study were the only levels of interest to the analyst. As a result, we were restricted to making statistical inferences only over these specific levels. For example, the transistor gain study from Example 2.1 examined the impact of emitter drive-in time and emitter dose on transistor gain for an integrated circuit. The levels for drive-in time ranged from 195 to 255, and the levels for dose ranged from 4.00 to 4.72. These levels were not randomly selected from a much larger population.

The key notion underlying fixed effects is that the levels we use in the study are not randomly selected. Almost always, quantitative regressors are fixed effects. Categorical regressors are fixed effects when the levels used are the only ones available, and the analyst is content to making inferences over these specific levels.

The levels used in a study for **random effects** represent a random sample from a much larger population of possible levels. For example, patients in a biomedical study are often random effects. The analyst selects the patients for the study from a large population of possible people. The focus of all statistical inference is not on the specific patients selected; rather, the focus is on the population of all possible patients. The key point underlying all random effects is this focus on the population and not on the specific units selected for the study. Random effects are almost always categorical.

Many studies involve mixed effects models where some regressors are fixed effects and some are random effects. Mixed effects models are useful for a wide array of study types. For instance, in medical studies, a subject's response to some treatment is monitored over time (a longitudinal study). Subjects in these studies are selected at random from a population of possible subjects. In

Generalized Linear Models, Second Edition, by Myers, Montgomery, Vining, and Robinson

wildlife studies, the abundance of some species is determined for each of several observation stations (transects) over several years. Observation stations such as transects are selected at random from an entire study area. In industrial split plot designs, several sub plot observations are taken for each whole plot unit. The whole plot units in the study are selected at random from a population of possible whole plot units. The common thread through these applications is that the data are grouped into *clusters* according to the levels of one or more classification factors (by subjects, transects, whole plot units, etc.). Unlike classical linear regression models (Chapter 2), nonlinear regression models (Chapter 3), and generalized linear models (Chapters 4 and 5), mixed effects models are built to accommodate the inherent correlation that exists among observations within the same cluster. These models also enable the user to consider the clusters of observations (subjects, transects, whole plots, etc.) as random samples from a common probability distribution, thus enabling the user to make more general interpretations.

This chapter discusses the extension of linear regression models to linear mixed effect models (Section 7.1) and generalized linear models to generalized linear mixed models (Section 7.2). Section 7.3 outlines a Bayesian approach to generalized linear mixed models. The emphasis of this section is on some of the advantages afforded by such an approach.

7.1 LINEAR MIXED EFFECTS MODELS

7.1.1 Linear Regression Models

To begin our discussion, assume that we wish to model some response, y, as a function of a set of k regressors, $x_1,x_2,\ldots,x_k$. Also, assume that the structure of the data involves m clusters of observations, each with n_i observations ($i = 1, 2,\ldots m$). One approach would allow for separate intercepts and slopes for each of the clusters of observations. As an example, suppose that the response is a function of a single, continuous regressor x. A possible linear model is

$$y = \beta_0 + \beta x + \sum_{j=1}^{m-1} \gamma_j z_j + \sum_{j=1}^{m-1} (\beta\gamma)_j z_j x + \varepsilon \tag{7.1}$$

where z_j denotes the indicator variable for the jth cluster, γ_j is the intercept term for the jth cluster, $(\beta\gamma)_j$ is the slope term for the jth cluster, and the ε are assumed normal with mean zero and constant variance, σ_ε^2. In matrix notation, the Equation (7.1) is

$$\mathbf{y} = \mathbf{X}\boldsymbol{\beta} + \boldsymbol{\varepsilon} \tag{7.2}$$

where $\mathbf{y}$ is the $(n \times 1)$ vector of responses with $n = \sum_{j=1}^{m} n_j$, $\mathbf{X}$ is the the $(n \times p)$ model matrix with $p = 2+2(m-1)$, $\boldsymbol{\beta}$ is the associated $(p \times 1)$ vector of model parameters, and $\boldsymbol{\varepsilon}$ is the $(n \times 1)$ vector of model errors. Equation (7.2) is similar to the linear model discussed in Chapter 2 except the vector of model errors now is a multivariate normal with $E(\boldsymbol{\varepsilon}) = \mathbf{0}$, Var $(\boldsymbol{\varepsilon}) = \mathbf{S}$, and $\mathbf{S} = \sigma_\varepsilon^2 \mathbf{I}_n$. The regression parameters in Equation (7.2) are the parameters in $\boldsymbol{\beta}$ as well as the γ_j's and the $(\beta\gamma)_j$'s. The parameters β_0 and β_1 represent the intercept and slope for the reference station, and γ_j and $(\beta\gamma)_j$ represent the intercept and slope deviation, respectively, from the reference for the jth station.

Example 7.1. Kukupa Counts. Westbrooke and Robinson (2009) reported on monitoring efforts of the Kukupa, New Zealand's only native pigeon, in a study area exposed to intensive pest control. The nesting success rates of the Kukupa, which are slow breeders, have been hampered by the presence of stoats and rats. The pest control effort was designed to reduce the populations of stoats and rats in this study area. Biologists noted average Kukupa counts at 10 observation stations over 12 years from 1995 to 2006. The 10 observation stations were randomly chosen locations in a large study area.

Given the fact that each station was repeatedly observed over the study period, we can structure the data in clusters by observation station. A portion of the data are provided in Table 7.1. In this case, the data are *balanced*; that is, there are equal numbers of observations (i.e., $n_i = 12$) for each of the 10 observation stations, yielding a total of $n = 120$ points in the data set. A goal of the study is to determine if there is a significant trend in the average counts (*Avg_Count*) over time. The entire data set is in *Kukupu.xls* at ftp://ftp.wiley.com/public/sci_tech_med/generalized_linear.

A naïve approach to this problem fits the multiple linear regression model from Equation (7.1) in order to allow for different intercepts and slope for each of the observation stations. Such an approach requires the estimation of a total of $p = 20$ regression parameters since there are 10 observation stations. Figure 7.1 shows the scatterplots and corresponding model fits for each of the 10 observation stations.

Fitting the model in Equation (7.1) enables the user to determine if there is a significant effect across time on an observation station by observation station basis with $(\beta_0 + \gamma_j)$ and $(\beta + (\beta\gamma)_j)$ denoting the intercept and slope, respectively, for the jth station ($j = 1, 2, \ldots, 9$). Figure 7.1 suggests that Kukupa numbers are increasing over time for the stations observed in this study. □

Although useful in a preliminary sense, fitting the model in Equation (7.1) is problematic on two fronts: (1) the repeated observations taken at each station over time are not independent; and perhaps more importantly, (2) it is not the desire of the biologist to make inferences only regarding these stations. Instead, the biologist needs to determine (1) if there is a significant time effect for the entire study area, and (2) how variable the time effects are from one station to the next. The analyst can broaden the scope of inference by treating the station variable as a *random* effect whose levels correspond to the different station numbers.

Table 7.1 Data Structure for Kukupa Data Set

Station	Average Count (Avg_Count)	Year	Total Noise (Tot_Noise)
81	0.000	1995	2.333
81	0.167	1996	1.833
⋮	⋮	⋮	⋮
80	0.833	2004	0.500
80	0.500	2005	0.833
80	2.167	2006	0.500

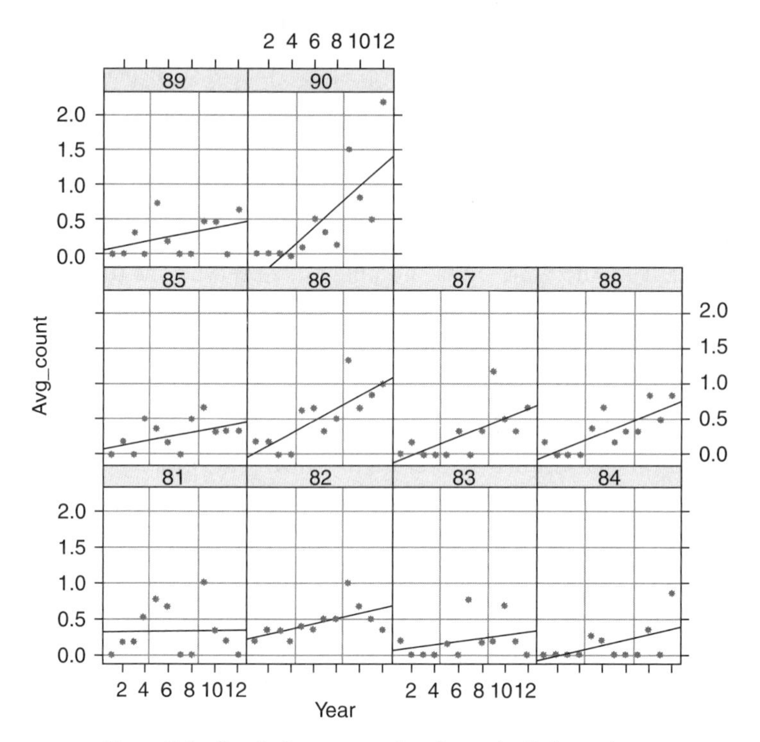

Figure 7.1 Simple linear regression fits to the Kukupa data.

7.1.2 General Linear Mixed Effects Models

To begin our discussion of the general linear mixed model, we return to the scenario of having m clusters of data where we are interested in modeling the response as a function of a single regressor variable. The linear *mixed* model is

$$y = \beta_0 + \beta x + \sum_{j=1}^{m} \delta_{0,j} z_j + \sum_{j=1}^{m} \delta_{1,j} z_j x + \varepsilon \qquad (7.3)$$

where z_j denotes the indicator variable for the jth cluster, $\delta_{0,j}$ is the *random intercept* term for the jth cluster with $\delta_{0,j} \sim N(0, \sigma^2_{\delta_0})$, $\delta_{1,j}$ is the *random slope* term for the jth cluster with $\delta_{1,j} \sim N(0, \sigma^2_{\delta_1})$, and the ε are as assumed in Equation (7.1). The variance component $\sigma^2_{\delta_0}$ represents how variable the intercepts are from one cluster to the next, and likewise, $\sigma^2_{\delta_1}$ represents how the slopes vary from one cluster to the next. The regression parameters β_0 and β_1 denote the average intercept and slope across all clusters in the population.

Although Equations (7.1) and (7.3) look similar in that they both accommodate a separate intercept and slope for each of the m clusters, there are several important differences. We compare Equations (7.1) and (7.3) using the Kukupa example as a reference. When treating the stations as fixed indicators as in Equation (7.1) the relationship of $E(y)$ to x for the reference station is $E(y) = \beta_0 + \beta_1 x$. The relationship of y to x for the jth non reference station is $E(y) = (\beta_0 + \gamma_j) + (\beta + (\beta\gamma)_j)x$. Thus γ_j and $(\beta\gamma)_j$ represent how the intercept and slope for the jth station deviates from the intercept and slope of the *reference station*. On the other hand, Equation (7.3) presents the general linear mixed model. In this case β_0 and β_1 denote the average intercept and slope across the population from which the stations are drawn (e.g., the entire study areas). We then can attribute any deviations that exist from station to station as random fluctuations. For instance, $\delta_{0,j}$ and $\delta_{1,j}$ represent random deviations in the intercept and slope, respectively, for the jth station. Both models allow for differences in the intercepts and slopes across the clusters. Equation (7.1) only allows the analyst to compare the specific m clusters in the study. Equation (7.3) focuses not on the specific clusters in the study but rather on the population from which the clusters are drawn. As such, the $\delta_{0,j}$'s and $\delta_{1,j}$'s are known as *random effects* since they correspond to the jth cluster that is randomly drawn from a population of clusters. The proper analysis based on Equation (7.3) centers on $\sigma^2_{\delta_0}$ and $\sigma^2_{\delta_1}$, since they describe the entire population of interest.

Another difference in the two models is that the linear mixed model implies that observations made on the same cluster are correlated whereas no correlation is implied by Equation (7.1). Specifically, in Equation (7.3), the covariance between any two observations y_{ji} and y_{jk} $(i \neq k)$, in the same cluster, is $\mathrm{Cov}(y_{ji}, y_{jk}) = \sigma^2_{\delta_0} + \sigma_{\delta_0\delta_1} x_{jk} + \sigma_{\delta_0\delta_1} x_{ji} + \sigma^2_{\delta_1} x_{jk} x_{ji}$. If the cluster to cluster variability is negligible (i.e., $\sigma^2_{\delta_0}, \sigma^2_{\delta_1} = 0$), then observations within the same cluster are independent.

General Linear Mixed Model in Matrix Form

Before discussing parameter estimation for the general linear mixed model, it is helpful to express the general linear mixed model in matrix notation. We write the general linear mixed model as

$$\mathbf{y} = \mathbf{X}\boldsymbol{\beta} + \mathbf{Z}\boldsymbol{\delta} + \boldsymbol{\varepsilon} \tag{7.4}$$

where $\mathbf{X}$, $\boldsymbol{\beta}$, and $\boldsymbol{\varepsilon}$ are as defined in Equation (7.2), $\mathbf{Z}$ is the $(n \times qm)$ matrix of predictor variables for the random effects, and $\boldsymbol{\delta}$ is the associated $(qm \times 1)$

vector containing the m levels of each of the q random effects. Furthermore, $\boldsymbol{\delta}$ is assumed multivariate normal with $E(\boldsymbol{\delta}) = \mathbf{0}$ and Var $(\boldsymbol{\delta}) = \mathbf{D}$ with $\mathbf{D}$ being a $(qm \times qm)$ positive definite matrix. Typically, the model errors and the random effects are assumed to be independent (i.e., Cov $(\boldsymbol{\varepsilon},\boldsymbol{\delta}) = \mathbf{0}$). Relating the fixed effects portion of Equation (7.4) to the fixed effects in Equation (7.3),

$$\mathbf{X} = \begin{bmatrix} \mathbf{1}_{n_1} & \mathbf{x}_{n_1} \\ \mathbf{1}_{n_2} & \mathbf{x}_{n_2} \\ \vdots & \vdots \\ \mathbf{1}_{n_{m-1}} & \mathbf{x}_{n_{m-1}} \\ \mathbf{1}_{m} & \mathbf{x}_{n_m} \end{bmatrix}, \quad \mathbf{y} = \begin{bmatrix} \mathbf{y}_{n_1} \\ \mathbf{y}_{n_2} \\ \vdots \\ \mathbf{y}_{n_{m-1}} \\ \mathbf{y}_{n_m} \end{bmatrix}, \quad \text{and } \boldsymbol{\beta} = \begin{bmatrix} \beta_0 \\ \beta_1 \end{bmatrix} \tag{7.5}$$

with $\mathbf{1}_{n_j}$ being a $(n_j \times 1)$ vector of ones, $\mathbf{x}_{n_j}$ is the $(n_j \times 1)$ vector of regressor values for the jth cluster, $\mathbf{y}_{n_j}$ is the $(n_j \times 1)$ vector of responses for the jth cluster, and β_0 and β_1 denote the population averaged intercept and slope, respectively. Exemplifying this notation with the Kukupa example, $\mathbf{1}_{n_1}, \mathbf{x}_{n_1}, \mathbf{y}_{n_1}$, and $\boldsymbol{\beta}$ are

$$\mathbf{1}_{n_1} = \begin{bmatrix} 1 \\ 1 \\ 1 \\ 1 \\ 1 \\ 1 \\ 1 \\ 1 \\ 1 \\ 1 \\ 1 \\ 1 \end{bmatrix} \quad \mathbf{x}_{n_1} = \begin{bmatrix} 1 \\ 2 \\ 3 \\ 4 \\ 5 \\ 6 \\ 7 \\ 8 \\ 9 \\ 10 \\ 11 \\ 12 \end{bmatrix}, \quad \mathbf{y}_{n_1} = \begin{bmatrix} 0.00 \\ 0.17 \\ 0.17 \\ 0.50 \\ 0.75 \\ 0.67 \\ 0.00 \\ 0.00 \\ 1.00 \\ 0.33 \\ 0.17 \\ 0.00 \end{bmatrix}, \quad \text{and } \boldsymbol{\beta} \begin{bmatrix} \beta_0 \\ \beta_1 \end{bmatrix} \tag{7.6}$$

respectively, where $\mathbf{x}_{n_1}$ represents the fixed effects vector of study years for which observations were conducted at station 1 and $\mathbf{y}_{n_1}$ is the corresponding vector of average counts (Avg-count) at station 1. The fixed effects β_0 and β_1 represent the intercept and slope parameters, respectively, for the line that relates the mean Kukupa count, across the population of observation stations, to time.

Relating the random effects portion of Equation (7.4) to that of Equation (7.3),

$$\mathbf{Z} = \begin{bmatrix} \mathbf{1}_{n_1} & \mathbf{0}_{n_1} & \cdots & \mathbf{0}_{n_1} & \mathbf{x}_{n_1} & \mathbf{0}_{n_1} & \cdots & \mathbf{0}_{n_1} \\ \mathbf{0}_{n_2} & \mathbf{1}_{n_2} & \cdots & \mathbf{0}_{n_2} & \mathbf{0}_{n_2} & \mathbf{x}_{n_2} & \cdots & \mathbf{0}_{n_2} \\ \cdots & \vdots & \cdots & \vdots & \vdots & \vdots & & \vdots \\ \mathbf{0}_{n_{m-1}} & \mathbf{0}_{n_{m-1}} & \cdots & \mathbf{0}_{n_{m-1}} & \mathbf{0}_{n_{m-1}} & \mathbf{0}_{n_{m-1}} & \cdots & \mathbf{0}_{n_{m-1}} \\ \mathbf{0}_{n_m} & \mathbf{0}_{n_m} & \cdots & \mathbf{1}_{n_m} & \mathbf{0}_{n_m} & \mathbf{0}_{n_m} & \cdots & \mathbf{x}_{n_m} \end{bmatrix} \text{ and } \boldsymbol{\delta} = \begin{bmatrix} \delta_{0,1} \\ \delta_{0,2} \\ \vdots \\ \delta_{0,m} \\ \delta_{1,1} \\ \delta_{1,2} \\ \vdots \\ \delta_{1,m} \end{bmatrix} \tag{7.7}$$

In thinking of the contribution of the random effects, it is often helpful to partition $\mathbf{Z}$ and $\boldsymbol{\delta}$ according to the number of random factors. Partitioning $\mathbf{Z}$ and $\boldsymbol{\delta}$ from Equation (7.7), we have

$$\mathbf{Z} = \begin{bmatrix} \mathbf{Z}_1^* & \vdots & \mathbf{Z}_2^* \end{bmatrix} \text{ and } \boldsymbol{\delta} = \begin{bmatrix} \boldsymbol{\delta}_1^* \\ \cdots \\ \boldsymbol{\delta}_2^* \end{bmatrix}$$

where

$$\mathbf{Z}_1^* = \begin{bmatrix} \mathbf{1}_{n_1} & \mathbf{0}_{n_1} & \cdots & \mathbf{0}_{n_1} \\ \mathbf{0}_{n_2} & \mathbf{1}_{n_2} & \cdots & \mathbf{0}_{n_2} \\ \vdots & \vdots & \cdots & \vdots \\ \mathbf{0}_{n_{m-1}} & \mathbf{0}_{n_{m-1}} & \cdots & \mathbf{0}_{n_{m-1}} \\ \mathbf{0}_{n_m} & \mathbf{0}_{n_m} & \cdots & \mathbf{1}_{n_m} \end{bmatrix}, \; \mathbf{Z}_2^* = \begin{bmatrix} \mathbf{x}_{n_1} & \mathbf{0}_{n_1} & \cdots & \mathbf{0}_{n_1} \\ \mathbf{0}_{n_2} & \mathbf{x}_{n_2} & \cdots & \mathbf{0}_{n_2} \\ \vdots & \vdots & \cdots & \vdots \\ \mathbf{0}_{n_{m-1}} & \mathbf{0}_{n_{m-1}} & \cdots & \mathbf{0}_{n_{m-1}} \\ \mathbf{0}_{n_m} & \mathbf{0}_{n_m} & \cdots & \mathbf{x}_{n_m} \end{bmatrix} \tag{7.8}$$

and

$$\boldsymbol{\delta}_1^* = \begin{bmatrix} \delta_{0,1} \\ \delta_{0,2} \\ \vdots \\ \delta_{0,m} \end{bmatrix}, \; \boldsymbol{\delta}_2^* = \begin{bmatrix} \delta_{1,1} \\ \delta_{1,2} \\ \vdots \\ \delta_{1,m} \end{bmatrix} \tag{7.9}$$

with $\boldsymbol{\delta}_1^*$ and $\boldsymbol{\delta}_2^*$ representing the vectors containing the m levels of the random intercept and slope terms, respectively. Using this notion of partitions, the general linear mixed model in Equation (7.4) is sometimes written as

$$\mathbf{y} = \mathbf{X}\boldsymbol{\beta} + \sum_{l=1}^{q} \mathbf{Z}_l^* \boldsymbol{\delta}_l^* + \boldsymbol{\varepsilon} \tag{7.10}$$

where $\boldsymbol{\delta}_l^*$ is the vector whose values comprise the levels of the lth random effect and $\mathbf{Z}_l^*$ is the associated matrix of model terms.

For the Kukupa example, if stations differed from one another only in terms of the intercept, the appropriate general linear mixed model is

$$\mathbf{y} = \mathbf{X}\boldsymbol{\beta} + \mathbf{Z}_1^*\boldsymbol{\delta}_1^* + \boldsymbol{\varepsilon} \tag{7.11}$$

where $\mathbf{X}$ is as defined in Equation (7.5), $\boldsymbol{\beta} = \begin{bmatrix} \beta_0 \\ \beta_1 \end{bmatrix}$, and $\mathbf{Z}_1^*$ and $\boldsymbol{\delta}_1^*$ are as defined in Equations (7.8) and (7.9), respectively, with $n_j = 12$ and $j = 1, 2, \ldots, 10$. If, however, the stations differ from one another due to random fluctuations in both the intercept and slope parameters, the appropriate general linear mixed model is

$$\mathbf{y} = \mathbf{X}\boldsymbol{\beta} + \mathbf{Z}_1^*\boldsymbol{\delta}_1^* + \mathbf{Z}_2^*\boldsymbol{\delta}_2^* + \boldsymbol{\varepsilon} \tag{7.12}$$

where $\mathbf{X}$, $\boldsymbol{\beta}$, $\mathbf{Z}_1^*$, and $\boldsymbol{\delta}_1^*$ are identical to what is written in Equation (7.11) and $\mathbf{Z}_2^*$ and $\boldsymbol{\delta}_2^*$ are defined in Equations (7.8) and (7.9), respectively.

7.1.3 Covariance Matrix, V

Recall that the classical multiple linear regression model assumes that all the observations have equal variances and that the observations are uncorrelated. In such a case, the variance-covariance matrix of $\mathbf{y}$, $\text{Var}(\mathbf{y}) = \mathbf{V} = \sigma_\varepsilon^2\mathbf{I}_n$. When the data structure is such that observations are clustered and random effects are present in the model, the clusters themselves are assumed to be independent, but observations within the same cluster are correlated. Thus the variance-covariance matrix of $\mathbf{y}$ is no longer diagonal but *block diagonal.* The variance-covariance matrix of $\mathbf{y}$ can be derived by taking the variance operator through the general linear mixed model from Equation (7.4). Thus

$$\text{Var}(\mathbf{y}) = \mathbf{V} = \text{Var}(\mathbf{X}\boldsymbol{\beta} + \mathbf{Z}\boldsymbol{\delta} + \boldsymbol{\varepsilon})$$

Since the model errors and the random effects are assumed to be uncorrelated (i.e., $\text{Cov}(\boldsymbol{\varepsilon},\boldsymbol{\delta}) = \mathbf{0}$),

$$\mathbf{V} = \text{Var}(\mathbf{X}\boldsymbol{\beta}) + \text{Var}(\mathbf{Z}\boldsymbol{\delta}) + \text{Var}(\boldsymbol{\varepsilon})$$

Taking advantage of the fact that $\mathbf{X}\boldsymbol{\beta}$ is the fixed effects portion of our model, $\text{Var}(\mathbf{X}\boldsymbol{\beta}) = \mathbf{0}$. In addition, $\mathbf{Z}$ is a matrix of constants; thus we can simplify $\mathbf{V}$ to

$$\mathbf{V} = \mathbf{Z}\text{Var}(\boldsymbol{\delta})\mathbf{Z}' + \text{Var}(\boldsymbol{\varepsilon}) \tag{7.13}$$

We see that the covariance matrix of the responses in a general linear mixed model is composed of two parts: (1) the residual variation that is manifested in the dispersion among the model errors (i.e., the *ε's*) and (2) the random effect variation given by $\mathbf{Z}\text{Var}(\boldsymbol{\delta})\mathbf{Z}'$. The variation among the model errors is equivalent to the variance of the conditional response, that is,

$$\text{Var}(\boldsymbol{\varepsilon}) = \text{Var}(\mathbf{y}|\boldsymbol{\delta}) = \mathbf{S} \tag{7.14}$$

which is often referred to as the *within-subjects* variation. Classical linear regression assumes that $\mathbf{S}$ is diagonal with $\mathbf{S} = \sigma_\varepsilon^2\mathbf{I}_n$. Let $\mathbf{D}$ denote Var $(\boldsymbol{\delta})$. Since the random effects are assumed to follow normal distributions, we have $\boldsymbol{\delta} \sim MVN\,(\mathbf{0},\mathbf{D})$. The diagonal elements of $\mathbf{D}$ tell us about how the levels of the random factors vary from subject to subject, after adjusting for the fixed effect covariates. As a result, we say that $\mathbf{D}$ models the *between-subjects* variation since it reflects the natural dispersion from one cluster (or one subject) to another. The covariance matrix of the responses in a general linear mixed model is then

$$\mathbf{V} = \mathbf{ZDZ}' + \mathbf{S} \tag{7.15}$$

The analysis of the general linear model makes various assumptions about the structures of $\mathbf{D}$ and $\mathbf{S}$, depending on the problem at hand. When the variation among model errors (the *ε's*) is due to sampling error or just random fluctuations, they are typically taken to be independently and identically distributed with constant variance, σ_ε^2. However, in situations where the observations within a cluster have a clear ordering or structure, we should assume that the correlation among the *ε's* is non zero and varies in a systematic fashion. For instance, when observations are taken at equally spaced points in time, a common assumption is a first-order autoregressive structure (AR(1)) for the model errors within each of the *j* clusters. As an example, suppose there are four observations in each cluster. In this case, the covariance matrix of the model errors in the *j*th cluster is

$$\text{Var}(\boldsymbol{\varepsilon}_j) = \mathbf{S}_j = \sigma_\varepsilon^2\mathbf{R}_j = \sigma_\varepsilon^2\begin{bmatrix} 1 & \rho & \rho^2 & \rho^3 \\ \rho & 1 & \rho & \rho^2 \\ \rho^2 & \rho & 1 & \rho \\ \rho^3 & \rho^2 & \rho & 1 \end{bmatrix} \tag{7.16}$$

where ρ denotes the correlation among any two adjacent model errors in the cluster. Here, $\mathbf{R}_j$ denotes the correlation matrix associated with observations within cluster *j*. The structure of $\mathbf{R}_j$ is generally specified by the user. When the data are balanced, the $\mathbf{R}_j$ are all identical such that $\mathbf{R}_j = \mathbf{R}$ for $j = 1,2,\ldots,m$. In many situations, observations are not ordered based on time or space, and an exchangeable (also known as compound symmetric) correlation structure is assumed where

$$\mathbf{R}_j = \begin{bmatrix} 1 & \rho & \cdots & \rho \\ \rho & 1 & \cdots & \rho \\ \vdots & \vdots & \ddots & \vdots \\ \rho & \rho & \cdots & \rho \end{bmatrix}.$$

The exchangeable (compound symmetric) structure assumes that observations within the cluster are equally correlated. Split plot experiments often assume such a structure.

Example 7.2. Industrial Split Plot Plastic Strength Experiment. Kowalski, Parker, and Vining (2007) consider a response surface approach to an industrial split plot experiment involving the strength of a plastic, y. Four factors are identified as potentially important: 1—baking temperature (*temp*), 2—additive percentage (*add*), 3—agitation rate (*rate*), and 4—processing time (*time*). Each factor has two levels: low = -1 and high = 1. Temperature is a much harder to change factor than the other three factors. Consequently, the experiment randomly assigns the eight combinations of the other factors within each setting of temperature. Table 7.2 summarizes the data. Note that there are four whole plots (clusters) and within each whole plot there are eight sub plot units. Since the sub plot treatment combinations are randomly assigned within each whole plot, observations on sub plots within a whole plot are correlated. There is no natural time or space ordering among each of the eight observations within a whole plot. One approach to modeling this data is the marginal model

$$\mathbf{y} = \mathbf{X}\boldsymbol{\beta} + \boldsymbol{\varepsilon} \tag{7.17}$$

where $\mathbf{X}$ is the assumed model matrix involving the intercept and the four experimental factors, along with their interactions and it is assumed that $E(\boldsymbol{\varepsilon}) = \mathbf{0}$ and $\text{Var}(\boldsymbol{\varepsilon}) = \mathbf{S}$.

Assuming an exchangeable (compound symmetric) structure, the variance–covariance matrix of the responses is given by

$$\mathbf{V} = \mathbf{S} = \sigma_\varepsilon^2 \begin{bmatrix} \mathbf{R}_1 & & & \mathbf{0} \\ & \mathbf{R}_2 & & \\ & & \mathbf{R}_3 & \\ \mathbf{0} & & & \mathbf{R}_4 \end{bmatrix}$$

Table 7.2 Industrial Split Plot Data for Plastic Experiment

Oven Setting	Temperature	Additive	Rate	Time	Strength
1	1	−1	1	1	68.5
1	1	1	−1	1	66.8
1	1	−1	−1	−1	58.5
1	1	1	1	1	70.8
1	1	−1	1	−1	61.3
1	1	1	−1	−1	51.8
1	1	−1	−1	1	58.5
1	1	1	1	−1	66.2
2	−1	1	−1	−1	57.4
2	−1	1	−1	1	57.5
2	−1	−1	1	−1	56.5
2	−1	1	1	1	63.8
2	−1	−1	1	1	56.4
2	−1	1	1	−1	58.1
2	−1	−1	−1	1	53.2
2	−1	−1	−1	−1	58.5
3	−1	−1	−1	−1	66.6
3	−1	−1	−1	1	63.8
3	−1	1	1	−1	62.6
3	−1	1	1	1	63.2
3	−1	−1	1	−1	56.1
3	−1	1	−1	1	63.3
3	−1	−1	1	1	62.7
3	−1	1	−1	−1	65
4	1	−1	−1	−1	58.5
4	1	1	1	−1	64
4	1	−1	1	1	68
4	1	1	−1	−1	65.6
4	1	−1	1	−1	58.6
4	1	1	1	1	73.3
4	1	1	−1	1	61.5
4	1	−1	−1	1	64.2

where each of the $\mathbf{R}_j$ ($j = 1,2,3,4$) is an (8×8) exchangeable correlation matrix. Equation (7.17) is a *marginal* model since it involves only fixed effect regression coefficients and the correlation among the sub plot observations within the same whole plot is handled solely through the variance–covariance of the model errors.

As with Example 7.1, it may be of interest not only to model the relationship between the response and the experimental factors but also the variability from one whole plot to another. Having separate estimates of whole plot variation and subplot variation is often useful in terms of quality improvement.

Kowalski, Parker, and Vining fit the following first-order plus interactions mixed effects model for this design:

$$\begin{aligned} y &= \beta_0 + \beta_1 temp + \delta_{wp} + \beta_2 add + \beta_3 rate + \beta_4 time + \\ &\beta_{23} add * temp + \beta_{24} add * time + \beta_{34} rate * time + \\ &\beta_{12} temp * add + \beta_{13} temp * rate + \beta_{14} temp * time + \varepsilon \end{aligned} \tag{7.18}$$

where $\delta_{wp} \sim N(0, \sigma_\delta^2)$ denotes the random effect associated with the random oven settings and ε is the model error term. In matrix notation, the model is

$$\mathbf{y} = \mathbf{X}\boldsymbol{\beta} + \mathbf{Z}\boldsymbol{\delta} + \boldsymbol{\varepsilon} \tag{7.19}$$

with $\mathbf{y}$ denoting the (32×1) vector of responses, $\mathbf{X}$ is the (32×11) model matrix, $\boldsymbol{\beta}$ is the associated (11×1) vector of fixed effects model parameters, $\mathbf{Z}$ is the (32×32) matrix

$$\mathbf{Z} = \begin{bmatrix} \mathbf{1}_8 & \mathbf{0}_8 & \mathbf{0}_8 & \mathbf{0}_8 \\ \mathbf{0}_8 & \mathbf{1}_8 & \mathbf{0}_8 & \mathbf{0}_8 \\ \mathbf{0}_8 & \mathbf{0}_8 & \mathbf{1}_8 & \mathbf{0}_8 \\ \mathbf{0}_8 & \mathbf{0}_8 & \mathbf{0}_8 & \mathbf{1}_8 \end{bmatrix}$$

and the single random effect, $\boldsymbol{\delta}$ has levels given by the vector

$$\boldsymbol{\delta} = \begin{bmatrix} \delta_1 \\ \delta_2 \\ \delta_3 \\ \delta_4 \end{bmatrix}$$

The model errors are represented by the (32×1) vector $\boldsymbol{\varepsilon} = \begin{bmatrix} \boldsymbol{\varepsilon}_1 \\ \boldsymbol{\varepsilon}_2 \\ \boldsymbol{\varepsilon}_3 \\ \boldsymbol{\varepsilon}_4 \end{bmatrix}$ with $\boldsymbol{\varepsilon}_j$ $(j = 1,2,3,4)$ denoting the (8×1) vector of model errors in the jth whole plot (i.e. j^{th} cluster). Note that $\mathbf{Z}$ has the identical format as $\mathbf{Z}_1^*$ in equation (7.8). Since the subplot treatment level combinations of *add*, *rate*, and *time* are randomly assigned within a fixed level of *temp*, the model errors could logically be assumed to be independent for a fixed whole plot. Consequently, taking the variance operator through (7.19), we can write the variance–covariance matrix of the responses as

$$\begin{aligned} \mathbf{V} &= \sigma_\delta^2 \mathbf{Z}\mathbf{Z}' + \mathbf{S} \\ &= \sigma_\delta^2 \mathbf{Z}\mathbf{Z}' + \sigma_\varepsilon^2 \mathbf{I}_n \end{aligned} \tag{7.20}$$

Thus for the jth cluster,

$$\mathrm{Var}(\mathbf{y}_j) = \sigma_\delta^2 \begin{bmatrix} 1 & 1 & \cdots & 1 \\ 1 & 1 & \cdots & 1 \\ \vdots & \vdots & \ddots & \vdots \\ 1 & 1 & \cdots & 1 \end{bmatrix} + \sigma_\varepsilon^2 \mathbf{I}_8 \tag{7.21}$$

The covariance among any two observations, y_{ji} and y_{jk} $(i \neq k)$, within the same whole plot j is $\mathrm{Cov}(y_{ji}, y_{jk}) = \sigma_\delta^2$. The variance–covariance structure given in Equation (7.20) for the mixed model is an exchangeable structure with $\rho = \dfrac{\sigma_\delta^2}{\sigma_\varepsilon^2 + \sigma_\delta^2}$ Thus whether one utilizes the marginal model in Equation (7.17) or the subject-specific model in Equation (7.19) to model the data, the variance–covariance matrix of the responses for both models has an exchangeable structure. The difference is that, in the marginal model, the correlation is treated as a nuisance, but in Equation (7.19) the cluster (i.e., whole plot) variance is of specific interest and is thus explicitly modeled. We return to the analysis of the model in Equation (7.19) later.

Just as the structure of **S** is an important decision in the analysis of a general linear mixed model, the structure of **D** is also important. Recall that the diagonal elements of **D** tell us how much the individual random effects vary from cluster to cluster. In many applications, the random effects can be assumed to be independent, and thus **D** is a diagonal matrix. However, in situations that require random coefficient models, such as in Example 7.1, one may need to allow for correlation between the random intercept and the random slopes. In these situations, the correlation is assumed to occur for coefficients describing the same cluster, but coefficients on different clusters are assumed uncorrelated. In the Kukupa example, where we are allowing for a random intercept and a random slope, the covariance matrix of the random effect vector at the jth station would be

$$\mathrm{Var}(\boldsymbol{\delta}_j) = \mathbf{D}_j = \mathrm{Var}\left(\begin{bmatrix} \delta_{0,j} \\ \delta_{1,j} \end{bmatrix}\right) = \begin{bmatrix} \sigma_{\delta_0}^2 & \mathrm{Cov}(\delta_{0,j}, \delta_{1,j}) \\ \mathrm{Cov}(\delta_{0,j}, \delta_{1,j}) & \sigma_{\delta_1}^2 \end{bmatrix} \tag{7.22}$$

Regarding the notation in Equation (7.22), $\sigma_{\delta_0}^2$ denotes the variance associated with the random intercepts, $\sigma_{\delta_1}^2$ denotes the variance among the random slopes, and Cov $(\delta_{0,j}, \delta_{1,j})$ denotes the correlation of the random intercept and slope at station j. If we assume independence among the random intercept and slope at station j, Cov $(\delta_{0,j}, \delta_{1,fj}) = 0$. It may not be immediately obvious when one needs to allow for a non zero correlation between $\delta_{0,j}$ and $\delta_{1,j}$. Since the response in the Kukupa example involves average Kukupa counts, it is possible that a station where large counts were observed at the beginning of the study suggests a favorable habitat at that particular station. Thus given the fact that the study area was observed during a predator reduction period, counts may increase over time at a greater rate for stations with favorable habitats than for stations with less favorable habitats. Large counts at the beginning of the study would result in a larger intercept, and these stations may have steeper slopes over time. In such a setting, the random intercept and slope would be positively correlated. This phenomenon does not seem to occur in our example as observed in Figure 7.1. It thus may be more appropriate to assume Cov $(\delta_{0,j}, \delta_{1,j}) = 0$. Just as the user must specify the structure of **S** in the software, the user also must specify the structure of **D**. The default structures of **S** and **D** are diagonal, which corresponds to independent model errors and independent random effects, respectively.

It is important to keep in mind that different assumptions regarding the structures of $\mathbf{S}_j$ and $\mathbf{D}_j$ influence the number of parameters that one must estimate. For instance, in Example 7.1, if the user were to assume independence and equal variance among the model errors and independent random intercepts and slopes, only three variance parameters are estimated. Specifically, σ_ε^2, since $\mathbf{S}_j = \sigma_\varepsilon^2 \mathbf{I}_{n_j}$, $\sigma_{\delta_0}^2$ (random intercept variance), and $\sigma_{\delta_1}^2$ (random slope variance). If, however, the user assumes that the random intercepts and slopes are correlated, four parameters must be estimated. Specifically, three parameters in $\mathbf{D}(\sigma_{\delta_0}^2, \sigma_{\delta_1}^2,$ and $\sigma_{\delta_0,\delta_1})$ and one parameter in **S**, the model error variance, σ_ε^2. In SAS, the user specifies the structure of the $\mathbf{S}_j$ in the REPEATED statement in PROC MIXED. The structure of $\mathbf{D}_j$ is specified using the RANDOM statement in PROC MIXED. We illustrate the specification of various structures using both SAS and R when we revisit the examples. It is important to keep in mind that the structures for $\mathbf{S}_j$ and $\mathbf{D}_j$ are assumed to be consistent for every cluster.

7.1.4 Parameter Estimation in the General Linear Mixed Model

For the responses in the general linear mixed model, the marginal distribution of the response vector is

$$\mathbf{y} \sim MVN(\mathbf{X}\boldsymbol{\beta}, \mathbf{V}) \tag{7.23}$$

As discussed in Section 2.5, the ordinary least squares estimator $\mathbf{b} = (\mathbf{X}'\mathbf{X})^{-1}\mathbf{X}'\mathbf{y}$ of the regression parameters is no longer optimal when $\mathbf{V} \neq \sigma_\varepsilon^2 \mathbf{I}_n$. Instead, the optimal solution is the generalized least squares estimate

$$\mathbf{b}^{\mathrm{GLS}} = \left(\mathbf{X}'\mathbf{V}^{-1}\mathbf{X}\right)^{-1}\mathbf{X}'\mathbf{V}^{-1}\mathbf{y} \tag{7.24}$$

where **V** is assumed known. When **V** is known, this estimate is also the maximum likelihood estimate of $\boldsymbol{\beta}$.

Clearly, in practice, the covariance matrix **V** is not known and must be estimated from the data. One of the most commonly used methods to estimate

V is restricted maximum likelihood (REML), an approach developed by Patterson and Thompson (1971). REML addresses the bias that is present in the maximum likelihood estimates of the variance components. Essentially, REML estimates the variance components by maximizing the likelihood function of a set of error contrasts given by $\mathbf{C} = \mathbf{Hy}$, where $\mathbf{H}$ is a $((n-p) \times n)$ matrix whose columns are orthogonal to the fixed effects model matrix $\mathbf{X}$. For the general linear mixed model, $\mathbf{C}$ follows a normal distribution with mean $\mathbf{0}$ and covariance matrix $\mathbf{HVH}'$. Working with a set of error contrasts produces a likelihood that does not depend on the fixed effects parameters in $\boldsymbol{\beta}$. Instead, the resulting likelihood is only a function of the variance components in $\mathbf{V}$. Since the likelihood used for variance component estimation is based on this *restricted* set of error contrasts, estimates obtained from it are known as the *restricted maximum likelihood estimates.* Littell et al. (2006) show that the log-likelihood of $\mathbf{C}$ can be written as

$$l_{\text{REML}} \propto -\frac{1}{2}\ln|\mathbf{V}| - \frac{1}{2}\ln|\mathbf{X}'\mathbf{V}\mathbf{X}| - \frac{1}{2}\left(\mathbf{y} - \mathbf{X}\mathbf{b}^{\text{GLS}}\right)'\mathbf{V}^{-1}\left(\mathbf{y} - \mathbf{X}\mathbf{b}^{\text{GLS}}\right) \qquad (7.25)$$

Let $\boldsymbol{\alpha}$ denote the $((q+1) \times 1)$ vector of unknown parameters in the covariance matrix so that $\mathbf{V} = \mathbf{V}(\boldsymbol{\alpha})$. Also, let $\mathbf{l}'_{\text{REML}}(\boldsymbol{\alpha}) = \frac{\partial}{\partial\boldsymbol{\alpha}} l_R(\boldsymbol{\alpha})$ and $l''_{\text{REML}}(\boldsymbol{\alpha}) = \dfrac{\partial^2}{\partial\boldsymbol{\alpha}\partial\boldsymbol{\alpha}'} l_{\text{REML}}(\boldsymbol{\alpha})$. Note that $\boldsymbol{l}'_{\text{REML}}$ is a $((q+1) \times 1)$ vector and that $\boldsymbol{l}'_{\text{REML}}$ is a $((q+1) \times (q+1))$ matrix. The REML estimate is then calculated using an iterative procedure with the value at the $(s+1)$st iteration given by

$$\boldsymbol{\alpha}^{(s+1)} = \boldsymbol{\alpha}^{(s)} + \mathbf{M}^{-1}\left(\boldsymbol{\alpha}^{(s)}\right) l'_{\text{REML}}\left(\boldsymbol{\alpha}^{(s)}\right) \qquad (7.26)$$

where $\mathbf{M}\,(\boldsymbol{\alpha}^{(s)})$ is a matrix evaluated at $\boldsymbol{\alpha}^{(s)}$. The Scoring procedure uses $\mathbf{M}\,(\boldsymbol{\alpha}^{(s)}) = -E[\boldsymbol{l}''_{\text{REML}}(\boldsymbol{\alpha}^{(s)})]$, while the Newton–Raphson procedure sets $\mathbf{M}\,(\boldsymbol{\alpha}^{(s)}) = -\boldsymbol{l}''_{\text{REML}}(\boldsymbol{\alpha}^{(s)})$. The former quantity is the information matrix. Wolfinger et al. (1993) indicate that the Scoring algorithm in SAS drops second-order derivatives involving $\mathbf{V}(\boldsymbol{\alpha})$ and is less computationally demanding than the Newton–Raphson approach. Generally, the procedure continues until the change between $\boldsymbol{\alpha}^{(s)}$ and $\boldsymbol{\alpha}^{(s+1)}$ is sufficiently small. The resulting REML estimate of $\boldsymbol{\alpha}$ is taken to be $\hat{\boldsymbol{\alpha}} = \boldsymbol{\alpha}^{(s+1)}$ and $\hat{\mathbf{V}} = \mathbf{V}(\hat{\boldsymbol{\alpha}})$.

For instance, in Example 7.2, the variance–covariance matrix for the responses in the *j*th whole plot, written in Equation (7.21), would be estimated by

$$V_j(\hat{\boldsymbol{\alpha}}) = \begin{bmatrix} \hat{\sigma}_\varepsilon^2 + \hat{\sigma}_\delta^2 & \hat{\sigma}_\delta^2 & \cdots & \hat{\sigma}_\delta^2 \\ \hat{\sigma}_\delta^2 & \hat{\sigma}_\varepsilon^2 + \hat{\sigma}_\delta^2 & \cdots & \hat{\sigma}_\delta^2 \\ \vdots & \vdots & \ddots & \vdots \\ \hat{\sigma}_\delta^2 & \hat{\sigma}_\delta^2 & \cdots & \hat{\sigma}_\varepsilon^2 + \hat{\sigma}_\delta^2 \end{bmatrix}$$

where $\hat{\alpha} = \begin{bmatrix} \hat{\sigma}_\varepsilon^2 \\ \hat{\sigma}_\delta^2 \end{bmatrix}$ is the vector of converged REML variance component estimates. Once the variance parameters are estimated, the fixed effects model coefficients are estimated by

$$\mathbf{b} = \left(\mathbf{X}'\hat{\mathbf{V}}^{-1}\mathbf{X}\right)^{-1}\mathbf{X}'\mathbf{V}^{-1}\mathbf{y} \tag{7.27}$$

The estimated variance–covariance matrix of **b**, is

$$\widehat{\text{Var}}(\mathbf{b}) = \left(\mathbf{X}'\hat{\mathbf{V}}^{-1}\mathbf{X}\right)^{-1} \tag{7.28}$$

The estimated standard errors of the regression coefficients are the square roots of the diagonal elements of Equation (7.28) and are used in hypothesis tests and confidence intervals involving the coefficients.

7.1.5 Statistical Inference on Regression Coefficients and Variance Components

The general linear mixed model, like the generalized linear model, lends itself to both Wald and likelihood inference. As observed in our discussions about Wald inference in Sections 4.2.4, 5.7, and 5.12, a possible test statistic for the hypothesis H_0: $\beta_j = 0$ is $\left(\frac{b_j - \beta_j}{\widehat{se}(b_j)}\right)^2$ which is asymptotically χ_1^2 under H_0. However, Kenward and Roger (1997) and Searle, Casella, and McCulloch (1992) point out that the estimated standard errors obtained from Equation (7.28) may underestimate the true standard errors of the coefficient estimates because they do not take into account the variability introduced by estimating the variance components in $\boldsymbol{\alpha}$. To resolve this, some analysts approximate the distribution of $\dfrac{(b_j - \beta_j)}{\widehat{se}(b_j)}$ by appropriate t- or F-statistics. In general, the t-statistic takes on the form

$$t_{df} = \frac{b_j - \beta_j}{\widehat{se}(b_j)} \tag{7.29}$$

where the estimated standard errors can be obtained from Equation (7.28). If one is interested in testing a contrast involving the vector of fixed effect coefficients, that is, H_0: $\mathbf{C}\boldsymbol{\beta} = \mathbf{0}$, where $\mathbf{C}$ is a $(c \times k)$ matrix of contrasts, the general form of the F-statistic is

$$F_{\text{df}_C,\text{df}_{\text{error}}} = \frac{(\mathbf{b} - \boldsymbol{\beta})'\mathbf{C}'\left[\mathbf{C}\left(\mathbf{X}'\hat{\mathbf{V}}^{-1}\mathbf{X}\right)^{-1}\mathbf{C}'\right]^{-1}\mathbf{C}(\mathbf{b} - \boldsymbol{\beta})}{\text{rank}(\mathbf{C})} \tag{7.30}$$

In practice, there are several methods for obtaining the appropriate error degrees of freedom for these tests. A popular choice within SAS PROC MIXED is a Satterthwaite type of approximation. For other methods and for a more detailed discussion of this matter see Little et al. (2006) and Verbeke and Molenberghs (1997, Section 3.5.2). Kenward and Roger (1997) propose corrected *t*- and F-statistics, which incorporate bias-adjusted standard errors instead of those obtained from Equation (7.28). SAS PROC MIXED also implements the Kenward – Roger approach. Wulff and Robinson (2009) use a simulation study to compare various standard error estimates for the general linear mixed model. They show that the Kenward – Roger approach is quite robust to a range of possible covariance structures.

Often, an important question is whether the model should include a specific random effect. Under certain circumstances, if the lth variance component is not significantly different from 0, the analyst reduces the covariance matrix in Equation (7.15) by removing the associated random component (i.e., $\boldsymbol{\delta}_l^*$) from the general linear mixed model in Equation (7.10). Any decision to remove a variance component must take into account how the data were collected, especially with regard to the randomization. Typically, we conduct a REML likelihood ratio test.

Let $l_{\text{REML}}^{\text{Full}}$ denote the restricted log-likelihood given in Equation (7.25) for the full model. In this case, by full we mean the model with the more complicated covariance structure. Similarly, let $l_{\text{REML}}^{\text{Red}}$ denote the restricted log-likelihood in Equation (7.25) for the reduced model. The REML likelihood ratio test statistic is

$$\hat{\lambda} = \frac{2l_{\text{REML}}^{\text{Full}}}{2l_{\text{REML}}^{\text{Red}}} = -2l_{\text{REML}}^{\text{Red}} + 2l_{\text{REML}}^{\text{Full}} \tag{7.31}$$

The distribution of the REML likelihood ratio test statistic depends on whether or not any of the parameters being tested fall on the boundary of the variance parameter space as defined by Self and Liang (1987) and Verbeke and Molenberghs (2000, Section 6.3.4). Although different scenarios of model comparisons exist, three general cases are common. We outline the distribution of the likelihood ratio statistic given in Equation (7.31) for each scenario:

Scenario 1. If v variance parameters are tested and none of the parameters specified under H_0 as well as any nuisance parameters lie on the boundary, the p-value is

$$p = Pr\left(\chi_v^2 > \hat{\lambda}\right) \tag{7.32}$$

This situation occurs when the user specifies a random intercept and a random slope and wants to see if there is a covariance between the two random effects. In this situation, $H_0 : \sigma_{\delta_0,\delta_1} = 0$, where δ_0 and δ_1 denote the two random effects and $v = 1$.

Scenario 2. One parameter is specified under H_0, and it falls on the boundary. If no other parameters are on the boundary, p is calculated using a mixture of a χ_0^2 and χ_1^2 distribution as

$$p = 0.5\,\Pr\left(\chi_0^2 > \hat{\lambda}\right) + 0.5\,\Pr\left(\chi_1^2 > \hat{\lambda}\right) \tag{7.33}$$

This situation occurs when one compares a model with both a random intercept and random slope, independent of one another, versus a model with only a random intercept. In this situation, $H_0 : \sigma_{\delta_1}^2 = 0$, where δ_1 denotes the random slope term. Verbeke and Molenberghs (2003) extend this situation to the case that tests the removal of j random effects from $j+k$ uncorrelated random effects.

Scenario 3. When testing the removal of a random effect from an unstructured **D** matrix, where **D** is the variance–covariance matrix of random effects, p is calculated as

$$p = 0.5Pr\left(\chi_k^2 > \hat{\lambda}\right) + 0.5Pr\left(\chi_{k-1}^2 > \hat{\lambda}\right) \tag{7.34}$$

where k is the number of random effects in the full model. This situation occurs when one tests whether a model that includes a correlated random intercept and random slope is significantly different from one that includes only a random intercept. Here, $H_0 : \sigma_{\delta_1}^2 = \sigma_{\delta_0,\delta_1} = 0$ and $k = 2$.

All tests used to compare covariance structures require the full fixed effects vector, $\boldsymbol{\beta}$. Thus one must determine the appropriate covariance structure before conducting significance tests for the fixed effects terms. We illustrate the use of the REML likelihood ratio tests when the examples are revisited later.

The restricted log-likelihood is also a popular statistic for comparing model fit among models containing the same fixed effects: the smaller the value of $-2l_{\text{REML}}^{\text{model}}$ the better the fit for the *model*. It should be noted, however, that one can always make this value smaller by adding more parameters to the model. Two alternative information criteria for assessing model fit are Akaike's Information Criterion (AIC) (Akaike, 1974) and the Bayesian Information Criterion (BIC) (Schwarz, 1978). The AIC is

$$\text{AIC} = -2l_{\text{REML}}^{\text{model}} + 2q^* \tag{7.35}$$

where q^* denotes the number of parameters in the variance–covariance matrix of the responses, **V**. In some software packages, such as R, q^* denotes the total number of parameters in the model, that is, the number of fixed effects parameters plus the number of parameters in **V**. The BIC is

$$\text{BIC} = -2l_{\text{REML}}^{\text{model}} + q^* ln(n) \tag{7.36}$$

where m is the number of clusters. In R, q^* once again denotes the total number of parameters in the model and $m = N$, the total number of observations in the data set.

7.1.6 Conditional and Marginal Means

As we proceed with mixed models, it is important to understand the difference in the conditional response mean and the marginal response mean. By conditional, we mean the average response conditioned on the random effects. In general, the conditional mean is

$$E(\mathbf{y}|\boldsymbol{\delta}) = \mathbf{X}\boldsymbol{\beta} + \mathbf{Z}\boldsymbol{\delta} \tag{7.37}$$

More specifically, the conditional mean for the jth cluster is

$$E\left(\mathbf{y}_{n_j}\middle|\boldsymbol{\delta}_j\right) = \mathbf{X}_j\boldsymbol{\beta} + \mathbf{Z}_j\boldsymbol{\delta}_j \tag{7.38}$$

where $\mathbf{y}_{n_j}$ is the vector of responses for the jth cluster, $\mathbf{X}_j$ is the $(n_j \times \text{p})$ matrix of fixed effect model terms associated with the jth cluster, $\boldsymbol{\beta}$ is the corresponding $(p \times 1)$ vector of fixed effect regression coefficients, δ_j is the $(q \times 1)$ vector of random factor levels associated with the jth cluster, and $\mathbf{Z}_j$ is the corresponding matrix of predictors for the jth cluster. Note the difference in notation between the δ_j in Equation (7.38) and the $\boldsymbol{\delta}_l^*$ in Equation (7.10). The $\boldsymbol{\delta}_l^*$ represent each of the q random factors whereas δ_j denotes the levels of the q random factors at the jth cluster. So, for the jth cluster, $\boldsymbol{\delta}_j = \begin{bmatrix} \delta_{1,j} \\ \delta_{2,j} \\ \vdots \\ \delta_{q,j} \end{bmatrix}$. Similarly, the $\mathbf{Z}_l^*$ represent the matrix predictors for each of the q random effects and the $\mathbf{Z}_j$ are the matrices of predictors for each of the j clusters. It is helpful to illustrate this notation using the Kukupa example. The model for the conditional mean at observations station 1, for instance, is

$$E\left(\mathbf{y}_{n_1}\middle|\boldsymbol{\delta}_1\right) = \mathbf{X}_1\boldsymbol{\beta} + \mathbf{Z}_1\boldsymbol{\delta}_1 \tag{7.39}$$

where $\mathbf{y}_{\mathbf{n}_1}$ is as defined in Equation (7.6), $\text{X}_1 = [\mathbf{1}_{\text{n}_1}\mathbf{x}_{\text{n}_1}]$ with $\mathbf{1}_{n_1}$ and $\mathbf{x}_{n_1}$ defined in Equation (7.6), $\delta_1 = \begin{bmatrix} \delta_{0,1} \\ \delta_{1,1} \end{bmatrix}$ with $\delta_{0,1}$ and $\delta_{1,1}$ denoting the levels of the random intercept and random slope factors, respectively, for station 1, and $\mathbf{Z}_1 = [\mathbf{1}_{n_1}\ \mathbf{x}_{n_1}]$.

The mean response across all clusters is

$$E(\mathbf{y}) = E[E(\mathbf{y}_{n_j}|\boldsymbol{\delta}_j)] = \mathbf{X}\boldsymbol{\beta}. \tag{7.40}$$

The expressions in Equations (7.39) and (7.40) are commonly referred to as the *subject-specific* and *population-averaged* means, respectively. From Equations (7.39) and (7.40) it is once again intuitive to see why the $\boldsymbol{\delta}_j$ are referred to as random effects. They are *effects* since $\mathbf{Z}_j\boldsymbol{\delta}_j$ represents the deviation in the conditional mean at cluster j and the mean across all clusters. The effects represented by the parameters in $\boldsymbol{\delta}_j$ are *random* because they correspond directly to the the jth cluster and the jth cluster is assumed to be a randomly selected cluster from the population of clusters. □

7.1.7 Estimation of the Random Coefficients

The coefficients for the random effects in Equation (7.10), the δ_l^* 's, are assumed to have normal distributions with zero means. Their specific values for a given cluster are just realizations from normal distributions. For instance, in the Kukupa example, $\delta_{0,1}$and $\delta_{1,1}$ are considered realizations of the random intercept and random slope, respectively, at station 1. Although the random effects are assumed to each have a mean of zero, it is possible to obtain predicted values for the realizations associated with each cluster. We can obtain the predicted values for the δ_l^*'s in Equation (7.10) by maximizing the joint likelihood of the distribution of the random effects and the distribution of the conditional response. When we write the conditional response, $\mathbf{y}_{n_j}|\boldsymbol{\delta}_j$, we mean the responses within the jth cluster. The levels of the random effects are fixed by the *conditioning*; thus the only stochastic component now is the model errors for the jth cluster given by ε_j. From Equations (7.38) and (7.14) we have that

$$\mathbf{y}|\boldsymbol{\delta} \sim N(\mathbf{X}\boldsymbol{\beta} + \mathbf{Z}\boldsymbol{\delta}, \mathbf{S}) \tag{7.41}$$

Thus the likelihood for the conditional response is

$$L(\boldsymbol{\beta}, \boldsymbol{\alpha}_S; \mathbf{y}) \propto |\mathbf{S}^{-1}| \exp\left\{-\frac{1}{2}(\mathbf{y} - \mathbf{X}\boldsymbol{\beta} - \mathbf{Z}\boldsymbol{\delta})'\mathbf{S}^{-1}(\mathbf{y} - \mathbf{X}\boldsymbol{\beta} - \mathbf{Z}\boldsymbol{\delta})\right\} \tag{7.42}$$

where $\boldsymbol{\alpha}_S$ is the vector whose elements comprise the unknown variance and covariance parameters in $\mathbf{S}$. Since the random effects are assumed to have mean $\mathbf{0}$ and variance–covariance matrix $\mathbf{D}$, the likelihood of the random effects is

$$\mathscr{L}(\boldsymbol{\alpha}_D; \boldsymbol{\delta}) \propto |\mathbf{D}^{-1}| \exp\left\{-\frac{1}{2}\boldsymbol{\delta}'\mathbf{D}^{-1}\boldsymbol{\delta}\right\} \tag{7.43}$$

where $\boldsymbol{\alpha}_D$ is the vector whose elements comprise the unknown variance and covariance parameters in $\mathbf{D}$. Utilizing the assumption that Cov ($\boldsymbol{\varepsilon}$, $\boldsymbol{\delta}$) = $\mathbf{0}$ from

Section 7.1.3, the joint likelihood is obtained by multiplying the two likelihood expressions in (7.42) and (7.43). Upon taking the log of the joint likelihood we have

$$\ln[\mathscr{L}(\boldsymbol{\beta}, \boldsymbol{\alpha}_S, \boldsymbol{\delta}, \boldsymbol{\alpha}_D; \mathbf{y})] = -\frac{1}{2}\{\ln|\mathbf{S}| + (\mathbf{y} - \mathbf{X}\boldsymbol{\beta} - \mathbf{Z}\boldsymbol{\delta})'\mathbf{S}^{-1}(\mathbf{y} - \mathbf{X}\boldsymbol{\beta} - \mathbf{Z}\boldsymbol{\delta}) + \ln|\mathbf{D}| + \boldsymbol{\delta}'\mathbf{D}^{-1}\boldsymbol{\delta}\} + \text{constant} \tag{7.44}$$

The predicted values for the random effects, commonly referred to as the *best linear unbiased predictors* or *BLUP* are then found by differentiating the log-likelihood in Equation (7.44) with respect to $\boldsymbol{\delta}$ and then solving. It can be shown that this process yields

$$\hat{\boldsymbol{\delta}} = \mathbf{D}\mathbf{Z}'\mathbf{V}^{-1}(\mathbf{y} - \mathbf{X}\boldsymbol{\beta}) \tag{7.45}$$

where $\mathbf{V} = \mathbf{Z}\mathbf{D}\mathbf{Z}' + \mathbf{S}$. The estimated generalized least squares estimates of $\boldsymbol{\beta}$, $\mathbf{b} = (\mathbf{X}'\hat{\mathbf{V}}^{-1}\mathbf{X})^{-1}\mathbf{X}'\hat{\mathbf{V}}^{-1}\mathbf{y}$, are then supplied for $\boldsymbol{\beta}$ in the expression above and the REML estimates of the variance–covariance matrices ($\hat{\mathbf{D}}$ and $\hat{\mathbf{S}}$, and thus $\hat{\mathbf{V}}$) are also substituted in the above. Once the random coefficients are estimated, then the vector of estimated conditional means is

$$\widehat{E}(\mathbf{y}|\boldsymbol{\delta}) = \hat{\boldsymbol{\mu}}|\boldsymbol{\delta} = \mathbf{X}\mathbf{b} + \mathbf{Z}\hat{\boldsymbol{\delta}} \tag{7.46}$$

A $(1 - \alpha)$ 100% confidence interval on the conditional mean, $E(y|\ \delta)$ is

$$\hat{\mu}|\delta \pm z_{1-\alpha/2}\widehat{se}(\mu|\delta) \tag{7.47}$$

where $\widehat{se}(\mu|\delta)$ denotes the estimated standard error of the estimated conditional mean. SAS PROC MIXED and R provide the lower and upper confidence bounds for the conditional mean at user specified locations. In a similar fashion, the vector of estimated marginal means is

$$\hat{E}(\mathbf{y}) = \widehat{\boldsymbol{\mu}} = \mathbf{X}\mathbf{b} \tag{7.48}$$

A $(1 - \alpha)$100% confidence interval on the marginal mean, $E(y)$, is

$$\widehat{\mu} \pm z_{1-\alpha/2}\widehat{se}(\widehat{\mu}) \tag{7.49}$$

where $\widehat{se}(\hat{\mu})$ denotes the estimated standard error of the estimated marginal mean. SAS PROC MIXED and R provide the lower and upper confidence bounds for the marginal mean at user specified locations. Estimation of the random effects, and subsequently the conditional and marginal means, are illustrated with the Kukupa example in the next section.

7.1.8 Examples Revisited

Example 7.3. Kukupa Data Revisited. Consider again Example 7.1 in which averaged Kukupa counts are observed for each of ten stations over a 12-year period. The biologist seeks to determine whether or not Kukupa numbers have changed significantly over the study period.

Based on some exploratory analyses, a square root transformation of the response was deemed to be appropriate. Also, to provide meaningful results for the intercept, the *Year* variable was centered as *Year-6.5*.

The first step in a general linear mixed model analysis is to decide on the appropriate covariance structure of the data. Figure 7.1, which plots the raw data along with separate linear model fits for each station, suggests that there is more variability among the slopes than the intercepts. We consider five potential linear mixed models and the results from each are given in Table 7.3.

Each of the models fit in Table 7.3 contains the single fixed effect, Year, but different covariance patterns are fit. The first model does not use random effects to model the correlation among observations within the same cluster. Instead, it specifies a first-order auto regressive (AR(1)) structure for the repeated observations. An unstructured variance–covariance matrix was originally specified for the repeated observations but convergence was an issue. The model is

$$\mathbf{y} = \mathbf{X}\boldsymbol{\beta} + \boldsymbol{\varepsilon}$$

Table 7.3 Estimated Parameters for Models Fit to Kukupa Data Using Different Covariance Structures

	No Random Effects + AR(1)	Random Intercepts + Independence	Random Slopes + Independence	Independent Random Slopes and Intercepts + Independence	*Dependent Random Slopes and Intercepts + Independence*
Intercept	0.4417	0.4420	0.4420	0.4420	0.4420
Year	0.0521	0.0525	0.0525	0.0525	0.0525
σ^2_ε	0.1008	0.0888	0.0934	**0.0807**	0.0807
ρ	0.2311	—	—	—	—
$\sigma^2_{\delta_0}$	—	0.0121	—	0.0128	0.0128
$\sigma^2_{\delta_1}$	—	—	0.0006	0.0007	0.0007
$\sigma_{\delta_0,\delta_1}$	—	—	—	—	0.0015
−2 log L	69.4	69.9	73	66.7	**65.7**
AIC	73.4	73.9	77	**72.7**	73.7
BIC	74.0	74.5	77.6	**73.6**	75.0

where we take the variance–covariance of the responses in the jth cluster to be given by $V_j = S_j = \sigma_\varepsilon^2 \begin{bmatrix} 1 & \rho & \cdots & \rho^{11} \\ \rho & 1 & \cdots & \rho^{10} \\ \vdots & \vdots & \ddots & \vdots \\ \rho^{11} & \rho^{10} & \cdots & 1 \end{bmatrix}$. Although we designate the response by y, keep in mind that for the analyses presented y denotes the square root of the counts.

In the second model, we allow for random intercepts among stations and assume independent model errors for the observations across time within a cluster. The model is

$$\mathbf{y} = \mathbf{X}\boldsymbol{\beta} + \mathbf{Z}_1^*\boldsymbol{\delta}_1^* + \boldsymbol{\varepsilon}$$

with $\mathbf{Z}_1^*$ defined as in Equation (7.8) and $\boldsymbol{\delta}_1^*$ defined as in Equation (7.9), $m = 10$, and the variance–covariance matrix of responses is

$$\mathrm{V} = \sigma_{\delta_0}^2 \mathbf{Z}_1^*\mathbf{Z}_1'^* + \sigma_\varepsilon^2 \mathbf{I}_n$$

In the third model, we allow only for random slopes and independent model errors. The resulting model is

$$\mathbf{y} = \mathbf{X}\boldsymbol{\beta} + \mathbf{Z}_2^*\boldsymbol{\delta}_2^* + \boldsymbol{\varepsilon}$$

with $\mathbf{Z}_2^*$ defined as in Equation (7.8) and $\boldsymbol{\delta}_2^*$ defined as in Equation (7.9). The variance–covariance matrix of responses is

$$\mathrm{V} = \sigma_{\delta_1}^2 \mathbf{Z}_2^*\mathbf{Z}_2'^* + \sigma_\varepsilon^2 \mathrm{I}_n$$

The fourth model allows for independent random intercepts and random slopes with independent model errors. The model is explicitly

$$\mathbf{y} = \mathbf{X}\boldsymbol{\beta} + \mathbf{Z}_1^*\delta_1^* + \mathbf{Z}_2^*\boldsymbol{\delta}_2^* + \boldsymbol{\varepsilon}$$

and the resulting variance–covariance matrix takes the form

$$\mathrm{V} = \sigma_{\delta_0}^2 \mathbf{Z}_1^*\mathbf{Z}_1'^* + \sigma_{\delta_1}^2 \mathbf{Z}_2^*\mathbf{Z}_2'^* + \sigma_\varepsilon^2 \mathrm{S}$$

Here, $\mathbf{Z}_1^*$, $\mathbf{Z}_2^*$, $\boldsymbol{\delta}_1^*$, and $\boldsymbol{\delta}_2^*$ are as defined for models 2 and 3.

The final model specifies random intercepts and random slopes; however, it also allows for the two random effects to be correlated. Thus the response model is the same as it is for model 4 but the variance–covariance matrix is now

$$\mathbf{V} = \mathbf{ZDZ}' + \sigma_{\varepsilon}^{2}\mathbf{S}$$

where **Z** is as defined in Equation (7.7) and **D**, the variance–covariance matrix of the two random effects, is

$$\mathbf{D} = \begin{bmatrix} \sigma_{\delta_0}^2 & \sigma_{\delta_0,\delta_1} \\ \sigma_{\delta_0,\delta_1} & \sigma_{\delta_1}^2 \end{bmatrix}$$

Model 5 fits best in terms of $-2l_{\text{REML}}^{\text{model}}$ and model 4, which is a more parsimonious model in terms of covariance structure, fits best in terms of AIC and BIC. The next best fitting model is model 1, which uses no random effects. Instead, it models the covariance structure solely through the model errors.

One can construct a likelihood ratio test to compare formally the covariance structures in models 4 and 5. Such a test takes the difference of the $-2l_{\text{REML}}^{\text{model}}$ values from both models, resulting in

$$\hat{\lambda} = -2l_{\text{REML}}^{\text{model 4}} + 2l_{\text{REML}}^{\text{model 5}} = 66.7 - 65.7 = 1.0$$

As pointed out in Section 7.1.5, one needs to be careful when thinking about the null distribution of this statistic. Here, $H_0 : \sigma_{\delta_0,\delta_1} = 0$ and from Section 7.1.5 one can easily see that the test of interest here falls under scenario 1. Thus the p-value is computed as

$$p = Pr\left(\chi_{\nu}^{2} > \hat{\lambda}\right) = Pr\left(\chi_{1}^{2} > 1\right) = 0.3173$$

which suggests that one cannot reject H_0. The implication here is that there is no need to model a covariance term for the two random effects.

One can take a similar approach to determine if a more parsimonious model fits as well as model 4. Consider model 2 where only the random intercept is specified. Now, the null hypothesis is $H_0 : \sigma_{\delta_1}^2 = 0$. The REML ratio test statistic is

$$\hat{\lambda} = -2l_{\text{REML}}^{\text{model 2}} + 2l_{\text{REML}}^{\text{model 4}} = 69.9 - 66.7 = 3.2$$

In this case, our test falls under scenario 2. The p-value is then

$$p = 0.5Pr\left(\chi_{0}^{2} > 3.2\right) + 0.5Pr\left(\chi_{1}^{2} > 3.2\right)$$

which is essentially one-half of the p-value from a χ_1^2 distribution. We have then $Pr\left(\chi_1^2 > 3.2\right) = 0.074$. Thus the p-value of our test is approximately 0.037, and we conclude that the variance associated with the random slope is significantly different from zero.

Table 7.4 SAS PROC MIXED Code for Running the Analyses of Models 1 and 4 Specified for the Kukupa Data

SAS Code Kukupa Data:

```
*Model 1 CS and no random effects;

proc mixed data=Kukupa;
class station Cal_year;
model sqcount = year / corrb s;
repeated Cal_year / type=cs subject=station r;
ods output covparms=cov;
run;

*Model 4 random intercept and random slope and independent errors;

ods html;
proc mixed data=Kukupa;
class station;
model sqcount = year / ddfm=Satterthwaite residual s
outpred=outpred outpredm=outpredm;
random int year / subject=station G;
run;
ods output solutionR=random; *produces data set of blups;
ods html close;
```

We can proceed with the analysis as we would with any other regression model once we have selected the appropriate covariance structure. If we have multiple regressor variables, we may need to reduce the model for the mean so as to obtain a more parsimonious one based on tests for the fixed effects. Table 7.4 gives the appropriate SAS PROC MIXED code to fit model 4. Table 7.5 summarizes the results. The first section of code fits model 1, the marginal model with the exchangeable correlation structure.

The information in the *Dimensions* section provides the number of covariance parameters, which is 3: $\sigma_\varepsilon^2, \sigma_{\delta_0}^2$, and $\sigma_{\delta_1}^2$. The two columns in **X** correspond to the overall intercept and overall slope corresponding to the time trend. The two columns in **Z** correspond to the random intercept and random slope. The material in the *Estimated G Matrix* section is a result of the *G* option in the Random statement and provides the estimated variance–covariance matrix of random effects for station 81. Since we only have two random effects, this is a (2×2) matrix with the diagonal entries being $\hat{\sigma}_{\delta_0}^2 = 0.01281$ and $\hat{\sigma}_{\delta_1}^2 = 0.000681$. The estimate for σ_ε^2 is reported adjacent to *Residual* in the next set of output labeled as *Covariance Parameter Estimates*.

Table 7.5 Output for Model 4 Analysis of Kukupa Data

The Mixed Procedure

Model Information

Data Set	WORK.BHEAD
Dependent Variable	sqcount
Covariance Structure	Variance Components
Subject Effect	Station
Estimation Method	REML
Residual Variance Method	Profile
Fixed Effects SE Method	Model-Based
Degrees of Freedom Method	Satterthwaite

Class Level Information

Class	Levels	Values
Station	10	81 82 83 84 85 86 87 88 89 90

Dimensions

Covariance Parameters	3
Columns in X	2
Columns in Z Per Subject	2
Subjects	10
Max Obs Per Subject	12

Number of Observations

Number of Observations Read	120
Number of Observations Used	120
Number of Observations Not Used	0

Iteration History

Iteration	Evaluations	-2 Res Log Like	Criterion
0	1	75.07378128	
1	1	66.66645496	0.00000000

Estimated G Matrix

Row	Effect	Station	Col1	Col2
1	Intercept	81	0.01281	
2	year2	81		0.000681

Covariance Parameter Estimates

Cov Parm	Subject	Estimate
Intercept	Station	0.01281
year2	Station	0.000681
Residual		0.08072

Fit Statistics

-2 Res Log Likelihood	66.7
AIC (smaller is better)	72.7
AICC (smaller is better)	72.9
BIC (smaller is better)	73.6

(Continued)

Table 7.5 *Continued*

Solution for Fixed Effects

Effect	Estimate	Standard Error	DF	t Value	Pr>|t|
Intercept	0.4420	0.04420	9	10.00	<.0001
year2	0.05247	0.01116	9	4.70	0.0011

Solution for Random Effects

Effect	Station	Estimate	Std Err Pred	DF	t Value	Pr > |t|
Intercept	81	-0.00614	0.07246	15.1	-0.08	0.9336
year	81	-0.03016	0.01860	8.56	-1.62	0.1411
Intercept	82	0.1296	0.07246	15.1	1.79	0.0938
year	82	-0.01325	0.01860	8.56	-0.71	0.4951
Intercept	83	-0.08934	0.07246	15.1	-1.23	0.2364
year	83	-0.01345	0.01860	8.56	-0.29	0.4888
Intercept	84	-0.1588	0.07246	15.1	-2.19	0.0445
year	84	-0.00541	0.01860	8.56	-0.29	0.7780
Intercept	85	0.004781	0.07246	15.1	0.07	0.9483
year	85	-0.00510	0.01860	8.56	-0.27	0.7904
Intercept	86	0.1250	0.07246	15.1	1.73	0.1049
year	86	0.01347	0.01860	8.56	0.29	0.4882
Intercept	87	-0.03060	0.07246	15.1	-0.42	0.6787
year	87	0.01339	0.01860	8.56	0.72	0.4907
Intercept	88	0.03436	0.07246	15.1	0.47	0.6422
year	88	0.01085	0.01860	8.56	0.58	0.5747
Intercept	89	-0.06676	0.07246	15.1	-0.92	0.3714
year	89	-0.00877	0.01860	8.56	-0.47	0.6491
Intercept	90	0.05786	0.07246	15.1	0.80	0.4370
year	90	0.03844	0.01860	8.56	2.07	0.0704

The *Solution for Fixed Effects* section is a result of the *S* option in the Model statement and provides the generalized least squares estimates of the parameters in $\boldsymbol{\beta}$. The estimated standard errors of **b** provided in this section are the square roots of the diagonal elements of Equation (7.28). In this example, the Kenward–Rogers adjustment does not result in different standard errors for the fixed effects estimates. The degrees of freedom are calculated using the Satterthwaite approximation due to the specification of `ddfm=Satterthwaite` as an option in the Model statement. Note that average Kukupa counts are increasing over time. The estimated marginal relationship of average Kukupa counts to *Year* is

$$\hat{y} = 0.442 + 0.0525 \times (Year - 6.5) = 0.1008 + 0.0525 \times Year$$

The estimated marginal means are output to the *outpredm* data set by specifying `outpredm=outpredm` as an option in the Model statement. The estimated profile for the jth station is obtained by taking

$$
\begin{aligned}
\hat{y}|\hat{\delta}_{0,j}, \hat{\delta}_{1,j} &= b_0 + \hat{\delta}_{0,j} + b_1 \times (Year - 6.5) + \hat{\delta}_{1,j} \times (Year - 6.5) \\
&= \left(b_0 + \hat{\delta}_{0,j}\right) + \left(b_1 + \hat{\delta}_{1,j}\right) \times (Year - 6.5)
\end{aligned}
$$

where $\hat{\delta}_{0,j}$ and $\hat{\delta}_{1,j}$ denote the random intercept and random slope BLUPs, respectively, for the *j*th station. The estimated conditional means are output to the *outpred* data set using the `outpred=outpred` option in the Model statement. The BLUPs are found in the section "Solution for Random Effects". As an illustration, the estimated profile for station 81 (station 81 is the first station) is

$$
\begin{aligned}
\hat{y} \mid \hat{\delta}_{0,1}, \hat{\delta}_{1,1} &= (0.442 - 0.0061) + (0.0525 - 0.0302) \times (Year - 6.5) \\
&= 0.2909 + 0.0223 \times Year.
\end{aligned}
$$

Figure 7.2 gives the profiles for each of the stations along with the overall mean square root count (dark profile) versus. Year produced by MINITAB after exporting the *outpred* (data set of conditional means) and the *outpredm* (data set of marginal means).

7.1.9 Diagnostics

The general linear mixed model in Equation (7.4) makes two main sets of assumptions. The first set involves the model errors, the ε's. Recall that the model errors reflect the natural dispersion among observations *within* a cluster. Thus the model errors describe how the conditional responses, the $y|\delta$'s, vary around their cluster means. We assume (1) the model errors are normally

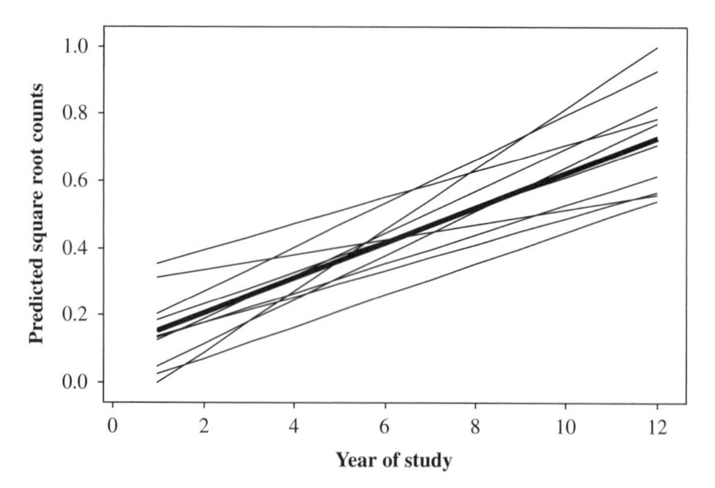

Figure 7.2 Plot of conditional mean square root counts for each station over time along with the marginal mean square root count over time (dark profile).

distributed with mean zero and constant variance; (2) their variance–covariance matrix follows a specified structure, **S**; and (3) they are independent of the random effects. Since the model errors are

$$\boldsymbol{\varepsilon} = \mathbf{y} - \mathbf{X}\boldsymbol{\beta} - \mathbf{Z}\boldsymbol{\delta}$$

the proper residuals to investigate assumptions regarding the ε's are

$$\mathbf{e}_c = \mathbf{y} - \widehat{E}(\mathbf{y}|\boldsymbol{\delta}) = \mathbf{y} - \mathbf{X}\mathbf{b} - \mathbf{Z}\hat{\boldsymbol{\delta}}$$

We use the notation $\mathbf{e_c}$ to denote the fact that these residuals are calculated using the conditional mean, and we refer to them as the conditional residuals. Gregoire, Schabenberger, and Barrett (1995) suggest using studentized residuals for investigating assumptions regarding the model errors where the studentized residual for the jth observation in the ith cluster is

$$e_{ij}^{c,\text{stud}} = e_{ij}^{c} \Big/ \sqrt{\widehat{\text{Var}}(e_{ij}^{c})} \tag{7.50}$$

where $\sqrt{\widehat{\text{Var}}(e_{ij}^{c})}$ is the estimated standard error of e_{ij}^{c}. The studentized conditional residuals are output to the *outpred* data set by specifying `residual` as an option in the Model statement. Figure 7.3 gives the boxplots of the studentized e_{ij}^{c} for each of the stations after exporting the data set *outpred* to MINITAB. It appears that the studentized conditional residuals are in fact centered at zero with approximately constant variance. Figure 7.4 gives the normal Q-Q plot of the studentized conditional residuals produced by

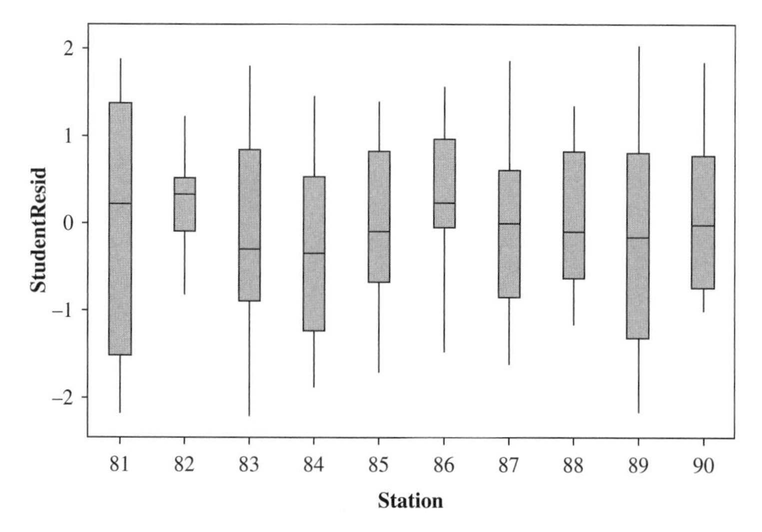

Figure 7.3 Boxplots of studentized, conditional residuals for each station.

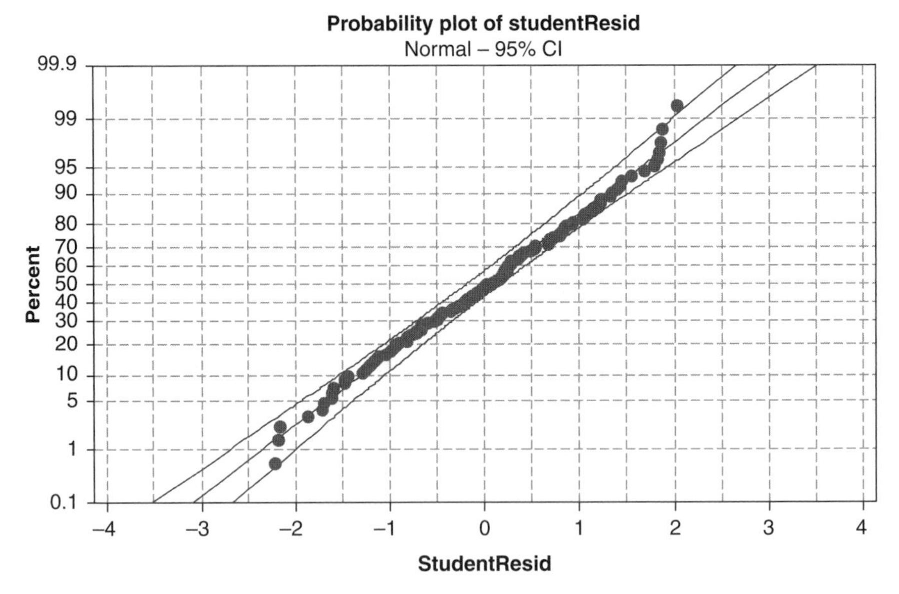

Figure 7.4 Normal Q-Q plot of studentized, conditional residuals.

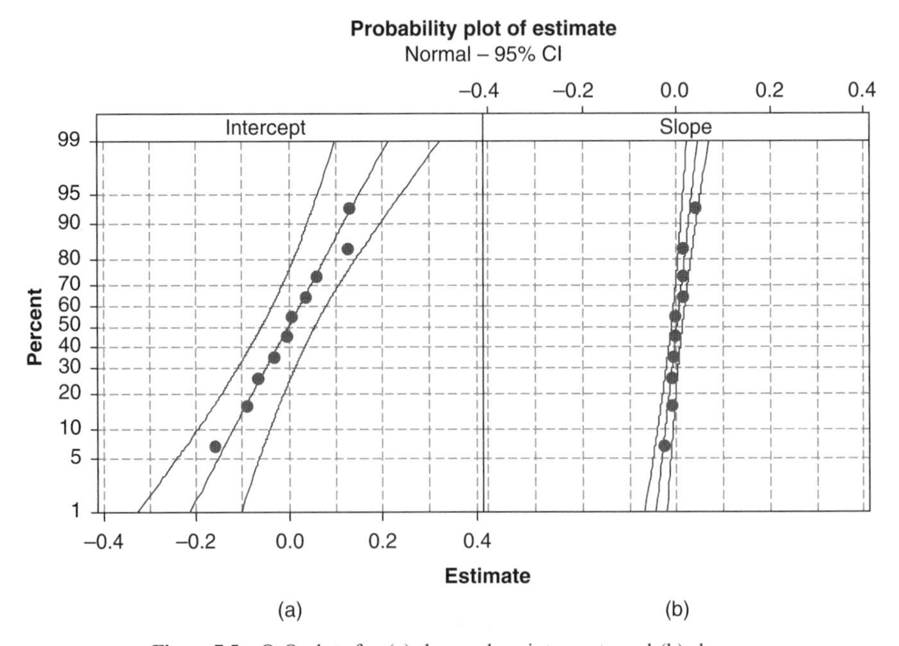

Figure 7.5 Q-Q plots for (a) the random intercepts and (b) slopes.

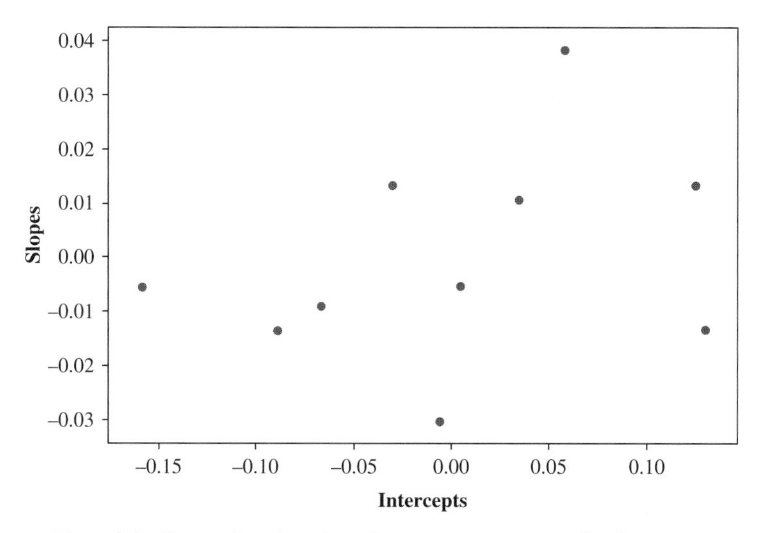

Figure 7.6 Scatterplot of random slopes versus corresponding intercepts.

MINITAB to assess the normality assumption. This plot does not indicate any obvious violation of the normality assumption of the model errors.

The next set of assumptions concerns the random effects, the δ's. The random effects are assumed to each be normally distributed with mean zero and constant variance. We can check for normality by doing Q-Q plots for each of the random effects. The `ods output solutionR=random` statement produces the BLUPs associated with the random intercept and slope terms. Figure 7.5a is the normal Q-Q plot for the random intercepts. Figure 7.5b is the normal Q-Q plot for both the random slopes. The normality assumptions for the random effects seem OK.

Model 4 also assumes that the random intercept and slope effects are independent; that is,

$$\mathrm{Var}(\delta_j) = \mathrm{Var}\left(\begin{bmatrix} \delta_{0,j} \\ \delta_{1,j} \end{bmatrix}\right) = \mathbf{D}_j$$

is a diagonal matrix for $j = 1, 2, \ldots, m$. To check this assumption, we use the scatterplot of the estimated random intercepts and slopes given in Figure 7.6. With the exception of one of the stations, the scatter appears quite random, suggesting independence between the two factors. □

Example 7.4. Industrial Split Plot Plastic Experiment Revisited. Example 7.2 considers the relationship between the strength of plastic, y, and four factors: baking temperature (*temp*), additive percentage (*add*), agitation rate (*rate*), and processing time (*time*). Each factor has two levels: low = −1 and high = 1. The factor *temp* is a hard-to-change, whole plot factor. There are four whole plots (clusters), representing the randomized levels of *temp*. Within each

Table 7.6 Model Fit Statistics and Estimated Variance Components From Example 7.4

	No Random (Independent **V**)	Random Intercept
σ^2_ε	14.213	9.782
$\sigma^2_{\delta 0}$	—	5.802
−2 log L	153.40	149.11
AIC	177.44	175.11
BIC	189.97	188.69

whole plot, the eight combinations of the other factor levels are randomly assigned. Unlike the Kukupa example, there is no natural time ordering among the eight observations within a cluster (a whole plot). Thus an exchangeable correlation structure seems sensible. To fit a compound symmetric correlation structure, the user can specify a random effect whose levels correspond to the individual whole plots, like a random intercept model, the user can specify the compound symmetric structure through **S**. However, one of the engineer's interest is to ascertain the magnitude of the whole plot variability for quality improvement purposes. As a result, a mixed model where the whole plot variability is explicitly modeled is more appropriate. Equation (7.18) gives the expression for the model. For comparison purposes, we also fit the marginal model in Equation (7.17); however, we assume that $\text{Var}(\boldsymbol{\varepsilon}) = \mathbf{S} = \sigma^2_\varepsilon \mathbf{I}_{32}$, which assumes independence among sub plot observations within each whole plot.

This specific experimental design is OLS-GLS equivalent (see Vining, Kowalski, and Montgomery, 2005). One consequence of OLS-GLS design is that the fixed effects coefficient estimates do not depend on the variance structure. As a result, the fixed effects estimates are the same for both models. Table 7.6 summarizes the variance component estimates and model fit statistics. The random intercept model, implying an exchangeable covariance structure, is better than the model assuming an independent covariance structure in terms of AIC and BIC. This result is no surprise given the experimental protocol.

One can conduct a REML likelihood ratio test of $H_0 : \sigma^2_{\delta_0} = 0$ versus $H_1 : \sigma^2_{\delta_0} \neq 0$ by taking the difference of the $-2l^{\text{model}}_{\text{REML}}$ values from both models as follows:

$$-2l^{\text{reduced}}_{\text{REML}} - \left(-2l^{\text{full}}_{\text{REML}}\right) = 153.4 - 149.11 = 4.29$$

As pointed out in the analysis of Example 7.1 as well as in Section 7.1.5, the null distribution of the statistic above is a mixture of χ^2_1 and χ^2_0 distributions. The p-value is essentially one-half of $Pr\left(\chi^2_1 > 4.29\right) = 0.038$. Thus the p-value of our test is approximately 0.019, implying that $\sigma^2_{\delta_0} \neq 0$. The random intercept is indeed needed and the full model is preferred.

The analysis of the model in Equation (7.18) is done here in R. Table 7.7 gives the appropriate R code. Table 7.8 provides the corresponding results. The first set of output in Table 7.8 consists of the model fit statistics AIC, BIC, and

Table 7.7 R Code for the Mixed Effects analysis of Data in Example 7.4

R Code Example 7.2Data:

```
#plastics is the name of the data set

plastics <- read.table("plastics.csv", header=TRUE, sep=",",
na.strings="NA", dec=".", strip.white=TRUE)

model1<-lme(Strength~ Temp + Add + Rate + Time + I(Temp*Add)+
                      I(Temp*Rate)+ I(Temp*Time)+I(Add*Rate)+
                      I(Add*Time)+I(Rate*Time),
                      random = ~ 1|OvenSet, data=plastics)
summary(model1)

#obtain the blups from model 1

blupmod1 <- ranef(model1)
eblupmod1

#extract the estimated V matrix from model 1 for first whole plot
#Need to install the mgcv package to do this

library(mgcv)
extract.lme.cov2(model1,plastics,start.level=1)

# standardized residuals vs fitted is from first plot statement
# Normal Q-Q plot is the second plot and is from qqnorm statement
# Code for producing Figure 7.7

par(mfrow=c(2,1))
plot(residuals(model1)~ fitted(model1), xlab="Fitted values",
           ylab="Standardized residuals")
abline(h=0)

qqnorm(residuals(model1), xlab = "Normal quantiles",
                      ylab="Standardized residuals", main="")
qqline(residuals(model1))

#Figure 7.9 produces Normal Q-Q plot for blups

qqnorm(ranef(model1)[,1], xlab="Normal quantiles",
          ylab="Estimated BLUPS", main="")
```

the REML log-likelihood. Recall that in likelihood ratio testing, the log-likelihood should be multiplied by −2. The material under *Random Effects* provides the estimated standard deviation of the random intercept under *(Intercept)* and the estimated standard deviation of the model errors, $\hat{\sigma}_\varepsilon$, under *Residual*. The squared values of these quantities appear in Table 7.6 Next in the

Table 7.8 R Output for Analysis of Strength in Plastics Data in Example 7.4

```
        Linear Mixed-Effects Model Fit by REML
                  Data: Plastics
              AIC          BIC          logLik
              175.11       188.69       -74.553

                  Random Effects:
              Formula: ~1 | OvenSet
                        (Intercept)    Residual
            StdDev:       2.4087         3.1276

 Fixed Effects: Strength ~ Temp + Add + Rate + Time + I(Temp * Add) +
  I(Temp * Rate) + I(Temp * Time) + I(Add * Rate) + I(Add * Time) +
                          I(Rate * Time)
                   Value    Std.Error  DF   t-value   p-value
  (Intercept)      62.003   1.32520    19   46.788    0.0000
  Temp             1.634    1.32520    2    1.233     0.3427
  Add              1.191    0.55289    19   2.153     0.0443
  Rate             1.134    0.55289    19   2.052     0.0542
  Time             1.541    0.55289    19   2.786     0.0118
  I(Temp * Add)    0.184    0.55289    19   0.333     0.7424
  I(Temp * Rate)   1.566    0.55289    19   2.832     0.0107
  I(Temp * Time)   1.397    0.55289    19   2.526     0.0206
  I(Add * Rate)    0.934    0.55289    19   1.690     0.1074
  I(Add * Time)    0.303    0.55289    19   0.548     0.5899
  I(Rate * Time)   1.172    0.55289    19   2.120     0.0474

                  Estimated BLUPS
                    (Intercept)
                  1     -0.57815
                  2     -2.11129
                  3      2.11129
                  4      0.57815

      Estimated Covariance Matrix for First Whole Plot
     1       2       3       4       5       6       7       8
1 15.5839 5.8019  5.8019  5.8019  5.8019  5.8019  5.8019  5.8019
2 5.8019  15.5839 5.8019  5.8019  5.8019  5.8019  5.8019  5.8019
3 5.8019  5.8019  15.5839 5.8019  5.8019  5.8019  5.8019  5.8019
4 5.8019  5.8019  5.8019  15.5839 5.8019  5.8019  5.8019  5.8019
5 5.8019  5.8019  5.8019  5.8019  15.5839 5.8019  5.8019  5.8019
6 5.8019  5.8019  5.8019  5.8019  5.8019  15.5839 5.8019  5.8019
7 5.8019  5.8019  5.8019  5.8019  5.8019  5.8019  15.5839 5.8019
8 5.8019  5.8019  5.8019  5.8019  5.8019  5.8019  5.8019  15.5839
```

output is the summary table of the analysis of the fixed effect coefficients. R does not provide the Satterthwaite adjustments and the Kenward–Rogers adjustments. Apparently, those in charge of the lme routine, have issues with both approaches. All of the output produced to this point is a result of the `summary(model1)` statement in the R code. The material in *Estimated BLUPS* is a result of the statement *ranef(model1)* and summarizes the estimated values of $\hat{\delta}_{0,1}, \hat{\delta}_{0,2}, \hat{\delta}_{0,3}$, and $\hat{\delta}_{0,4}$, which are the estimated levels of the random intercept corresponding to whole plots 1 through 4, respectively. The *Estimated Covariance Matrix For First Whole Plot* is a result of `extract.lme.cov2(model1,plastics,start.level=1)`. This section of output is the first (8×8) block of the estimated (32×32) variance–covariance matrix of responses, $\mathbf{V}$. The (1,1) entry of this first block, $\mathbf{V}_1$, denotes $\widehat{\text{Var}}\,(y_{11})$, where

$$\widehat{\text{Var}}(y_{11}) = \hat{\sigma}^2_{\delta_0} + \hat{\sigma}^2_{\varepsilon} = 2.4087^2 + 3.1276^2 = 15.5837$$

In this case $\mathbf{V}_1$ has the compound symmetric structure described in Section 7.1.3.

Figure 7.7a gives the plot of the studentized conditional residuals as defined by Equation (7.50) versus the predicted values for the conditional mean. Figure 7.7b gives the normal Q-Q plot of the conditional residuals. Other than a couple of unusually large standardized residuals, the assumptions

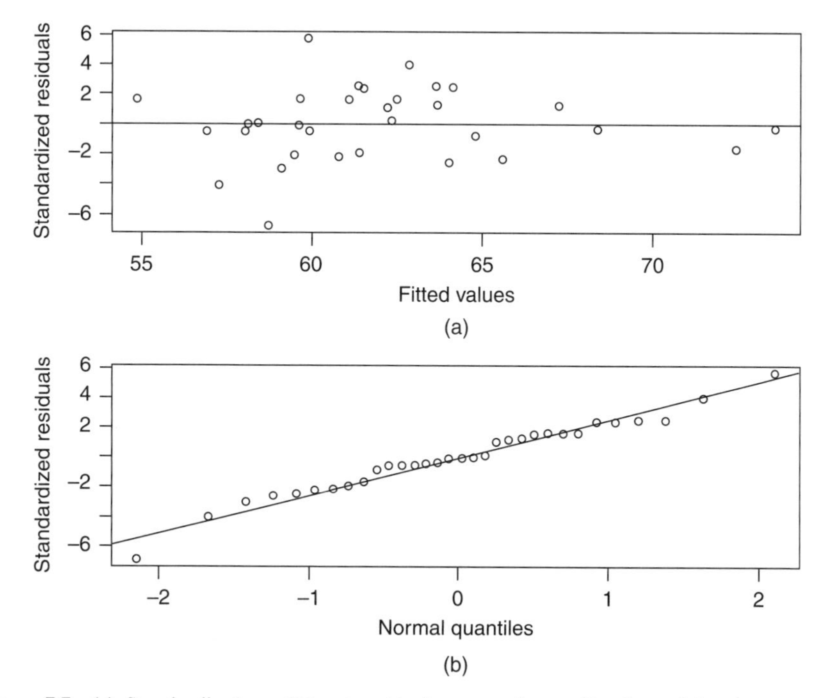

Figure 7.7 (a) Standardized conditional residuals versus the predicted conditional means and (b) normal probability plot of conditional residuals.

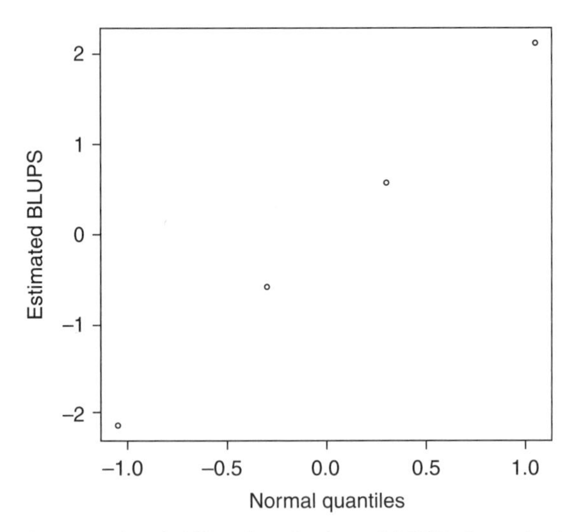

Figure 7.8 Normal probability plot of estimated BLUPs from plastics data.

regarding the model errors appear to be met. Figure 7.8 gives the normal Q-Q plot of the estimated BLUPs. This plot suggests that the levels of the random intercepts follow a normal distribution. Documentation in the R code in Table 7.7 specifies the code used to produce the plots in Figures 7.7 and 7.8 □

7.2 GENERALIZED LINEAR MIXED MODELS

Chapters 4 and 5 develop a flexible class of models for non normal data known as generalized linear models. Inherent to this development is the assumption of independent data. As pointed out in Section 7.1, many situations exist in which there is a need for models that involve both fixed and random effects. Generalized linear models that contain random effects are known as *generalized linear mixed models* (GLMMs). These models, like linear mixed effects models, are useful for a wide breadth of study types.

Example 7.5. Epileptic Seizure Data. Thall and Vail (1990) present data from a clinical trial of 59 epileptics (see Exercise 6.7). This study records for each patient the number of epileptic seizures experienced every 2 weeks for a period of 8 weeks. Half of the patients receive an anti-epileptic drug, progabide, and the other half receive a placebo. The goals of the study are to study the rate of seizures over the course of the study for both the placebo and treatment group and to characterize the patient-to-patient variability in terms of the seizure rates over time for both treatment groups. Since the observations at each time point are counts, a Poisson distribution is a natural choice for modeling. Although patients are independent of one another, the multiple responses on each patient over time are likely correlated. One can address this correlation by introducing one or more random effects to the Poisson generalized linear model. □

Example 7.6. Film Manufacturing Experiment. Robinson et al. (2004) consider a film manufacturing experiment involving both mixture and process variables. The film manufacturer produces rolls of film using a screw extruder and wishes to investigate the affect of six factors on film quality. In manufacturing, three mixture components (x_1, x_2, and x_3) are melted and mixed in an extruder, eventually producing a solid roll of film. This particular experiment contains five distinct *recipes* involving the combinations of the three mixture variables, some of which are replicated, to give a total of 13 batches. Once a roll of film is produced, the roll is cut into pieces and the pieces are randomly assigned to the levels of the process variables (p_1, p_2, and p_3) according to a 2^{3-1} design. For practical purposes, two separate randomizations take place. The first level of randomization involves the order in which the mixture combinations are used for film formulation. The second level of randomization occurs after the roll is produced and involves the levels of the process variables being randomly assigned to cut pieces of film. Figure 7.9 shows schematically the process. Figure 7.10 gives the mixture design region. The primary goal of the experiment is to determine factor levels that optimize the amount of polarized light that passes through the film. Engineering knowledge of the

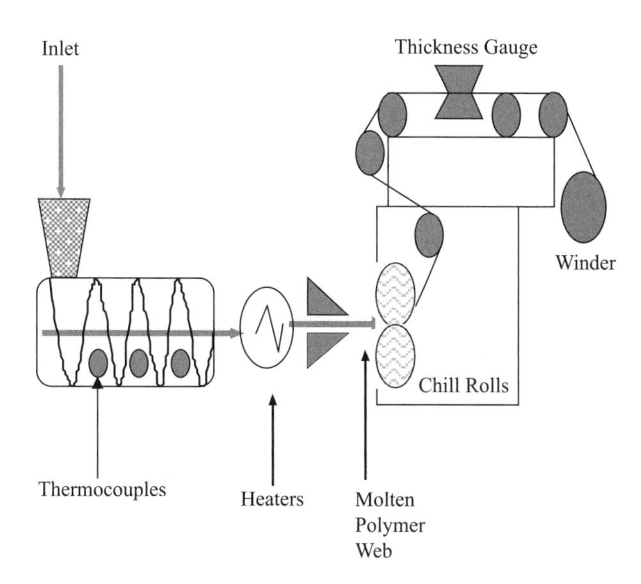

Figure 7.9 Schematic of film manufacturing process.

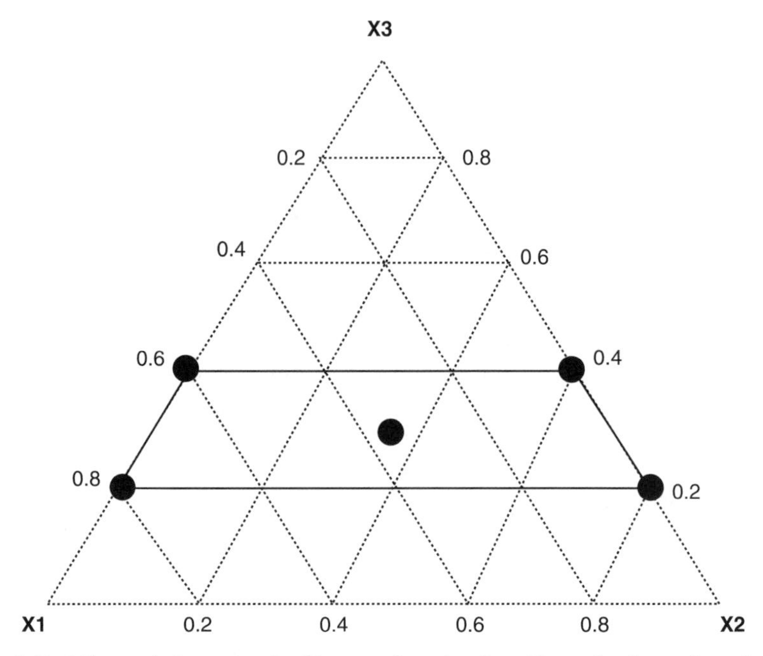

Figure 7.10 Mixture design region for film manufacturing data. For each mixture formulation, a 2^{3-1} design is run in the process variables.

Table 7.9 Data for Film Manufacturing Example 7.6 in Coded Variables

Mixture Formulation		Process Variables (p_1, p_2, p_3)			
Batch Number	(x_1, x_2, x_3)	(1,−1, −1)	(−1,−1,1)	(1,1,1)	(−1,1,−1)
1	(0.35,0.35,0.3)	397.88	127.41	356.35	70.93
4	(0.8,0,0.2)	136.3	92.78	406.3	284.71
5	(0.6,0,0.4)	56.17	113.02	74.26	119.99
8	(0,0.8,0.2)	125.89	48.29	33.59	7.19
9	(0,0.6,0.4)	5.57	17.11	12.35	4.52
17	(0.8,0,0.2)	230.29	114.39	242.39	92.74
21	(0,0.8,0.2)	91.14	26.22	77.67	11.85
24	(0,0.6,0.4)	74.89	11.06	45.79	8.98
29	(0.35,0.35,0.3)	367.15	133.64	175.24	89.27
32	(0.8,0,0.2)	146.03	38.16	169.1	81.97
33	(0.6,0,0.4)	53.77	50.11	71.82	124.26
36	(0,0.8,0.2)	83.52	20.99	51.58	7.01
37	(0,0.6,0.4)	14.77	7.22	38.32	3.21

process suggests that the response follows a gamma distribution. Observed values of the response on sub plots (pieces of film) within a given whole plot are correlated. One can account for this correlation by introducing a random effect into the gamma generalized linear model. Table 7.9 provides the data. □

7.2.1 The Generalized Linear Mixed Model

It is natural to begin by considering a model for the conditional mean because the observations are grouped into clusters. The generalized linear mixed model (GLMM) relates the conditional mean for the jth cluster to the fixed and random effects as follows:

$$E\left(\mathbf{y}_{n_j}|\boldsymbol{\delta}_j\right) = g^{-1}(\boldsymbol{\eta}_j) = g^{-1}(\mathbf{X}_j\boldsymbol{\beta} + \mathbf{Z}_j\boldsymbol{\delta}_j) \tag{7.51}$$

where $\mathbf{y}_{n_j}$ is the vector of responses at the jth cluster, g is a differentiable monotonic link function, $\boldsymbol{\eta}_j$ is the linear predictor given by $\boldsymbol{\eta}_j = \mathbf{X}_j\boldsymbol{\beta} + \mathbf{Z}_j\boldsymbol{\delta}_j$, $\mathbf{X}_j$ is the $(n_j \times p)$ matrix of fixed effects model terms associated with the jth cluster, $\boldsymbol{\beta}$ is the corresponding $(p \times 1)$ vector of fixed effects regression coefficients, $\boldsymbol{\delta}_j$ is the $(q \times 1)$ vector of random factor levels associated with the jth cluster, and $\mathbf{Z}_j$ is the corresponding matrix of predictors for the jth cluster. There are n_j observations in the jth cluster for a total of $n=\sum_{j=1}^{m} n_j$ observations. The conditional response, $y|\delta$, is assumed to have an exponential family member distribution. Each of the q random effects are assumed normally distributed with mean zero. The variance–covariance matrix of the vector of random effects in the jth cluster is denoted $\mathbf{D}_j$. The $\mathbf{D}_j$ are typically taken to be the same for each cluster. For the linear mixed model discussed in Section 7.1, the link function, g, is simply the identity link. As we discuss estimation and other topics in this section, we refer to the general expression of the conditional mean given by

$$E(\mathbf{y}|\boldsymbol{\delta}) = g^{-1}(\boldsymbol{\eta}) = g^{-1}(\mathbf{X}\boldsymbol{\beta} + \mathbf{Z}\boldsymbol{\delta}) \tag{7.52}$$

The variance–covariance matrix of the vector of conditional responses is Var $(\mathbf{y}|\boldsymbol{\delta}) = \mathbf{S}$, where

$$\mathbf{S} = \mathbf{A}^{1/2}(\boldsymbol{\eta})\mathbf{R}\mathbf{A}^{1/2}(\boldsymbol{\eta}) \tag{7.53}$$

with $\mathbf{A}(\boldsymbol{\eta})$ being the diagonal matrix that contains the variance functions associated with the assumed probability distribution of the response. See Section 6.4.1 regarding specific forms of $\mathbf{A}(\boldsymbol{\eta})$. The variance functions are evaluated at the linear predictor, $\boldsymbol{\eta}$. $\mathbf{R}$ is the user specified correlation matrix, which is common to all clusters.

To illustrate the model notation, consider the epilepsy data set. Suppose that we wish to model the number of seizures over time for the control group patients (i.e., trt = 0). We specify a random patient effect because the patients in the study are a random sample of patients from a population of possible patients. We consider both a random intercept for each cluster as well as a random slope for each cluster, like the Kukupa data set in Example 7.1. We begin with a random intercept model. Since we are interested in modeling a count, we assume that the number of seizures for a given patient, $y|\delta$, follows a Poisson distribution. Using a log link to relate $E\,(\mathbf{y}|\boldsymbol{\delta})$ to time, $\mathbf{x}$, the random intercept model for the conditional mean is

$$E(\mathbf{y} \mid \boldsymbol{\delta}_1^*) = g^{-1}(\boldsymbol{\eta}) = \exp(\mathbf{X}\boldsymbol{\beta} + \mathbf{Z}\boldsymbol{\delta}_1^*)$$

where the fixed effects components are

$$\mathbf{X} = \begin{bmatrix} \mathbf{1}_{n_1} & \mathbf{x}_{n_1} \\ \mathbf{1}_{n_2} & \mathbf{x}_{n_2} \\ \vdots & \vdots \\ \mathbf{1}_{n_{27}} & \mathbf{x}_{n_{27}} \\ \mathbf{1}_{n_{28}} & \mathbf{x}_{n_{28}} \end{bmatrix}, \mathbf{y} = \begin{bmatrix} \mathbf{y}_{n_1} \\ \mathbf{y}_{n_2} \\ \vdots \\ \mathbf{y}_{n_{27}} \\ \mathbf{y}_{n_{28}} \end{bmatrix}, \text{ and } \boldsymbol{\beta} = \begin{bmatrix} \beta_0 \\ \beta_1 \end{bmatrix} \tag{7.54}$$

Regarding notation, $\mathbf{1}_{n_j}$ is an $(n_j \times 1)$ vector of ones, $\mathbf{x}_{n_j}$ is the $(n_j \times 1)$ vector of time points for the jth cluster, $\mathbf{y}_{n_j}$ is the $(n_j \times 1)$ vector of responses for the jth cluster, and $n_j = 5$ for each of the $j = 1, 2, \ldots, 28$ clusters in this data set. The parameters β_0 and β_1 denote the population averaged intercept and slope, respectively. Regarding the random effects,

$$\mathbf{Z} = \begin{bmatrix} \mathbf{1}_{n_1} & \mathbf{0}_{n_1} & \cdots & \mathbf{0}_{n_1} \\ \mathbf{0}_{n_2} & \mathbf{1}_{n_2} & \cdots & \mathbf{0}_{n_2} \\ \vdots & \vdots & \cdots & \vdots \\ \mathbf{0}_{n_{27}} & \mathbf{0}_{n_{27}} & \cdots & \mathbf{0}_{n_{27}} \\ \mathbf{0}_{n_{28}} & \mathbf{0}_{n_{28}} & & \mathbf{1}_{n_{28}} \end{bmatrix} \text{ and } \boldsymbol{\delta}_1^* = \begin{bmatrix} \delta_{0,1} \\ \delta_{0,2} \\ \vdots \\ \delta_{0,28} \end{bmatrix}$$

where the levels of $\boldsymbol{\delta}_1^*$ correspond to the number of patients in the control group. In summary, the conditional response $y|\delta$ is assumed to follow a Poisson distribution with mean $E(y|\delta_1^*) = g^{-1}(\eta) = \exp(\beta_0 + \beta_1 x + \delta_1^*)$. The variability from patient to patient is expressed by the random intercept δ_1^*. It is this random intercept term that accounts for the correlation among the repeated observations for each patient.

For the industrial split plot film manufacturing example, prior data and an engineering understanding of the process suggests a gamma response. Thus we have that $y_{ji}|r, 1/\lambda_{ji} \sim Gamma(r, 1/\lambda_{ji})$, where y_{ji} denotes the film quality for the ith piece of film from the jth batch (roll). Section 5.5.1 uses the following parameterization of the gamma density for $y \sim Gamma(r, 1/\lambda)$,

$$f_y(y) = \frac{\left(1/\lambda\right)^r y^{r-1} e^{-y/\lambda}}{\Gamma(r)} \tag{7.55}$$

Robinson, Myers, and Montgomery (2004) choose a log link [i.e., $g(\mu_{ji}|\delta_j) = \ln(\mu_{ji}|\delta_j)$] to model the response as a function of the mixture and process variables. Thus the linear predictor is

$$\eta_{ji}|\delta_j = \ln\left(\mu_{ji}|\delta_j\right) = \mathbf{x}_{ji}'\boldsymbol{\beta} + \delta_j \tag{7.56}$$

where

$$\mathbf{x}_{ji}'\boldsymbol{\beta} = \sum_{b=1}^{3} \beta_b x_{b,ji} + \beta_{12} x_{1,ji} x_{2,ji} + \sum_{c=1}^{3} \gamma_{c,ji} p_{c,ji} + \sum_{b=1}^{2}\sum_{c=1}^{3} \psi_{bc} x_{b,ji} p_{c,ji} \tag{7.57}$$

and δ_j is the random effect associated with the jth batch. The random batch effects, the δ_j's, are assumed to be i.i.d. $Normal\left(0, \sigma_\delta^2\right)$. In this equation, the β's are the first- and second- order mixture terms, the γ's are the linear process terms, and the ψ's represent the mixture-by-process interactions. The linear predictor does not include an intercept since we are fitting a second-order Scheffé model for mixture components. Relating the model in Equation (7.56) to the general formulation in Equation (7.52), we have that $g(\cdot) = \ln(\cdot)$, $\mathbf{y}$ is the (52×1) vector of observations starting with the four observations from the first roll and so on, $\mathbf{Z}$ is a 52×13 classification matrix of ones and zeros where the klth entry is a one if the kth observation ($k = 1,\ldots,52$) belongs to the lth roll ($l = 1,\ldots,13$). The rows of the 52×13 model matrix $\mathbf{X}$ are formed by $\mathbf{x}_{ij}'$ for each observation. In summary, we assume that the conditional distribution of $y_{ji}|\delta_j$ is gamma and the conditional mean, $\mu_{ji}|\delta_j$, depends on r, $\mathbf{x}_{ji}'$, $\boldsymbol{\beta}$, and δ_j.

The major difference between generalized linear mixed models (GLMMs) and generalized linear models (GLMs) is the presence of the random effects in the GLMMs. When we use GLMMs, the data consists of clusters (e.g., patients, whole plots) and the levels of the random effects correspond to the clusters. Recall from Chapter 6 that we also can use GEE models for data structured in clusters. GEE models are most useful when the interest is solely in the marginal mean. We can use GLMMs to model both the conditional and marginal means. GLMMs have the added advantage of explicitly modeling variance components. In GLMs the data are not structured into clusters and the observations are assumed independent of one another. In GLMMs, we assume that the conditional response (i.e., the response conditioned on a fixed setting of the random effects) follows a probability distribution that is part of the exponential family. For GLMs, however, we assume that the unconditional response, y, follows an exponential family member distribution.

7.2.2 Parameter Estimation in the GLMM

Sections 4.2.2 and 4.3 presented the details of maximum likelihood estimation of the parameters in the linear predictor, the β's, for binary and Poisson responses, respectively. Section 5.3 outlines maximum likelihood equations for any member of the exponential family. Section 7.1 points out that we can base our inferences on maximizing the *marginal* likelihood of the response for the linear mixed model. Section 7.1 also pointed out that we typically base our inferences on maximizing the marginal residual likelihood of the data, REML. When the random effects and the data are normal, and we have an identity link, we have the standard linear mixed model

$$\mathbf{y} = \mathbf{X\beta} + \mathbf{Z\delta} + \boldsymbol{\varepsilon}$$

The marginal distribution of $\mathbf{y}$ is easily obtained by taking the expectation and variance operators through the model, producing $E(\mathbf{y}) = \mathbf{X\beta}$ and $\text{Var}(\mathbf{y}) = \mathbf{ZDZ'} + \text{S}$. We know that $\mathbf{y} \sim N(\mathbf{X\beta}, \mathbf{ZDZ'} + \text{S})$ since a linear combination of normal variables is also normal. For GLMMs, however, we assume a non normal conditional response, $y|\delta$, and normal random effects. In these cases, obtaining the marginal distribution of $\mathbf{y}$ is a more challenging task.

There are three general approaches for coming up with the marginal likelihood in GLMM settings:(1) linearizing the conditional mean and then repeatedly applying linear mixed models techniques to the approximated model; (2) using numerical methods to approximate the integrals involved in the marginal likelihood and developing a set of estimating equations based on this approximation; and (3) using a Bayesian approach. In this section we outline two linearization methods: pseudo-likelihood (PQL) and marginal quasi-likelihood (MQL) for parameter estimation. These methods are broadly applicable and have intuitive appeal. PQL is the default method in the SAS procedure GLIMMIX. Despite their utility, PQL and MQL can produce biased estimates in situations where the sample size is small, particularly for binary data. As a result, an increasingly popular method for the analysis of GLMMs numerically approximates the integrals involved in the marginal likelihood. The procedure then sets up the estimating equations for parameter estimation based on the approximated integrals. Breslow and Lin (1995) and Pinheiro and Chao (2006) discuss the limitations of PQL and MQL. A user can specify the integral approximation methods in PROC GLIMMIX through the METHOD = option where the user may request the Laplace or the quadrature approximations. Schabenberger (2007) gives details and examples of the use of these methods. Section 7.3 outlines the Bayesian approach.

Pseudo-Likelihood Parameter Estimation for GLMMs

GLMMs use the conditional response, $y|\delta$, to model the variation among observations within the same cluster. Variation among the clusters is attributed to the random effects in $\boldsymbol{\delta}$. Thus the joint density of $\mathbf{y}|\boldsymbol{\delta}$ and $\boldsymbol{\delta}$ describes the total variability in the data. Under the assumption that the model errors are independent of the random effects in mixed models, this joint density is

$$f_{\mathbf{y},\boldsymbol{\delta}}(\mathbf{y}|\boldsymbol{\delta},\boldsymbol{\delta}) = f_{\mathbf{y}|\boldsymbol{\delta}}(\mathbf{y}|\boldsymbol{\delta}) * f_{\boldsymbol{\delta}}(\boldsymbol{\delta}) \tag{7.58}$$

where $f_{\mathbf{y}|\boldsymbol{\delta},\boldsymbol{\delta}}(\cdot)$ denotes the joint density of $\mathbf{y}|\boldsymbol{\delta}$ and $\boldsymbol{\delta}$, $f_{\mathbf{y}|\boldsymbol{\delta}}(\cdot)$ denotes the conditional distribution of the response given the random effects, and $f_{\boldsymbol{\delta}}(\cdot)$ denotes the distribution of the random effects. We obtain the marginal distribution of the response by integrating over the distribution of the q random effects as follows:

$$f_{\mathbf{y}}(\mathbf{y}) = \int f_{\mathbf{y}|\boldsymbol{\delta}}(\mathbf{y}|\boldsymbol{\delta}) * f_{\boldsymbol{\delta}}(\boldsymbol{\delta})\, d\boldsymbol{\delta} \tag{7.59}$$

Unfortunately, this integral rarely exists in closed form; hence the need for the Laplace and quadrature methods to approximate this integral for the purpose of setting up parameter estimation equations. The linearization methods such as PQL and MQL do not work with the marginal distribution in Equation (7.59) directly. Instead, they work with the *pseudo* response, proposed by Breslow and Clayton (1993) and Wolfinger and O'Connell (1993).

A Subject-Specific Approach

The PQL linearization method first constructs a *pseudo-model* involving *pseudo-data* based on a first-order Taylor series approximation of the conditional mean. The conditional mean is

$$\begin{aligned} E(\mathbf{y}|\boldsymbol{\delta}) &= g^{-1}(\mathbf{X}\boldsymbol{\beta} + \mathbf{Z}\boldsymbol{\delta}) \\ &= g^{-1}(\boldsymbol{\eta}) = \boldsymbol{\mu}|\boldsymbol{\delta} \end{aligned}$$

A first-order Taylor series approximation of $\boldsymbol{\mu}|\boldsymbol{\delta}$ about given estimates of $\boldsymbol{\beta}$ and $\boldsymbol{\delta}$, $\mathbf{b}$ and $\hat{\boldsymbol{\delta}}$, results in

$$g^{-1}(\boldsymbol{\eta}) \approx g^{-1}(\hat{\boldsymbol{\eta}}) - \boldsymbol{\Delta}^{g^{-1}}_{\boldsymbol{\eta}=\hat{\boldsymbol{\eta}}}(\boldsymbol{\eta} - \hat{\boldsymbol{\eta}}) \tag{7.60}$$

where $\boldsymbol{\Delta}^{g^{-1}}_{\boldsymbol{\eta}=\hat{\boldsymbol{\eta}}}$ is an $n \times n$ diagonal matrix of derivatives with the (i,i) entry $(\delta g^{-1}(\boldsymbol{\eta}_i)/\delta\boldsymbol{\eta}_i)$ evaluated at $\hat{\boldsymbol{\eta}} = \mathbf{X}\mathbf{b} + \mathbf{Z}\hat{\boldsymbol{\delta}}$. Multiplying both sides by $\left[\boldsymbol{\Delta}^{g^{-1}}_{\boldsymbol{\eta}=\hat{\boldsymbol{\eta}}}\right]^{-1}$ and re arranging terms, we obtain

$$\left[\boldsymbol{\Delta}^{g^{-1}}_{\boldsymbol{\eta}=\hat{\boldsymbol{\eta}}}\right]^{-1}(g^{-1}(\boldsymbol{\eta}) - g^{-1}(\hat{\boldsymbol{\eta}})) + \mathbf{X}\mathbf{b} + \mathbf{Z}\hat{\boldsymbol{\delta}} \approx \mathbf{X}\boldsymbol{\beta} + \mathbf{Z}\boldsymbol{\delta} \tag{7.61}$$

Replacing $g^{-1}(\boldsymbol{\eta})$ with the vector of responses, $\mathbf{y}$, we have a vector of pseudo-responses, $\tilde{\mathbf{y}}$,

$$\tilde{\mathbf{y}} = \left[\boldsymbol{\Delta}^{g^{-1}}_{\boldsymbol{\eta}=\hat{\boldsymbol{\eta}}}\right]^{-1}(\mathbf{y} - g^{-1}(\hat{\boldsymbol{\eta}})) + \mathbf{Xb} + \mathbf{Z}\hat{\boldsymbol{\delta}} \tag{7.62}$$

It is easy to see that $\mathbf{E}(\tilde{\mathbf{y}})$ in Equation (7.62), assuming fixed $\mathbf{b}$ and $\hat{\boldsymbol{\delta}}$, is just the left-hand side of Equation (7.61). The variance–covariance matrix of $\tilde{\mathbf{y}}|\boldsymbol{\delta}$ is

$$\begin{aligned}\mathrm{Var}(\tilde{\mathbf{y}}|\boldsymbol{\delta}) &= \left[\boldsymbol{\Delta}^{g^{-1}}_{\boldsymbol{\eta}=\hat{\boldsymbol{\eta}}}\right]^{-1}\mathrm{Var}(\mathbf{y}|\boldsymbol{\delta})\left[\boldsymbol{\Delta}^{g^{-1}}_{\boldsymbol{\eta}=\hat{\boldsymbol{\eta}}}\right]^{-1} \\ &= \left[\boldsymbol{\Delta}^{g^{-1}}_{\boldsymbol{\eta}=\hat{\boldsymbol{\eta}}}\right]^{-1}\mathbf{A}^{1/2}(\boldsymbol{\eta})\mathbf{R}\mathbf{A}^{1/2}(\boldsymbol{\eta})\left[\boldsymbol{\Delta}^{g^{-1}}_{\boldsymbol{\eta}=\hat{\boldsymbol{\eta}}}\right]^{-1}\end{aligned}$$

where $\mathbf{R}$ is a user specified correlation matrix common to all clusters as discussed in Section 6.4. Thus one can consider estimation of model parameters using the linear mixed model

$$\tilde{\mathbf{y}} = \mathbf{X}\boldsymbol{\beta} + \mathbf{Z}\boldsymbol{\delta} + \boldsymbol{\varepsilon} \tag{7.63}$$

where $\tilde{\mathbf{y}}$ is the vector of pseudo-responses calculated using Equation (7.62), $\boldsymbol{\beta}$ is the vector of fixed effect parameters, $\boldsymbol{\delta}$ is the vector of random effects whose variance–covariance matrix is given by $\mathbf{D}$, while the variance–covariance matrix of the model errors, $\mathrm{Var}(\boldsymbol{\varepsilon}) = \mathrm{Var}(\tilde{\mathbf{y}}|\boldsymbol{\delta})$. The variance–covariance matrix of the pseudo-responses is

$$\mathbf{V}_{\tilde{\mathbf{y}}} = \mathbf{ZDZ}' + \left[\boldsymbol{\Delta}^{g^{-1}}_{\boldsymbol{\eta}=\hat{\boldsymbol{\eta}}}\right]^{-1}\mathbf{S}\left[\boldsymbol{\Delta}^{g^{-1}}_{\boldsymbol{\eta}=\hat{\boldsymbol{\eta}}}\right]^{-1} \tag{7.64}$$

where $\mathbf{S} = \mathbf{A}^{1/2}(\boldsymbol{\eta})\mathbf{R}\mathbf{A}^{1/2}(\boldsymbol{\eta})$. The model for the pseudo-response in (7.63) is identical to the linear mixed model given by Equation (7.4) in Section 7.1 except for an additional step where the vector of pseudo-responses, $\tilde{\mathbf{y}}$, depends on estimates of $\mathbf{D}$, $\mathbf{R}$, $\boldsymbol{\beta}$, and $\boldsymbol{\delta}$.

The iterative procedure that fits the model in Equation (7.63) initially uses the raw data as the value of $\tilde{\mathbf{y}}$ and the identity matrix $\mathbf{I}$ for $\mathbf{S}$. The procedure then fits a normal linear mixed model using a method such as REML to estimate the variance components in $\mathbf{D}$. It then substitutes these estimates into Equation (7.64) to obtain an estimated variance–covariance matrix $\hat{\mathbf{V}}_{\tilde{\mathbf{y}}}$. An estimated value for $\boldsymbol{\beta}$, $\mathbf{b}$, can then be calculated using

$$\mathbf{b} = \left(\mathbf{X}'\hat{\mathbf{V}}_{\tilde{\mathbf{y}}}^{-1}\mathbf{X}\right)^{-1}\mathbf{X}'\hat{\mathbf{V}}_{\tilde{\mathbf{y}}}^{-1}\tilde{\mathbf{y}} \tag{7.65}$$

which has an estimated variance–covariance matrix

$$\widehat{\mathrm{Var}}(\mathbf{b}) = \left(\mathbf{X}'\hat{\mathbf{V}}_{\tilde{\mathbf{y}}}^{-1}\mathbf{X}\right)^{-1} \tag{7.66}$$

The estimate of the random effect vector is

$$\hat{\boldsymbol{\delta}} = \hat{\mathbf{D}}\mathbf{Z}\hat{\mathbf{V}}_{\tilde{\mathbf{y}}}^{-1}(\tilde{\mathbf{y}} - \mathbf{X}\mathbf{b}) \tag{7.67}$$

and the pseudo-response vector is then re calculated from these estimates and a second linear mixed model is fit. The process continues until estimates of the fixed effects and variance components converge. It is important to note that the estimated regression coefficients in $\mathbf{b}$ and the estimated BLUPs in $\hat{\boldsymbol{\delta}}$ have asymptotic normal distributions.

Since the estimation method described above is based on a linearization of the conditional expectation as a function of $\hat{\boldsymbol{\eta}} = \mathbf{X}\mathbf{b} + \mathbf{Z}\hat{\boldsymbol{\delta}}$, calculation of the pseudo-response involves the vector of estimated random effects $\hat{\boldsymbol{\delta}}$. As a result the value of $\hat{\boldsymbol{\delta}}$ affects the estimates of $\boldsymbol{\beta}$. Since the levels of $\hat{\boldsymbol{\delta}}$ correspond to individual clusters or subjects, this methodology is commonly referred to in the literature as a *subject-specific* estimation procedure. Substituting the results from Equations (7.65) and (7.67) into Equation (7.52) yields the estimated conditional mean,

$$\widehat{\boldsymbol{\mu}}|\boldsymbol{\delta} = g^{-1}\left(\mathbf{X}\mathbf{b} + \mathbf{Z}\hat{\boldsymbol{\delta}}\right) \tag{7.68}$$

Plots of the profiles of the estimates of the conditional means for each of the clusters enable the user to visualize the variability from one cluster to the next.

A Population-Averaged Approach

In many applications, the user is more interested in estimating the unconditional or the marginal mean of the response than in estimating the conditional mean. For the general linear mixed model discussed in Section 7.1, where $E(\mathbf{y}|\boldsymbol{\delta}) = \mathbf{X}\boldsymbol{\beta} + \mathbf{Z}\boldsymbol{\delta}$, one easily can find the estimate of the marginal mean by taking the expectation of the conditional mean with respect to the random effects as follows:

$$E(\mathbf{y}) = E\left[E(\mathbf{y}|\boldsymbol{\delta})\right] = \mathbf{X}\boldsymbol{\beta}$$

Obtaining the marginal mean for a GLMM is more cumbersome since it involves the evaluation of the integral in Equation (7.59). In general, unless the link function is the identity, or $\boldsymbol{\delta} = \mathbf{0}$, in which case the GLMM is actually a GLM, one must resort to approximations. Approximate expressions for the marginal mean and variance involve linearizing the conditional mean, given in Equation (7.52), via a first-order Taylor series expansion about $E(\boldsymbol{\eta}) = \mathbf{X}\boldsymbol{\beta}$. The expansion here is about the average value of the linear predictor over the random effects.

A first-order Taylor series expansion of $E(\mathbf{y}|\boldsymbol{\delta})$ about the expectation of the $E(\boldsymbol{\eta}) = \mathbf{X}\boldsymbol{\beta}$ yields

$$E(\mathbf{y}|\boldsymbol{\delta}) = g^{-1}(\boldsymbol{\eta}) \approx g^{-1}(\mathbf{X}\boldsymbol{\beta}) + \boldsymbol{\Delta}^{g^{-1}}_{\boldsymbol{\eta}=\mathbf{X}\boldsymbol{\beta}}(\boldsymbol{\eta} - \mathbf{X}\boldsymbol{\beta}) \tag{7.69}$$

where $\boldsymbol{\Delta}^{g^{-1}}_{\boldsymbol{\eta}=\mathbf{X}\boldsymbol{\beta}}$ is an $n \times n$ diagonal matrix of derivatives with the (i,i) entry $(\delta g^{-1}(\boldsymbol{\eta}_i)/\delta\boldsymbol{\eta}_i)$ evaluated at $\mathbf{X}\boldsymbol{\beta}$. Taking the expectation of this expression gives the following approximation of the unconditional mean:

$$E(\mathbf{y}) = E\,[E(\mathbf{y}|\boldsymbol{\delta})] \approx g^{-1}(\mathbf{X}\boldsymbol{\beta}) \tag{7.70}$$

This approximation is exact for a linear link function and is more accurate when the variance components associated with $\boldsymbol{\delta}$ are close to zero. When the variance components are not close to zero, second-order Taylor series expansions may be preferable.

The approximate unconditional variance–covariance matrix of $\mathbf{y}$ can be determined from the relation

$$\text{Var}(\mathbf{y}) = \text{Var}[E(\mathbf{y}|\boldsymbol{\delta})] + E[\text{Var}(\mathbf{y}|\boldsymbol{\delta})] \tag{7.71}$$

The operations over the square brackets are calculated over the distribution of the random effects. The unconditional variance, Var $[E\,(\mathbf{y}|\boldsymbol{\delta})]$, is the sum of the variation in the average response across different batches. The expected conditional variance is E [Var $(\mathbf{y}|\boldsymbol{\delta})$]. Using Equation (7.71), the unconditional variance can be approximated by

$$\text{Var}(\mathbf{y}) \approx \boldsymbol{\Delta}^{g^{-1}}_{\boldsymbol{\eta}=\boldsymbol{\eta}_0}\mathbf{Z}\mathbf{D}\mathbf{Z}'\boldsymbol{\Delta}^{g^{-1}}_{\boldsymbol{\eta}=\boldsymbol{\eta}_0} + \mathbf{S}_{\boldsymbol{\eta}_0} \tag{7.72}$$

where the variance of the conditional expectation is obtained by taking the variance operator across the expression in Equation (7.69). The expectation of the conditional variance is found by taking the expectation of the expression in Equation (7.53). The linear predictor $\boldsymbol{\eta}$ is evaluated here at $\boldsymbol{\eta}_0 = \mathbf{X}\boldsymbol{\beta}_0$ for a given value $\boldsymbol{\beta}_0$. The matrix $\mathbf{D}$ is the same variance–covariance matrix associated with the specified random effects that was discussed in Section 7.1.3.

Wolfinger and O'Connell (1993) develop an approach known as marginal quasi-likelihood (MQL) for estimating the fixed effects regression parameters when the interest primarily focuses on the unconditional mean. This method is similar to the pseudo-likelihood method described for the *subject-specific* approach except that the linearization in MQL is conducted about $\hat{\boldsymbol{\eta}}_0 = \mathbf{X}\mathbf{b}$, the estimated expectation of the linear predictor. Since the linearization in MQL occurs about the population average of the linear predictor, the approach is known as a *population-averaged* estimation procedure.

Performing this linearization, we have

$$g^{-1}(\boldsymbol{\eta}) \approx g^{-1}(\hat{\boldsymbol{\eta}}_0) - \boldsymbol{\Delta}^{g^{-1}}_{\boldsymbol{\eta}=\hat{\boldsymbol{\eta}}}(\boldsymbol{\eta} - \hat{\boldsymbol{\eta}}_0) \tag{7.73}$$

where $\Delta^{g^{-1}}_{\eta=\hat{\eta}_0}$ is an $n \times n$ diagonal matrix of derivatives with the *(i,i)* entry $\delta g^{-1}(\eta_i)/\delta\eta_i$ evaluated at $\hat{\eta}_0 = \mathbf{Xb}$. Multiplying both sides by $\left[\Delta^{g^{-1}}_{\eta=\hat{\eta}_0}\right]^{-1}$ and rearranging terms, we obtain

$$\left[\Delta^{g^{-1}}_{\eta=\hat{\eta}_0}\right]^{-1}\left(g^{-1}(\eta) - g^{-1}(\hat{\eta}_0)\right) + \mathbf{Xb} \approx \mathbf{X}\beta + \mathbf{Z}\delta$$

The population-averaged pseudo-responses, $\tilde{\mathbf{y}}_{\mathrm{pa}}$, are

$$\tilde{\mathbf{y}}_{\mathrm{pa}} = \left[\Delta^{g^{-1}}_{\eta=\hat{\eta}_0}\right]^{-1}\left(\mathbf{y} - g^{-1}(\hat{\eta}_0)\right) + \mathbf{Xb} \tag{7.74}$$

Consider the linear mixed model

$$\tilde{\mathbf{y}}_{\mathrm{pa}} = \mathbf{X}\beta + \mathbf{Z}\delta + \varepsilon \tag{7.75}$$

for estimation of the model parameters. The expressions in Equations (7.73)–(7.75) are the same as those in Equations (7.60), (7.62), and (7.63), respectively, except now they are functions of $\hat{\eta}_0$, rather than $\hat{\eta}$. Also, the pseudo-response is no longer a function of $\hat{\delta}$; however, the right-hand side of the equation is still a function of δ. As a result there is no need to calculate $\hat{\delta}$ as was done in the *subject-specific* method when computing the pseudo-response.

The estimates in **b** now are

$$\mathbf{b} = \left(\mathbf{X}'\hat{\mathbf{V}}^{-1}_{\tilde{\mathbf{y}}_{\mathrm{pa}}}\mathbf{X}\right)^{-1}\mathbf{X}'\hat{\mathbf{V}}^{-1}_{\tilde{\mathbf{y}}_{\mathrm{pa}}}\tilde{\mathbf{y}}_{\mathrm{pa}} \tag{7.76}$$

which are impacted by the random effects only through the estimated variance–covariance matrix, $\hat{\mathbf{V}}^{-1}_{\tilde{\mathbf{y}}}$. Like the estimates in **b** for PQL, these estimates also have asymptotic normal distributions. The variance–covariance matrix of these estimates is

$$\widehat{\mathrm{Var}}(\mathbf{b}) = \left(\mathbf{X}'\hat{\mathbf{V}}^{-1}_{\tilde{\mathbf{y}}_{\mathrm{pa}}}\mathbf{X}\right)^{-1} \tag{7.77}$$

with

$$\mathbf{V}_{\tilde{\mathbf{y}}_{\mathrm{pa}}} = \mathbf{ZDZ}' + \left[\Delta^{g^{-1}}_{\eta=\hat{\eta}_0}\right]^{-1}\mathbf{S}\left[\Delta^{g^{-1}}_{\eta=\hat{\eta}_0}\right]^{-1} \tag{7.78}$$

In general, the estimated fixed effect parameters, the *b's*, obtained via the *subject-specific* method, are not the same as those obtained via the *population-averaged* method. Once the final estimates of β and $\mathbf{V}_{\tilde{\mathbf{y}}_{\mathrm{pa}}}$ are obtained from MQL, the estimated marginal (population-averaged) mean response is

$$\hat{\mu} = g^{-1}(\mathbf{Xb}) \tag{7.79}$$

where **b** is given by Equation (7.76).

7.2.3 Statistical Inference on Regression Coefficients and Variance Components

Inference involving the regression coefficients and variance components in GLMMs proceeds similarly to what was in Section 7.1 for the linear mixed model. An appropriate test on the variance component associated with a particular random effect determines if that effect belongs in the model, which in turn allows the analysts to reduce the covariance matrix in Equation (7.64) or (7.78). One can compare two covariance structures through a REML pseudo-likelihood ratio test (LRT). In SAS 9.2, the REML pseudo-likelihood ratio test for comparing two covariance structures uses the COVTEST statement in SAS PROC GLIMMIX. This approach fits the full model, which is the model with the most complex covariance structure of interest, and then takes the pseudo-data from the last iteration to calculate the associated REML pseudo-likelihood value. Let $l_{\text{FULL}}^{\text{R}}$ denote this restricted log-pseudo-likelihood value. The procedure then uses the same pseudo-data to fit the reduced model and computes the corresponding pseudo-likelihood. Let $l_{\text{RED}}^{\text{R}}$ denote this restricted log-pseudo-likelihood value. The appropriate likelihood ratio test statistic, $\hat{\lambda}$, is

$$\hat{\lambda} = l_{\text{FULL}}^{\text{R}} - l_{\text{RED}}^{\text{R}} \tag{7.80}$$

Once again, the distribution of this test statistic depends on whether any of the parameters fall on the boundary of the variance parameter space as defined by Self and Liang (1987) and Verbeke and Molenberghs (2000, Section 6.3.4). The SAS GLIMMIX procedure allows for the Scenarios 1–3 outlined in Section 7.1.5 when computing p-values. See SAS/STAT 9.2 User's Guide *Statistical Inference for Covariance Parameters* for more details. Similar to what was done for linear mixed models, the tests used to compare covariance structures for GLMMs are conducted with the full fixed effects vector, $\boldsymbol{\beta}$. We illustrate tests for comparing covariance structures in the next section.

Once the form of the covariance matrix is determined, it is necessary to test the significance of the fixed effects for the model in Equation (7.75). Again, we can utilize Wald tests. To test the hypothesis $H_0 : \beta_i = 0$ versus $H_a : \beta_i \neq 0$, where β_i is the ith element of $\boldsymbol{\beta}$, the Wald test statistic is

$$t_i = \frac{b_i}{\widehat{se}(b_i)} \tag{7.81}$$

where b_i is the corresponding element in $\mathbf{b}$ and $\widehat{se}(b_i)$ is the square root of the corresponding diagonal element of $\widehat{\text{var}}(\mathbf{b})$ given in Equation (7.66) for subject-specific inference and in Equation (7.77) for population-averaged inference. The null distribution of this test statistic is a t-distribution.

Quite often, there is also interest in conducting inference on the conditional and/or marginal mean response at a specific point, $\mathbf{x}_0$. Since the estimated conditional and marginal means are nonlinear functions, inference focuses on the linear predictor. Taking the inverse of the link function provides the corresponding inference for the means. For instance, from Equation (7.68) the estimated conditional mean for cluster j at $\mathbf{x}_0$ is

$$\hat{\mu}(\mathbf{x}_0)|cluster_j = g^{-1}\left(\mathbf{x}_0'\mathbf{b} + \sum_{l=1}^{q}\hat{\delta}_{lj}\right) \tag{7.82}$$

where the δ_{lj} represent the BLUPs for the jth subject associated with each of the q random effects. The linear predictor is then

$$\hat{\eta}(\mathbf{x}_0)|cluster_j = \mathbf{x}_0'\mathbf{b} + \sum_{l=1}^{q}\hat{\delta}_{lj} \tag{7.83}$$

Since the parameter estimates in $\hat{\eta}|cluster_j$ are asymptotically normal, $\hat{\eta}|cluster_j$ is also asymptotically normal. The approximate $(1 - \alpha)100\%$ confidence interval on $\eta|cluster_j$ is

$$\hat{\eta}|cluster_j \pm z_{1-\alpha/2}\widehat{se}(\hat{\eta}|cluster_j) \tag{7.84}$$

GLIMMIX and R both provide estimates of the standard error of the conditional linear predictor, $se(\hat{\eta}|cluster_j)$. The corresponding $(1 - \alpha)100\%$ confidence interval on $\mu(\mathbf{x}_0) \mid cluster_j$ is

$$\left[g^{-1}\left(\hat{\eta}|cluster_j - z_{1-\alpha/2}\widehat{se}(\hat{\eta}|cluster_j)\right), g^{-1}\left(\hat{\eta}|cluster_j + z_{1-\alpha/2}\widehat{se}(\hat{\eta}|cluster_j)\right)\right] \tag{7.85}$$

Similarly, from Equation (7.79), the estimated marginal mean at $\mathbf{x}_0$ is

$$\hat{\mu}(\mathbf{x}_0) = g^{-1}(\mathbf{x}_0'\mathbf{b}) \tag{7.86}$$

The corresponding estimated linear predictor is

$$\hat{\eta}(\mathbf{x}_0) = \mathbf{x}_0'\mathbf{b} \tag{7.87}$$

where $\mathbf{b}$ is defined in Equation (7.76). Like the conditional linear predictor in Equation (7.82), the marginal linear predictor in Equation (7.87) has an asymptotic normal distribution. The approximate $(1 - \alpha)100\%$ confidence interval on η is

$$\hat{\eta} \pm z_{1-\alpha/2}\widehat{se}(\hat{\eta}) \tag{7.88}$$

GLIMMIX and R both provide estimates for the standard error of the marginal linear predictor, $\widehat{se}(\hat{\eta})$. The corresponding $(1 - \alpha)100\%$ confidence interval on $\mu(\mathbf{x}_0)$ is

$$\left[g^{-1}(\hat{\eta} - z_{1-\alpha/2}\widehat{se}(\hat{\eta})), g^{-1}(\hat{\eta} + z_{1-\alpha/2}\widehat{se}(\hat{\eta}))\right] \tag{7.89}$$

We illustrate the construction of confidence limits for both the conditional and marginal means later with the epilepsy data.

7.2.4 Subject-Specific Versus Population-Averaged Prediction Models

As previously mentioned, prediction models for GLMMs depend on the scope of inference desired by the practitioner. Subject-specific models are particularly useful in repeated measures studies where the focus is the individual profiles of subjects across time. From Equation (7.68) these models yield estimates of the mean that are conditional on the levels of the random effects. When interest is in estimating general trends across the entire population of random effects, the population-averaged models are more useful. For instance, in the film manufacturing example, the engineer is likely more interested in the average film quality across a population of rolls of film rather than the individual prediction models for each specific roll of film. A popular approach for estimating the marginal mean using the subject-specific predictions given by Equation (7.68) is to simply set $\hat{\boldsymbol{\delta}} = \mathbf{0}$ since $E(\boldsymbol{\delta}) = \mathbf{0}$. However, this estimate of the marginal mean differs from the estimate of the mean produced by the population-averaged modeling approach.

To illustrate this point, consider the following GLMM

$$E(y|\delta) = \exp\{1 + X + \delta\} \tag{7.90}$$

where we assume that $\delta \sim N(0, 0.65^2)$ and that X takes on values between -1.0 and 1.0. We also assume a gamma distribution for the conditional response, $y|\delta$. We generated five random values for the δ. The dotted plots in Figure 7.11 are the profiles corresponding to each of the five levels of δ. Recall that the levels of the random effect, δ, correspond to specific clusters of observations. By generating five random normal variates for δ, we are simulating the presence of five clusters of observations. The distance between the conditional profiles is a function of the magnitude of $\text{Var}(\delta)$ and represents the cluster to cluster variability. The larger $\text{Var}(\delta)$ is, the more spread out the profiles are. Note the histogram of the gamma distribution to the right of Figure 7.11. While the normal distribution represents the natural variation from one cluster to another, the gamma distribution represents the natural variation among observations within the same cluster.

We can provide an estimate for the marginal mean by simply taking the average of the five random normal variates generated for δ and substituting

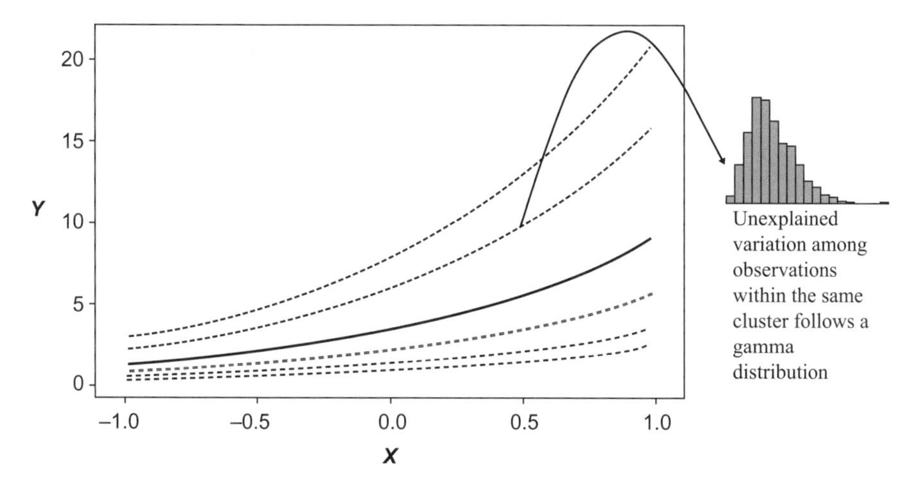

Figure 7.11 Profiles of the conditional mean, $E(y|\delta) = \exp\{1 + X + \delta\}$, for five random normal realizations of δ ($\delta = -0.28, -1.08, 1.05, -0.745$, and 0.774; profiles represented by dotted curves). Profile of the estimated marginal mean when $\delta = \bar{\delta}$ (represented by bold dashed curve) and for the true marginal mean, $E(y) = \exp\{1.211 + X\}$ (represented by bold solid curve).

this average for δ into Equation (7.90). The bold, dashed curve in Figure 7.11 is the profile for this estimate of the marginal mean. We obtain the *true* average profile by integrating out the δ's from Equation (7.90) as follows:

$$E[E(y|\delta)] = \int_{-\infty}^{\infty} \exp\{1+X+\delta\} f(\delta) \quad (7.91)$$

where $f(\cdot)$ denotes the normal density centered at 0 with variance 0.65^2. Note that $E[E(y|\delta)] = E(y)$ is the quantity actually modeled in population-averaged GLMM. Evaluating this integral we have

$$E(y) = \exp\{1.211 + X\} \quad (7.92)$$

The bold solid curve in Figure 7.11 is the profile corresponding to the true marginal mean. Please note the difference between the approximate marginal mean obtained by substituting the average of the deltas into Equation (7.90) (the bold dashed curve) and the true marginal mean from Equation (7.92) (the bold solid curve). The practical implication of this difference is that any estimate of the marginal mean based on the subject-specific prediction model is not the same as the estimate of the marginal mean based on the population-averaged prediction model. If interest is in the marginal mean, one should use a GLMM that specifically models the marginal mean.

7.2.5 Examples Revisited

Example 7.7. Epilepsy Data Revisited. Consider again Example 7.5. One goal of the study is to examine the rate of seizures over the course of the study for both the placebo and treatment groups. The number of seizures at the first time point (year = 0) is the baseline number of seizures experienced 8 weeks. The seizure counts at each of the other time points are observed over a 2 week period. We let *treat* represent the indicator variable for treatment where

$$treat = \begin{cases} 0 \text{ for placebo subjects} \\ 1 \text{ for progabide subjects} \end{cases}$$

Following Thall and Vail (1990) we define *lage* = ln (*age*) and *lbase* = ln (*baseline* / 4). We divide *baseline* by 4 to put it on the same scale for seizure counts as the counts observed at each of the other time periods. Two patients, 207 and 227, have what appears to be very unusual data. Consequently, we omit them from the analysis. Thus there are a total of 57 subjects in the modeled data. The main goals are to determine if the average number of seizures for those on progabide is less than for those on the placebo and to see if the seizure rates decline over the time periods (*period*). Interest also lies in determining if seizure rates vary from one subject to another. Since the patients in the study are a random sample of patients, random slopes and intercepts are of interest if seizure rates vary from one subject to another. As a result, we utilize the subject-specific PQL approach to the analysis of the Poisson GLMM.

The first step is to decide on the appropriate covariance structure of the data. We consider four potential GLMMs. Table 7.10 summarizes the results for each. Each of the four models contains four fixed effects: *treat, period, lage*, and *lbase*. However, we fit different covariance structures to each model. This analysis assumes a log link.

The first model does not use random effects to model the correlation among observations within the same cluster, where each subject is a cluster. Instead, this model specifies a first-order auto regressive (AR(1)) structure for the

Table 7.10 Summary of Model Fits to the Epilepsy Data

	No Random Effects + AR(1) **R**	Random Intercepts + Independent **R**	Random Slopes + Independent **R**	*Independent Random Intercepts and Slopes + Independent* **R**
Intercept	−2.2218	−0.8359	−2.4033	−1.2264
treat	−0.1900	−0.3039	−0.2333	−0.3154
period	−0.0613	−0.0663	−0.1111	−0.0733
lage	0.7450	0.3551	0.8384	0.4751
lbase	0.9573	0.9038	0.9330	0.9094
$\hat{\sigma}^2_{\delta_0}$	—	0.2415	—	0.2296
$\hat{\sigma}^2_{\delta_1}$	—	—	0.0285	0.0083
ϕ	3.2190	—	—	—
ρ	0.4145	—	—	—
-2 *Res* $\ln(PL)$	515.10	506.77	565.79	504.06

repeated observations. An unstructured correlation structure for $\mathbf{R}$ turns out not to be feasible due to convergence issues, which is something that often occurs with an unstructured specification. Directly specifying the covariance structure of the responses in a GLMM and not using random effects to account for the correlation structure is said to be modeling with *R-side* effects. These models are strictly marginal models. GLMMs specified using only R-side effects to account for the correlation directly parallel what is done in GEE models.

In the linear mixed model the R-side covariance is the covariance structure of the residuals. There are applications when one may be interested in modeling with both random effects as well as R-side effects. Schabenberger (2007) provides more details. The first model under consideration is

$$E(\mathbf{y}) = g^{-1}(\boldsymbol{\eta}) = \exp(\mathbf{X}\boldsymbol{\beta}) \tag{7.93}$$

Regarding notation,

$$\mathbf{X} = \begin{bmatrix} \mathbf{1}_{n_1} & \mathbf{x}_{1n_1} & \mathbf{x}_{2n_1} & \mathbf{x}_{3n_1} & \mathbf{x}_{4n_1} \\ \mathbf{1}_{n_2} & \mathbf{x}_{1n_2} & \mathbf{x}_{2n_2} & \mathbf{x}_{3n_2} & \mathbf{x}_{4n_2} \\ \vdots & \vdots & \vdots & \vdots & \vdots \\ \mathbf{1}_{n_{58}} & \mathbf{x}_{1n_{56}} & \mathbf{x}_{2n_{56}} & \mathbf{x}_{3n_{56}} & \mathbf{x}_{4n_{56}} \\ \mathbf{1}_{n_{59}} & \mathbf{x}_{1n_{57}} & \mathbf{x}_{2n_{57}} & \mathbf{x}_{3n_{57}} & \mathbf{x}_{4n_{57}} \end{bmatrix}, \quad \mathbf{y} = \begin{bmatrix} \mathbf{y}_{n_1} \\ \mathbf{y}_{n_2} \\ \vdots \\ \mathbf{y}_{n_{56}} \\ \mathbf{y}_{n_{57}} \end{bmatrix}, \text{ and } \boldsymbol{\beta} = \begin{bmatrix} \beta_0 \\ \beta_1 \\ \beta_2 \\ \beta_3 \\ \beta_4 \end{bmatrix} \tag{7.94}$$

where each of the $\mathbf{1}_{nj}$ is a (4×1) vector of ones, $\mathbf{x}_{1n_j}$ is a (4×1) vector of ones if the jth subject is taking the drub progabide and is a (4×1) vector of zeros if on the placebo, $\mathbf{x}_{2n_j}$ is the (4×1) vector of periods, and $\mathbf{x}_{3n_j}$ and $\mathbf{x}_{4n_j}$ are the (4×1) vectors representing log age and log baseline number of seizures divided by four, respectively, for the jth subject. Finally, $\mathbf{y}_{n_j}$ is the (4×1) vector of responses for the jth subject with $n_j = 4$ for each of the $j = 1, 2, \ldots, 57$ subjects in this data set. The parameters β_0, β_1, β_2, β_3, and β_4 denote the population averaged intercept and slopes, respectively. If our only interest is to make inferences on the population-averaged parameters, β_0, β_1, β_2, β_3, and β_4, a marginal analysis, such as MQL, is more appropriate. In our situation, we seek to determine the range of the period effect across the subjects. The variance–covariance matrix of the responses for the jth subject is

$$\text{Var}(\mathbf{y}_j) = \mathbf{S}_j = \mathbf{A}_j^{1/2}(\boldsymbol{\eta})\mathbf{R}\mathbf{A}_j^{1/2}(\boldsymbol{\eta}) \tag{7.95}$$

This expression for the variance–covariance matrix differs from the $\mathbf{S}_j$ that compose $\mathbf{S}_{\eta_0}$ in Equation (7.72). Equation (7.95) has no component associated with the random effects since it only uses the R-side effects to account for the correlation structure. For Equation (7.95),

$$\mathbf{A}_j(\boldsymbol{\eta}) = \begin{bmatrix} a_{j1} & & & \phi \\ & a_{j2} & & \\ & & a_{j3} & \\ \phi & & & a_{j4} \end{bmatrix} \tag{7.96}$$

where a_{ji} is the component of the variance delivered by the Poisson distribution for the ith observation on the jth subject. So, here, $a_{ji} = \mu_{ji} = \exp(\beta_0 + \beta_1 x_{1ji} + \beta_2 x_{2ji} + \beta_3 x_{3ji} + \beta_4 x_{4ji})$ with x_{1ji} denoting whether subject j took the placebo or progabide, x_{2ji} denoting the ith period of the study, x_{3ji} is the log age, and x_{4ji} is the log baseline number of seizures divided by four, respectively, for the jth subject. As discussed in Section 6.4, ϕ is the natural scale parameter for the Poisson distribution. The matrix $\mathbf{R}$ is the common correlation matrix for each subject. This model uses a first-order autoregressive structure.

Models 2–4 use random effects to account for the correlation among observations taken on the same subject. Model 2 allows for random intercepts among the subjects. It also assumes an common identity correlation matrix for all the subjects. The model is

$$E(\mathbf{y}) = g^{-1}(\boldsymbol{\eta}) = \exp\left(\mathbf{X}\boldsymbol{\beta} + \mathbf{Z}_1^*\boldsymbol{\delta}_1^*\right) \tag{7.97}$$

where $\mathbf{X}$ and $\boldsymbol{\beta}$ are as defined in Equation (7.94), $\mathbf{Z}_1^*$ is defined as in Equation (7.8), and $\boldsymbol{\delta}_1^*$ is defined as in Equation (7.9). For the expressions in Equations (7.8) and (7.9), m, the number of clusters is 57. The only difference in this model and model 2 of the Kukupa example from Table 7.3 is that we have a Poisson model with log link here and the Kukupa example assumed normal data with an identity link. For model 2 here, the variance–covariance matrix of responses, given generally in Equation (7.72), is approximately

$$\mathrm{Var}(\mathbf{y}) \approx \sigma_{\delta_0}^2 \boldsymbol{\Delta}^{g^{-1}}_{\boldsymbol{\eta}=\boldsymbol{\eta}_0} \mathbf{Z}_1^* \mathbf{Z}_1^{*\prime} \boldsymbol{\Delta}^{g^{-1}}_{\boldsymbol{\eta}=\boldsymbol{\eta}_0} + \mathbf{A}(\boldsymbol{\eta}_0) \tag{7.98}$$

Model 3 fits a random slope without a random intercept. The model is

$$E(\mathbf{y}) = g^{-1}(\boldsymbol{\eta}) = \exp\left(\mathbf{X}\boldsymbol{\beta} + \mathbf{Z}_2^*\boldsymbol{\delta}_2^*\right) \tag{7.99}$$

Here, $\mathbf{X}$ and $\boldsymbol{\beta}$ are defined the same as in Equation (7.97), $\boldsymbol{\delta}_2^*$ is defined exactly as in Equation (7.9), and $\mathbf{Z}_2^*$ is

$$\mathbf{Z}_2^* = \begin{bmatrix} \mathbf{x}_{2n_1} & \mathbf{0}_{n_1} & \cdots & \mathbf{0}_{n_1} \\ \mathbf{0}_{n_2} & \mathbf{x}_{2n_2} & \cdots & \mathbf{0}_{n_2} \\ \vdots & \vdots & \cdots & \vdots \\ \mathbf{0}_{n_{56}} & \mathbf{0}_{n_{56}} & \cdots & \mathbf{0}_{n_{56}} \\ \mathbf{0}_{n_{57}} & \mathbf{0}_{n_{57}} & & \mathbf{x}_{2n_{57}} \end{bmatrix} \tag{7.100}$$

with $\mathbf{x}_{2n_j}$ denoting the n_j time points for subject j. The variance–covariance matrix of responses, given generally in Equation (7.72), is approximately

$$\text{Var}(\mathbf{y}) \approx \sigma^2_{\delta_1} \boldsymbol{\Delta}^{g^{-1}}_{\boldsymbol{\eta}=\boldsymbol{\eta}_0} \mathbf{Z}^*_2 \mathbf{Z}^{*\prime}_2 \boldsymbol{\Delta}^{g^{-1}}_{\boldsymbol{\eta}=\boldsymbol{\eta}} + \mathbf{A}(\boldsymbol{\eta}_0) \tag{7.101}$$

Model 4 allows for both random intercepts as well as random slopes, where we assume that these two random effects are independent of one another. For the R-side effects we take $\mathbf{R}$ to be the identity matrix, as we did for models 2 and 3. The resulting model is

$$E(\mathbf{y}) = g^{-1}(\boldsymbol{\eta}) = \exp\left(\mathbf{X}\boldsymbol{\beta} + \mathbf{Z}^*_1\boldsymbol{\delta}^*_1 + \mathbf{Z}^*_2\boldsymbol{\delta}^*_2\right) \tag{7.102}$$

with $\mathbf{X}$, $\boldsymbol{\beta}$, $\mathbf{Z}^*_1$, $\mathbf{Z}^*_2$, $\boldsymbol{\delta}^*_1$, and $\boldsymbol{\delta}^*_2$ defined exactly as in Equations (7.97) and (7.99). The variance–covariance matrix of responses is then approximately

$$\text{Var}(\mathbf{y}) \approx \boldsymbol{\Delta}^{g^{-1}}_{\boldsymbol{\eta}=\boldsymbol{\eta}_0} \left(\sigma^2_{\delta_0} \mathbf{Z}^*_1 \mathbf{Z}^{*\prime}_1 + \sigma^2_{\delta_1} \mathbf{Z}^*_2 \mathbf{Z}^{*\prime}_2\right) \boldsymbol{\Delta}^{g^{-1}}_{\boldsymbol{\eta}=\boldsymbol{\eta}_0} + \mathbf{A}(\boldsymbol{\eta}_0) \tag{7.103}$$

Although convergence was an issue, we also considered a model that fit random intercepts and slopes; however, an allowance was made for the two random effects to be correlated. Theoretically, the model is the same as Equation (7.102) with the variance–covariance matrix of responses given by

$$\text{Var}(\mathbf{y}) \approx \boldsymbol{\Delta}^{g^{-1}}_{\boldsymbol{\eta}=\boldsymbol{\eta}_0} (\mathbf{Z}\mathbf{D}\mathbf{Z}') \boldsymbol{\Delta}^{g^{-1}}_{\boldsymbol{\eta}=\boldsymbol{\eta}_0} + \mathbf{A}(\boldsymbol{\eta}_0) \tag{7.104}$$

where $\mathbf{D}$ is

$$\mathbf{D} = \begin{bmatrix} \sigma^2_{\delta_0} & \sigma_{\delta_0,\delta_1} \\ \sigma_{\delta_0,\delta_1} & \sigma^2_{\delta_1} \end{bmatrix}$$

and

$$\mathbf{Z} = \begin{bmatrix} \mathbf{1}_{n_1} & \mathbf{0}_{n_1} & \cdots & \mathbf{0}_{n_1} & \mathbf{x}_{2n_1} & \mathbf{0}_{n_1} & & \mathbf{0}_{n_1} \\ \mathbf{0}_{n_2} & \mathbf{1}_{n_2} & \cdots & \mathbf{0}_{n_2} & \mathbf{0}_{n_2} & \mathbf{x}_{2n_2} & & \mathbf{0}_{n_2} \\ \vdots & \vdots & \cdots & \vdots & \vdots & \vdots & & \\ \mathbf{0}_{n_{56}} & \mathbf{0}_{n_{56}} & \cdots & \mathbf{0}_{n_{56}} & \mathbf{0}_{n_{56}} & \mathbf{0}_{n_{56}} & & \mathbf{0}_{n_{56}} \\ \mathbf{0}_{n_{57}} & \mathbf{0}_{n_{57}} & \cdots & \mathbf{1}_{n_{57}} & \mathbf{0}_{n_{57}} & \mathbf{0}_{n_{57}} & & \mathbf{x}_{2n_{57}} \end{bmatrix} \text{ and } \boldsymbol{\delta} = \begin{bmatrix} \delta_{0,1} \\ \delta_{0,2} \\ \vdots \\ \delta_{0,57} \\ \delta_{1,1} \\ \delta_{1,2} \\ \vdots \\ \delta_{1,57} \end{bmatrix}$$

In order to decide on the best of the three random effects models (models 2–4 in Table 7.10), we use the REML pseudo-likelihood ratio test discussed in Section 7.2.4. Table 7.11 gives the SAS GLIMMIX code used to fit each of the models.

Since model 4 involves the more complex covariance structure, we consider it to be the full model. We test $H_0 : \sigma^2_{\delta_1} = 0$ to determine if the covariance structure from the random intercept only model (model 2) fits as well as the covariance structure from the full model. We test $H_0 : \sigma^2_{\delta_0} = 0$ to determine if the covariance structure from the random slope only model (model 3) fits as well as the covariance structure from the full model. Recall from Section 7.1.5 that these testing scenarios fall under scenario 2. Thus the p-values for either of these tests is calculated using a mixture of a χ^2_0 and χ^2_1 distribution as

$$p = 0.5Pr\left(\chi^2_0 > \hat{\lambda}\right) + 0.5Pr\left(\chi^2_1 > \hat{\lambda}\right)$$

Table 7.11 SAS GLIMMIX Code for Epilepsy Data

SAS Code Epilepsy Data:

```
ods html;
ods graphics on;

*method=rspl for subject-specific, method=rmpl for marginal
analysis;
proc glimmix data=epinew method=rspl plots=(studentpanel(blup));
class patient;
model seizure = treat period lage lbase / solution ddfm=satterth
link=log dist=poisson;
random int period / subject = patient solution g;

*Next, Model 4 vs. Model 2 using REML pseudo-likelihood ratio test;
covtest . 0;

*Next, Model 4 vs. Model 3 using REML pseudo-likelihood ratio test;
covtest 0 .;

output out=sresult pred(blup ilink)=pred pred(blup) = lpred
            stderr(blup)=selpred lcl(blup) = lowlink
ucl(blup) = uplink lcl(blup ilink)=slower ucl(blup ilink)=supper
            resid(blup ilink)=resid resid(blup)=invresid;
run;

ods output solutionR=random; *produces data set of blups;
ods graphics off;
ods html close;
```

where $\hat{\lambda} = l^{R}_{FULL} - l^{R}_{RED}$. In GLIMMIX, l^{R}_{FULL} denotes the restricted log-pseudo-likelihood value corresponding to the pseudo-data from the last iteration of the full model fit. Using the same pseudo-data and fitting the reduced model (model 2 or model 3), l^{R}_{RED} is the corresponding restricted log-pseudo-likelihood value. This value for l^{R}_{RED} in general is different from the restricted log-psuedo likelihood value for the reduced model fit to the raw data. This analysis uses the COVTEST statement in GLIMMIX to conduct these tests. Table 7.12 gives the full model (model 4) GLIMMIX analysis of the epilepsy data. In the section of output entitled *Fit Statistics* $l^{R}_{FULL} = -2$ Res Log Pseudo-Likelihood = 504.06, which is the restricted log-pseudo-likelihood value corresponding to the pseudo-data from the last iteration of the model 4 fit. Consequently, to test if model 2, the random intercept model, fits as well as model 4 (i.e., $H_0 : \sigma^2_{\delta_1} = 0$), we construct the test statistic

$$\hat{\lambda} = l^{R}_{FULL} - l^{R}_{RED} = 509.34 - 504.06 = 5.28$$

where the value of l^{R}_{RED} is found in the section of output entitled *Tests of Covariance Parameters Based on the Residual Pseudo-Likelihood.* This value of l^{R}_{RED} differs from the -2 Res Log Pseudo-Likelihood value for model 2 that appears in the last line of the second column of Table 7.10, due to how SAS conducts this test as described earlier in this Chapter. The corresponding p-value is approximately one-half the p-value from a χ^2_1 distribution. More specifically, the p-value $\approx \mathrm{P}(\chi^2_1 > 5.28) = 0.5{*}0.0216 = 0.0108$. Judging from the p-values in *Tests of Covariance Parameters Based on the Residual Pseudo-Likelihood* for $H_0 : \sigma^2_{\delta_1} = 0\,(0.0108)$ and $H_0 : \sigma^2_{\delta_0} = 0\,(<0.0001)$, we conclude that the full model, model 4, is best.

Now that the appropriate covariance structure has been selected, we proceed with an explanation of the rest of the output in Table 7.12. The material in *Dimensions* yields relevant information for the matrix form of our model given explicitly in Equation (7.102). The number of *G-side Cov. Parameters* represent the number of unknowns in the variance–covariance matrix of the random effects, which we have denoted as $\mathbf{D}$. Clearly, the two unknowns in $\mathbf{D}$ are $\sigma^2_{\delta_0}$ and $\sigma^2_{\delta_1}$. The fixed effects model matrix, $\mathbf{X}$, has five columns since there are four fixed effects plus the intercept, and the random effects model matrix, $\mathbf{Z} = \left[\mathbf{Z}^*_1 \mid \mathbf{Z}^*_2\right]$, has two columns corresponding to the two random effects $\boldsymbol{\delta}^*_1$ and $\boldsymbol{\delta}^*_2$ in (7.102). Since there are 57 subjects, the variance–covariance matrix of responses, $\mathbf{V}$, is a (57×57) block-diagonal matrix. The *Iteration History* shows the changes in the restricted log-pseudo-likelihood values.

In *Fit Statistics*, the ratio of the generalized chi-square statistic and its associated degrees of freedom is close to 1 (actual value is 1.39). This is an indication that there is little to no residual overdispersion in the data, suggesting that the Poisson distribution is adequate. If overdispersion were present here, the negative binomial distribution would be a logical alternative. The material in *Estimated G Matrix* suggests that $\hat{\sigma}^2_{\delta_0} = 0.2296$ and $\hat{\sigma}^2_{\delta_1} = 0.0083$. It is important

Table 7.12 Model 4 GLIMMIX Analysis of Epilepsy Data

Model Information

Data Set	WORK.EPINEW
Response Variable	Seizure
Response Distribution	Poisson
Link Function	Log
Variance Function	Default
Variance Matrix Blocked By	Patient
Estimation Technique	Residual PL
Degrees of Freedom Method	Satterthwaite

Class Level Information

Class	Levels	Values
patient	57	101 102 103 104 106 107 108 110 111 112 113 114 116 117 118 121 122 123 124 126 128 129 130 135 137 139 141 143 145 147 201 202 203 204 205 206 208 209 210 211 213 214 215 217 218 219 220 221 222 225 226 228 230 232 234 236 238

Number of Observations Read	228
Number of Observations Used	228

Dimensions

G-side Cov. Parameters	2
Columns in X	5
Columns in Z per Subject	2
Subjects (Blocks in V)	57
Max Obs per Subject	4

Iteration History

Itera-tion	Restarts	Subiter-ations	Objective function	Change	Max Gradient
0	0	6	438.35229416	0.51422975	9.916E-6
1	0	5	496.25152187	0.24640954	0.009429
2	0	5	503.78517582	0.02029656	7.572E-6
3	0	3	504.05640172	0.00066869	6.177E-7
4	0	2	504.06054074	0.00001142	2.625E-7
5	0	1	504.06060667	0.00000141	0.000812
6	0	1	504.06061107	0.00000121	6.465E-6
7	0	1	504.06060731	0.00000131	0.000851
8	0	1	504.06060349	0.00000128	3.124E-6
9	0	0	504.06060746	0.00000000	7.736E-6

Convergence criterion (PCONV = 1.11022E-8) satisfied.

(Continued)

Table 7.12 *Continued*

Fit Statistics

-2 Res Log Pseudo-Likelihood	504.06
Generalized Chi-Square	310.08
Gener. -Square / DF	1.39

Estimated G Matrix

Effect	Row	Col1	Col2
Intercept	1	0.2296	
Period	2		0.008301

Covariance Parameter Estimates

Cov Parm	Subject	Estimate	Standard Error
Intercept	patient	0.2296	0.06470
period	patient	0.008301	0.005020

Solutions for Fixed Effects

Effect	Estimate	Standard Error	DF	t Value	Pr > |t|
Intercept	-1.2264	1.2037	46.57	-1.02	0.3136
treat	-0.3154	0.1530	46.29	-2.06	0.0449
period	-0.07334	0.02752	78.53	-2.66	0.0094
lage	0.4751	0.3444	45.84	1.38	0.1744
lbase	0.9094	0.1087	45.7	8.37	<.0001

Type III Tests of Fixed Effects

Effect	Num DF	Den DF	F Value	Pr > F
treat	1	46.29	4.25	0.0449
period	1	78.53	7.10	0.0094
lage	1	45.84	1.90	0.1744
lbase	1	45.7	70.03	<.0001

Solution for Random Effects

Effect	Subject	Estimate	Std Err Pred	DF	t Value	Pr > |t|
Intercept	patient 101	0.05289	0.2681	57.06	0.20	0.8443
period	patient 101	-0.01509	0.07278	13.12	-0.21	0.8390
Intercept	patient 102	0.008613	0.2555	61.37	0.03	0.9732
period	patient 102	-0.01512	0.07579	11.26	-0.20	0.8454
Intercept	patient 103	-0.3192	0.3214	78.39	-0.99	0.3237
period	patient 103	-0.02451	0.08280	7.991	-0.30	0.7748
Intercept	patient 104	0.09182	0.2919	74.92	0.31	0.7540
period	patient 104	-0.00550	0.08033	8.995	-0.07	0.9469
Intercept	patient 106	0.07885	0.2918	74.36	0.27	0.7878
period	patient 106	0.008066	0.08024	9.038	0.10	0.9221

Tests of Covariance Parameters
Based on the Residual Pseudo-Likelihood

Label	DF	-2 Res Log P-Like	ChiSq	Pr > ChiSq	Note
Parameter list	1	509.34	5.28	0.0108	MI
Parameter list	1	555.56	51.50	<.0001	MI

MI: P-value based on a mixture of chi-squares.

to note that these variance estimates are on the log-scale. Although the standard errors for the estimated variance components are found in the section *Covariance Parameter Estimates*, the results here should not be used to conduct Wald inference on variance components unless none of the parameters being tested lie on the boundary of the parameter space.

The solutions for the fixed intercept and slope coefficients are found in *Solutions for Fixed Effects*. Given the parameterization of the *treat* variable, we have the following expressions for the mean number of seizures, apart from the estimated random effects, for the progabide treatment group and the placebo group:

$$\hat{\mu}_{\text{drug}} = \exp(b_0 + b_1 treat + b_2 period + b_3 lage + b_4 lbase)$$

and

$$\hat{\mu}_{\text{placebo}} = \exp(b_0 + b_2 period + b_3 lage + b_4 lbase)$$

respectively. Taking the ratio of the two means, we have

$$\hat{\mu}_{\text{drug}} \Big/ \hat{\mu}_{\text{placebo}} = \exp(b_1) = \exp(-0.3154) = 0.729$$

Thus for subjects of the same age and baseline number of seizures, at a fixed time period in the study, subjects on progabide experience on average 27.1% fewer seizures than subjects on the placebo. The groups are significantly different at the 0.0449 level. In terms of the population-averaged change in average seizures from period t to the next period $(t+1)$ holding fixed age, baseline, and treatment group,

$$\hat{\mu}_{\text{period}(t+1)} \Big/ \hat{\mu}_{\text{period}(t)} = \exp(b_2) = \exp(-0.0733) = 0.929$$

Thus, on average, the number of seizures declines by approximately 7.1% every 2 weeks. The trend across time is significant at the 0.0094 level. It appears that adjusting for the log of the baseline number of seizures is important ($p<0.0001$).

It is also of interest to see how the period effect varies from patient to patient. For the jth patient, the estimated mean number of seizures is

$$\hat{\mu}|patient_j = \exp\left(b_0 + b_1 treat + b_2 period + b_3 lage + b_4 lbase + \hat{\delta}_{0j} + \hat{\delta}_{1j} period\right) \tag{7.105}$$

where $\hat{\delta}_{0j}$ and $\hat{\delta}_{1j}$ are the random intercept and slope BLUPs, respectively, for the jth patient. Thus, for the jth patient, the average change in the number of seizures from period t to period $(t+1)$ is

$$\hat{\mu}_{period(t+1)}|patient_j \Big/ \hat{\mu}_{period(t)}|patient_j = \exp\left(b_2 + \hat{\delta}_{1j}\right)$$

The estimated BLUPs for the random intercept and slope for patients 101–106 are provided in the section *Solutions for Random Effects*. So, for patient 101, the average change in seizures from period t to period $(t+1)$ is

$$\hat{\mu}_{period(t+1)}|101 \Big/ \hat{\mu}_{period(t)}|101 = \exp(-0.0733 - 0.015) = 0.915$$

or an 8.5% decline every 2 weeks. Figure 7.12 shows the range of period effects across patients separated by treatment groups. Period effects for each patient are calculated just as we did above for patient 101 and output to the SAS data set *random* using the ods output solutionR = random statement in Table 7.11. Figure 7.12 gives the boxplots generated by MINITAB based on the data exported from SAS. For both treatment groups, the percent change ranges from a little over a 15% decline to just under a 5% increase in seizures every 2 weeks.

As discussed in Section 7.2.4, the $(1 - \alpha)100\%$ confidence interval for the conditional mean for a given setting of the regressors is

$$\left[g^{-1}\left(\hat{\eta}|cluster_j - z_{1-\alpha/2}\widehat{se}\left(\hat{\eta}|cluster_j\right)\right), g^{-1}\left(\hat{\eta}|cluster_j + z_{1-\alpha/2}\widehat{se}\left(\hat{\eta}|cluster_j\right)\right)\right].$$

So, for patient 101 during the first period, we have

$$\begin{aligned}\hat{\eta}|101 &= (-1.2264 - 0.3154(1) + -0.0733(1) \\ &\quad + 0.4751(2.89) + 0.9094(2.944) + 0.053 - 0.015) \\ &= 2.474 \end{aligned} \tag{7.106}$$

In the OUTPUT statement from GLIMMIX in Table 7.11, the keyword pred(blup) = lpred produces values for the estimated linear predictors, stderr(blup) = selpred produces standard errors of estimated linear predictors, lcl(blup) = lowlink and ucl(blup) = uplink produce the lower and upper confidence limits for the linear predictors, respectively, pred(blup ilink) = pred

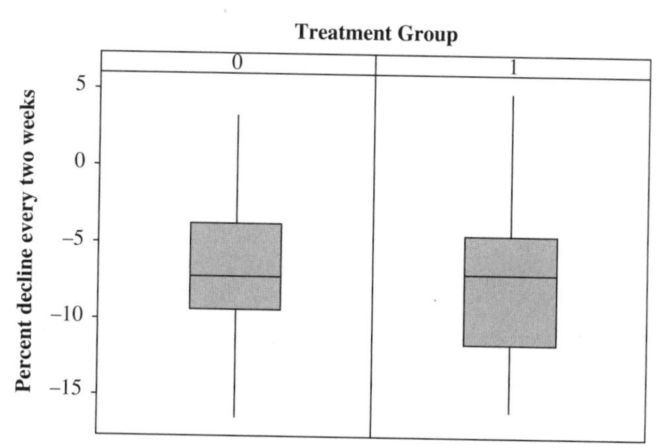

Figure 7.12 Boxplots showing average percent change from period t to period $(t + 1)$ for patients in both treatment groups.

yields the estimated conditional means, and lcl(blup ilink) = slower and ucl(blup ilink) = supper yield the lower and upper confidence limits on the conditional means, respectively. A variety of residual types can also be requested and we describe some of these later. Table 7.13 provides the aforementioned statistics as produced by GLIMMIX, along with some others, related to patient 101. The value of the variable *sslpred* in the first row gives the calculated estimate of the linear predictor based on Equation (7.106). The variable *pred* is the estimated conditional mean and is simply $g^{-1}(\hat{\eta}|101) = \exp(2.474) = 11.869$. The variables *lowlink* and *uplink* give the 95% confidence limits on the linear predictor for patient 101. So, for patient 101 during period 1, the 95% confidence limits based on Equation (7.84) are

$$\hat{\eta}|101 \pm z_{1-\alpha/2}\,\widehat{se}(\hat{\eta}|101)$$

$$2.474 \pm 1.96^{*}0.174$$

$$[2.132, 2.815]$$

The corresponding confidence interval for the conditional mean is then [exp (2.132), exp (2.815)] = [8.431,16.693]. Table 7.13 gives these limits in the first row under *slower* and *supper*, respectively. This confidence interval reflects the uncertainty in the conditional mean in terms of the hypothetical sense of patient 101 repeatedly going through this clinical trial. Note that this uncertainty is a *replication* type of uncertainty. In many observational studies, such as this one, the practical meaning is somewhat abstract. The interval does not, however, reflect the uncertainty in the mean that exists across all subjects. To develop an interval on the population-averaged mean, one must use a method such as marginal quasi-likelihood as discussed in *A Population-Averaged Approach* in Section 7.2.3. PROC GLIMMIX can run this method by specifying the method = rmpl option in the GLIMMIX procedural statement. Although the details of this analysis are not shown here, the resulting marginal mean estimate is

$$\hat{\mu} = \exp(b_0 + b_1 treat + b_2 period + b_3 lage + b_4 lbase)$$

where the b's are estimated by Equation (7.76). The variable *margmn* is this result in Table 7.13. The variables *marglcl* and *margucl* give the corresponding 95% confidence limits on the marginal mean. For 18 year olds (i.e. $e^{2.89}$) who report 19 (i.e. $e^{2.944}$) biweekly seizures as baseline and take progabide, we are 95% confident that the average number of seizures after 2 weeks on treatment is between 0.795 and 190.836. Note the vast difference in widths between the intervals on the conditional mean and the marginal mean. One would expect a wider interval for the marginal mean than for the conditional mean since the interval for the marginal mean reflects the uncertainty from patient to patient as well as the uncertainty existing from patients repeatedly going through the clinical trial.

Table 7.13 Estimated Conditional Mean Number of Seizures (Link and Inverse Link Scales), Estimated Population-Averaged Number of Seizures (Inverse Link Scale), and Associated 95% Confidence Limits for Patient 101 in Epilepsy Data

patient	*seizure*	*treat*	*period*	*lage*	*lbase*	*lpred*	*pred*	*selpred*	*lowlink*	*uplink*	*slower*	*supper*	*margmn*	*marglcl*	*margucl*
101	11	1	1	2.890	2.944	2.474	11.869	0.174	2.132	2.815	8.431	16.693	12.320	0.795	190.836
101	14	1	2	2.890	2.944	2.385	10.861	0.149	2.091	2.679	8.092	14.578	11.571	0.749	178.784
101	9	1	3	2.890	2.944	2.297	9.942	0.160	1.982	2.612	7.257	13.621	10.867	0.674	175.274
101	8	1	4	2.890	2.944	2.208	9.101	0.200	1.811	2.606	6.115	13.544	10.206	0.581	179.449

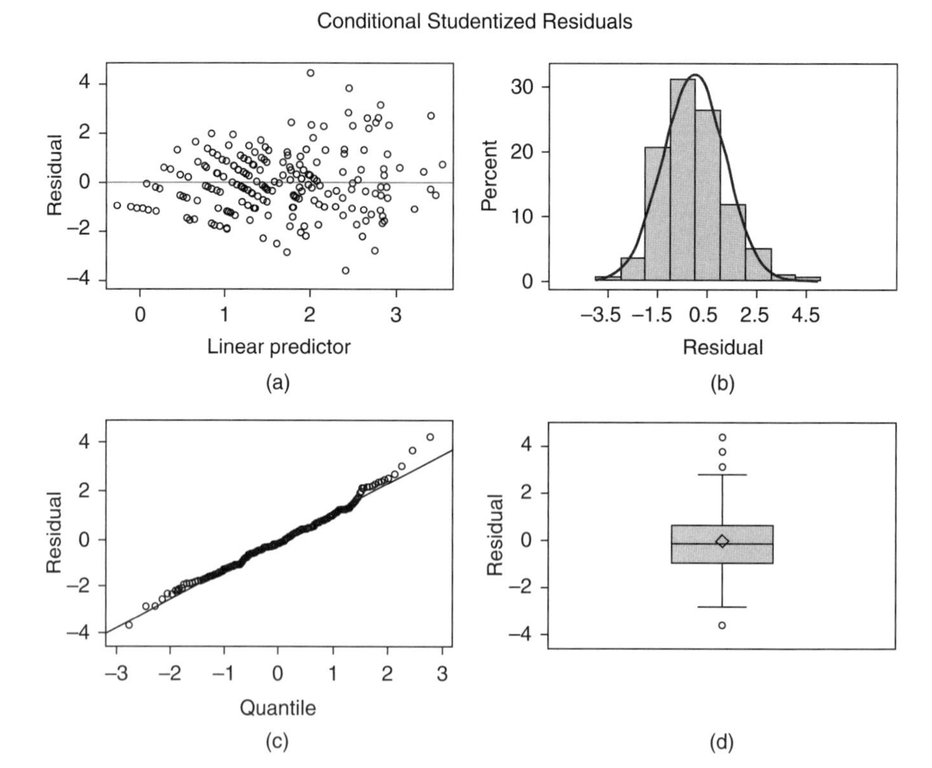

Figure 7.13 Studentized residual plots for the model 4 analysis of the epilepsy data: (a) studentized residuals (pseudo-data—linear predictor) versus linear predictor, (b) histogram of studentized residuals, (c) normal Q-Q plot of studentized residuals, and (d) boxplot of studentized residuals.

7.2.6 Diagnostics

Equation (7.51) requires two sets of assumptions: one set for the conditional responses, the $y|\delta$'s, and another set for the random effects, the δ's. We require different diagnostics for each set of assumptions.

First, consider the conditional responses. In general, we assume that (1) the data follow an exponential family member distribution; (2) a specified link function, g, relates the conditional mean linearly to the regression parameters $g(\boldsymbol{\mu}|\boldsymbol{\delta}'s) = \mathbf{x}'\boldsymbol{\beta} + \sum_{l=1}^{q} \delta_l$ (3) the variance–covariance matrix of the conditional responses follows a specified structure, $\mathbf{S} = \mathbf{A}^{1/2}(\boldsymbol{\eta})\mathbf{R}\mathbf{A}^{1/2}(\boldsymbol{\eta})$ with $\mathbf{R}$ as the common correlation matrix for each cluster; and (4) the conditional responses are independent of the random effects.

In the epilepsy data, the conditional responses are counts of observed seizures per 2 week period. Again, the Poisson distribution is a natural choice for count data. From Table 7.10 in the section entitled *Fit Statistics*, we observe that the ratio of the generalized chi-square and its degrees of freedom is 1.39, which is quite close to 1. Such a value implies that there are no issues with overdispersion. The suitability of the specified link can be examined by plotting the studentized conditional residuals on the link scale versus the linear predictor. The vector of studentized residuals is

$$\mathbf{e}_c = \frac{\tilde{\mathbf{y}} - \hat{\boldsymbol{\eta}}}{\widehat{se}(\tilde{\mathbf{y}} - \hat{\boldsymbol{\eta}})}$$

where $\tilde{\mathbf{y}}$ is the vector of pseudo-responses given in (7.62), $\hat{\boldsymbol{\eta}} = \mathbf{Xb} + \mathbf{Z}\hat{\boldsymbol{\delta}}$, the estimated linear predictor, and $\widehat{se}(\cdot)$ denotes the estimated standard error. Figure 7.13 a shows the studentized residuals from model 4 plotted by the estimated linear predictor for the epilepsy data. The random scatter suggests no violation of the assumed log link. Recall from the analysis discussion in Section 7.2.3, *A Subject-Specific Approach,* the pseudo-response model in Equation (7.63) is a normal, linear mixed model. Figures 7.13b, 7.13c, and 7.13d, produced by the plots = (studentpanel(blup)) option in the GLIMMIX procedural statement, examine this normality assumption via a histogram, normal Q-Q plot, and boxplot of the studentized residuals. None of these plots suggests any violation of the normality assumption of the pseudo-responses.

We next consider the assumptions surrounding the random effects. The random effects (random intercept and random slope in the epilepsy example) are each assumed to be normally distributed with mean zero and constant variance. We can check for normality by doing Q-Q plots for the random intercepts and slopes. Figures 7.14a and 7.14b display the Q-Q plots for the random intercepts and random slopes, respectively. Neither plot indicates any

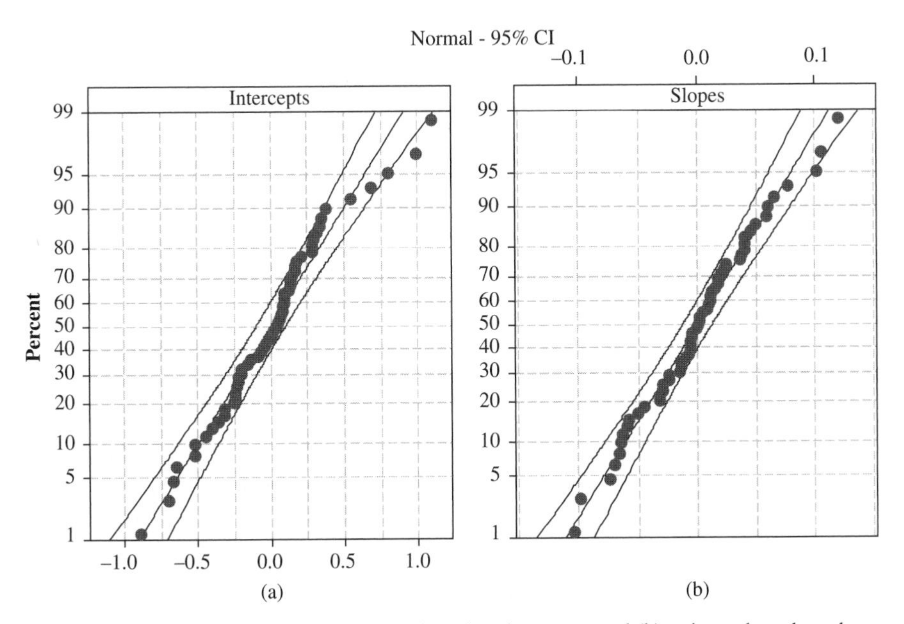

Figure 7.14 Normal Q-Q plots: (a) estimated random intercepts and (b) estimated random slopes from model 4 fit of the epilepsy data.

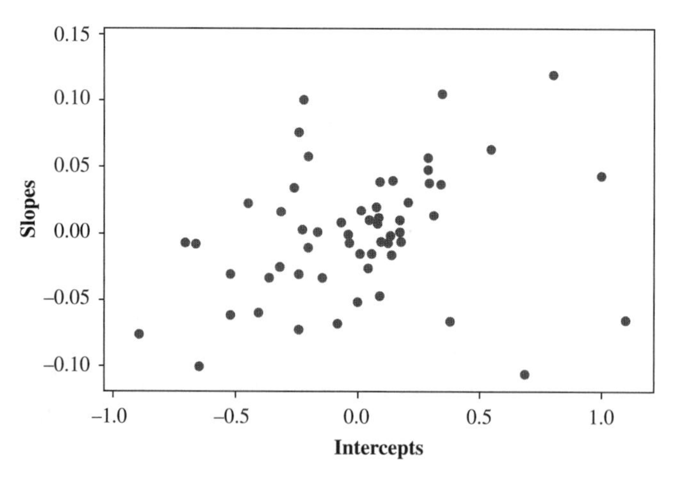

Figure 7.15 Scatterplot of estimated slopes versus estimated intercepts from model 4 fit to the epilepsy data.

issues with the normality assumptions. Finally, for model 4 we also assume independent random intercepts and slopes. Figure 7.15 allows us to check this assumption via a scatterplot of the estimated slopes versus the estimated intercepts. The observed random scatter implies independence between the random intercept and random slope parameters.

Example 7.8. Film Manufacturing Experiment Revisited. In Example 7.6, the film manufacturer models the response, a measure of film quality, as a function of six factors. Three of the factors, x_1, x_2, and x_3, are mixture components and the other three factors, p_1, p_2, and p_3, are process factors. Their model assumes a gamma response with a log link based on previous knowledge. The split plot structure of the experiment requires a random effect corresponding to the batches to account for the correlation among the sub plots within each whole plot. The linear predictor is

$$\eta_{ji}|\delta_j = \ln\left(\mu_{ji}|\delta_j\right) = \mathbf{x}'_{ji}\boldsymbol{\beta} + \delta_j \tag{7.107}$$

where

$$\mathbf{x}'_{ji}\boldsymbol{\beta} = \sum_{b=1}^{3} \beta_b x_{b,ji} + \beta_{12}x_{1,ji}x_{2,ji} + \sum_{c=1}^{3} \gamma_{c,ji}p_{c,ji} + \sum_{b=1}^{2}\sum_{c=1}^{3} \psi_{bc}x_{b,ji}p_{c,ji} \tag{7.108}$$

and δ_j is the random effect associated with the jth batch. As with all GLMMs, the random effects are assumed to be i.i.d. Normal$\left(0, \sigma_\delta^2\right)$. In this case the linear predictor does not include a fixed effect intercept since we are fitting a Scheffé model for mixture experiments.

We use R to fit a gamma GLMM with log link and the linear predictor given in Equation (7.107) using the *lmer* function from the lme4 library. The lme4 library is not part of the default installation of R; however, it is easy to download it from

Table 7.14 R Code for lmer PQL Analysis of Film Data

R Code Film Data:

```
film <- read.csv("film_data.csv")
names(film)

#Next line loads the lme4 library
library(lme4)

#m1 is a gamma GLMM with log link and linear predictor as given in
(7.105)

m1 <- lmer(y ~ -1 + x1 + x2 + x3 + x1x2 + p1 + p2 + p3 + x1p1
          + x1p2 + x1p3 + x2p1 + x2p2 + x2p3 + (1|batch),
          data=film, method="PQL",family=Gamma(link="log"))

summary(m1)

preds <- fitted(m1)

#Code below does Q-Q plot of the estimated random intercepts

blups <- ranef(m1)
qqnorm (unlist(blups))
```

the R-project website. Table 7.14 gives the R commands for a PQL analysis. The function lmer from the lme4 R library also incorporates the integral approximation methods (e.g., Laplace and quadrature) for parameter estimation that were briefly mentioned in Section 7.2.3. For consistency of presentation, we demonstrate the PQL analysis here. Table 7.15 summarizes the output from the lmer analysis. To specify the use of the quadrature methods, replace the phrase method = "PQL" by method = "LaPlace" or method = "Quad" in the lmer sentence. The random batch effect is specified within the parentheses using (1|batch). The summary(.) command provides the basic output from the model. The lmer analysis only provides t-statistics for the model coefficients and not the corresponding p-values. The estimated variance components are given in the section *Random Effects*. The preds object contains the estimated conditional means. The ranef(.) function extracts the BLUPs that are useful for checking the normality assumption of the random effects. Figure 7.16 provides the normal Q-Q plot for the estimated random intercepts.

Unlike SAS PROC GLIMMIX, R at the present time does not produce confidence limits for the conditional and marginal means. Residual diagnostics can be obtained using the glmmPQL command from the MASS library. Table 7.16 provides the R commands for the glmmPQL analysis of the full model. Since the glmmPQL analysis is identical to that produced by lmer, the output from glmmPQL is not provided in a separate table. The advantage of lmer

Table 7.15 Output from Lmer Analysis of Film Data

```
Generalized Linear Mixed Model Fit Using PQL

Formula: y ~ -1 + x1 + x2 + x3 + x1x2 + p1 + p2 + p3 + x1p1 + x1p2 + x1p3 +
x2p1 + x2p2 + x2p3 + (1 | batch)

Data: Film
Family: Gamma(log link)
AIC BIC logLik deviance
47.78 75.1 -9.89 19.78

                     Random Effects:
          Groups     Name           Variance    Std.Dev.
          batch      (Intercept)    0.052548    0.22923
                     Residual       0.162660    0.40331

           number of obs: 52, groups: batch, 13
Fixed Effects:

                    Estimate   Std. Error   t value
          x1          6.1132     0.3062     19.966
          x2          3.9847     0.3190     12.490
          x3          1.2413     0.6648      1.867
          x1x2       10.7063     1.9181      5.582
          p1         -0.6767     0.4372     -1.548
          p2          0.3587     0.4372      0.820
          p3          0.2911     0.4372      0.666
          x1p1        1.1695     0.6111      1.914
          x1p2       -0.2134     0.6111     -0.349
          x1p3       -0.4696     0.6111     -0.768
          x2p1        1.9483     0.6251      3.117
          x2p2       -1.0219     0.6251     -1.635
          x2p3       -0.1283     0.6251     -0.205
```

over glmmPQL is that glmmPQL only allows for the PQL GLMM analysis; whereas the lmer analysis allows for PQL, Laplace, and the quadrature methods of estimation. However, glmmPQL offers features such as the production of residuals and fitted values on both the link and inverse link scales. Figure 7.17 shows the diagnostics for the conditional residuals on the link scale. In Figure 7.17a we observe random scatter when the conditional residuals are plotted against the linear predictor, which implies that there are no violations of the assumed log link. Recall from the analysis discussion in Section 7.2.3, "A Subject-Specific Approach," the pseudo-response model in Equation (7.63) assumes a normal, linear mixed model. Figures 7.17b–d suggest that this assumed normality is reasonable. □

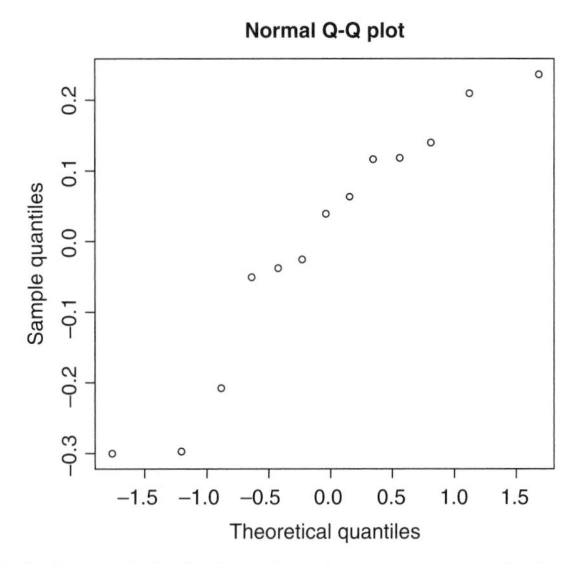

Figure 7.16 Normal Q-Q plot for estimated random intercepts in film example.

Table 7.16 R Code for glmmPQL Analysis of Film Data

R Code Film Data:

```
#For residual diagnostics, need to re-analyze using glmmPQL
function from MASS library

library(MASS)

m2 <- glmmPQL(y ~ -1 + x1 + x2 + x3 + x1x2 + p1 + p2 + p3 +
            x1p1 + x2p1 + x3p1 + x2p1 + x2p2 + x2p3, random = ~1
         |batch,data=film,family = Gamma("log"))

summary(m2)

#lresids are residuals to the pseudo-data
lresids <- residuals(m4)

#lpreds contains estimated linear predictor
lpreds <- predict(m4,type="link")

#residual diagnostic plots produced in lines below

par(mfrow=c(2,2))
plot (predsm4, resids,xlab="Linear Predictor",ylab="Residuals")
hist(resids,freq=FALSE,main="",xlab="Residuals",ylab="Percent")
qqnorm(resids,xlab="Quantile",ylab="Residuals")
boxplot(resids,ylab="Residuals")
```

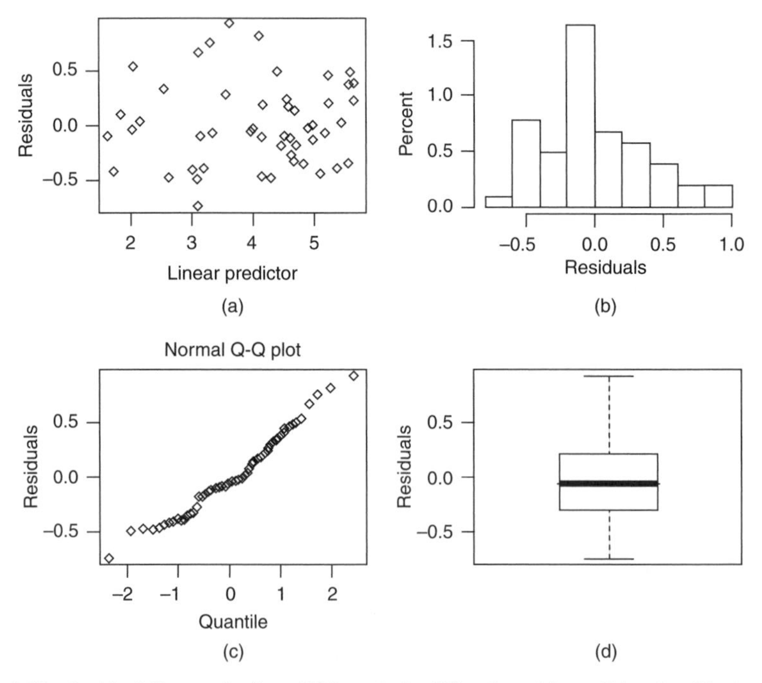

Figure 7.17 Residual diagnostics from PQL analysis of film data: (a) conditional residuals on link scale versus linear predictor, (b) histogram of conditional residuals on link scale, (c) normal Q-Q plot of conditional residuals on link scale, and (d) boxplot of conditional residuals on link scale.

7.3 GENERALIZED LINEAR MIXED MODELS USING BAYESIAN METHODS

7.3.1 Model Formulation

Another important approach for modeling correlated, non normal data is hierarchical generalized linear models (HGLMs) (Lee and Nelder, 1996). Hierarchical generalized linear models include GLMMs as a special case but do not constrain the random effects to have a normal distribution. HGLMs can be implemented via the GenStat procedures written by Lee and Nelder.

In this section, we consider a Bayesian approach to modeling exponential family data with random effects. While the GLMM, HGLM, and GEE approaches tend to focus primarily on inference for the population mean, the Bayesian approach offers a great deal of flexibility in terms of the types of distribution characteristics. We begin by defining an extended class of generalized linear models with random effects, which consists of two parts:

1. The vector of responses, conditional on the jth cluster, follows a GLM family in which

$$E(\mathbf{y}_{n_j}|\boldsymbol{\delta}_j) = g^{-1}(\boldsymbol{\eta}_j) = g^{-1}(\mathbf{X}_j\boldsymbol{\beta} + \mathbf{Z}_j\boldsymbol{\delta}_j) \tag{7.109}$$

Once again, g is a differentiable monotonic link function, and the parameters in Equation (7.109) are defined exactly as those in Equation (7.51).

2. $\boldsymbol{\delta}_j$ and $\mathbf{Z}_j$ are defined exactly as in Equation (7.51) except now the $\boldsymbol{\delta}_j$ are assumed to be independent and to follow a *specified* probability distribution.

Note that the model in Equation (7.109) is the same as that for GLMMs given in Equation (7.51) when the random effects in $\boldsymbol{\delta}_j$ are assumed i.i.d. normal with mean $\mathbf{0}$ and diagonal variance–covariance matrix $\mathbf{D}_j$. HGLMs are included in this framework when one assumes that the $\boldsymbol{\delta}_j$ follow a distribution conjugate to a GLM family. HGLMs also assume that the linear predictor takes the form $\boldsymbol{\eta}_j = \mathbf{X}_j\boldsymbol{\beta} + \mathbf{Z}_j\boldsymbol{\delta}_j^*$, where $\boldsymbol{\delta}_j^* = g(\boldsymbol{\delta}_j)$.

As discussed in Section 7.2.4, statistical inference for GLMMs requires asymptotic theory involving likelihoods and pseudo-likelihoods. The same is true regarding HGLMs. Provided one is comfortable with the sample size assumptions being met, point and interval estimates as well as a wide array of hypothesis test results involving the conditional and marginal means are available as output from SAS PROC GLIMMIX or R.

This section outlines an approach that does not rely on asymptotics; instead, it builds on assumptions regarding prior densities on the model parameters. The Bayesian approach is useful not only for conducting inference on the conditional and/or marginal mean but also for other characteristics of the response such as quantiles and exceedance probabilities. For instance, in the film manufacturing example, suppose that pieces of film whose response exceeds 150 are considered high quality and can be sold as premium. A natural question of the manufacturer might be "What proportion of film pieces (across all rolls) are of premium grade?" Since this question involves all rolls of film produced, it involves an exceedance probability associated with the marginal distribution of the response. Technically speaking, the question involves estimation of $P(y > 150)$ where

$$\begin{aligned} P(y > 150) &= \int_{150}^{\infty}\int_{-\infty}^{\infty} f_{y|\delta,\delta}(y,\delta)\,d\delta\,dy \\ &= \int_{150}^{\infty}\int_{-\infty}^{\infty} f_y(y|\delta)f_\delta(\delta)d\delta\,dy \end{aligned} \tag{7.110}$$

and $f_{y|\delta,\delta}(y)$ is the marginal density defined in matrix-vector notation in Equation (7.58). For the film manufacturing example, $f_{y|\delta,\delta}(y)$ is a mixture of gamma densities for the individual rolls integrated over the normal distribution of random effects δ. As a result, it would be difficult, at best, to ascertain the distribution of any estimate of $P(y > 150)$. The Bayesian approach, however, offers a straightforward method of conducting inference characteristics such as the one given in Equation (7.110).

7.3.2 Bayesian Inference

The Bayesian approach to inference combines prior information about the model parameters with the information in the data. Let $\mathbf{\Theta}$ denote the vector of model parameters and $\pi(\mathbf{\Theta})$ represent the joint density of the model parameters before any data are observed. In Example 7.4, there are 28 unknowns in the theoretical GLMM. Thus, $\mathbf{\Theta}' = (r, \boldsymbol{\delta}', \boldsymbol{\beta}', \sigma_\delta)$, where r is the gamma shape parameter from Equation (7.55), $\boldsymbol{\delta}'$ is the $1x13$ vector of random roll effects, $\boldsymbol{\beta}'$ denotes the $1x13$ vector of regression coefficients in Equation (7.108), and σ_δ denotes the standard deviation of the random effect distribution. In frequentist approaches, such as the PQL approach described in Section 7.2, the vector of model parameters is assumed to be fixed, and information regarding the model parameters comes solely from the data. Bayesian approaches, on the other hand, assume prior probability densities on each of the model parameters. The Bayesian approach uses any additional information regarding model parameters from the data to update the assumed prior densities. If little is known a priori regarding the model parameters, most analysts choose diffuse prior densities, which allow for the possibility of a wide range of values for the parameters.

For Example 7.6, we assume that little is known about the model parameters; hence, we choose diffuse proper prior distributions. Specifically, for the regression coefficients (the parameters in $\boldsymbol{\beta}$) and for r, we use Normal(0,1000^2) and Uniform(0,100) for $f_{\beta_\kappa}(\cdot)$ ($k = 1,\ldots,13$) and $f_\alpha(\cdot)$, respectively. For σ_δ, we use the diffuse prior distribution Uniform(0,100) for $f_{\sigma_\delta}(\cdot)$. The diffuse prior Uniform(0,100) is suggested by Gelman (2006) for the standard deviation of the random effects distribution in hierarchical models. Figures 7.18 a-c illustrate the shape of each of these prior densities. It is important to point out that the choice of parameters to specify diffuse prior distributions is problem specific; for example, in Example 7.6, we expect the β's to be at most in the 10's; however, we choose prior distributions that allow for the possibility that the β's are a couple orders of magnitude larger.

The foundation of the Bayesian analysis is the prior distribution for the parameters. As a result, this approach requires additional assumptions for the distributional forms of these priors, which makes many frequentists quite uncomfortable. On the other hand, frequentist approaches often rely on asymptotic assumptions for estimates of uncertainty on the quantities of interest. Bayesian methods do not require such asymptotic theory. In the final analysis the overall number of assumptions for both analyses, frequentist and Bayesian, seems to be relatively similar.

The data sampling model, $f_{\mathbf{y}}(\mathbf{y}|\mathbf{\Theta})$, commonly known as the likelihood, captures the information in the data. For Example 7.6, the likelihood $f_{\mathbf{y}}(\mathbf{y}|\mathbf{\Theta})$ has the form

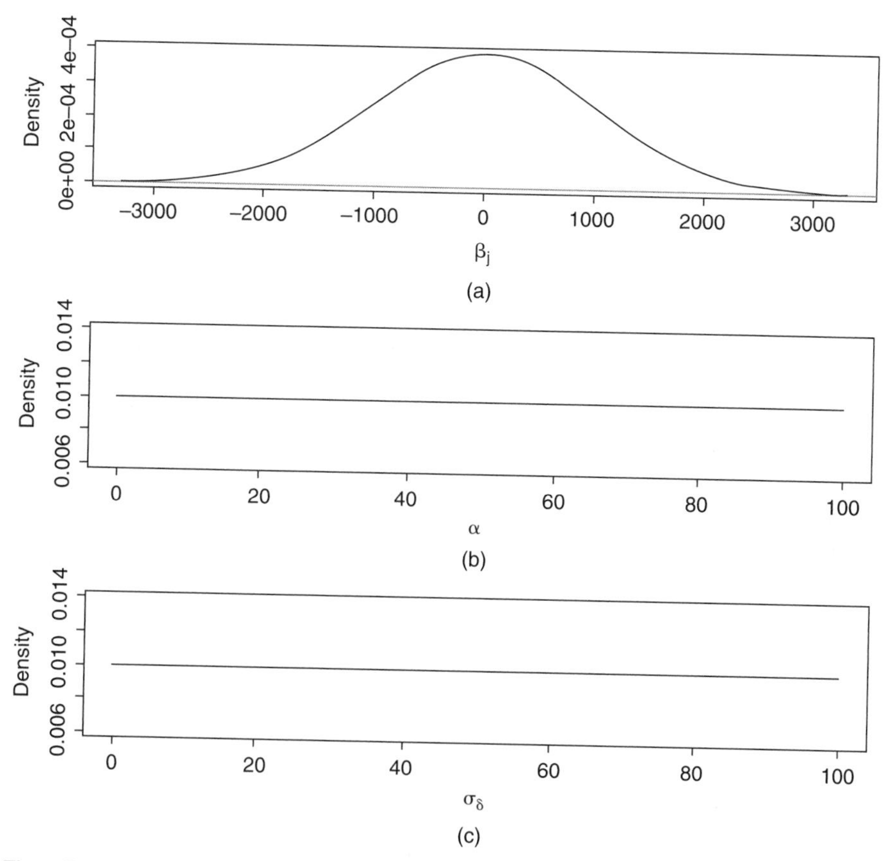

Figure 7.18 Assumed prior densities for the parameters in Θ for the Bayesian modeling of the film data (a) top, prior density for each of the β's, (b) middle, prior density for r, and (c) lower, prior density for σ_δ

$$f_{\mathbf{y}}(\mathbf{y}|\mathbf{\Theta}) = \prod_{j=1}^{13}\prod_{i=1}^{4} f_y\left(y_{ji}|r, \lambda_{ji} = {}^{\mu_{ji}}/_{r}\right) \tag{7.111}$$

with μ_{ji} given in Equation (7.107) where μ_{ji} depends on $\mathbf{x}'_{ji}$, $\boldsymbol{\beta}$, and δ_j. The posterior density, π $(\mathbf{\Theta}|\mathbf{y})$, describes the combined information from the data and the specified prior distributions. The posterior density is then evaluated using Bayes' Theorem (Degroot (1970) p. 28) as

$$\pi(\mathbf{\Theta}|\mathbf{y}) \propto f(\mathbf{y}|\mathbf{\Theta})\pi(\mathbf{\Theta}).$$

The prior density π $(\mathbf{\Theta})$ has the form

$$\pi(\mathbf{\Theta}) \propto \left(\prod_{j=1}^{13} f_{\delta_j}(\delta_j|\sigma_\delta^2)\right)\left(\prod_{k=1}^{13} f_{\beta_k}(\beta_k)\right) f_\alpha(\alpha) f_{\sigma_\delta}(\sigma_\delta), \tag{7.112}$$

where $f_\delta(\cdot)$ is the Normal(0, σ_δ^2) density assumed for the random batch effects (i.e. the δj's).

A popular approach to the Bayesian analysis is to approximate the form of the posterior distribution, $\pi(\mathbf{\Theta}|\mathbf{y})$, using recent advances in Bayesian computing such as Markov chain Monte Carlo algorithms (see Gelfand and Smith (1990), Casella and George (1992), Chib and Greenberg (1995) for details). Essentially, Markov chain Monte Carlo (MCMC) algorithms produce samples from the joint posterior distribution of $\mathbf{\Theta}$ by sequentially updating each model parameter, conditional on the current values of the other model parameters. These samples of the posterior distribution of $\mathbf{\Theta}$ are easy to work with in evaluating various performance criteria, which are functions of $\mathbf{\Theta}$. A popular software package for conducting the MCMC analysis is the freely downloadable (http://www.mrc-bsu.cam.ac.uk/bugs/) program, WinBUGS (Spiegelhalter, Thomas, Best, and Lunn (2004)). Table 7.17 provides the WinBUGS code for the analysis of Example 7.6. The structure of the WinBUGS code is similar to SAS PROC GLIMMIX as well as what is used in lmer and glmmPQL in R. The model statement specifies the fixed effects first. A separate statement specifies the random effects (like the RANDOM statement in GLIMMIX and (1|Batch) in lmer and Random = in glmmPQL).

In WinBUGS, the parameterization for the gamma density is $y \sim Gamma(r,\xi)$ where

$$f_y(y) = \frac{\xi^r y^{r-1} e^{-\xi y}}{\Gamma(r)}. \tag{7.113}$$

The parameterization in Equation (7.113) is identical Equation (7.55) with $\xi = {}^1/_\lambda$. In producing the MCMC samples (or draws) from the posterior, WinBUGS uses the first 4000 draws of $\mathbf{\Theta}$ (referred to as burn-in) to tune the algorithm for subsequent draws. After the burn-in, it is recommended to have WinBUGS generate enough MCMC samples in order to *thin* the samples so as to have a final set of independent samples. In this analysis, we generated 1,000,000 samples then systematically selected every 100th draw for a sample of 10,000 independent draws of $\mathbf{\Theta}$. The final samples of 10,000 draws are essentially empirical representations of the posterior distributions for each of the parameters in $\mathbf{\Theta}$. The draws are easily exported to a text file from WinBUGS, which can then be imported into a software package, such as R, for further analysis. Using R, the medians from each sample of 10,000 draws (the empirical posteriors) are calculated for each of the parameters in $\mathbf{\Theta}$. Table 7.18 gives the fixed effect parameter estimates, and Table 7.19 gives the random effect parameter estimates, along with the median of the posteriors for r and σ_δ. Both tables also give the corresponding estimates from PQL. The Bayesian and PQL summaries are similar in both tables. The major difference between the Bayesian analysis and the frequentist analysis is that the Bayesian approach can work directly with the posteriors of each of the parameters for doing inference

Table 7.17 WinBUGS Code for the Analysis of Data from Example 7.6

WINBUGS Code Film Data:

```
Model {

        For(i in 1:52) {
                y[i] ~dgamma(alpha,theta[i])
                #gamma mean is alpha/theta[i]
                #var is alpha/theta[i]^2
                theta[i]<-alpha/mu[i]
                mu[i]<-exp(
                beta[1]*x1[i]+beta[2]
                *x2[i]+beta[3]*x3[i]+ beta
                [4]*x1x2[i]+0*x1x3[i]
                +0*x2x3[i]+ beta [5]*p1[i]
                + beta [6]*p2[i]
                + beta [7]*p3[i]+ beta
                [8]*x1p1[i]
                + beta [9]*x2p1[i]+0*x3p1[i]+ beta [10]*x1p2[i]
                + beta [11]*x2p2[i]+0*x3p2[i]+ beta [12]*x1p3[i]
                + beta [13]*x2p3[i]+0*x3p3[i]+0*p1p2[i]
                +0*p1p3[i]+0*p2p3[i] + delta[roll[i]])

        }

        for (j in 1:13) {
                delta[j]~dnorm(0,taudelta)
                #taudelta is a whole plot precision
        }

#priors
taudelta <-
1/pow(sigmadelta,2)

sigmadelta~dunif(0,100)
for(k in 1:13){
                        beta[k]~ dnorm(0.0,1.0E-6)
                        #1.0E-6 is a precision
                }
alpha~dunif(0,100)
```

while the frequentist analysis (e.q., PQL) must rely on asymptotics. For instance, in the Bayesian approach, if one wishes to provide a 95% confidence interval for, say, β_1, one need only take the 250th smallest and 9750th largest observations from the 10,000 empirical values from the posterior of β_1.

Table 7.18 Fixed Effect Estimates and Standard Errors for Bayesian and PQL Analyses

	Bayesian Analysis		GLMM PQL Analysis	
Effect	Estimate	Standard Deviation	Estimate	Standard Error
x_1	6.146	0.408	6.113	0.306
x_2	4.022	0.407	3.984	0.319
x_3	1.332	0.878	1.241	0.665
$x_1{}^*x_2$	10.510	2.472	10.706	1.918
p_1	−0.697	0.584	−0.677	0.437
p_2	0.272	0.533	0.359	0.437
p_3	0.224	0.597	0.291	0.437
$x_1{}^*p_1$	1.190	0.827	1.170	0.611
$x_1{}^*p_2$	−0.098	0.753	−0.213	0.611
$x_1{}^*p_3$	−0.380	0.840	−0.469	0.611
$x_2{}^*p_1$	1.981	0.820	1.948	0.625
$x_2{}^*p_2$	−0.897	0.758	−1.022	0.625
$x_2{}^*p_3$	−0.036	0.843	−0.128	0.625

Table 7.19 Parameter Estimates for the Gamma Shape Parameter (r), Random Effects (δ_j's), and Random Effect Variance (σ_δ^2)

	Bayesian Analysis		GLMM PQL Analysis	
Effect	Estimate[a]	Standard Deviation	Estimate	Standard Error[b]
α	4.2580	1.1300	4.7214	1.2190
δ_1	0.0174	0.2442	0.0408	0.2410
δ_2	0.0710	0.2209	0.1298	0.2165
δ_3	0.1350	0.2352	0.2299	0.2179
δ_4	0.0663	0.2124	0.1273	0.2165
δ_5	−0.2095	0.2648	−0.3346	0.2117
δ_6	−0.0108	0.2087	−0.0268	0.2165
δ_7	0.0806	0.2169	0.1540	0.2165
δ_8	0.1526	0.2370	0.2606	0.2117
δ_9	−0.0205	0.2443	−0.0408	0.2410
δ_{10}	−0.1953	0.2564	−0.3290	0.2165
δ_{11}	0.0347	0.2165	0.0715	0.2179
δ_{12}	−0.0286	0.2082	−0.0553	0.2165
δ_{13}	−0.1298	0.2303	−0.2274	0.2117
σ_δ^2	0.0572	0.1098	0.0837	0.0659

[a] Bayesian estimates are medians of posteriors.

[b] Note that the standard error of the estimated gamma shape parameter from GLMM was estimated using the delta method since SAS does not provide the standard error for the shape parameter but rather the reciprocal of the shape parameter.

7.3.3 Inference on Response Distribution Characteristics

It is relatively straightforward to extend this analysis to making inference on response distribution characteristics. Consider both the mean and an exceedance probability associated with the marginal distribution of the response. Robinson, Anderson-Cook, and Hamada (2009) discuss examples of Bayesian inference on characteristics of the conditional response distribution as well as other details of Bayesian inference.

Once again, consider Example 7.6. Suppose we need to make inferences for the setting on characteristics of the marginal response distribution at $x_1 = 0.35$, $x_2 = 0.35$, and $x_3 = 0.30$ and the process variable levels at $p_1 = 1$, $p_2 = -1$, and $p_3 = -1$. In thinking of inference, it is helpful to recall the sources of uncertainty, which in this case are a function of parameter estimation, roll-to-roll variability, and the variation due to the response distribution. The uncertainty due to parameter estimation is reflected in the joint posterior distribution of $\mathbf{\Theta}$. Roll-to-roll variation is summarized by the parameter σ_δ^2 and the conditional response distribution. Here, we use the gamma distribution to describe the uncertainty that exists in the response from one piece of film to another for a given roll.

Instead of deriving an explicit expression for the marginal distribution, a more straightforward approach uses simulation by exploiting the posterior draws which make up the empirical posterior of $\mathbf{\Theta}$. For each realization in $\mathbf{\Theta}$, one can generate a large number (say, 1000) of δ_i realizations from a normal distribution with mean 0 and variance σ_δ^2 to represent the population of rolls. Table 7.20 provides the first three realizations of the posterior distributions of $\beta_1,\beta_2,\ldots,\beta_{13}$, σ_δ, and r. Here, β_{13} denotes the 13th model parameter in the linear predictor in Equation (7.108), so $\beta_{13} = \psi_{23}$. From Table 7.20, the first generation of 1000 δ_i realizations used a normal distribution with mean 0 and standard deviation $\sigma_\delta = 0.15$. Next, for each of the 1000 δ_i realizations, a large number of film pieces (say, 1000) are generated from the gamma density with parameters r and $\lambda_{ji} = E(y_{ji}|\delta_j)/r$. For the first realization in $\mathbf{\Theta}$ (i.e., first row in Table 7.20), an

Table 7.20 First Three Realizations of the Posterior Distributions[a] of $\beta_1,\beta_2,\ldots,\beta_{13}$, σ_δ, and r Obtained from WinBuGS

β_1	β_1	...	β_{13}	α	σ_δ
5.64	3.74	...	−0.20	4.20	0.15
5.80	4.46	...	−0.30	6.58	0.31
5.84	3.92	...	0.11	4.21	0.15
⋮	⋮	...	⋮	⋮	⋮

[a](Note that β_{13} denotes the 13th model parameter in the linear predictor in expression (7.108), so $\beta_{13} = \psi_{23}$).

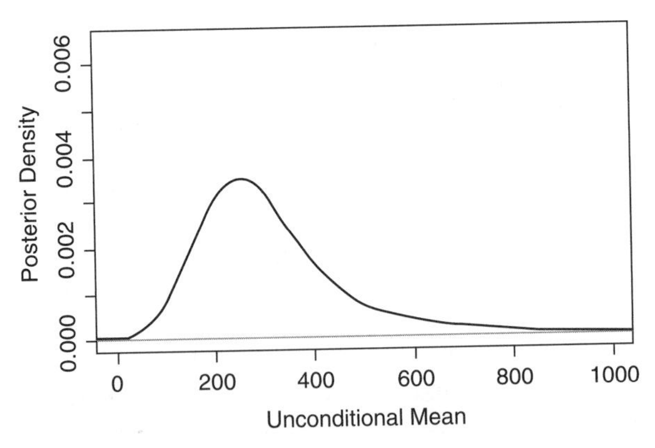

Figure 7.19 Empirical posterior of marginal mean.

entire population of roll effects are simulated (i.e., the 1000 δ_j's), and for each roll, an entire population of film pieces are simulated (i.e., 1000 gamma random variates using r and $\lambda_{ji} = E(y_{ji}|\delta_j)/r$; here, r and $E(y_{ji}|\delta_j)$ are determined by the first row in Table 7.20 along with the specified link function). A realization of the marginal mean posterior distribution is simply the mean of these $1000 \times 1000 = 1{,}000{,}000$ observations. Similarly, a realization of posterior of $P(y > 150)$ is the proportion of these 1,000,000 observations that exceed 150. By conducting this exercise for all 10,000 draws in the posterior distributions of $\boldsymbol{\Theta}$, one essentially has an empirical representation of the posterior distributions for the marginal mean and the proportion of film pieces that exceed 150. Figure 7.19 is a plot of the empirical posterior of the marginal mean. We do not provide the empirical posterior distribution of $P(y > 150)$; however, it is straightforward to obtain. Finally, if one wishes to obtain, say, a 95% credible interval on the marginal mean or the proportion of film pieces of film whose response exceeds 150, one needs only to choose the 250th and 9,750th order statistics of the appropriate posteriors for the lower and upper bounds of the credible interval.

Not only does the Bayesian approach allow the user to easily make inference on a variety of characteristics of the conditional and marginal response distributions, it is flexible enough to be used with any exponential family response distribution. Unlike the frequentist approach to GLMMs, the Bayesian method can easily be used for other assumed distributions for the random effects. While the Bayesian approach does not rely on asymptotic distributional assumptions, it does rely on assumptions regarding the priors for the model parameters. Although not presented here, basic diagnostic checks using summaries from the posterior distributions easily allow for exploration of the residuals to see if the assumptions are appropriate. For a more detailed treatment of the Bayesian approach to generalized linear models with non normal responses, see Robinson, Anderson-Cook, and Hamada (2009).

EXERCISES

7.1 Jensen, Birch, and Woodall (2008) considered profile monitoring of a calibration data set in which the data consists of 22 calibration samples. One of the purposes of the experiment was to determine the relationship between an absorbance measure (absorbance) of a chemical solution to the volume at which the solution was prepared (volume). The raw data are provided in *calibration.xls*.

(a) Graphically investigate the relationship between absorbance and volume in a sample specific manner.

(b) Determine which model—(1) random intercept model, (2) random slope but common intercept, or (3) random intercept plus random slope model—is most appropriate for these data, shown in the following table. Use hypothesis testing with a significance level of 0.05 to make your decision.

(c) Based on the model in part (b), estimate the common variance–covariance matrix for each sample.

(d) Provide an expression for the marginal mean relationship between absorbance and volume.

(e) Provide an expression for the conditional mean relationship between absorbance and volume.

Data for Exercise 7.1 (Calibration Data)

Sample	Volume	Absorbance	Repeat
1	0	1	1
2	0	4	1
3	0	3	1
4	0	4	1
5	0	−9	1
6	0	3	1
7	0	3	1
8	0	2	1
9	0	−6	1
10	0	2	1
11	0	1	1
12	0	−7	1
13	0	5	1
14	0	3	1
15	0	−8	1
16	0	4	1
17	0	2	1
18	0	2	1
19	0	0	1
20	0	1	1

Sample	Volume	Absorbance	Repeat
21	0	−9	1
22	0	−8	1
1	50	104	1
2	50	104	1
3	50	105	1
4	50	104	1
5	50	92	1
6	50	107	1
7	50	104	1
8	50	105	1
9	50	95	1
10	50	104	1
11	50	103	1
12	50	94	1
13	50	105	1
14	50	106	1
15	50	94	1
16	50	104	1
17	50	105	1
18	50	104	1
19	50	101	1
20	50	104	1
21	50	92	1
22	50	95	1
1	100	206	1
2	100	206	1
3	100	207	1
4	100	206	1
5	100	195	1
6	100	209	1
7	100	207	1
8	100	208	1
9	100	196	1
10	100	206	1
11	100	205	1
12	100	198	1
13	100	210	1
14	100	208	1
15	100	196	1
16	100	207	1
17	100	206	1
18	100	206	1
19	100	203	1
20	100	206	1
21	100	194	1
22	100	195	1
1	150	307	1
2	150	308	1

Sample	Volume	Absorbance	Repeat
3	150	311	1
4	150	308	1
5	150	296	1
6	150	311	1
7	150	311	1
8	150	310	1
9	150	297	1
10	150	311	1
11	150	309	1
12	150	298	1
13	150	313	1
14	150	311	1
15	150	299	1
16	150	311	1
17	150	308	1
18	150	309	1
19	150	305	1
20	150	311	1
21	150	298	1
22	150	298	1
1	200	409	1
2	200	412	1
3	200	414	1
4	200	411	1
5	200	397	1
6	200	412	1
7	200	414	1
8	200	412	1
9	200	401	1
10	200	413	1
11	200	412	1
12	200	404	1
13	200	415	1
14	200	411	1
15	200	400	1
16	200	415	1
17	200	410	1
18	200	414	1
19	200	409	1
20	200	410	1
21	200	400	1
22	200	401	1
1	0	3	2
2	0	2	2
3	0	2	2
4	0	2	2
5	0	−8	2
6	0	3	2

Sample	Volume	Absorbance	Repeat
7	0	2	2
8	0	2	2
9	0	−7	2
10	0	4	2
11	0	2	2
12	0	−7	2
13	0	7	2
14	0	2	2
15	0	−6	2
16	0	6	2
17	0	4	2
18	0	0	2
19	0	1	2
20	0	4	2
21	0	−10	2
22	0	−8	2
1	50	104	2
2	50	103	2
3	50	104	2
4	50	104	2
5	50	95	2
6	50	105	2
7	50	105	2
8	50	104	2
9	50	94	2
10	50	105	2
11	50	104	2
12	50	96	2
13	50	107	2
14	50	104	2
15	50	95	2
16	50	106	2
17	50	106	2
18	50	103	2
19	50	102	2
20	50	106	2
21	50	92	2
22	50	95	2
1	100	206	2
2	100	204	2
3	100	207	2
4	100	207	2
5	100	197	2
6	100	207	2
7	100	208	2
8	100	208	2
9	100	197	2
10	100	207	2

Sample	Volume	Absorbance	Repeat
11	100	206	2
12	100	199	2
13	100	208	2
14	100	207	2
15	100	199	2
16	100	210	2
17	100	208	2
18	100	206	2
19	100	206	2
20	100	208	2
21	100	194	2
22	100	199	2
1	150	308	2
2	150	307	2
3	150	309	2
4	150	312	2
5	150	299	2
6	150	308	2
7	150	308	2
8	150	309	2
9	150	300	2
10	150	310	2
11	150	307	2
12	150	301	2
13	150	315	2
14	150	308	2
15	150	302	2
16	150	310	2
17	150	310	2
18	150	308	2
19	150	307	2
20	150	309	2
21	150	297	2
22	150	301	2
1	200	412	2
2	200	413	2
3	200	411	2
4	200	413	2
5	200	400	2
6	200	410	2
7	200	410	2
8	200	412	2
9	200	401	2
10	200	412	2
11	200	411	2
12	200	402	2
13	200	415	2
14	200	414	2

Sample	Volume	Absorbance	Repeat
15	200	404	2
16	200	413	2
17	200	413	2
18	200	409	2
19	200	411	2
20	200	414	2
21	200	398	2
22	200	403	2

7.2 Robinson, Wulff, Montgomery, and Khuri (2006) consider a wafer etching process in semi conductor manufacturing. During the etching process, some of the variables are not perfectly controllable and the net effect is that wafers produced on any given day (i.e., within the same batch) may be different from wafers produced on another day (i.e., wafers produced in different batches). Variation due to time is designed into the experimentation process by using test wafers chosen at random across several days. If there are significant interactions between any of the control variables and time of production (consider batch as a proxy for time of production), then it may be possible to minimize the impact of the variation due to time by manipulating the levels of these control variables. Resistivity is the response of interest and the following control variables were considered: gas flow rate (x_1), temperature (x_2), and pressure (x_3). A resistivity of 350 is desirable and process engineers hope to obtain operating conditions that result in minimal prediction variance. Previous experience with the process suggests that resistivity follows a gamma distribution. Batches of wafers from 11 different days are used in the experiment and the levels of the control variables are manipulated according to a central composite design with four center runs. The data set is provided in the table below.

(a) Fit the full, marginal gamma mixed model with log link, where batches are considered random. Be sure to fit the main effects, the two-factor interactions, and the pure quadratics in the control variables. Also fit the interactions between all of the fixed effects and the random batch effect.

(b) Simplify the response variance–covariance matrix via hypothesis testing.

(c) Obtain an explicit expression for the marginal mean response in terms of the significant control factor terms.

(d) Obtain an explicit expression for the estimated prediction variance of the marginal mean response.

(e) Propose a method for finding the optimal control factor settings such that the marginal mean is constrained to be 350 while minimizing the variance of the estimated marginal mean.

Coded Settings			Response										
x_1	x_2	x_3	Batch1	Batch2	Batch3	Batch4	Batch5	Batch6	Batch7	Batch8	Batch9	Batch10	Batch11
−1	−1	−1	186.56	274.03	323.62	185.80	386.65	321.98	59.39	404.40	208.24	186.52	366.65
1	1	1	378.06	374.53	379.89	336.44	373.96	589.72	219.17	504.36	328.44	367.63	438.12
−1	−1	−1	196.32	219.53	244.17	268.18	187.78	185.34	65.73	327.87	145.12	219.77	204.44
1	1	1	265.31	294.17	362.81	179.73	256.60	402.19	62.26	444.72	279.84	266.38	403.77
−1	−1	−1	442.54	301.11	224.29	159.76	367.49	266.03	587.50	145.87	352.47	233.24	110.84
1	1	1	327.29	321.55	197.20	278.44	566.95	276.37	546.48	153.76	288.26	293.87	75.35
−1	−1	−1	128.22	212.22	146.31	89.72	231.17	262.77	304.92	101.37	180.55	87.66	62.30
1	1	1	205.31	178.25	130.01	78.59	273.14	102.83	329.75	87.67	175.30	126.09	77.12
−1.6818	0	0	232.07	248.85	274.92	139.97	264.33	305.75	189.35	192.82	157.59	141.47	174.31
1.6818	0	0	293.59	252.31	330.11	192.82	368.85	297.27	189.42	271.05	266.92	241.99	288.50
0	−1.6818	0	381.57	229.30	289.35	177.20	401.81	310.84	255.30	255.23	368.72	256.91	305.37
0	1.6818	0	138.67	227.13	110.53	119.70	211.23	153.19	135.20	128.78	138.06	100.39	155.58
0	0	−1.6818	215.31	517.09	477.31	289.59	394.24	642.83	31.03	873.86	364.32	330.26	468.26
0	0	1.6818	300.00	332.85	252.49	90.27	428.11	221.10	719.12	73.41	354.05	206.78	25.52
0	0	0	400.39	296.12	324.95	228.89	303.66	153.87	240.72	276.95	269.38	231.07	226.14
0	0	0	340.11	237.53	189.76	206.69	416.08	260.55	129.83	174.99	250.79	166.19	228.51
0	0	0	381.70	205.66	267.14	188.08	375.61	291.52	256.51	232.91	191.50	170.39	189.15
0	0	0	397.16	262.74	289.68	226.40	572.07	307.89	223.15	272.82	310.37	347.19	261.68

7.3 Schall (1991) considers data from an experiment to measure the mortality of cancer cells under radiation. Four hundred cells were placed on a dish, and three dishes were irradiated at a time, or occasion. The zero-dose data provided here were obtained from OzDASL and are found in cell_survive.xls.

(a) Consider occasion of irradiation as a fixed effect and assume that the three dishes irradiated at a given time represent a random sample of possible dishes of cells that could be irradiated at a given time. Fit an appropriate logistic regression model to determine if there is a significant difference among the occasions.

(b) Now assume that both occasion and dish are random effects and re-fit your logistic model to determine if there is a difference between the occasions.

(c) How does the interpretation of your hypothesis test in part (b) differ from the interpretation of the hypothesis test in part (a)?

(d) Provide an estimate of the overall proportion of cells that survive using the model in part (a) as well as the model in part (b).

Data for Exercise 7.3

Occasion	Survived
1	178
1	193
1	217
2	109
2	112
2	115
3	66
3	75
3	80
4	118
4	125
4	137
5	123
5	146
5	170
6	115
6	130
6	133
7	200
7	189
7	173
8	88
8	76
8	90
9	121
9	124
9	136

7.4 Weiss (2005, p. 353) describes a patient controlled analgesia study presented by Henderson and Shimakura (2003). The number of self-administered doses in a 4-hour period was recorded for each patient for each of 12 consecutive periods. Two groups of patients were considered: a 1-milligram (mg) per dose group and a 2 milligram. The variables in the data set are *id* (subject id), *group1mg* = 1 for subjects in the 1-mg group, 0 in the 2-mg group, *count* (# of self-administered dosages), and *time* (dose period). The data is found in selfdose.xls.

(a) Plot *count* and ln(*count*) versus *time* and comment on which transformation of *count* appears to be most appropriate. Does there appear to be a linear relationship of ln(*count*) with *time*?

(b) When considering a model with both categorical and continuous regressors, the interpretation of model coefficients is often aided by centering the continuous regressors (here *time* could be considered continuous). After centering *time*, determine the most appropriate Poisson mixed effects model.

(c) Overlay plots of the estimated marginal mean number of self-administered doses for the 1-mg group and 2-mg groups over time and interpret the plot.

7.5 In Exercise 6.2 a process for steel normalization was discussed.

(a) Propose an appropriate mixed model for the data described.

(b) Provide an estimate of the marginal mean strength of normalized steel and compare this estimate to the GEE estimate.

(c) What advantages can you see to analyzing this data using a mixed model versus the GEE approach?

7.6 In Exercise 6.8, an arthritis clinical trial was described.

(a) Assume the patients represent a random sample from a population of possible patients and refit the data using an appropriate logistic mixed effects model (be sure to argue whether a model with random intercepts, random slopes, or both should be used). In fitting the logistic model, let $y = 1$ if the self-assesment was 'good' and $y = 0$ if 'good' was not indicated.

(b) Determine if there is a significant treatment effect.

(c) Determine if there is a significant treatment by time interaction.

(d) Suppose we would like to estimate the difference in the proportion of patients who felt "poor" at baseline and those who felt "poor" after 3 months in the study. Fit an appropriate mixed effects model for estimating this difference.

(e) Provide a 90% confidence interval on the difference in the proportion of patients who felt "poor" at baseline and those who felt "poor" after 3 months in the study.

7.7 In Example 6.5, a milling process for corn was discussed in terms of yield of grits.

(a) Assume *Batch* is a random effect and fit an appropriate linear mixed model.

(b) Produce an appropriate set of residual diagnostic plots and comment on whether or not you feel there are any unusual observations.

(c) Produce a normal Q-Q plot for the random *Batch* effect and comment on whether or not you feel that the assumed distribution for the *Batch* effect is appropriate.

7.8 Somner (1982) reported on an Indonesian Children's Health Study designed to determine the effects of vitamin A deficiency in preschool children. The investigators were particularly interested in whether children with vitamin A deficiency were at increased risk of developing respiratory infection. Each child was measured at six times: 0, 3, 6, 9, 12, and 15 months. The following is a description of the variabes in the subset of the actual data set which is considered here: *bage* (baseline age in years), *gender* (gender = 1 if female and 0 if male), *vita* (vitamin A deficient = 1 if yes and = 0 if not), *infect* (respiratory infection = 1 if yes and 0 if not), and *time* (time of examination). The data is found in Somner.xls.

(a) Compute appropriate summary statistics for the response across the levels of the regressors and comment on some preliminary relationships that appear to exist.

(b) Fit a random intercept logistic model along with a random intercept and random slope logistic model and determine which is most appropriate via hypothesis testing.

(c) Produce an interaction plot involving *vita* and *time* and interpret.

(d) After removing nonsignificant terms, compute the estimated odds of a child developing a respiratory infection when he/she is vitamin A deficient. Do the same for children who are not vitamin A deficient and then construct the estimated odds ratio comparing the two groups. Interpret this estimated odds ratio.

(e) Characterize the children who seem to have the greatest chance of developing a respiratory infection.

7.9 McKnight and van den Eeden (1993) analyzed a multi period, two-treatment crossover experiment to establish whether the artificial sweetener aspartame caused headaches. The trial involved randomly assigning 27 patients to different sequences of placebo and aspartame. The variables in headache.xls are *Case* (subject id number), *Week* (week of measurement), *Aspartame* (patient took aspartame = 1 if not = 0), *Headaches* (number of reported headaches), and *Exp-days* (the number days for which headaches counted).

(a) Many times with Poisson models, the counts reported should be analyzed as rates instead of the raw count due to unequal amounts of effort for each subject (e.g., here, some patients had headaches counted over differing numbers of days compared to other patients). The amount of time for which a count was recorded is commonly referred to as an *offset* and one models the count/time. For the log-link, the offset is then ln(time). Fit an appropriate random effects Poisson model with *Exp-days* specified as the offset variable.

(b) Does there appear to be a relationship between the number of reported headaches and whether or not a person took aspartame? Explain.

(c) Interpret the coefficients from the final model.

(d) Provide an expression for the marginal mean number of headaches and interpret.

CHAPTER 8

Designed Experiments and the Generalized Linear Model

8.1 INTRODUCTION

We have illustrated applications of the GLM in a variety of settings, including data from both unplanned or observational studies and designed experiments. GLMs arise in the context of designed experiments in two ways. First, one of the responses in a multiresponse experiment may be nonnormal and require the use of a GLM to obtain a satisfactory model. This response may not necessarily be the primary response of interest when the experiment is designed, or the experimenter may not suspect that it is nonnormal and requires the use of the GLM. In this situation, a factorial design, or perhaps a response surface design, or even an optimal design may have been selected. Second, the experimenter may be very interested in this response and know at the outset of the study that the response is nonnormal. In this situation, the experimenter should design the experiment with this information in mind.

This chapter presents several topics that involve designed experiments and GLMs. We discuss designed experiments for the GLM and show one approach to constructing optimal designs for the GLM using the popular D-optimality criterion. This technique can be useful in the situation described above where the experimenter knows in advance that he/she requires the GLM to obtain an adequate model for the response. We also present an approach to evaluating the efficiency of standard designs employed with a response that should be modeled with the GLM. This provides some insight on the utility of standard designs when the need for a GLM arises after a standard experiment has been conducted. We also provide some guidance on how GLMs can be used in the analysis of screening experiments, which are often fractional factorial designs.

8.2 EXPERIMENTAL DESIGNS FOR GENERALIZED LINEAR MODELS

It should be clear to the reader by now that when regressor variables can be controlled in a situation with nonnormal responses then the selection of proper settings of the variables through design of experiments may be very important. Even in the case of a single variable dose–response situation the selection of the number of drug doses and location of the doses can have an impact on quality of the inference, namely, on the estimation of model coefficients. Before we embark on a discussion of difficulties associated with design choice, we review some important and well-known design concepts for the case of linear models with normal responses.

8.2.1 Review of Two-Level Factorial and Fractional Factorial Designs

Two-level factorial and fractional factorial designs for linear models that contain first-order and interaction cross-product terms possess the property of **orthogonality**. For example, see Box and Hunter (1961a, b), Montgomery (2009), and Myers, Montgomery, and Anderson-Cook (2009). Consider the model matrix **X** in the model formulation

$$\mathbf{y} = \mathbf{X}\boldsymbol{\beta} + \boldsymbol{\varepsilon}$$

that was used in the linear regression review in Chapter 2. The property of orthogonality implies that the columns of the **X** matrix are mutually orthogonal. For example, if the coded levels of the design variables are at ± 1, then in the case of a 2^2 factorial with model terms x_1, x_2, x_1x_2, and a constant term, the **X** matrix is given by

$$\mathbf{X} = \begin{array}{c} \begin{matrix} \quad & x_1 & x_2 & x_1x_2 \end{matrix} \\ \begin{bmatrix} 1 & -1 & -1 & 1 \\ 1 & -1 & 1 & -1 \\ 1 & 1 & -1 & -1 \\ 1 & 1 & 1 & 1 \end{bmatrix} \end{array}$$

and the orthogonality of the columns of **X** gives that $\mathbf{X}'\mathbf{X} = 4\mathbf{I}_4$ and thus $(\mathbf{X}'\mathbf{X})^{-1} = \frac{1}{4}\mathbf{I}_4$. Here the standard errors of model coefficients are given by $\sigma/\sqrt{4} = \sigma/2$, where σ is the population standard deviation. It turns out that **on a per run basis** these standard errors are as small as possible given the restriction that design levels remain within the [−1,+1] ranges. This extends to any **orthogonal design** with levels at ± 1 for models containing main-effect (x_i)

terms and interaction ($x_i x_j$) terms for $i \pm j$. A proof of this result is found in Myers, Montgomery, and Anderson-Cook (2009).

The orthogonal design for models described above is often termed the **variance optimal** design, and it relies on the orthogonality and the placing of all points in the design to the ± 1 extremes. Indeed, this variance optimal characterization extends to alphabetic optimality such as D and G-optimality. Many of these alphabetic optimality criteria are due to Kiefer (1959) and Kiefer and Wolfowitz (1959). Here we only define D-optimality. (For a practical discussion of alphabetic optimality, see Myers, Khuri, and Carter, 1989.) A D-optimal design is one that minimizes the **generalized variance** of the estimated coefficients. Thus the design is chosen for which

$$D = \left|\frac{\mathbf{X}'\mathbf{X}}{n}\right| \tag{8.1}$$

is maximized. The matrix $(\mathbf{X}'\mathbf{X})/n$ is often called the **moment matrix**. From a practical interpretation viewpoint, a D-optimal design results in a simultaneous confidence region on the model parameters that has the smallest possible volume.

While we do not discuss the practical utility of D-optimality further here, we should point out that D-optimality is used widely in industry as a design selection criterion. In addition, any design criterion that has **variance of coefficients** as the basis of its motivation involves the information matrix and, of course, for linear models with i.i.d. $N(0, \sigma^2)$ errors the information matrix is proportional to $\mathbf{X}'\mathbf{X}$. D-optimality is merely one criterion whose characterization involves maximizing the **determinant of the information matrix** (put in moment matrix form so that one is maximizing information on a per run basis). The notion of choosing the experimental plan to maximize some norm on the information certainly makes intuitive sense, since it offers a plan that is most efficient, with the type of efficiency depending on what norm on the information matrix is used. Good fortune abounds for this approach in the case where the model is linear in the parameters because $\mathbf{X}'\mathbf{X}$ is **parameter free**. As a result optimal experimental designs can be found that require **only the knowledge of the form of the model**. Thus even if the model contains terms that are additional to the x_i and $x_i x_j$ for $i \neq j$, the determination of an optimal design is not a major task.

There is one further intuitive point that should be made that provides some more justification of the property of orthogonality. The reader who is not entirely familiar with the literature on industrial design of experiments may have some understanding of the linear regression topic *multicollinearity*. This area deals with the study of diagnostics that determine the extent of any linear dependencies among the regressor variables. Eigenvalues of the correlation matrix as well as **variance inflation factors** are used. If the design is orthogonal there are no linear dependencies among the design variables. In this case the correlation matrix is the identity matrix with all eigenvalues and variance inflation factors equal to unity.

8.2.2 Finding Optimal Designs in GLMs

As in the case of normal error linear models it would seem reasonable that optimal designs for GLMs should also be derived from the information matrix. Recall from Chapters 4 and 5 that for the most general of the GLMs, the information matrix is given by (apart from the constant $a(\phi)$) the matrix

$$\mathbf{X}'\boldsymbol{\Delta}\mathbf{V}\boldsymbol{\Delta}\mathbf{X} = \mathbf{X}'\mathbf{W}\mathbf{X} \tag{8.2}$$

where $\mathbf{W}$ is the **Hessian weight matrix**. Now we know that for the case of the identity link and normal errors, the information matrix is given by $\mathbf{X}'\mathbf{X}$ (apart from the divisor $1/\sigma^2$). However, for other GLMs we know that the diagonal elements in $\mathbf{V}$ are functions of the means and hence the $\mathbf{x}'\boldsymbol{\beta}$ (i.e., the unknown model parameters), and the diagonal elements of $\boldsymbol{\Delta}$ are also a function of the unknown parameters apart from the case of the canonical link in which $\boldsymbol{\Delta} = \mathbf{I}$. As a result, for many generalized linear models, the optimal design cannot be found without knowledge of the parameters. For many years there have been few attempts to obtain a practical solution to the design problem for obvious reasons.

Despite the practical difficulty in the implementation of an optimal design in generalized linear models, there has been some attention given to the problem, particularly in the case of logistic regression. However, the majority of the research in this area has been confined to the case of a single design variable. Some of the contributions in this area include Abdelbasit and Plackett (1983), Sitter (1992), Heise and Myers (1996), Myers, Myers, and Carter (1994), Minkin (1987), and Kalish and Rosenberger (1978). For the two-variable problem, there are contributions in both Poisson and logistic regression models. See, for example, Branden, Vidmar, and McKean (1988), Jia and Myers (2001), Van Mullekom and Myers (2001), and Sitter and Torsney (1995). Minkin (1993) has dealt with designs for the single-variable Poisson regression model where a special application involving impairment of reproduction applies.

Because the information matrix contains unknown parameters, the experimenter has to have an estimate or guessed values of these parameters in order to find an optimal design. Some useful references include Dror and Steinberg (2006), Ford, Kitsos, and Titterington (1989), Khuri, Mukherjee, Sinh and Ghosh (2006). As should be obvious, the requirement of knowledge of parameters, or rather parameter guesses, breeds even more difficulty when one moves beyond a single design variable. However, break-throughs have been made in special types of applications. The impairment of reproduction situation mentioned above finds applications where responses are Poisson random variables that come from a system (cancer cell colonies or important organisms in our ecosystem) and the design variables represent doses of combinations of drugs/pollutants that impair the growth of the system. In these applications simplifications allow estimated optimal designs to come from parameter knowledge that has as its source guessed information easily derivable from characteristics of individual

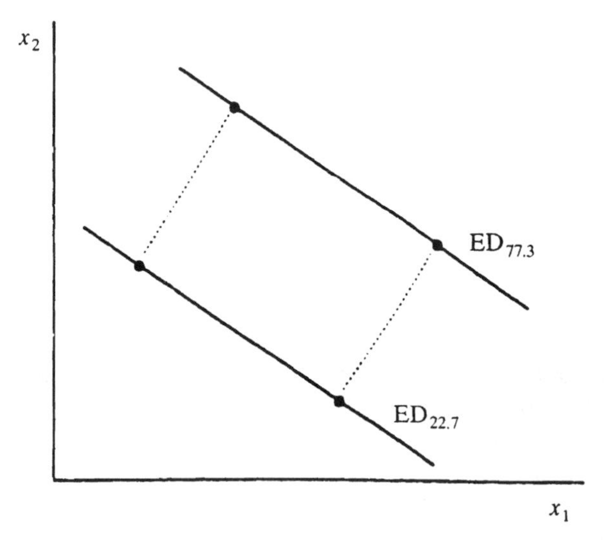

Figure 8.1 D-optimal design for two variables using logistic regression.

drugs or pollutants. In these applications and certain others, the use of design optimality can be combined with either Bayesian approaches or two-stage procedures to produce efficient designs or robust designs. In fact, in the impairment of reproduction application discussed above, efficient factorial or even fractional factorial designs can be developed for any number of drugs/pollutants.

The requirement of parameter guessing has as its foundation the placing of points which accommodates the variance function involved, since in the case of members of the experimental family the variance is a function of the mean. For example, in a single variable logistic regression model the D-optimal design is to place $\frac{1}{2}$ of the experimental units at the $ED_{17.6}$ and the other half at the $ED_{82.4}$. The guess of the design level, or drug dose, that produces the probabilities 0.176 and 0.824 must be made. Now, of course, in a practical setting a two-point design is rarely embraced by a practitioner. In the case of two design variables there is greater difficulty that surrounds the guessing. The points must be placed on ED **contours**, once again with an attempt to account for the changing variance structure in the design space. For example, any parallelogram bounded by the lines of constant $ED_{22.7}$ and $ED_{77.3}$ as in Figure 8.1 is indeed the D-optimal design. Jia and Myers (2001) established that the D-criterion is invariant to the location and angle in the parallelogram. Figure 8.2 illustrates this point. However, if the design becomes larger because of wider ranges the D criterion becomes larger, and thus the design is more efficient. Here we have an analog to factorial experiments in which increased ranges decrease variances of parameter estimates. Points must, however, accommodate the binomial variance structure by being on the special ED lines (see Figure 8.3). Similar situations occur in Poisson regression particularly in the impairment of reproduction application.

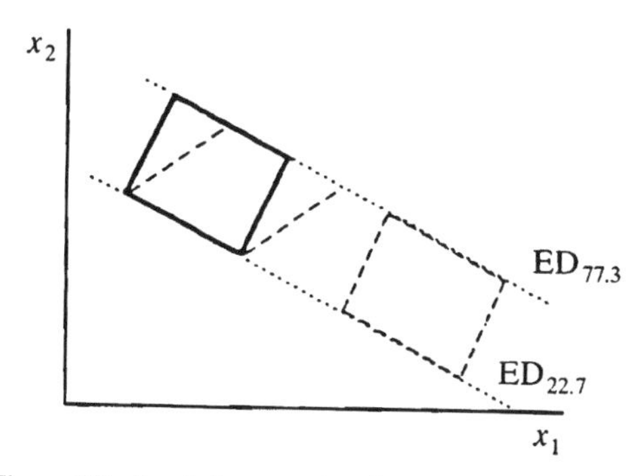

Figure 8.2 Logistic regression: invariance to location.

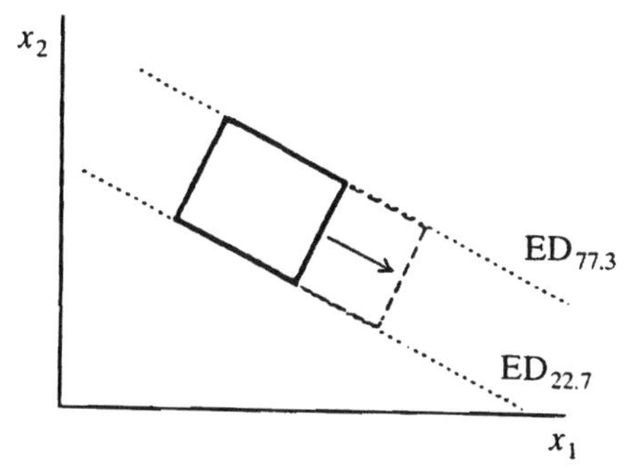

Figure 8.3 Logistic regression: design is improved with larger range.

One approach to the optimal design problem for a GLM is to guess or estimate values of the unknown parameters. In this case the information matrix is a function of the design points only and a D-optimal design can be found. More properly, this design should be called a **conditional D-optimal design**, because it depends on the estimated values of the unknown parameters. A two-stage or **sequential** procedure could also be employed, where the initial design is run and preliminary estimates of the model parameters obtained, then the initial design is augmented with additional runs to complete the experiment.

A third alternative is to employ a **Bayesian** approach. This would involve assessing a prior distribution on the unknown parameters and integrating them out of the information matrix. If the prior distribution is $f(\boldsymbol{\beta})$, then the criterion that we will use is the Bayesian D-optimality criterion developed by Chaloner and Larntz (1989):

$$\phi = \int \ln(|\mathbf{I}(\boldsymbol{\beta}|)f(\boldsymbol{\beta})d\boldsymbol{\beta} \tag{8.3}$$

where $\mathbf{I}(\boldsymbol{\beta})$ is the information matrix. Also see Chaloner and Verdinelli (1995). Finding a design that maximizes this criterion is a very computationally intensive process as typically it requires that a multidimensional integral be evaluated many times. Gotwalt, Jones, and Steinberg (2009) develop a clever quadrature scheme that provides extremely accurate values of this integral and is very computationally efficient. JMP uses this procedure with a coordinate exchange procedure developed by Meyer and Nachtsheim (1995) that can be used to find designs for a variety of nonlinear models, including GLMs. We now present several examples of this procedure.

Example 8.1. Logistic Regression with Two Predictor Variables. An experimenter wants to fit a logistic regression model in two variables x_1 and x_2, where both variables are in coded units in the range $-1 \leq x_i \leq 1$. Therefore the model we plan to fit is

$$E(y) = \frac{e^{\beta_0+\beta_1 x_1+\beta_2 x_2}}{1 + e^{\beta_0+\beta_1 x_1+\beta_2 x_2}}$$

Our prior information about the model suggests that reasonable ranges for the model parameters are

$$\begin{aligned} 1 &\leq \beta_0 \leq 3 \\ 1.5 &\leq \beta_1 \leq 4.5 \\ -1 &\leq \beta_2 \leq -3 \end{aligned}$$

Table 8.1 JMP Bayesian D-Optimal Design for Example 8.1

Nonlinear Design

Parameters

Name	Role	Values	
b0	Continuous	1	3
b1	Continuous	1.5	4.5
b2	Continuous	-1	-3

Design

Run	**x1**	**x2**
1	-0.57255	1
2	-0.44157	-1
3	-1	-0.02109
4	-0.57255	1
5	-1	-1
6	-1	-1
7	-0.57255	1
8	0.402077	1
9	-1	-1
10	0.402077	1
11	0.402077	1
12	1	1

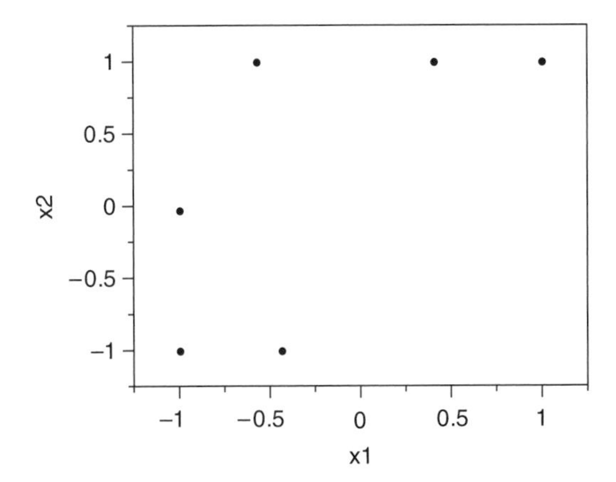

Figure 8.4 JMP Bayesian D-optimal design for Example 8.1.

The experimenter believes that a normal distribution defined over these ranges constitutes a reasonable summary of the prior information. A design with $n = 12$ runs is desired.

Table 8.1 and Figure 8.4 present the Bayesian D-optimal design found by JMP. Notice that there are six distinct design points and the model has three parameters. Runs 2, 3, and 12 from Table 8.1 are not replicated, while runs (5,6, 9), (1, 4, 7), and (8, 10, 11) comprise the replicates for the other three distinct design points. □

Example 8.2. The Impact of Changing the Prior Distribution. In Example 8.1 we used a normal prior distribution to summarize the information about the unknown model parameters in the logistic regression model. We now consider the impact of using a different prior. JMP allows us to use the normal, lognormal, uniform, and exponential distributions as priors. Suppose that we use exactly the same information as we did in Example 8.1 but now select the uniform distribution as the prior. That is, the model parameters are uniformly distributed over the ranges

$$
\begin{aligned}
1 &\leq \beta_0 \leq 3 \\
1.5 &\leq \beta_1 \leq 4.5 \\
-1 &\leq \beta_2 \leq -3
\end{aligned}
$$

Table 8.2 and Figure 8.5 present the Bayesian D-optimal design found by JMP. This design is slightly different from the design that we found using the normal prior. It has five distinct design points rather than six. Runs 8 and 9 from Table 8.2 are unreplicated, while runs (1, 6, 10, 12), (2, 3, 4), and (5, 7, 11) are the replicates for the other three distinct design points. While there is some impact of the choice of prior distribution, it is not dramatic in this example. □

Table 8.2 JMP Bayesian D-Optimal Design for Example 8.2, Uniform Prior

Nonlinear Design

Parameters

Name	Role	Values	
b0	Continuous	1	3
b1	Continuous	1.5	4.5
b2	Continuous	-1	-3

Design

Run	x1	x2
1	-1	-1
2	-0.50204	1
3	-0.50204	1
4	-0.50204	-1
5	0.56504	1
6	-1	-1
7	0.56504	1
8	-1	0.253568
9	0.336435	1
10	-1	-1
11	0.56504	1
12	-1	-1

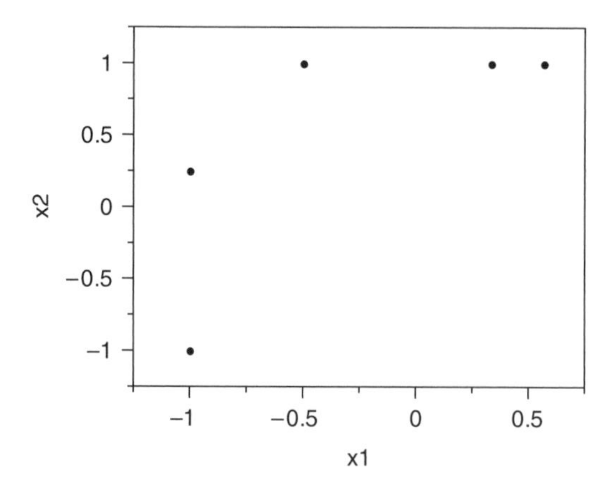

Figure 8.5 JMP Bayesian D-optimal design for Example 8.2.

Example 8.3. Logistic Regression with Interaction. We now reconsider the logistic regression model from Example 8.1 but include an interaction in the linear predictor. Therefore the model is now

$$E(y) = \frac{e^{\beta_o + \beta_1 x_1 + \beta_2 x_2 + \beta_{12} x_1 x_2}}{1 + e^{\beta_o + \beta_1 x_1 + \beta_2 x_2 + \beta_{12} x_1 x_2}}$$

As in Example 8.1, we assume that a normal prior distribution is appropriate. The likely ranges for the model parameters are

$$1 \leq \beta_0 \leq 3$$
$$1.5 \leq \beta_1 \leq 4.5$$
$$-1 \leq \beta_2 \leq -3$$
$$-0.5 \leq \beta_{12} \leq -1.5$$

The ranges for β_0, β_1, and β_2 are identical to those chosen in Example 8.1. We want to use a design with 12 runs.

Table 8.3 and Figure 8.6 present the Bayesian D-optimal design. This design is rather different from the designs for the no-interaction model. It has seven distinct design points for a model with four parameters with runs 3, 6, and 10 from Table 8.3 unreplicated and runs (1, 12), (2, 11), (4, 5, 8), and (7, 9) forming the other four points. □

Table 8.3 Bayesian D-Optimal Design From JMP for a Logistic Regression Model with Two Predictors and an Interaction Term

Nonlinear Design

Parameters

Name	Role	Values	
b0	Continuous	1	3
b1	Continuous	1.5	4.5
b2	Continuous	-1	-3
b12	Continuous	-0.5	-1.5

Design

Run	**x1**	**x2**
1	1	1
2	-1	1
3	-0.39451	-1
4	-1	-1
5	-1	-1
6	0.261768	1
7	-0.03877	-0.15756
8	-1	-1
9	-0.03877	-0.15756
10	-0.24769	1
11	-1	1
12	1	1

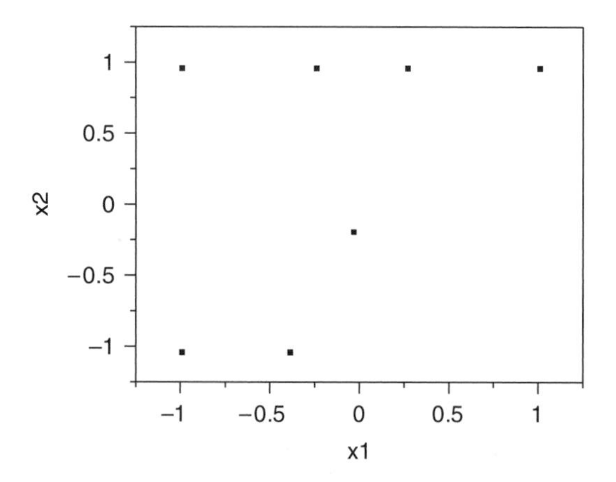

Figure 8.6 JMP Bayesian D-optimal design for Example 8.3.

Example 8.4. Logistic Regression with a Quadratic Model in the Linear Predictor. Suppose that we want to fit a logistic regression model in two predictor variables but that a quadratic model in the linear predictor should be used. That is, the model is

$$E(y) = \frac{e^{\beta_o + \beta_1 x_1 + \beta_2 x_2 + \beta_{12} x_1 x_2 + \beta_{11} x_1^2 + \beta_{22} x_2^2}}{1 + e^{\beta_o + \beta_1 x_1 + \beta_2 x_2 + \beta_{12} x_1 x_2 + \beta_{11} x_1^2 + \beta_{22} x_2^2}}$$

Suppose that the prior information about the model parameters can be summarized with a normal distribution and that

$$\begin{gathered} 1 \leq \beta_0 \leq 3 \\ 1.5 \leq \beta_1 \leq 4.5 \\ -1 \leq \beta_2 \leq -3 \\ -0.5 \leq \beta_{12} \leq -1.5 \\ 1.5 \leq \beta_{11} \leq 4.5 \\ -2 \leq \beta_{22} \leq -6 \end{gathered}$$

The model that we want to fit has six unknown parameters. Suppose we decide to use a 12-run design.

The JMP Bayesian D-optimal design is shown in Table 8.4 and Figure 8.7. Notice that all 12 runs are distinct; that is, there is no replication. There are eight levels of both factors in the design.

Table 8.4 12-Run Bayesian D-Optimal Design From JMP for a Logistic Regression Model with a Full Quadratic in the Linear Predictor

Nonlinear Design

Parameters

Name	Role	Values	
b0	Continuous	1	3
b1	Continuous	1.5	4.5
b2	Continuous	-1	-3
b11	Continuous	1.5	4.5
b12	Continuous	-0.5	-1.5
b22	Continuous	-2	-6

Design

Run	**x1**	**x2**
1	-0.05222	-1
2	0.466732	-1
3	0.959903	1
4	-0.09415	0.1796
5	-1	0.25415
6	-1	-1
7	-1	-0.62031
8	-1	1
9	1	0.796762
10	-0.15693	0.539652
11	0.560026	1
12	-1	0.680569

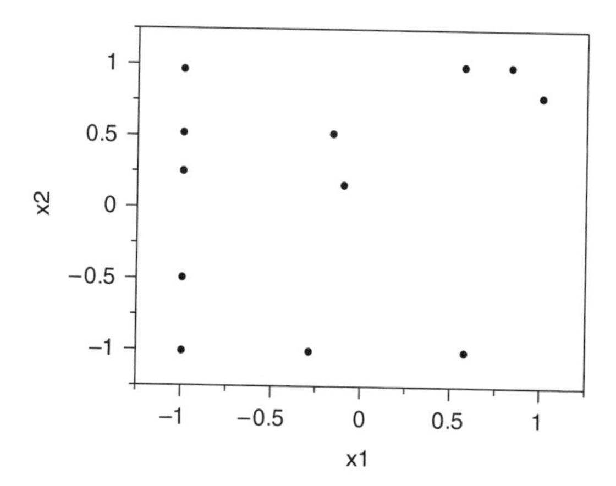

Figure 8.7 12-Run JMP Bayesian D-optimal design for Example 8.4.

Table 8.5 16-Run Bayesian D-Optimal Design From JMP for a Logistic Regression Model with a Full Quadratic in the Linear Predictor

Nonlinear Design

Parameters

Name	Role	Values	
b0	Continuous	1	3
b1	Continuous	1.5	4.5
b2	Continuous	-1	-3
b11	Continuous	1.5	4.5
b12	Continuous	-0.5	-1.5
b22	Continuous	-2	-6

Design

Run	x1	x2
1	1	0.776367
2	-0.19592	0.562149
3	0.050948	-1
4	-1	0.366591
5	-1	-1
6	-0.48048	-1
7	0.411704	1
8	-0.19592	0.562149
9	1	0.776367
10	-0.16357	0.059825
11	-1	0.366591
12	0.050948	-1
13	-1	-0.53623
14	0.797396	1
15	0.620068	-1
16	-1	1

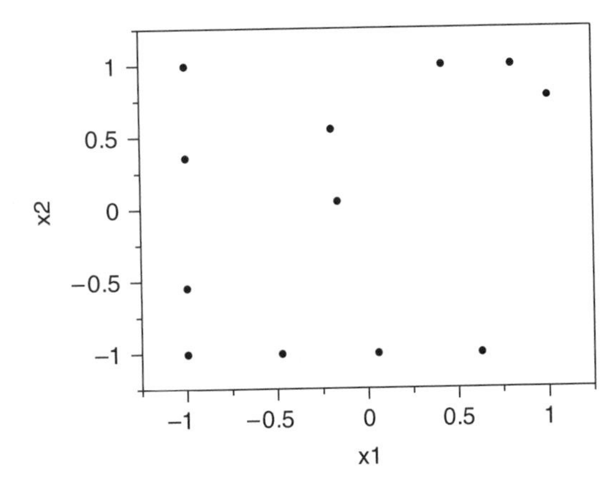

Figure 8.8 16-Run JMP Bayesian D-optimal Design for Example 8.4.

Table 8.5 and Figure 8.8 show what happens when we increase the number of runs to 16. There are still 12 distinct design points, but now there are nine levels of x_1 and eight levels of x_2. Runs (1, 9), (2, 8), (3, 12), and (4, 11) from Table 8.5 are replicates. □

Example 8.5. Poisson Regression with a Full Quadratic in Two Predictor Variables. An experimenter wants to fit a Poisson regression model in two variables x_1 and x_2, where both variables are in coded units in the range $-1 \leq x_i \leq 1$. It is suspected that a quadratic model in the linear predictor is required to provide an adequate model of the response. Therefore the model is

$$E(y)=e^{\beta_0+\beta_1 x_1+\beta_2 x_2+\beta_{12} x_1 x_2+\beta_{11} x_1^2+\beta_{22} x_2^2}$$

The prior information about the model parameters can be summarized with a normal distribution where the ranges of the parameters are

$$1 \leq \beta_0 \leq 3$$
$$0.25 \leq \beta_1 \leq 0.75$$
$$-0.1 \leq \beta_2 \leq -0.3$$
$$-0.5 \leq \beta_{12} \leq -1.5$$
$$0.45 \leq \beta_{11} \leq 1.35$$
$$-0.2 \leq \beta_{22} \leq -0.6$$

Notice that the experimenter is anticipating a stronger effect in the x_1 direction than in the x_2 direction. The model has six parameters that must be estimated and a design with 14 runs is desired.

Table 8.6 and Figure 8.9 present the Bayesian D-optimal design from JMP. There are nine distinct design points; the replicates are runs (2, 8), (3, 4), (4, 7), and (5, 9, 11). There are four levels of x_1 and seven levels of x_2. □

8.2.3 The Use of Standard Designs in Generalized Linear Models

It is clear from the discussion in Section 8.2.2 that in general (quite apart from special applications) the use of design optimality requires either sequential experimental design or Bayesian design. These approaches may not be available to all experimenter. This suggests the question, What about classic standard designs that are used in conjunction with ordinary least squares estimation with linear models? Suppose that are consider the

Table 8.6 14-Run Bayesian D-Optimal Design From JMP for a Poisson Regression Model with a Full Quadratic in the Linear Predictor

Nonlinear Design

Parameters

Name	Role	Values	
b0	Continuous	1	3
b1	Continuous	0.25	0.75
b2	Continuous	-0.1	-0.3
b11	Continuous	0.45	1.35
b12	Continuous	-0.5	-1.5
b22	Continuous	-0.2	-0.6

Design

Run	x1	x2
1	0	0
2	1	1
3	1	-0.22794
4	1	-1
5	-1	1
6	-1	-1
7	1	-1
8	1	1
9	-1	1
10	0.495294	-1
11	-1	1
12	-1	-0.92313
13	-1	0.005825
14	1	-0.22688

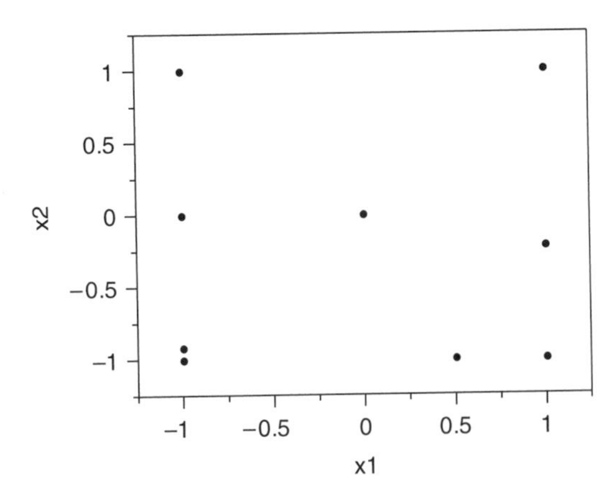

Figure 8.9 14-Run JMP Bayesian D-optimal design for the poisson regression model in Table 8.5.

two-level factorial or fractional factorial designs. As we indicated earlier, these designs were developed and intended for standard conditions, quite unlike the nonlinear structure in GLMs and the concomitant complication of nonhomogeneous variance. And yet the examples given in Chapters 4, 5, and 6 gave evidence that under certain circumstances they can be very efficient.

We know that the property of orthogonality discussed in Section 8.2.1 is the key that produces the *minimum variance* property possessed by these designs under standard conditions. In addition, this orthogonality property has its roots in the conditioning of the information matrix. In Section 8.2.2 we wrote the information matrix for the general GLM as

$$\mathbf{I}(\mathbf{b})=\mathbf{X}'\mathbf{W}\mathbf{X} \tag{8.1}$$

where $\mathbf{W} = \Delta\mathbf{V}\Delta$ is called the Hessian weight matrix, a diagonal matrix that contains weights $w_1, w_2, \ldots, w_n$, which are functions of parameters. This form of the information is discussed at length in Chapters 4 and 5. Suppose that we define

$$\mathbf{Z}=\mathbf{W}^{1/2}$$

where $\mathbf{W}^{1/2}$ is diagonal with $_i$th diagonal element $w_i^{1/2}$. As a result Equation (8.1) can be written

$$\mathbf{I}(\mathbf{b})=\mathbf{Z}'\mathbf{Z} \tag{8.2}$$

where $\mathbf{Z}$ is $n \times p$. Here we have the information matrix in a form not unlike its appearance for the case of a linear model with the exception that $\mathbf{Z}$, in general, contains parameters and thus cannot be controlled. Now consider the diagonal and off-diagonal elements of $\mathbf{Z}'\mathbf{Z}$, **particularly for the case of a two-level design** with levels at ± 1. The diagonal elements are equal and they all take on the value

$$\mathbf{I}(\mathbf{b})_{ii}=\sum_{j=1}^{N} w_j, \quad i=1,2,\ldots,p \tag{8.3}$$

where w_j is the jth Hessian weight, that is, the jth diagonal element of $\mathbf{W}$. The question is: Is the two-level factorial or fractional design orthogonal for the GLM situation? Orthogonality here implies that the columns of the $\mathbf{Z}$ matrix are mutually orthogonal. This information matrix a diagonal and renders the model coefficient estimators asymptotically independent.

8.2.4 Orthogonal Designs in GLM: The Variance-Stabilizing Link

It should be emphasized that while the information matrix in Equation (8.2) does not look particularly imposing, the values in the information matrix depend on parameters that are functions of the distribution involved as well as the chosen link function. In fact, given a particular distribution, the choice of certain link functions can certainly alter the properties of the designs through the information matrix.

One important type of link function corresponds to the function that is applied as a transformation to the raw response and stabilizes the variance when the transformed response is fit with a linear model. For example, for a Poisson distribution it is well known that a linear model that regresses $y^{1/2}$ against the regressors stabilizes the residual variance. In Chapter 5 we discuss the difference between transforming response and the use of generalized linear models. For the case of the former, where the response is Poisson, the homogeneous variance model is

$$E(y^{1/2}) = \mathbf{x}'\boldsymbol{\beta} \tag{8.4}$$

whereas the GLM model, with a **square root link** is

$$\mu^{1/2} = \mathbf{x}'\boldsymbol{\beta} \tag{8.5}$$

For the model in (8.4) the data are being transformed, whereas the GLM model of Equation (8.5) features a transformation on the population mean itself. For the Poisson distribution the square root transformation in (8.5) is the **variance-stabilizing transformation.** This is illustrated easily through the use of the Taylor series expansion of $y^{1/2}$ around $y = \mu_y$ through first order:

$$y^{1/2} \cong \mu_y^{1/2} + \left[\frac{\partial(y^{1/2})}{\partial y}\right]_{y=\mu_y} (y - \mu_y) = \mu_y^{1/2} + \tfrac{1}{2}(\mu_y^{-1/2})(y - \mu_y)$$

Thus a Taylor series approximation of $\text{Var}(y^{1/2})$ is Var(y)/[$4\mu_y$]. Consequently, when $\text{Var}(y) = \mu_y$ as in the Poisson distribution, we have

$$\text{Var}(y^{1/2}) \cong \tfrac{1}{4} \tag{8.6}$$

Thus the **variance-stabilizing link** for the Poisson distribution is the square root link. Similarly, it can be shown that the variance-stabilizing link for the exponential distribution is the log link. The log link is also the variance-stabilizing link for the gamma distribution.

The reader should recall Example 5.1 in Section 5.8 involving the gamma distribution where a two-level factorial design was used. The variances of the parameter estimators were all equal when the variance stabilizing link was used. It should be emphasized that the term **variance-stabilizing link** does not imply that the variance of the distribution of y is constant. The Poisson, exponential, and other distributions certainly have variances that are functions of the mean. The variance-stabilizing aspect of the terminology simply means that the same function applied to the data, meaning applied to y, results in approximate stabilization of variance.

It turns out that a two-level factorial or fractional factorial of resolution of at least III is **an orthogonal design when the variance-stabilizing link is used**. This can be proved without difficulty and the proof is instructive, so we provide it here. Recall the Hessian weight matrix of Equation (8.2), that is, $\mathbf{W} = \mathbf{\Delta V \Delta}$. Using the notation and terminology introduced in Chapter 4, we have that the matrix $\mathbf{\Delta}$ is a diagonal matrix with

$$w_{ii} = \frac{\partial \theta_i}{\partial(\mathbf{x}_i'\boldsymbol{\beta})}, \quad i = 1, 2, \ldots, n \tag{8.7}$$

and $\mathbf{V}$ is diagonal with diagonal elements

$$v_i = \text{Var}(y_i) = \frac{\partial \mu_i}{\partial \theta_i} \tag{8.8}$$

apart from the scale parameter $a(\phi)$. At this point we drop $a(\phi)$, since it only remains as a proportionality constant.

Consider the link function $g(\mu) = \mathbf{x}'\boldsymbol{\beta}$. If this transformation were on y, then the model $E[g(y)] = \mathbf{x}'\boldsymbol{\beta}$ would need to be considered. Expanding $g(y)$ in a Taylor series around $y = \mu_y$, we have that

$$g(y) \cong g(\mu_y) + [g'(\mu_y)][y - \mu]$$

where $g'(\mu) = (\partial g/\partial y)_{y=\mu_y}$. Therefore an approximate expression for $\text{Var}[g(y)]$ is $[g'(\mu_y)]^2[\text{Var}(y)]$. Thus if $g(y)$ is a variance-stabilizing function, $[g'(\mu_y)]^2$ is proportional to $1/\text{Var}(y)$ with the constant of proportionality being independent of the mean μ_y. This is the condition that must be satisfied in order that $g(\mu)$ is the variance-stabilizing link function. Let us assume this condition and consider the information matrix given by Equation (8.2). The design is orthogonal if

$$\mathbf{\Delta V \Delta} = \mathbf{W} = k\mathbf{I}$$

where k is a constant, that is, if all of the Hessian weights are equal. Now consider that the ith diagonal element w_{ii} of $\mathbf{W}$ can be written

$$w_{ii} = \delta_i^2 \text{Var}(y_i)$$

where $\delta_i = \partial\theta_i/\partial(\mathbf{x}_i'\boldsymbol{\beta})$. Thus $w_{ii} = (\partial\theta_i/\partial\mathbf{x}_i'\boldsymbol{\beta})(\partial\mu_{y_i}/\partial\theta_i)(\partial\theta_i/\partial(\mathbf{x}_i'\boldsymbol{\beta})$, which can be written using the chain rule

$$w_{ii} = \left(\frac{\partial\theta_i}{\partial(\mathbf{x}_i'\boldsymbol{\beta})}\right)\left(\frac{\partial\mu_{y_i}}{\partial(\mathbf{x}_i'\boldsymbol{\beta})}\right) \tag{8.9}$$

Now assume that the variance-stabilizing link is used. This implies that

$$(g'(\mu_{y_i}))\cdot[\mathrm{Var}(y_i)] = k^* \tag{8.10}$$

where k^* is some constant. Since $g(\mu) = \mathbf{x}'\boldsymbol{\beta}$,

$$\left(\frac{\partial\mathbf{x}_i'\boldsymbol{\beta}}{\partial\mu_{y_i}}\right)\left(\frac{\partial(\mathbf{x}_i'\boldsymbol{\beta})}{\partial\mu_{y_i}}\right)\left(\frac{\partial\mu_{y_i}}{\partial\theta_i}\right) = k^*$$

and

$$\left[\frac{\partial(\mathbf{x}_i'\boldsymbol{\beta})}{\partial\mu_{y_i}}\right]\cdot\left(\frac{\partial(\mathbf{x}_i'\boldsymbol{\beta})}{\partial\theta_i}\right) = k^* \tag{8.11}$$

Now in light of Equations (8.11) and (8.9), we have that since the proportionality constant k^* does not involve model parameters,

$$w_{ii} = \left[\frac{\partial\theta_i}{\partial(\mathbf{x}_i'\boldsymbol{\beta})}\right]\cdot\left[\frac{\partial\mu_{y_i}}{\partial(\mathbf{x}_i'\boldsymbol{\beta})}\right] = \left\{\left[\frac{\partial(\mathbf{x}_i'\boldsymbol{\beta})}{\partial\mu_{y_i}}\right]\cdot\left[\frac{\partial(\mathbf{x}_i'\boldsymbol{\beta})}{\partial\theta_i}\right]\right\}^{-1} = \frac{1}{k^*} = k \quad \text{for all } i \tag{8.12}$$

a scalar constant. Clearly, all Hessian weights are equal, and the information matrix in Equation (8.2) is of the form

$$\mathbf{Z}'\mathbf{Z} = \mathbf{X}'\mathbf{W}\mathbf{X} = k(\mathbf{X}'\mathbf{X}) \tag{8.13}$$

Thus $(\mathbf{Z}'\mathbf{Z})^{-1} = (1/k)(\mathbf{X}'\mathbf{X})^{-1}$. So in the case of the variance-stabilizing link, a two-level design that is orthogonal for the linear model with ± 1 scaling is orthogonal for the GLM model with the variance-stabilizing link. Here, $(\mathbf{X}'\mathbf{X}) = N\mathbf{I}$, where N is the total number of runs, and thus the asymptotic variance of model coefficients are the diagonal elements of $(\mathbf{Z}'\mathbf{Z})^{-1}$ which are

$$\mathrm{Var}(b_j) = \frac{1}{kN}, \quad j = 0, 1, 2, \ldots \tag{8.14}$$

where k is the common Hessian weight given by Equation (8.12). Here the covariance is asymptotically zero.

Example 8.6. The Poisson Case. Consider the Poisson case where a square root link is used. The common Hessian weight can be determined easily by Equation (8.12). For the square root link, $\mu_y^{1/2} = \mathbf{x}'\boldsymbol{\beta}$ and $\theta = \ln\ \mu_y$. Thus $\mu_y = e^{\theta}$ and $\mathbf{x}'\boldsymbol{\beta} = e^{\theta/2}$. So $\partial(\mathbf{x}'\boldsymbol{\beta})/\partial\mu_y = 1/2\mu_y^{-1/2}$. In addition, $\partial\mathbf{x}'\boldsymbol{\beta}/\partial\theta = l^{\theta/2}/2$. But $e^{\theta/2} = \mu_y^{1/2}$. The common Hessian weight from Equation (8.12) is $(1/4)^{-1} = 4$. So all coefficients have asymptotic variances that are diagonals of $(\mathbf{X}'\mathbf{X})^{-1} \cdot 1/4$. This result is extremely intuitive, since a square root transformation on y would produce $\text{Var}(y^{1/2}) = 1/4$ from Equation (8.6). In Section 5.8 an example that makes use of resistivity data was discussed where the gamma distribution was used with a log link (variance-stabilizing link). The design was orthogonal. In this case the variance–covariance matrix of coefficient estimators was given by

$$\text{Var}(\mathbf{b}) = (\mathbf{X}'\mathbf{X})^{-1}\left(\frac{1}{r}\right)$$

where r is the scale parameter for the gamma distribution. This same result can be found using Equations (8.2) and (8.13). □

8.2.5 Use of Other Links

The use of the variance-stabilizing link in conjunction with a two-level factorial or fraction with design resolution greater than III produces results that are very pleasing. The orthogonality is maintained in the GLM via the information matrix, yielding asymptotically independent estimates of coefficients and equal variances. But what if other link functions are required? Does there remain a kind of robustness under which the design properties remain pleasing? There is certainly considerable evidence that this is the case in many instances. The factorial structure alone is a commonsense approach to experimentation that has scientific and intuitive appeal, even if standard assumptions do not hold. Indeed, there are situations in which the efficiency of a two-level factorial or fractional factorial design is quite high without a variance-stabilizing link. The whole idea of efficiency becomes a bit more difficult to conceptualize in GLM scenarios. Let us review the situation with the variance-stabilizing link. For the case of the variance-stabilizing link the information matrix in Equation (8.1) contains a Hessian weight matrix $\mathbf{W}$ in which $\mathbf{W}$ diag$\{w_1, w_2, \ldots, w_n)$ contains equal weights down the main diagonal, and hence

$$\mathbf{I}(\mathbf{b}) = \mathbf{X}'\mathbf{W}\mathbf{X} = k(\mathbf{X}'\mathbf{X})$$

Thus it is clear that if $\mathbf{X}'\mathbf{X}$ is diagonal from choice of design, then all parameter estimate variances are equal and are uncorrelated. Suppose, however, that the link chosen is not the variance-stabilizing link. Now the w_i are not equal, and $\mathbf{I}(\mathbf{b}) = \mathbf{Z}'\mathbf{Z}$, where $\mathbf{Z} = \mathbf{W}^{1/2}\mathbf{X}$. This is best illustrated by an example.

Suppose that we have a 2^3 factorial in three variables x_1, x_2, and x_3 and a GLM given by

$$g(\mu) = \beta_0 + \beta_1 x_1 + \beta_2 x_2 + \beta_3 x_3$$

Then the **Z** matrix is given by

$$\mathbf{Z} = \begin{bmatrix} w_1^{1/2} & w_1^{1/2} x_{11} & w_1^{1/2} x_{12} & w_1^{1/2} x_{13} \\ w_2^{1/2} & w_1^{1/2} x_{21} & w_1^{1/2} x_{22} & w_1^{1/2} x_{23} \\ \vdots & & & \\ w_8^{1/2} & w_8^{1/2} x_{81} & w_8^{1/2} x_{82} & w_8^{1/2} x_{83} \end{bmatrix}$$

where x_{ij} is the ith level of the jth variable. The columns of the **X** matrix here are orthogonal, but the columns of **Z** are orthogonal only if the Hessian weights are equal. As we indicated earlier, the diagonal elements of $\mathbf{I(b)} = \mathbf{Z'Z}$ are $\sum_{i=1}^{8} w_i$. Now the off-diagonal elements are **contrasts in the w's**, where the contrast constants are products of 1's and -1's that are mixed in sign unless factors are confounded, which they are not. In fact, the form of the information matrix is

$$\mathbf{I(b)} = \begin{bmatrix} \sum w & \text{contr}A & \text{contr}B & \text{contr}C \\ & \sum w & \text{contr}AB & \text{contr}AC \\ & & \sum w & \text{contr}BC \\ & & & \sum w \end{bmatrix}$$

In the above, "contr" is simply a contrast or effect of A **on the Hessian weights;** a similar definition holds for contr B and contr C. Contr AB is the contrast or AB interaction effect on the Hessian weights, and similarly for contr AC and contr BC. As a result the information matrix is well conditioned if these so-called effects are near zero. The interaction effects appear even though there are no interactions in the model. Thus the conditioning of the information matrix is good if the diagonal elements dominate. This depends on the nature of the Hessian weights. In most instances one expects the sum of Hessian weights to dominate contrasts in the Hessian weights. If there is no interaction in the model, one would expect interaction contrasts in these weights to be relatively small. One must keep in mind though that the weights are functions of the mean, and that in the case of the canonical link, the weights are equal to the variances. Consider the case of a binomial distribution. In most applications the binomial variances are quite unlikely to vary a great deal across the experiment, and thus the efficiency of a standard two-level design is usually expected to be quite good.

As a hypothetical example, consider a 2^2 factorial with 10 samples drawn at each data point and the following estimated probabilities. Assume a logit link.

$A(x_1)$	$B(x_2)$	$\hat{p}$	Hessian Weights
−1	−1	0.440	2.464
−1	1	0.750	1.875
1	−1	0.250	1.875
1	1	0.490	2.499

As a result of the above, the information matrix $\mathbf{I} = \mathbf{X}'\mathbf{V}\mathbf{X}$ is

$$\mathbf{I}(\mathbf{b}) = (\mathbf{X}'\mathbf{V}\mathbf{X}) = \begin{bmatrix} 8.713 & 0.035 & 0.035 \\ 0.035 & 8.713 & 1.213 \\ 0.035 & 1.213 & 8.713 \end{bmatrix}$$

Note that the value contr AB is much larger than either contr A or contr B, even though there is no detectable interaction between the two factors as far as the effect on the binomial response $\hat{p}$ is concerned. The asymptotic variance–covariance matrix is

$$\mathbf{I}(\mathbf{b})^{-1} = \begin{bmatrix} 0.1148 & -0.0004 & -0.0004 \\ -0.0004 & 0.1170 & -0.0163 \\ -0.0004 & -0.0163 & 0.1170 \end{bmatrix}$$

From this result it is clear that even though a variance-stabilizing link is not used, the information matrix is well conditioned, and thus the design is efficient. Further verification comes from observing the correlation matrix or any number of methods for computing design efficiencies. The correlation matrix is

$$\text{CORR} = \begin{bmatrix} 1 & -0.00345 & -0.00345 \\ -0.00345 & 1 & -0.1393 \\ -0.00345 & -0.1393 & 1 \end{bmatrix}$$

One interesting measure of design efficiency is to compare the variances of individual coefficients (or perhaps the determinant of the information matrix) with that obtained if in fact the same Hessian weights were observed in an

Table 8.7 A 2^4 Design with 20 Bernoulli Observations at each Design Point

A	B	C	D	y
−1	−1	−1	−1	4
1	−1	−1	−1	10
−1	1	−1	−1	5
1	1	−1	−1	7
−1	−1	1	−1	8
1	−1	1	−1	6
−1	1	1	−1	8
1	1	1	−1	10
−1	−1	−1	1	4
1	−1	−1	1	7
−1	1	−1	1	5
1	1	−1	1	14
−1	−1	1	1	7
1	−1	1	1	9
−1	1	1	1	7
1	1	1	1	7

orthogonal design. This gives a sense of level of ill conditioning of the information matrix. In this hypothetical illustration, the coefficient variance for an orthogonal design would be $1/8.713 = 0.1148$. Clearly, the efficiencies are 99.97%, 97.85%, and 97.85%, respectively, for the intercept and two linear coefficients. As a result the lack of use of a variance-stabilizing link does not produce a poor design. This is particularly true in the case of binomial experiments, since binomial variances are often relatively stable, except in rare extreme cases.

Consider a second example for illustrative purposes. The design is a 2^4 factorial in which 20 Bernoulli observations at each design point are simulated. The logit link is used, and thus the design is not orthogonal. There is no proper variance-stabilizing link for this case. The data are given in Table 8.7.

After model editing the parameter estimates, standard errors, Wald chi-square values, and observational statistics are listed in Table 8.8. Note the relative stability of the Hessian weights. The sum of the Hessian weights is 70.6392. As a result the efficiencies of the parameter estimates are

$$\begin{aligned} A&: \ 97.6\% \\ C&: \ 97.5\% \\ D&: \ 98.8\% \\ AC&: \ 98\% \\ AD&: \ 98.8\% \end{aligned}$$

Table 8.8 GLM for Data in Table 8.7

Analysis of Parameter Estimates

Parameter	DF	Estimate	Std Err	Chi Square	Pr > Chi
INTERCEPT	1	-0.5687	0.1204	22.3187	0.0001
A	1	0.0351	0.1204	6.4213	0.0113
C	1	0.1019	0.1203	0.7167	0.3972
D	1	0.2075	0.1197	3.0036	0.0831
AC	1	-0.2611	0.1203	4.7078	0.0300
AD	1	0.2075	0.1197	3.0036	0.0631
Scale	0	1.000	0.0000	-	-

Observation Statistics

Y	N	Pred	Xbeta	Std	HessWgt	Lower	Upper
4	20	0.2260	-1.2368	0.3198	3.4875	0.1343	0.3521
10	20	0.5770	0.3105	0.2791	4.8814	0.4411	0.7021
5	20	0.2250	-1.2368	0.3188	3.4875	0.1343	0.3521
7	20	0.3730	-0.5194	0.2832	4.6774	0.2546	0.5089
8	20	0.3780	-0.5108	0.2897	4.6875	0.2538	0.5142
6	20	0.3020	-0.8378	0.2928	4.2159	0.1960	0.4344
8	20	0.3750	-0.5108	0.2897	4.6875	0.2538	0.5142
10	20	0.4980	-0.007889	0.2788	4.9999	0.3657	0.6305
4	20	0.2250	-1.2388	0.3198	3.4876	0.1343	0.8521
7	20	0.3730	-0.5194	0.2832	4.6774	0.2546	0.5088
5	20	0.3250	-1.2368	0.3198	3.4875	0.1343	0.3521
14	20	0.5770	0.3106	0.2791	4.8814	0.4411	0.7021
7	20	0.3750	-0.5108	0.2897	4.6875	0.2538	0.5142
9	20	0.4980	-0.007959	0.2768	4.9999	0.3657	0.6806
7	20	0.3760	-0.6108	0.2897	4.6875	0.2538	0.5142
7	20	0.3020	-0.8378	0.2928	4.2150	0.1960	0.4544

Thus for this 2^4 factorial the logistic regression model yields very nearly an orthogonal design. Note how stable the standard errors of the coefficients are.

Example 8.7. Spermatozoa Survival. This example deals with a spermatozoa survival study in a sperm bank. The endpoints in the experiments are *survive* or *not survive* according to ability to impregnate. The spermatozoa are stored in sodium citrate and glycerol, and the amounts of these substances were varied along with equilibration time in a factorial array. Fifty samples of material were used in each design point. The purpose of the experiment is to assess the effects of the factors on proportion survival. Table 8.9 gives the data. The analysis included a logistic regression model that contains all main effects and two-factor interactions.

Table 8.9 Survival Data for Example 8.7

x_1 (Sodium Citrate)	x_2 (Glycerol)	x_3 (Equilibrium Time)	y (Number Surviving)
−1	−1	−1	34
1	−1	−1	20
−1	1	−1	8
1	1	−1	21
−1	−1	1	30
1	−1	1	20
−1	1	1	10
1	1	1	25

SAS PROC LOGISTIC is first used to fit the full model (main effects and two-factor interactions):

$$\hat{\pi} = \frac{1}{1 + e - (\beta_0 + \beta_1 x_1 + \beta_2 x_2 + \beta_3 x_3 + \beta_{12} x_1 x_2 + \beta_{13} x_1 x_3 + \beta_{23} x_2 x_3)}$$

One may initially test the overall model, which involves the hypotheses

$$H_0 : \boldsymbol{\beta} = \mathbf{0} \quad H_1 : \boldsymbol{\beta} \neq \mathbf{0}$$

where $\boldsymbol{\beta}$ contains all six parameters, eliminating the intercept. This is analogous to doing an F-test for the entire model in standard linear models. Using likelihood inference in this situation, we have for the reduced model $\pi = 1/\left(1 + e^{-\beta_0}\right)$,

$$-2 \ln \mathscr{L} = 544.234$$

whereas for the full model containing seven parameters

$$-2 \ln \mathscr{L} = 496.055$$

The appropriate likelihood ratio test statistic is a chi-square statistic, which is

$$\chi^2 = -2 \ln\left[\frac{\mathscr{L}(\text{reduced})}{\mathscr{L}(\text{full})}\right] = 544.234 - 496.055 = 48.179$$

with $7 - 1 = 6$ degrees of freedom. The value of the test statistic is significant at a level $p < 0.0001$, and thus we conclude that at least one model term impacts the probability of survival. The MLEs with Wald inference showing practical χ^2 values are given in Table 8.10.

It would appear that β_0, β_2, and β_{12} are the important model terms. If the researcher feels strongly that the adopted model must obey hierarchy, then β_1, would also be included.

Table 8.10 Maximum Likelihood Estimates and Wald Inference on Individual Coefficients for Data of Example 8.7

Model Term	df	Parameter Estimate	Standard Error	Wald χ^2	P-Value
Intercept	1	−0.3770	0.1101	11.7198	0.0006
x_1	1	0.0933	0.1101	0.7175	0.3970
x_2	1	−0.4632	0.1101	17.7104	0.0001
x_3	1	0.0259	0.1092	0.0563	0.8124
x_1x_2	1	0.5851	0.1101	28.2603	0.0001
x_1x_3	1	0.0544	0.1093	0.2474	0.6189
x_2x_3	1	0.1122	0.1088	1.0624	0.3027

The logistic procedure in SAS provide a stepwise procedure that is based on likelihood inference. Model terms enter according to the amount that log-likelihood is increased, much like the decrease in error sum of squares that is used in standard linear models. Suppose that initially the model contains an intercept. The model term that is placed in the model in the presence of the intercept is that which increased the log-likelihood by the largest amount. The results are:

Step 1. Variable x_1x_2 enters

	Intercept Only	**Intercept and Covariates**	χ^2
$-2 \ln \mathscr{L}$	544.234	516.116	28.118 with 1 df ($p < 0.0001$)

Step 2. Variable x_2 enters

	Intercept Only	**Intercept and Covariates**	χ^2
$-2 \ln \mathscr{L}$	544.234	498.287	45.947 with 2 df ($p < 0.0001$)

No additional variables met the 0.05 significance level for entry into the model. As in the case of linear models, the significance level for entry can be changed. The final analysis with Wald inference is:

Variable	df	Parameter Estimate	Standard Error	χ^2	P-Value
Intercept	1	−0.3637	0.1081	11.3116	0.0008
x_2	1	−0.4505	0.1084	17.2677	0.0001
x_1x_2	1	0.5747	0.1086	27.9923	0.0001

Thus the fitted model for estimating probability of survival is

$$\hat{\pi} = \frac{1}{1 + e^{-(-0.3637 - 0.4505\, x_2 + 0.5747\, x_1 x_2)}}$$

The model reflects a main effect for the amount of glycerol and an interaction between glycerol and the amount of sodium citrate. From the fitted model above it is easily established that when the sodium citrate level is low, the amount of glycerol has a negative effect on probability of survival. By that we mean that a low level of glycerol produces the desired result ($\hat{P}$ close to 1.0). On the other hand, if the sodium citrate level is high, the amount of glycerol has very little effect. The most desirable result occurs when both x_1 and x_2 are simultaneously at the low level; whereas the worst case scenario in terms of survival occurs when sodium citrate is low and glycerol is high. One must keep in mind that as in any designed experiment the conclusions drawn here are very much a function of the ranges chosen in the experiment. In addition, standard items like the use of residuals and other diagnostic information are covered in Chapters 5 and 7 as we deal with generalizing to the exponential family of distributions (generalized linear models). However, more should be presented at this point on the use of odds ratios and what they mean in this example.

The odds ratios for the model terms x_2 and x_1x_2 as given in SAS PROC LOGISTIC are

Variable	Odds Ratio
x_2	$0.637 = e^{b_2}$
x_1x_2	$1.777 = e^{b_{12}}$

The odds ratio for x_2 is calculated as $e^{b_2} = e^{-0.4505} = 0.637$. This can be viewed as the analog to an effect in standard linear models with a designed experiment. A value of 1.0 obviously implies no effect. Note that in general the odds, $\pi/(1 - \pi)$, of survival is

$$\frac{\pi}{1 - \pi} = e^{\mathbf{x}'\boldsymbol{\beta}}$$

which is given by $e^{\beta_0 + \beta_2 x_2 + \beta_{12} x_1 x_2}$ this example. If the model contains no interaction, then the value e^{β_2} is the **ratio of the odds of survival** at $x_2 = +1$ to the **odds of survival** at $x_2 = 0$. Thus $e^{2\beta_2}$ represents the ratio of odds at $x_2 = +1$ to that at $x_2 = -1$. Obviously, this is the case no matter how many other first-order terms appear in the model. However, the existence of interaction in the model produces a slight complication in the interpretation of an effect. In the case of this example with interaction, the odds ratio $e^{b_2} = 0.637$ is the ratio of odds of survival at $x_2 = 1$ to the odds at $x_2 = 0$

Table 8.11 Statistical Output for Example 8.7

Term	Coefficient	Standard Error	Observation	Hessian Weights
Intercept	−0.3770	0.1101	1	11.0149
x_1	0.0933	0.1101	2	12.0737
x_2	−0.4632	0.1101	3	6.9775
x_3	0.059	0.1092	4	12.1158
$x_1\ x_2$	0.5851	0.1101	5	11.9205
$x_1\ x_3$	0.0544	0.1093	6	11.9205
$x_2\ x_3$	0.1122	0.1088	7	7.7673
			8	12.4971

when $x_1 = 0$. Note that since the odds ratio is much Owalla than 1.0, the implication is that a high value for x_2, the amount of glycerol, is undesirable. Obviously, other odds ratios can be demonstrated. For example, the ratio of odds of survival at $x_2 = +1$ to that of $x_2 = 0$ for $x_1 = -1$ is $e^{b_2 - b_{12}}$ or $e^{b_2}/e^{b_{12}}$, which in this case is $0.637/1.777 \cong 0.36$. This number illustrates how the effect (far removed from 1.0) of glycerol is so much greater in reducing odds of survival at the high level when $x_1 = -1$ than at $x_1 = 0$. On the other hand, if we let $x_1 = +1$, the odds ratio for x_2 is $e^{b_2 + b_{12}} = e^{0.125}$, which is close to 1.0. So, even if interactions reside in the model, the odds ratio values computed by SAS can be manipulated to find interpretable effects.

Now consider the efficiency of the 2^3 design for this experiment. Table 8.11 gives the coefficients and standard errors again along with the Hessian weights. The sum of the Hessian weights is 86.2873. Thus the ideal variance of each coefficient, as obtained with an orthogonal design, is $(86.2873)^{-1} = 0.011589$. Thus the efficiency of estimation for the intercept, for example, is

$$\frac{0.01159}{(0.1101)^2} = 0.956$$

or 95.6%. The remaining efficiencies are

$$\begin{aligned} x_1 &: \ 95.6\% \\ x_2 &: \ 95.6\% \\ x_3 &: \ 97.2\% \\ x_1x_2 &: \ 95.6\% \\ x_1x_3 &: \ 97.1\% \\ x_2x_3 &: \ 97.9\% \end{aligned}$$

Thus in this example, as is often the case for logistic regression fit to data with a standard two-level design, the efficiencies when compared to that of an orthogonal design, are quite good. □

8.2.6 Further Comments Concerning the Nature of the Design

In most real-life situation a two-level factorial or fractional factorial design is at least reasonably efficient even if the variance-stabilizing link is not used. In fact many designs are very nearly orthogonal. The variances of coefficients are also very nearly equal in many cases, with the largest variation belonging to the variance of the estimated intercept term. It is interesting to see the patterns that are present in the information matrix and thus the variance–covariance matrix when interactions are present. For example, consider the survival data for Example 8.7, where the logit link was used with a binomial response. Initially, the logistic regression model contained the terms x_1, x_2, x_3, x_1x_2 and x_2x_3 in addition to the constant term in the linear predictor. Here, the off-diagonal elements in the information matrix proved very interesting. The contrasts in the Hessian weights that appeared are the *generalized interaction contrasts*. The generalized interactions are interactions between the terms that are present in the margins that jointly define the cell in the matrix. To illustrate, we have

$$\mathbf{I}(\mathbf{b}) = \begin{array}{c} \\ \left[\begin{array}{ccccccc} & x_1 & x_2 & x_3 & x_1x_2 & x_1x_3 & x_2x_3 \\ \sum w & \text{contr}x_1 & \text{contr}x_2 & \text{contr}x_3 & \text{contr}x_1x_2 & \text{contr}x_1x_3 & \text{contr}x_2x_3 \\ & \sum w & \text{contr}x_1x_2 & \text{contr}x_1x_3 & \text{contr}x_2 & \text{contr}x_3 & \text{contr}x_1x_2x_3 \\ & & \sum w & \text{contr}x_2x_3 & \text{contr}x_1 & \text{contr}x_1x_2x_3 & \text{contr}x_3 \\ & & & \sum w & \text{contr}x_1x_2x_3 & \text{contr}x_1 & \text{contr}x_2 \\ & & & & \sum w & \text{contr}x_2x_3 & \text{contr}x_1x_3 \\ & & & & & \sum w & \text{contr}x_1x_2 \\ \text{symm} & & & & & & \sum w \end{array}\right] \end{array}$$

Note how often there are repeat values in the information matrix. We saw this in the previous examples, and the same thing appears in an example later in this section. The existence of these off-diagonal terms does not imply that the designs in question are inefficient. The contrasts in the Hessian weights that appear on the off-diagonal elements often are small relative to the diagonal elements. In our experience, the highest efficiencies accompany examples with the binomial distribution, and lower efficiencies occur in the Poisson distribution with the log link when there are zero counts and/or extremely large effects.

Here, effects on the Hessian weights are effects on the mean itself, since the mean is equal to the variance. Ironically, large effects bring lack of efficiency here, but pragmatically, the latter may not be a problem because design efficiency is not needed to detect such large effects.

The efficiencies that we describe here in our examples in Section 8.2.5 are based on *fixed diagonal elements*. One must keep in mind that these important diagonal elements of the information matrix, namely, the sum of the Hessian weights, are very much a function of both the design and the link function. This relationship underscores again the crucial difference between controlling design performance for linear models and for nonlinear models. For linear models, the diagonal elements of $\mathbf{X}'\mathbf{X}$ are controlled by design scaling (± 1), and the designs, if orthogonal, are optimal conditioned on the ranges of the design variables. However, in the GLM situation one cannot control the diagonal elements of $\mathbf{I}(\mathbf{b}) = \mathbf{X}'\mathbf{W}\mathbf{X}$, since they depend heavily on unknown parameters. As a result, in the absence of guessing parameters and laboriously computing optimal designs, we are evaluating efficiencies based on a comparison of a given design (after data are collected) against one that exhibits the same Hessian weights but is orthogonal. In that way, the efficiency is an evaluation of the conditioning of the information matrix.

8.3 GLM ANALYSIS OF SCREENING EXPERIMENTS

The 2^k factorial and 2^{k-p} fractional factorial design are widely used as **screening experiments.** In such experiments the objective is to identify the subset of the experimental factors that has the greatest effect on the response. Usually the sparsity of effects principle prevails; that is, only a relatively small subset of the factors and their interactions is active. Montgomery (2009) describes the construction and analysis of these designs in detail.

A typical approach to the analysis of the 2^k (2^{k-p}) design in a screening experiment is to plot the effect estimates (or model coefficient estimates $\mathbf{b}$) on a normal probability plot. This approach is very effective because the estimates have equal variance and are uncorrelated. In the GLM, model coefficients are not generally uncorrelated and may not have equal variances. However, a normal probability plot of coefficient estimates divided by their standard errors often provides useful guidance in the selection of active factors, unless the correlations between the estimates is large. As noted in Section 8.2, use of the **variance-stabilizing links** results in uncorrelated parameter estimates. Thus the normal probability plot of these parameter estimates is very useful in identifying active effects.

Example 8.8. The Drill Experiment. The drill experiment was originally described by Daniel (1976). Four factors (x_i = drill load, x_2 = flow, x_3 = speed,

Table 8.12 Drill Data Design Matrix and Response Data for Example 8.8

	x_1	x_2	x_3	x_4	Advance Rate (Original Data)
1	−	−	−	−	1.68
2	+	−	−	−	1.98
3	−	+	−	−	3.28
4	+	+	−	−	3.44
5	−	−	+	−	4.98
6	+	−	+	−	5.70
7	−	+	+	−	9.97
8	+	+	+	−	9.07
9	−	−	−	+	2.07
10	+	−	−	+	2.44
11	−	+	−	+	4.09
12	+	+	−	+	4.53
13	−	−	+	+	7.77
14	+	−	+	+	9.43
15	−	+	+	+	11.75
16	+	+	+	+	16.30

x_4 = type of mud) were studied, and the response variable was the drill advance rate (y). The experiment, shown in Table 8.12, is a 2^4 factorial. Many authors have analyzed these data, including Box, Hunter, and Hunter (2005) and Montgomery (2009), who use a log transformation and find that three of the four mean effects (x_2, x_3, and x_4) are active.

We analyze the drill data using a GLM with a gamma distribution and the log link. Since this is the variance-stabilizing link, the model coefficients are uncorrelated with constant variance. This is confirmed by examining the covariance matrix in Table 8.13 and the correlation form of the covariance matrix in Table 8.14. The half-normal plot shown in Figure 8.10 indicates that factors x_2, x_3, and x_4 are clearly active. This conclusion conforms to previous work. Interpretation of this plot is straightforward since the parameter estimates are uncorrelated. □

Example 8.9. The Grille Defects Experiment. This example is taken from Bisgaard and Fuller (1994–1995). The experiment involves a 16 run two-level fractional factorial design in 9 factors. The purpose of this experiment is to screen out the insignificant factors. The generators for this 2^{9-5} resolution III design are $E = BD$, $F = BCD$, $G = AC$, $H = ACD$, and $J = AB$. The experimental responses in this case are counts of defects per grille, which is usually assumed to be a Poisson distributed variable. The design matrix, the count of defects, $\hat{c}$, and the aliasing pattern are shown in Table 8.15.

Bisgaard and Fuller analyzed these data by taking the square root of the counts and also by using the Freeman and Tukey modified transformation $(\sqrt{\hat{c}} + \sqrt{\hat{c}+1})/2$. Myers and Montgomery (1997) reanalyzed these data using the

Table 8.13 Covariance Matrix of Parameter Estimates for the Drill Experiment, Gamma Response, and Log Link

	b_0	b_1	b_2	b_3	b_4	b_{12}	b_{13}	b_{14}	b_{23}	b_{24}	b_{34}	b_{123}	b_{124}	b_{134}	b_{234}
b_0	3.73E	0.00E	0.00E	0.00E	0.00E	0.00E	0.00E	0.00E	0.00E	0.00E	0.00E	0.00E	0.00E	0.00E	0.00E
	−04	+00	+00	+00	+00	+00	+00	+00	+00	+00	+00	+00	+00	+00	+00
b_1	0.00E	3.73E	−1.03E	2.07E	2.07E	2.07E	2.07E	−6.21E	1.11E	−1.40E	1.03E	−2.07E	−2.07E	−2.07E	2.07E
	+00	−04	−20	−20	−20	−20	−20	−20	−35	−51	−20	−20	−20	−20	−20
b_2	0.00E	−1.03E	3.73E	1.03E	1.03E	2.07E	−1.03E	3.10E	−2.07E	2.07E	−2.07E	−2.07E	−2.07E	2.07E	2.07E
	+00	−20	−04	−20	−20	−20	−20	−20	−20	−20	−20	−20	−20	−20	−20
b_3	0.00E	2.07E	1.03E	3.73E	2.07E	−2.07E	4.59E	−2.07E	2.07E	1.46E	−5.69E	8.28E	4.14E	−6.32E	−2.58
	+00	−20	−20	−04	−20	−20	−36	−20	−20	−35	−20	−20	−20	−36	−20
b_4	0.00E	2.07E	1.03E	2.07E	3.73E	−8.11E	1.44E	4.14E	−2.07E	−5.17E	−1.03E	2.07E	−3.10E	−9.14E	−2.07E
	+00	−20	−20	−20	−04	−36	−36	−20	−20	−21	−20	−20	−20	−36	−20
b_{12}	0.00E	2.07E	2.07E	−2.07E	−8.11E	3.73E	−4.14E	2.07E	4.14E	2.59E	2.07E	−6.21E	3.10E	1.29E	2.07E
	+00	−20	−20	−20	−36	−04	−20	−20	−20	−20	−20	−20	−20	−35	−20
b_{13}	0.00E	2.07E	−1.03E	4.59E	1.44E	−4.14E	3.73E	3.10E	2.07E	−2.07E	4.14E	2.07E	4.14E	−2.07E	−4.14E
	+00	−20	−20	−36	−36	−20	−04	−20	−20	−20	−20	−20	−20	−20	−20
b_{14}	0.00E	−6.21E	3.10E	−2.07E	4.14E	2.07E	3.10E	3.73E	−2.07E	1.44E	−8.28E	1.42E	2.55E	−1.03E	5.69E
	+00	−20	−20	−20	−20	−20	−20	−04	−20	−36	−20	−35	−35	−20	−20
b_{23}	0.00E	1.11E	−2.07E	2.07E	−2.07E	4.14E	2.07E	−2.07E	3.73E	3.10E	1.03E	−5.95E	−2.34E	4.16E	−5.74E
	+00	−35	−20	−20	−20	−20	−20	−20	−04	−20	−20	−20	−35	−36	−37
b_{24}	0.00E	−1.40E	2.07E	1.46E	−5.17E	2.59E	−2.07E	1.44E	3.10E	3.73E	−2.07E	8.28E	−6.21E	−6.21E	−2.07E
	+00	−51	−20	−35	−21	−20	−20	−36	−20	−04	−20	−20	−20	−20	−20
b_{34}	0.00E	1.03E	−2.07E	−5.69E	−1.03E	2.07E	4.14E	−8.28E	1.03E	−2.07E	3.73E	−4.14E	−6.21E	−2.07E	4.14E
	+00	−20	−20	−20	−20	−20	−20	−20	−20	−20	−04	−20	−20	−20	−20
b_{123}	0.00E	−2.07E	−2.07E	8.28E	2.07E	−6.21E	2.07E	1.42E	−5.95E	8.28E	−4.14E	3.73E	1.03E	−6.21E	1.26E
	+00	−20	−20	−20	−20	−20	−20	−35	−20	−20	−20	−04	−19	−20	−35
b_{124}	0.00E	−2.07E	−2.07E	4.14E	−3.10E	3.10E	4.14E	2.55E	−2.34E	−6.21E	−6.21E	1.03E	3.73E	3.36E	8.28E
	+00	−20	−20	−20	−20	−20	−20	−35	−35	−20	−20	−19	−04	−20	−20
b_{134}	0.00E	−2.07E	2.07E	−6.32E	−9.41E	1.29E	−2.7E	−1.03E	4.16E	−6.21E	−2.07E	−6.31E	3.36E	3.73E	6.21E
	+00	−20	−20	−36	−36	−35	−20	−20	−36	−20	−20	−20	−20	−04	−20
b_{234}	−0.00E	2.07E	2.07E	−2.58E	−2.07E	2.07E	−4.14E	5.69E	−5.74E	−2.07E	4.14E	1.26E	8.28E	6.21E	3.73E
	+00	−20	−20	−36	−20	−20	−20	−20	−37	−20	−20	−35	−20	−20	−04

Table 8.14 Correlation Matrix of Parameter Estimates for the Drill Experiment, Gamma Response, and Log Link

	b_0	b_1	b_2	b_3	b_4	b_{12}	b_{13}	b_{14}	b_{23}	b_{24}	b_{34}	b_{123}	b_{124}	b_{134}	b_{234}
b_0	1.00	0.00	0.00	0.00	0.00	0.00	0.00	0.00	0.00	0.00	0.00	0.00	0.00	0.00	0.00
b_1	0.00	1.00	0.00	0.00	0.00	0.00	0.00	0.00	0.00	0.00	0.00	0.00	0.00	0.00	0.00
b_2	0.00	0.00	1.00	0.00	0.00	0.00	0.00	0.00	0.00	0.00	0.00	0.00	0.00	0.00	0.00
b_3	0.00	0.00	0.00	1.00	0.00	0.00	0.00	0.00	0.00	0.00	0.00	0.00	0.00	0.00	0.00
b_4	0.00	0.00	0.00	0.00	1.00	0.00	0.00	0.00	0.00	0.00	0.00	0.00	0.00	0.00	0.00
b_{12}	0.00	0.00	0.00	0.00	0.00	0.00	0.00	1.00	0.00	0.00	0.00	0.00	0.00	0.00	0.00
b_{13}	0.00	0.00	0.00	0.00	0.00	0.00	1.00	0.00	0.00	0.00	0.00	0.00	0.00	0.00	0.00
b_{14}	0.00	0.00	0.00	0.00	0.00	0.00	0.00	1.00	0.00	0.00	0.00	0.00	0.00	0.00	0.00
b_{23}	0.00	0.00	0.00	0.00	0.00	0.00	0.00	0.00	1.00	0.00	0.00	0.00	0.00	0.00	0.00
b_{24}	0.00	0.00	0.00	0.00	0.00	0.00	0.00	0.00	0.00	1.00	0.00	0.00	0.00	0.00	0.00
b_{34}	0.00	0.00	0.00	0.00	0.00	0.00	0.00	0.00	0.00	0.00	1.00	0.00	0.00	0.00	0.00
b_{123}	0.00	0.00	0.00	0.00	0.00	0.00	0.00	0.00	0.00	0.00	0.00	1.00	0.00	0.00	0.00
b_{234}	0.00	0.00	0.00	0.00	0.00	0.00	0.00	0.00	0.00	0.00	0.00	0.00	1.00	0.00	0.00
b_{234}	0.00	0.00	0.00	0.00	0.00	0.00	0.00	0.00	0.00	0.00	0.00	0.00	0.00	1.00	0.00
b_{234}	0.00	0.00	0.00	0.00	0.00	0.00	0.00	0.00	0.00	0.00	0.00	0.00	0.00	0.00	1.00

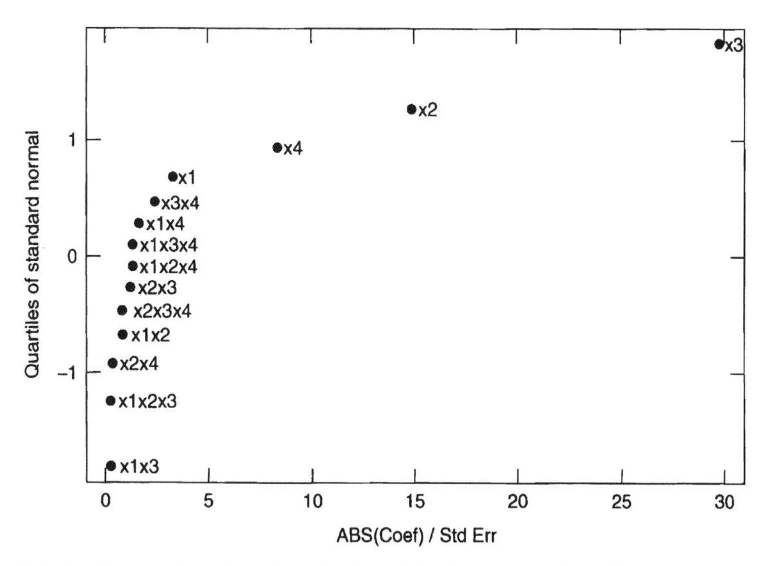

Figure 8.10 Half-normal probability plot for drill data using the GLM with the gamma distribution and log link.

Table 8.15 Design Matrix, the Responses, and Aliasing Pattern for Car Grille Data

Observations	A	B	C	D	E	F	G	H	J	$\hat{c}$	$\sqrt{\hat{c}}$	Freeman–Tukey Modification
1	−1	−1	−1	−1	1	−1	1	−1	1	56	7.48	7.52
2	1	−1	−1	−1	1	1	1	−1	−1	17	4.12	4.18
3	−1	1	−1	−1	1	−1	−1	1	−1	2	1.41	1.57
4	1	1	−1	−1	1	−1	−1	−1	1	4	2.00	2.12
5	−1	−1	1	−1	−1	1	−1	−1	1	3	1.73	1.87
6	1	−1	1	−1	−1	1	−1	1	−1	4	2.00	2.12
7	−1	1	1	−1	−1	−1	1	−1	−1	50	7.07	7.11
8	1	1	1	−1	−1	−1	1	1	1	2	1.41	1.57
9	−1	−1	−1	1	−1	−1	1	1	1	1	1.00	1.21
10	1	−1	−1	1	−1	−1	1	−1	−1	0	0.00	0.50
11	−1	1	−1	1	−1	1	−1	1	−1	3	1.73	1.87
12	1	1	−1	1	−1	1	−1	−1	1	12	3.46	3.53
13	−1	−1	1	1	1	−1	−1	−1	1	3	1.73	1.87
14	1	−1	1	1	1	−1	−1	1	−1	4	2.00	2.12
15	−1	1	1	1	1	1	1	−1	−1	0	0.00	0.50
16	1	1	1	1	1	1	1	1	1	0	0.00	0.50

Note:

$l_1 = A+BJ+CG$ $\quad l_9 = AD+CH+EJ$

$l_2 = B+AJ+DE$ $\quad l_{10} = E+BD+CF$

$l_3 = J+AB+FH$ $\quad l_{11} = CD+AH+BF$

$l_4 = C+AG+EF$ $\quad l_{12} = JD+AE+FG$

$l_5 = G+AC+DH$ $\quad l_{13} = H+DG+FJ$

$l_6 = BC+DF+GJ$ $\quad l_{14} = F+CE+HJ$

$l_7 = BG+CJ+EH$ $\quad l_{15} = AF+BH+EG$

$l_8 = D+BE+GH$

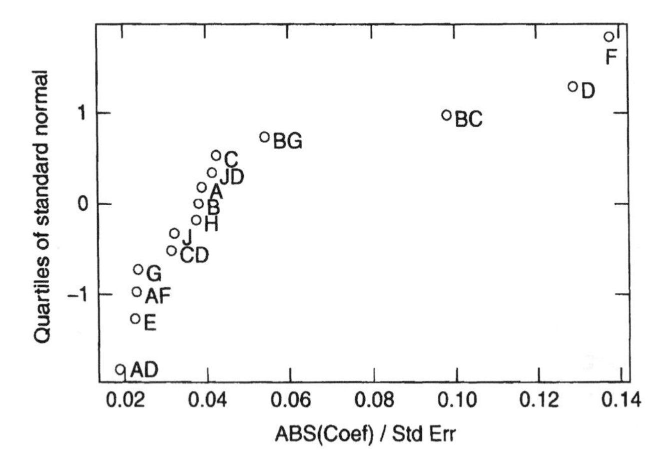

Figure 8.11 Half-normal probability plot for grille defects experiment using GLM with a Poisson response distribution and log link.

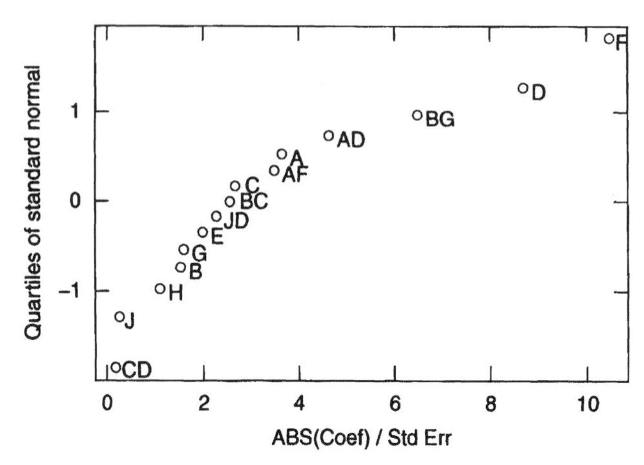

Figure 8.12 Half-normal probability plot for grille defects experiment using GLM with a Poisson response distribution and square root link.

generalized linear model with the Poisson distribution and the log link. They used the analysis of deviance to determine the significant factors.

In this example, the car grille data are again subjected to a GLM analysis with both log and square root (variance-stabilizing) links, and the half-normal probability plot is used to identify the active factors. The half-normal plot using the generalized linear model on Poisson responses with log and square root (variance-stabilizing) links are shown in Figures 8.11 and 8.12. The covariance

Table 8.16 Covariance Matrix Using GLM with Poisson Distribution and Square Root Link for the Grille Defects Example

	Intercept	*A*	*B*	*C*	*D*	*E*	*F*	*G*	*H*	*J*	*BC*	*BG*	*AD*	*CD*	*JD*	*AF*
Intercept	0.02	0.00	0.00	0.00	0.00	0.00	0.00	0.00	0.00	0.00	0.00	0.00	0.00	0.00	0.00	0.00
A	0.00	0.02	0.00	0.00	0.00	0.00	0.00	0.00	0.00	0.00	0.00	0.00	0.00	0.00	0.00	0.00
B	0.00	0.00	0.02	0.00	0.00	0.00	0.00	0.00	0.00	0.00	0.00	0.00	0.00	0.00	0.00	0.00
C	0.00	0.00	0.00	0.02	0.00	0.00	0.00	0.00	0.00	0.00	0.00	0.00	0.00	0.00	0.00	0.00
D	0.00	0.00	0.00	0.00	0.02	0.00	0.00	0.00	0.00	0.00	0.00	0.00	0.00	0.00	0.00	0.00
E	0.00	0.00	0.00	0.00	0.00	0.02	0.00	0.00	0.00	0.00	0.00	0.00	0.00	0.00	0.00	0.00
F	0.00	0.00	0.00	0.00	0.00	0.00	0.02	0.00	0.00	0.00	0.00	0.00	0.00	0.00	0.00	0.00
G	0.00	0.00	0.00	0.00	0.00	0.00	0.00	0.02	0.00	0.00	0.00	0.00	0.00	0.00	0.00	0.00
H	0.00	0.00	0.00	0.00	0.00	0.00	0.00	0.00	0.02	0.00	0.00	0.00	0.00	0.00	0.00	0.00
J	0.00	0.00	0.00	0.00	0.00	0.00	0.00	0.00	0.00	0.02	0.00	0.00	0.00	0.00	0.00	0.00
BC	0.00	0.00	0.00	0.00	0.00	0.00	0.00	0.00	0.00	0.00	0.02	0.00	0.00	0.00	0.00	0.00
BG	0.00	0.00	0.00	0.00	0.00	0.00	0.00	0.00	0.00	0.00	0.00	0.02	0.00	0.00	0.00	0.00
AD	0.00	0.00	0.00	0.00	0.00	0.00	0.00	0.00	0.00	0.00	0.00	0.00	0.02	0.00	0.00	0.00
CD	0.00	0.00	0.00	0.00	0.00	0.00	0.00	0.00	0.00	0.00	0.00	0.00	0.00	0.02	0.00	0.00
JD	0.00	0.00	0.00	0.00	0.00	0.00	0.00	0.00	0.00	0.00	0.00	0.00	0.00	0.00	0.02	0.00
AF	0.00	0.00	0.00	0.00	0.00	0.00	0.00	0.00	0.00	0.00	0.00	0.00	0.00	0.00	0.00	0.02

Table 8.17 Correlation Matrix Using GLM with Poisson Distribution and Square Root Link for the Grille Defects Example

	Intercept	*A*	*B*	*C*	*D*	*E*	*F*	*G*	*H*	*J*	*BC*	*BG*	*AD*	*CD*	*JD*	*AF*
Intercept	1.00	0.00	0.00	0.00	0.00	0.00	0.00	0.00	0.00	0.00	0.00	0.00	0.00	0.00	0.00	0.00
A	0.00	1.00	0.00	0.00	0.00	0.00	0.00	0.00	0.00	0.00	0.00	0.00	0.00	0.00	0.00	0.00
B	0.00	0.00	1.00	0.00	0.00	0.00	0.00	0.00	0.00	0.00	0.00	0.00	0.00	0.00	0.00	0.00
C	0.00	0.00	0.00	1.00	0.00	0.00	0.00	0.00	0.00	0.00	0.00	0.00	0.00	0.00	0.00	0.00
D	0.00	0.00	0.00	0.00	1.00	0.00	0.00	0.00	0.00	0.00	0.00	0.00	0.00	0.00	0.00	0.00
E	0.00	0.00	0.00	0.00	0.00	1.00	0.00	0.00	0.00	0.00	0.00	0.00	0.00	0.00	0.00	0.00
F	0.00	0.00	0.00	0.00	0.00	0.00	1.00	0.00	0.00	0.00	0.00	0.00	0.00	0.00	0.00	0.00
G	0.00	0.00	0.00	0.00	0.00	0.00	0.00	1.00	0.00	0.00	0.00	0.00	0.00	0.00	0.00	0.00
H	0.00	0.00	0.00	0.00	0.00	0.00	0.00	0.00	1.00	0.00	0.00	0.00	0.00	0.00	0.00	0.00
J	0.00	0.00	0.00	0.00	0.00	0.00	0.00	0.00	0.00	1.00	0.00	0.00	0.00	0.00	0.00	0.00
BC	0.00	0.00	0.00	0.00	0.00	0.00	0.00	0.00	0.00	0.00	1.00	0.00	0.00	0.00	0.00	0.00
BG	0.00	0.00	0.00	0.00	0.00	0.00	0.00	0.00	0.00	0.00	0.00	1.00	0.00	0.00	0.00	0.00
AD	0.00	0.00	0.00	0.00	0.00	0.00	0.00	0.00	0.00	0.00	0.00	0.00	1.00	0.00	0.00	0.00
CD	0.00	0.00	0.00	0.00	0.00	0.00	0.00	0.00	0.00	0.00	0.00	0.00	0.00	1.00	0.00	0.00
JD	0.00	0.00	0.00	0.00	0.00	0.00	0.00	0.00	0.00	0.00	0.00	0.00	0.00	0.00	1.00	0.00
AF	0.00	0.00	0.00	0.00	0.00	0.00	0.00	0.00	0.00	0.00	0.00	0.00	0.00	0.00	0.00	1.00

Table 8.18 Covariance Matrix Using GLM with Poisson Distribution and Log Link for the Grille Defects Example

	Intercept	*A*	*B*	*C*	*D*	*E*	*F*	*G*	*H*	*J*	*BC*	*BG*	*AD*	*CD*	*JD*	*AF*
Intercept	472.78	157.58	157.59	157.59	472.77	157.58	472.77	−157.58	−157.59	−157.59	472.77	157.59	157.58	157.59	−157.59	157.58
A	157.58	472.78	−157.59	−57.58	157.58	−157.59	157.58	157.59	157.59	157.59	157.59	472.77	472.77	−157.59	−157.58	472.77
B	157.59	157.59	472.78	472.77	157.58	472.77	157.59	157.59	157.58	157.58	157.59	−157.58	−157.59	472.77	157.58	−157.59
C	157.59	−157.58	472.77	472.78	157.59	472.77	157.58	157.58	157.58	157.59	157.59	−157.59	−157.59	472.77	157.58	−157.59
D	472.77	157.58	157.58	157.59	472.78	157.59	472.77	−157.59	−157.58	−157.59	472.77	157.58	157.58	157.59	−157.59	157.59
E	157.58	−157.59	472.77	472.77	157.59	472.78	157.59	157.58	157.59	157.58	157.59	−157.59	−157.59	472.77	157.58	−157.58
F	472.77	157.58	157.59	157.58	472.77	157.59	472.78	−157.59	−157.59	−157.59	472.77	157.58	157.59	157.59	−157.58	157.58
G	−157.58	157.59	157.59	157.58	−157.59	157.58	−157.59	472.78	472.77	472.77	−157.59	157.59	157.59	157.58	472.77	157.58
H	−157.59	157.59	157.58	157.58	−157.58	157.59	−157.59	472.77	472.78	472.77	−157.59	157.58	157.59	157.58	472.77	157.59
J	−157.59	157.59	157.58	157.59	−157.59	157.58	−157.59	472.77	472.77	472.78	−157.58	157.59	157.58	157.58	472.77	157.59
BC	472.77	157.59	157.59	157.59	472.77	157.59	472.77	−157.59	−157.59	−157.58	472.78	157.58	157.58	157.58	−157.59	157.58
BG	157.59	472.77	−157.58	−157.59	−157.58	−157.59	157.58	157.59	157.58	157.59	157.58	472.78	472.77	−157.59	157.59	472.44
AD	157.58	472.77	−157.59	−157.59	157.58	−157.59	157.59	157.59	157.59	157.58	157.58	472.77	472.78	−157.58	157.59	472.77
CD	157.59	−157.59	472.77	472.77	157.59	472.77	157.59	157.58	157.58	157.58	157.58	−157.59	−157.58	472.78	157.59	−157.59
JD	−157.59	157.58	157.58	157.58	−157.59	157.58	−157.58	472.77	472.77	472.77	−157.59	157.59	157.59	157.59	472.78	157.59
AF	157.58	472.77	−157.59	−157.59	157.59	−157.58	157.58	157.58	157.59	157.59	157.58	472.77	472.77	−157.59	157.59	472.78

Table 8.19 Correlation Matrix Using GLM with Poisson Distribution and Log Link for the Grille Defects Example

	Intercept	*A*	*B*	*C*	*D*	*E*	*F*	*G*	*H*	*J*	*BC*	*BG*	*AD*	*CD*	*JD*	*AF*
Intercept	1.00	0.33	0.33	0.33	1.00	0.33	1.00	−0.33	−0.33	−0.33	1.00	0.33	0.33	0.33	−0.33	0.33
A	0.33	1.00	−0.33	−0.33	0.33	−0.33	0.33	0.33	0.33	0.33	0.33	1.00	1.00	−0.33	0.33	1.00
B	0.33	−0.33	1.00	1.00	0.33	1.00	0.33	0.33	0.33	0.33	0.33	−0.33	−0.33	1.00	0.33	−0.33
C	0.33	−0.33	1.00	1.00	0.33	1.00	0.33	0.33	0.33	0.33	0.33	−0.33	0.33	1.00	0.33	−0.33
D	1.00	0.33	0.33	0.33	1.00	0.33	1.00	−0.33	−0.33	−0.33	1.00	0.33	0.33	0.33	−0.33	0.33
E	0.33	−0.33	1.00	1.00	0.33	1.00	0.33	0.33	0.33	0.33	0.33	−0.33	−0.33	1.00	0.33	−0.33
F	1.00	0.33	0.33	0.33	1.00	0.33	1.00	−0.33	−0.33	−0.33	1.00	0.33	0.33	0.33	−0.33	0.33
G	−0.33	0.33	0.33	0.33	−0.33	0.33	−0.33	1.00	1.00	1.00	−0.33	0.33	0.33	0.33	1.00	0.33
H	−0.33	0.33	0.33	0.33	−0.33	0.33	−0.33	1.00	1.00	1.00	−0.33	0.33	0.33	0.33	1.00	0.33
J	−0.33	0.33	0.33	0.33	−0.33	0.33	−0.33	1.00	1.00	1.00	−0.33	0.33	0.33	0.33	1.00	0.33
BC	1.00	0.33	0.33	0.33	1.00	0.33	1.00	−0.33	−0.33	−0.33	1.00	0.33	0.33	0.33	−0.33	0.33
BG	0.33	1.00	−0.33	−0.33	0.33	−0.33	0.33	0.33	0.33	0.33	0.33	1.00	1.00	−0.33	0.33	1.00
AD	0.33	1.00	−0.33	−0.33	0.33	−0.33	0.33	0.33	0.33	0.33	0.33	1.00	1.00	−0.33	0.33	1.00
CD	0.33	−0.33	1.00	1.00	0.33	1.00	0.33	0.33	0.33	0.33	0.33	−0.33	−0.33	1.00	0.33	−0.33
JD	−0.33	0.33	0.33	0.33	−0.33	0.33	−0.33	1.00	1.00	1.00	−0.33	0.33	0.33	0.33	1.00	0.33
AF	0.33	1.00	−0.33	−0.33	0.33	−0.33	0.33	0.33	0.33	0.33	0.33	1.00	1.00	−0.33	0.33	1.00

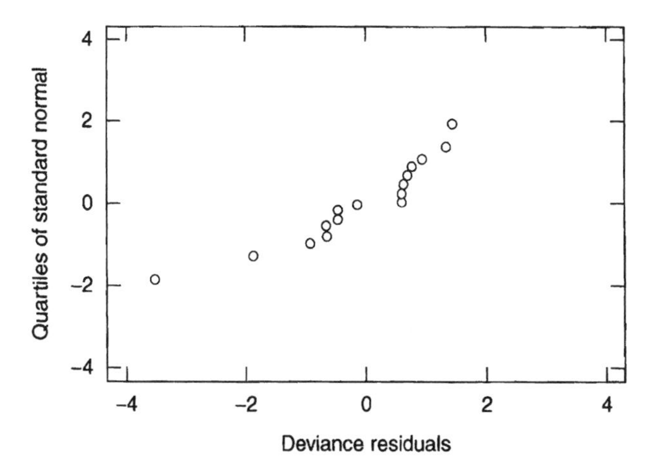

Figure 8.13 Normal probability plot for deviance residuals using GLM with factors *D*, *F*, *BC*, and *BG* and log link.

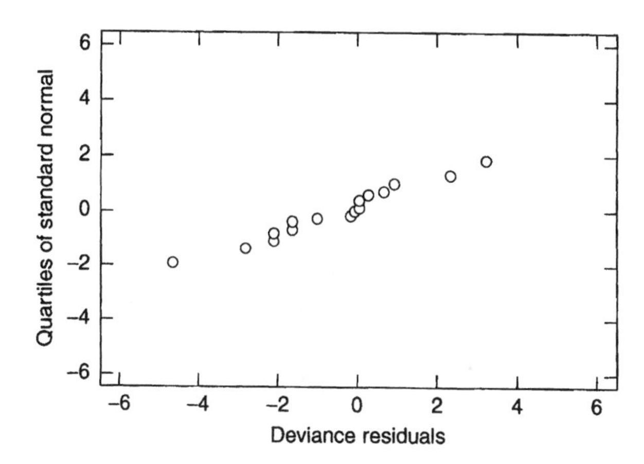

Figure 8.14 Normal probability plot for deviance residuals using GLM with factors *D*, *F*, and *BG* and square root link.

and correlation matrices using GLM with the square root and log link are also provided in Tables 8.16–8.19. Notice that when the log link is used, there are moderate to large correlations between the parameter estimates. However, when the variance-stabilizing square root link is used, the parameter estimates are uncorrelated.

Figure 8.11 shows that the main effects *D* and *F*, the interaction *BC*+*DF*+*GJ*, and possibly the interaction *BG*+*CJ*+*EH* are important when fitting the model with a log link. On the other hand, Figure 8.12 indicates only the main effects *D*,

Table 8.20 Comparison of 95% Confidence Intervals on the Mean Responses for Model Generated with Freeman–Tukey Modified Square Root Data Transformation GLM with Poisson and Square Root Link Defects for Example 8.9

	Using OLS with Freeman–Tukey Square Root				GLM Using Poisson Distribution and Square Root Link		GLM Using Poisson Distribution and Log Link		Length of 95% Confidence Intervals		
	Transformed		Untransformed								
Observation	Predicted Value	95% Confidence Intervals	Predicted Value	95% Confidence Intervals	Predicted Value	95% Confidence Intervals	Predicted Values	95% Confidence Intervals	Transformation	GLM Square Root	GLM Log
1	5.49	(4.13, 6.84)	29.64	(16.56, 46.29)	35.20	(29.39, 41.01)	50.85	(41.18, 60.53)	29.73	11.63	19.35
2	3.95	(2.60, 5.3)	15.11	(6.27, 27.59)	17.01	(12.97, 21.05)	11.65	(7.37, 15.92)	21.32	8.08	8.55
3	1.53	(0.18, 2.88)	1.87	(1.46, 7.80)	1.68	(0.41, 2.95)	1.21	(0.45, 1.98)	6.34	2.54	1.53
4	3.06	(1.71, 4.42)	8.87	(2.45, 19.04)	9.63	(6.59, 12.67)	5.29	(2.39, 8.19)	16.59	6.08	5.80
5	1.53	(0.18, 2.88)	1.87	(1.46, 7.80)	1.68	(0.41, 2.95)	1.21	(0.45, 1.98)	6.34	2.54	1.53
6	3.06	(1.71, 4.42)	8.87	(2.45, 19.04)	9.63	(6.59, 12.67)	5.29	(2.39, 8.19)	16.59	6.08	5.80
7	5.49	(4.13, 6.84)	29.64	(16.56, 46.29)	35.20	(29.39, 41.01)	50.85	(41.18, 60.53)	29.73	11.63	19.35
8	3.95	(2.60, 5.3)	15.11	(6.27, 27.59)	17.01	(12.97, 21.05)	11.65	(7.37, 15.92)	21.32	8.08	8.55
9	1.07	(−0.28, 2.42)	0.70	(*, 5.37)	0.47	(0.00, 1.14)	0.41	(0, 1.19)	*	1.14	1.57
10	−0.47	(−1.82, 0.88)	*	(*, 0.36)	1.27	(0.16, 2.37)	0.09	(0, 0.28)	*	2.21	0.36
11	1.95	(0.60, 3.31)	3.32	(0.03, 10.46)	2.91	(1.24, 4.58)	2.05	(0.97, 3.13)	10.43	3.34	2.16
12	3.49	(2.14, 4.84)	11.69	(4.09, 22.93)	12.34	(8.90, 15.78)	8.95	(5.15, 12.75)	18.84	6.89	7.60
13	1.95	(0.60, 3.31)	3.32	(0.03, 10.46)	2.91	(1.24, 4.58)	2.05	(0.97, 3.13)	10.43	3.34	2.16
14	3.49	(2.14, 4.84)	11.69	(4.09, 22.93)	12.34	(8.90, 15.78)	8.95	(5.15, 12.75)	18.84	6.89	7.60
15	1.07	(−0.28, 2.42)	0.70	(*, 5.37)	0.47	(0.00, 1.14)	0.41	(0, 1.19)	*	1.14	1.57
16	−0.47	(−1.82, 0.88)	*	(*, 0.36)	1.27	(0.16, 2.37)	0.09	(0, 028)	*	2.21	0.36

F, and the interaction BG (and its aliases) have large effects when using the square root as the link function. This result corresponds to Bisgaard and Fuller (1994–1995), and the Myers and Montgomery (1997) analysis. The straight-line part of the effects in Figure 8.11 may appear slightly straighter than in Figure 8.12. Nevertheless, both models include the main effect D, F, and the interaction BG. We use the normal plot of deviance residuals to check the adequacy of the model. Those plots are shown in Figures 8.13 and 8.14. The normal probability plot of the deviance residuals indicates that the model with factors D and F and the interactions BC and BG is not as good as the model based on the square root link, which implies that the interaction term $BC+DF+GJ$ is not important and should not be included in the model. However, the generalized linear model with main effects D and F and interactions BC and BG based on the log link is used in the comparison.

The model for the defect response using the Freeman – Tukey modified square root transformation is

$$\hat{c} = \left(\frac{4\mu^2 - 1}{4\mu}\right)^2$$

where $\mu = 2.513 - 0.996D - 1.21F - 0.772BG$. The reduced model using a generalized linear model with Poisson responses and log link with factors D and F and interactions BC and BG is

$$\hat{c} = e^{0.985-1.075D-1.338F-0.207BC-0.737BG}$$

The reduced model using generalized linear model with Poisson responses and square root link with factors D, F, and BG is

$$\hat{c} = (2.404 - 1.210D - 1.415F - 0.904BG)^2$$

Table 8.20 shows the predicted values at each design point for all three models. Since some of the predicted values in the transformed scale are negative, no untransformed prediction can be reliably made. Both GLMs produce reasonable predicted values. We also construct 95% confidence intervals on the mean response at each design point for all three models. The lengths of the confidence intervals for both GLMs are very comparable. Based on a simpler model and more attractive residuals, we conclude that the important factors are D, F, and BG. □

Example 8.10. The Wave Solder Experiment. This experiment deals with an electronic circuit card assembly and a wave-soldering process. The response is the number of defects in the solder joint. The process involves baking and

Table 8.21 2^{7-3} Factorial for the Wave Solder Experiment with Number of Defects

	Factor							y		
Run	*A*	*B*	*C*	*D*	*E*	*F*	*G*	1	2	3
1	−1	−1	−1	−1	−1	−1	−1	13	30	26
2	−1	−1	−1	+1	+1	+1	+1	4	16	11
3	−1	−1	+1	−1	−1	+1	+1	20	15	20
4	−1	−1	+1	+1	+1	−1	−1	42	43	64
5	−1	+1	−1	−1	+1	−1	+1	14	15	17
6	−1	+1	−1	+1	−1	+1	−1	10	17	16
7	−1	+1	+1	−1	+1	+1	−1	36	29	53
8	−1	+1	+1	+1	−1	−1	+1	5	9	16
9	+1	−1	−1	−1	+1	+1	−1	29	0	14
10	+1	−1	−1	+1	−1	−1	+1	10	26	9
11	+1	−1	+1	−1	+1	−1	+1	28	—	19
12	+1	−1	+1	+1	−1	+1	−1	100	129	151
13	+1	+1	−1	−1	−1	+1	+1	11	15	11
14	+1	+1	−1	+1	+1	−1	−1	17	11	17
15	+1	+1	+1	−1	−1	−1	−1	53	70	89
16	+1	+1	+1	+1	+1	+1	+1	23	22	7

preheating the circuit card and passing it through a solder wave by conveyor. Condra (1993) presented the results, which Hamada and Nelder (1997) later reanalyzed. The seven factors are (*A*) prebake condition, (*B*) flux density, (*C*) conveyor speed, (*D*) preheat condition, (*E*) cooling time, (*F*) ultrasonic solder agitator, and (*G*) solder temperature. Table 8.21 summarizes these results. Each factor is at two levels and the experimental design is a 2^{7-3} fractional factorial. Note that data point 11 has only two observations. A third observation was reported, but strong evidence suggested that it was an outlier (Hamada and Nelder, 1997).

Table 8.22 gives the PROC GENMOD output, which displays a full model for a Poisson response old she log link with all seven main effects and six interactions that were considered potentially important. Note that the mean deviance, which is deviance/$(n - p)$, exceeds one considerably, so there is an indication of overdispersion. To gain some insight on the impact of ignoring overdispersion, let us continue with the analysis. Recall that the danger here is the underestimation of standard errors of coefficients. Incidentally, note that the standard errors of the coefficients are similar but not identical, because the design is not orthogonal due to the missing observation. Note that all main effects are significant apart from *D* and *F*. The *DC*, *AD*, *BC*, and *BD* are also significant. Obviously, there may be a case here for including *D* in the edited model to preserve hierarchy.

Table 8.22 GENMOD Output for Full Model, Wave Solder Data

The GENMOD Procedure
Model Information

Description	Value
Data set	WORK.DEFECT
Distribution	POISSON
Link function	LOG
Dependent variable	Y
Observations used	47

Criteria for Assessing Goodness-of-Fit

Criterion	DF	Value	Value/DF
Deviance	33	139.7227	4.2340
Scaled deviance	33	139.7227	4.2340
Pearson chi-square	33	122.9190	3.7248
Scaled Pearson X2	33	122.9190	3.7248
Log-likelihood	–	3837.9339	–

Analysis of Parameter Estimates

Parameter	DF	Estimate	Std Err	Chi Square	Pr > Chi
INTERCEPT	1	3.0721	0.0344	7981.4934	0.0001
A	1	0.1126	0.0332	11.4710	0.0007
B	1	-0.1349	0.0344	15.3837	0.0001
C	1	0.4168	0.0345	146.1672	0.0001
D	1	-0.0577	0.0344	2.8118	0.0936
E	1	-0.0942	0.0334	7.9508	0.0048
F	1	-0.0176	0.0339	0.2696	0.6036
G	1	-0.3803	0.0343	122.9944	0.0001
AB	1	-0.0175	0.0334	0.2760	0.5993
AC	1	0.1741	0.0339	26.3925	0.0001
AD	1	0.0919	0.0332	7.6421	0.0057
BC	1	-0.0788	0.0343	5.2798	0.0216
BD	1	-0.2996	0.0344	75.8633	0.0001
CD	1	0.0466	0.0345	1.8275	0.1764
Scale	0	1.0000	0.0000	–	–

Note: The scle parameter was held fixed.

The PROC GENMOD output in Table 8.23 has the standard error adjusted for overdispersion as discussed previously. Note in Table 8.23 that the coefficients are the same as those obtained with the unadjusted analysis. However, the standard errors are larger by a factor of $\sqrt{\text{deviance}/33} = \sqrt{4.234} = 2.0577$. Now the model terms that are significant on the basis of χ^2 values are only C, G, AC, and BD. This conclusion is

Table 8.23 GENMOD Output for Wave Solder Data Adjusted for Overdispersion

Description	Value
Data set	WORK.DEFECT
Distribution	POISSON
Link function	LOG
Dependent variable	Y
Observations used	47

Criteria for Assessing Goodness-of-Fit

Criterion	DF	Value	Value/DF
Deviance	33	139.7227	4.2340
Scaled deviance	33	33.0000	1.0000
Pearson chi-square	33	122.9190	3.7248
Scaled Pearson X2	33	29.0313	0.8797
Log-likelihood	–	906.4510	–

Analysis of Parameter Estimates

Parameter	DF	Estimate	Std Err	Chi Square	Pr > Chi
INTERCEPT	1	3.0721	0.0708	1885.0853	0.0001
A	1	0.1126	0.0684	2.7092	0.0998
B	1	-0.1349	0.0708	3.6334	0.0566
C	1	0.4168	0.0709	34.5221	0.0001
D	1	-0.0577	0.0708	0.6641	0.4151
E	1	-0.0942	0.0687	1.8778	0.1706
F	1	-0.0176	0.0697	0.0637	0.8008
G	1	-0.3803	0.0706	29.0491	0.0001
AB	1	-0.0175	0.0687	0.0652	0.7985
AC	1	0.1741	0.0697	6.2334	0.0125
AD	1	0.0919	0.0684	1.8049	0.1791
BC	1	-0.0788	0.0706	1.2470	0.2641
BD	1	-0.2996	0.0708	17.9175	0.0001
CD	1	0.0466	0.0709	0.4316	0.5112
Scale	0	2.0577	0.0000	–	–

Note: The scle parameter was estimated by the square root of deviance/df.

quite different from the initial analysis where overdispersion was not taken into account. We used χ^2 tests on the parameter estimates. There certainly may be a case for t-tests derived from coeff/standard errors, since the diagonal elements of the variance–covariance matrix are multiplied by a scale parameter. However, consider the edited model with output shown in Table 8.24. To gain some insight on individual effects, consider factor G, which has a negative effect. If one were to change from the middle of the solder temperature (coded value = 0) to the high level, the number

Table 8.24 Edited Model for Wave Solder Data

Link function	LOG
Dependent variable	Y
Observations used	47

Criteria for Assessing Goodncss-of-Fit

Criterion	DF	Value	Value/DF
Deviance	42	241.7755	5.7566
Scaled deviance	42	42.0000	1.0000
Pearson chi-square	42	237.3981	5.6523
Scaled pearson X2	42	41.2396	0.9819
Log-likelihood	–	657.8421	–

Analysis of Parameter Estimates

Parameter	DF	Estimate	Std Err	Chi Square	Pr > Chi
INTERCEPT	1	3.0769	0.0825	1391.0260	0.0001
C	1	0.4405	0.0808	29.7429	0.0001
G	1	-0.4030	0.0808	24.8954	0.0001
AC	1	0.2821	0.0667	17.9039	0.0001
BD	1	-0.3113	0.0808	14.8557	0.0001
Scale	0	2.3993	0.0000	–	–

Note: The scale parameter was estimated by the square root of deviance/df.

of defects is reduced by a factor of $e^{-0.403} = 0.65$. obviously the individual effect of factor C cannot be evaluated without taking into account the interaction.

Consider using the square root link on these data. The analysis for the full model and the reduced model is shown in Tables 8.25 and 8.26, respectively.

The log link and the power link give lack-of-fit performances that are approximately the same. However, it is interesting that the conclusions regarding the roles of the variables are somewhat different, suggesting that the analyst may often wish to try more than one link. Figures 8.15–8.23 summarize the appropriate residual plots. These plots indicate some minor problems with the variability in the residuals, but not enough to significantly affect the analysis.

We see that exercising the d-scale option in SAS gives an estimated scale parameter. Notice that A, B, C, E, and G are significant although A, B, and E do not have large effects. The significant interactions are the same as found using the log link scale. The explanation is that the effects are now additive on the square root, so sizes of coefficients cannot be compared to those in the previous analysis. Notice

Table 8.25 Analysis for the Wave Solder Data Using Power (½) Link

The GENMOD Procedure
Model Information

Description	Value
Data set	WORK.DEFECT
Distribution	POISSON
Link function	POWER (0.5)
Dependent variable	Y
Observations used	47

Criteria for Assessing Goodness of Fit

Criterion	DF	Value	Value/DF
Deviance	33	148.2599	4.2927
Scaled deviance	33	33.0000	1.0000
Pearson chi-square	33	134.4467	4.0741
Scaled Pearson X2	33	29.9254	0.9068
Log-likelihood	—	853.3053	—

Analysis of Parameter Estimates

Parameter	DF	Estimate	Std Err	Chi Square	Pr > Chi
INTERCEPT	1	4.9953	0.1549	1040.1954	0.0001
A	1	0.4302	0.1552	7.6866	0.0056
B	1	-0.4041	0.1552	6.7831	0.0092
C	1	1.1957	0.1552	59.3566	0.0001
D	1	0.0096	0.1549	0.0038	0.9507
E	1	-0.3938	0.1552	6.4392	0.0112
F	1	0.0435	0.1552	0.0786	0.7792
G	1	-1.1105	0.1552	51.2249	0.0001
AB	1	-0.1292	0.1552	0.6937	0.4049
AC	1	0.5650	0.1552	13.2518	0.0003
AD	1	0.2621	0.1552	2.8525	0.0912
BC	1	-0.2802	0.1552	3.2603	0.0710
BD	1	-0.9461	0.1552	37.1730	0.0001
CD	1	0.2139	0.1552	1.9000	0.1681
Scale	0	2.1196	0.0000	—	—

Note: The scale parameter was estimated by the square root of deviance/df.

also that the standard errors are not quite equal. Recall that one of the observations was removed, rendering the design not quite orthogonal. Table 8.26 shows the analysis of the edited model. The standard errors are found from approximately the square roots of diagonals of $(\mathbf{X}'\mathbf{X})^{-1} \cdot (2.2078)^2$. □

Table 8.26 Edited Model for the Wave Solder Data

The GENMOD Procedure

Model Information

Description	Value
Data set	WORK.DEFECT
Distribution	POISSON
Link function	POWER(0.5)
Dependent variable	Y
Observation used	47

Criteria for Assessing Goodness of Fit

Criterion	DF	Value	Value/DF
Deviance	39	190.1069	4.8745
Scaled deviance	39	39.0000	1.0000
Pearson chi-square	39	173.1641	4.4401
Scaled Pearson X2	39	35.5242	0.9109
Log-likelihood	–	782.1753	–

Analysis of Parameter Estimates

Parameter	DF	Estimate	Std Err	Chi-Square	Pr > Chi
INTERCEPT	1	5.0178	0.1609	972.7881	0.0001
A	1	0.4315	0.1611	7.1719	0.0074
B	1	-0.4053	0.1615	6.2968	0.0121
C	1	1.1935	0.1612	54.8031	0.0001
E	1	-0.3864	0.1616	5.7193	0.0168
G	1	-1.1038	0.1614	46.7910	0.0001
AC	1	0.5833	0.1611	13.1014	0.0003
BD	1	-0.9587	0.1611	35.4251	0.0001
Scale	0	2.2078	0.0000	–	–

Note: The scale parameter was estimated by the square root of deviance/df.

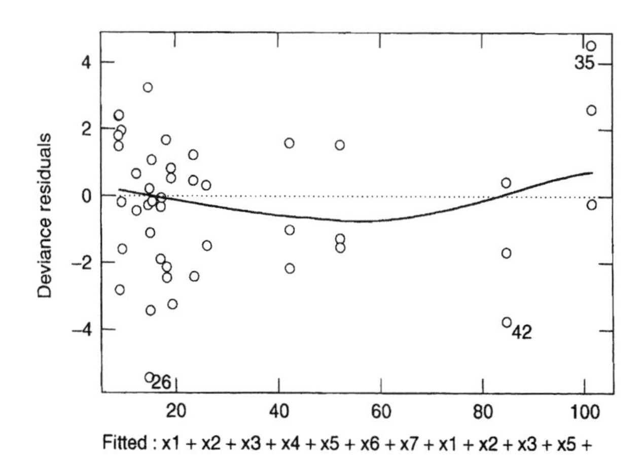

Figure 8.15 Deviance residuals plotted against the fitted values for the solder wave data.

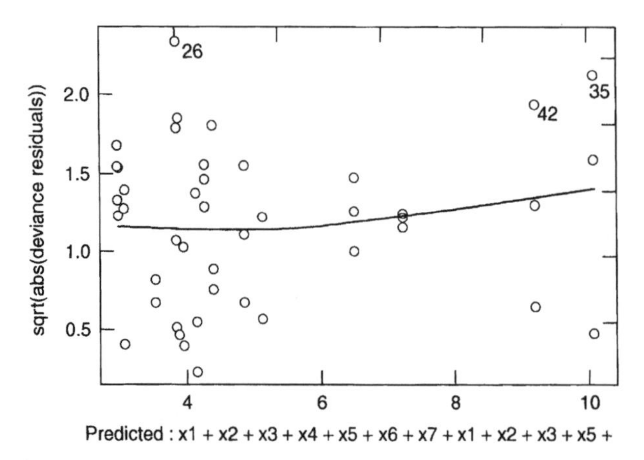

Figure 8.16 Signed deviance residuals plotted against the fitted values for the wave solder data.

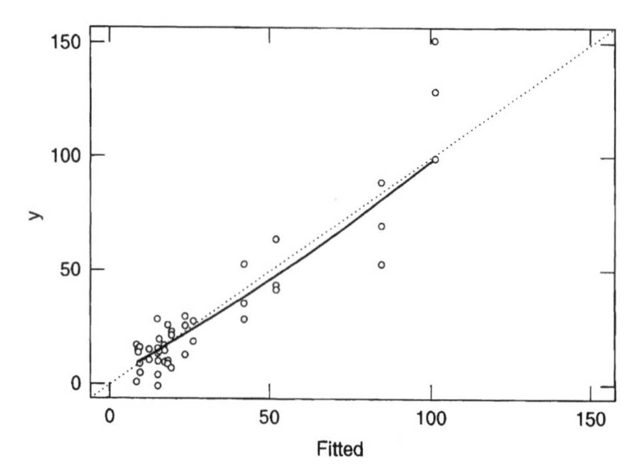

Figure 8.17 Plot of the observed values versus the fitted values for the wave solder data.

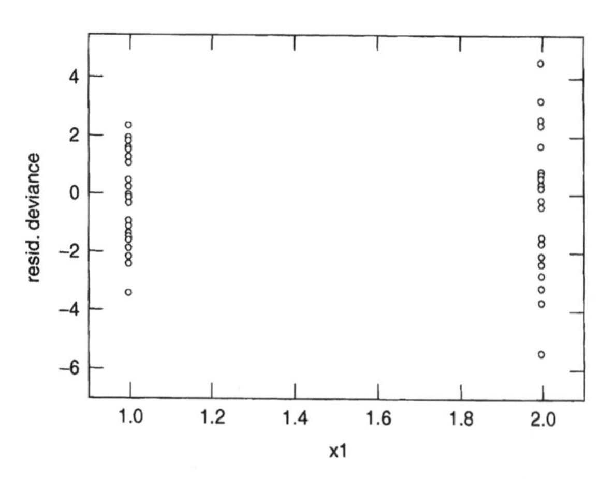

Figure 8.18 Plot of the signed deviance residuals versus x_1.

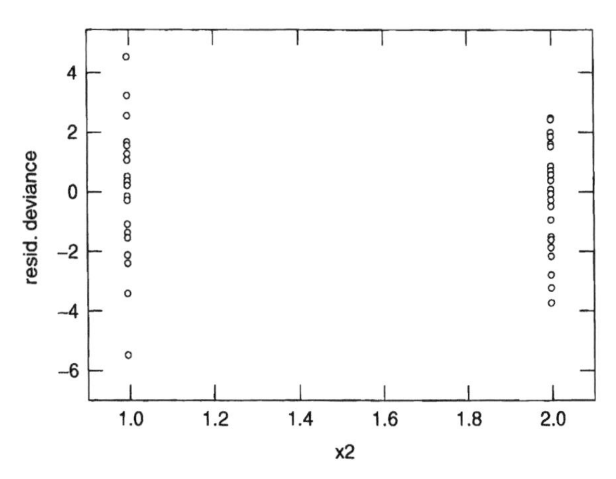

Figure 8.19 Plot of the signed deviance residuals versus x_2.

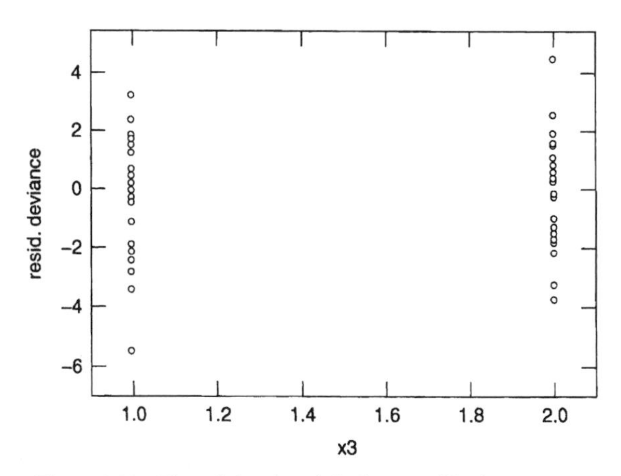

Figure 8.20 Plot of the signed deviance residuals versus x_3.

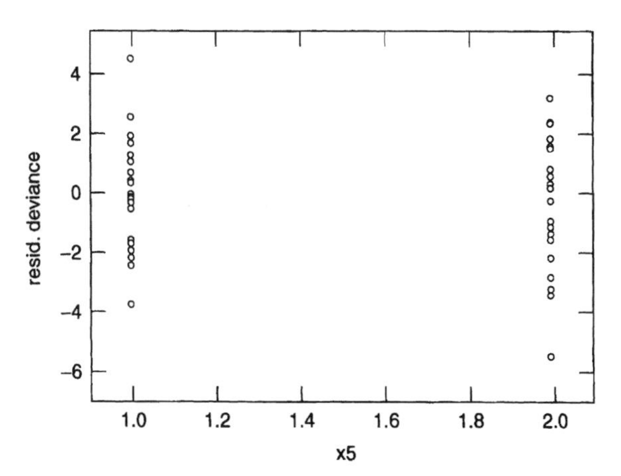

Figure 8.21 Plot of the signed deviance residuals versus x_5.

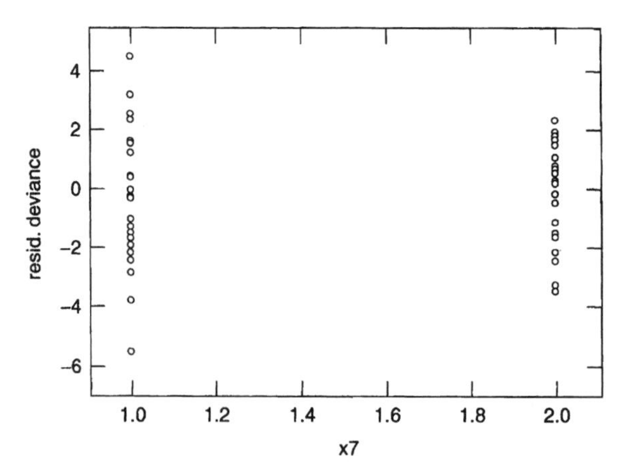

Figure 8.22 Plot of the signed deviance residuals versus x_7.

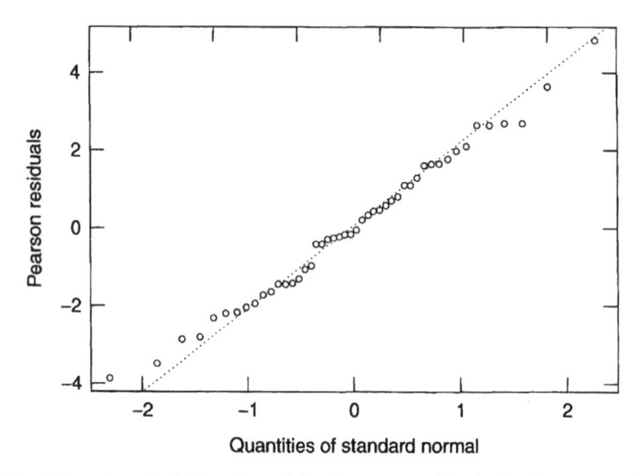

Figure 8.23 Normal probability plot of the Pearson residuals for the wave solder data.

EXERCISES

8.1 Suppose that you want to design an experiment to fit a logistic regression model in two predictors. A first-order model is assumed. Based on prior information about the experimental situation, reasonable ranges for the model parameters are

$$0.5 \leq \beta_0 \leq 5$$
$$2 \leq \beta_1 \leq 6$$
$$1 \leq \beta_2 \leq 4$$

and the normal distribution is a reasonable prior. Find a Bayesian D-optimal design with 10 runs for this experiment.

8.2 Reconsider the situation in Exercise 8.1. Suppose that a uniform prior distribution is selected. Find a Bayesian D-optimal design with 10 runs for this experiment. Compare this to the design found in Exercise 8.1.

8.3 Suppose that you want to design an experiment to fit a logistic regression model in two predictors. A first-order model with interaction is assumed. Based on prior information about the experimental situation, reasonable ranges for the model parameters are

$$0.5 \leq \beta_0 \leq 8$$
$$1 \leq \beta_1 \leq 5$$
$$1 \leq \beta_2 \leq 3$$
$$0.5 \leq \beta_{12} \leq 1.5$$

and the normal distribution is a reasonable prior. Find a Bayesian D-optimal design with 16 runs for this experiment.

8.4 Reconsider the situation in Exercise 8.3. Suppose that a uniform prior distribution is selected. Find a Bayesian D-optimal design with 16 runs for this experiment. Compare this to the design found in Exercise 8.3.

8.5 Consider the spermatozoa survival study in Table 8.7. Suppose that we want to construct a Bayesian D-optimal design for this experiment. We have some prior information about the anticipated results, which we have summarized with the following ranges of the parameters:

$$-1 \leq \beta_0 \leq 0.5$$
$$0 \leq \beta_1 \leq 0.5$$
$$-1 \leq \beta_2 \leq 0.5$$
$$0 \leq \beta_3 \leq 0.5$$
$$0.5 \leq \beta_{12} \leq 1.5$$
$$0 \leq \beta_{13} \leq 0.75$$
$$0 \leq \beta_{23} \leq 0.5$$

Assume that the normal distribution is a reasonable prior.

(a) Find a Bayesian D-optimal design with 8 runs for this experiment. How does this design compare to the 2^3 design that the experimenters actually used?

(b) Find a Bayesian D-optimal design with 16 runs for this experiment.

8.6 Rework Exercise 8.5 using a uniform prior. What difference does this make in the designs obtained?

8.7 Suppose that you want to design an experiment to fit a Poisson regression model in two predictors. A first-order model is assumed. Based on prior information about the experimental situation, reasonable ranges for the model parameters are

$$0.5 \le \beta_0 \le 3$$
$$1 \le \beta_1 \le 3$$
$$0.25 \le \beta_2 \le 0.75$$

and the normal distribution is a reasonable prior. Find a Bayesian D-optimal design with 16 runs for this experiment.

8.8 Reconsider the situation in Exercise 8.7. Suppose that a uniform prior distribution is selected. Find a Bayesian D-optimal design with 16 runs for this experiment. Compare this to the design found in Exercise 8.7.

8.9 Suppose that you want to design an experiment to fit a Poisson regression model in two predictors. A first-order model with interaction is assumed. Based on prior information about the experimental situation, reasonable ranges for the model parameters are

$$0.5 \le \beta_0 \le 3$$
$$1 \le \beta_1 \le 3$$
$$0.25 \le \beta_2 \le 0.75$$
$$0.1 \le \beta_{12} \le 0.5$$

and the normal distribution is a reasonable prior. Find a Bayesian D-optimal design with 16 runs for this experiment.

8.10 Reconsider the situation in Exercise 8.9. Suppose that a uniform prior distribution is selected. Find a Bayesian D-optimal design with 16 runs for this experiment. Compare this to the design found in Exercise 8.9.

8.11 Consider a 2^3 factorial design with ± 1 coding for the design variables. Assume that the response follows a gamma distribution, and suppose we wish to use the canonical link. Assume that the true values for the coefficients in the linear predictor are $\beta_0 = 1$, $\beta_1 = 3$, $\beta_2 = -4$, and $\beta_3 = 2$.

(a) What are the design efficiencies associated with the estimation of the four parameters?

(b) Give the values for the eight Hessian weights.

(c) Give the correlation matrix for the estimated parameters.

8.12 Consider a 2^4 factorial design with ± 1 coding for the design variables. Assume that the response follows a Poisson distribution, and suppose we wish to use the log link. Assume that the true values for the coefficients in the linear predictor are $\beta_0 = 1.5$, $\beta_1 = -1.0$, $\beta_2 = -3$, $\beta_3 = 1.5$, and $\beta_4 = 2.0$

(a) What are the design efficiencies associated with the estimation of the four parameters?

(b) Give the values for the sixteen Hessian weights.

(c) Give the correlation matrix for the estimated parameters.

8.13 Consider Exercise 8.11. Suppose now that the linear predictor also includes the x_1x_2 and the x_1x_3 interactions with $\beta_{12} = 0.5$ and $\beta_{13} = -2$. Calculate all the design efficiencies for the more complicated model.

8.14 Consider Exercise 8.11, where we use an identity link with a Poisson response. Assume the model

$$E(y) = 1.0 + 3x_1 + 2x_2 - x_3$$

Compute the design efficiencies and the correlation matrix of the estimated parameters.

8.15 Consider a 2^2 factorial experiment. Suppose that the response follows a Poisson distribution and that the linear predictor is $g(\mu) = 3.5 + 2.0x_1 + 1.5x_2$, where the design variables are coded as ± 1.

(a) Use a log link and generate three replicates of this experiment using a Poisson random number generator with the mean calculated from the link function and the linear predictor.

(b) Fit the GLM using the log link.

(c) Find the covariance and correlation matrix of the estimated parameters. Comment on the variances and the correlations.

(d) Calculate the fitted values at each design point.

(e) Compare your empirical results and then compare to the theoretical results.

(f) Rework parts (b), (c), and (d) using the square root link and comment on your results.

8.16 Consider the situation in Exercise 8.15. Suppose now that the linear predictor is $g(\mu) = 3.5+2.0x_1+1.5x_2+0.5x_1x_2$. Rework all the parts of Exercise 8.15 with this new model.

8.17 Consider a 2^3 factorial experiment. Suppose that the response follows an exponential distribution and that the linear predictor is $g(\mu) = 8.0 +2.0x_1+3.0x_2+x_3-1.5x_1x_2$, where the design variables are coded as ± 1.

(a) Use the inverse link and generate two replicates of this experiment using an exponential random number generator with the mean calculated from the link function and the linear predictor.

(b) Fit the GLM using the inverse link.

(c) Find the covariance and correlation matrix of the estimated parameters. Comment on the variances and the correlations.

(d) Calculate the fitted values at each design point.

(e) Compare your empirical results and then compare to the theoretical results.

(f) Rework parts (b), (c), and (d) using the log link and comment on your results.

8.18 Reconsider the situation in Exercise 8.17. Rework part (a) using the log link, and then rework parts (b) and (c) using the new data.

8.19 Consider the situation in Exercise 8.15. Use the linear predictor and log link to generate 500 samples of three replicates each.

(a) For each of the 500 samples fit the GLM assuming a Poisson response and log link.

(b) Calculate the predicted response at $x_1 = 1$ and $x_2 = -1$ for each of the 500 samples.

(c) Prepare histograms of the estimated model coefficients and the predicted values obtained in parts (a) and (b). Do these histograms look reasonable in light of the asymptotic properties of the parameter estimates?

(d) Compute 95% confidence intervals for β_1 and for the mean response when $x_1 = 1$ and $x_2 = 1$ for each of the 500 samples. How often do these intervals cover the true values for β_1 and the true mean response when $x_1 = 1$ and $x_2 = -1$?

8.20 Consider the situation in Exercise 8.17. Use the linear predictor and log link to generate 500 samples of two replicates each.

(a) For each of the 500 samples fit the GLM assuming an exponential response and log link.

(b) Calculate the predicted response at $x_1 = 1$, $x_2 = 1$, and $x_3 = 1$ for each of the 500 samples.

(c) Prepare histograms of the estimated model coefficients and the predicted values obtained in parts (a) and (b). Do these histograms look reasonable in light of the asymptotic properties of the parameter estimates?

(d) Compute 95% confidence intervals for β_1 and for the mean response when $x_1 = 1$, $x_2 = 1$, and $x_3 = 1$ for each of the 500 samples. How often do these intervals cover the true values for β_1 and the true mean response when at $x_1 = 1$, $x_2 = 1$, and $x_3 = 1$?

APPENDIX A.1

Background on Basic Test Statistics

We indicate that Y is a random variable that follows a normal distribution with mean μ and variance σ^2 by

$$Y \sim N(\mu, \sigma^2)$$

Central Distributions

1. Let $Y_1, Y_2, \ldots, Y_n$ be independent normally distributed random variables with $E(Y_i) = \mu_i$, and $\text{Var}(Y_i) = \sigma_i^2$. Let $a_1, a_2, \ldots, a_n$ be known constants. If we define the linear combination of the Y_i's by

   $$U = \sum_{i=1}^{n} a_i Y_i$$

   then

   $$U \sim N\left(\sum_{i=1}^{n} a_i \mu_i, \sum_{i=1}^{n} a_i^2 \sigma_i^2\right)$$

   The key point is that linear combinations of normally distributed random variables also follow normal distributions.

2. If $Y \sim N(\mu, \sigma^2)$, then

   $$Z = \frac{Y - \mu}{\sigma} \sim N(0, 1)$$

   where Z is the standard normal random variable.

---

3. Let $Z = (Y - \mu)/\sigma$. If $Y \sim N(\mu, \sigma^2)$, then Z^2 follows a χ^2 distribution with one degree of freedom, which we denote by

$$Z^2 \sim \chi_1^2$$

The key point is that the square of a standard normal random variable is a χ^2 random variable with one degree of freedom.

4. Let $Y_1, Y_2, \ldots, Y_n$ be independent normally distributed random variables with $E(Y_i) = \mu_i$ and $\text{Var}(Y_i) = \sigma_i^2$, and let

$$Z_i = \frac{Y_i - \mu_i}{\sigma_i}$$

then

$$\sum_{i=1}^{n} Z_i^2 \sim \chi_n^2$$

The key points are (a) the sum of n independent squared standard normal random variables follows a χ^2 distribution with n degrees of freedom, and (b) the sum of χ^2 random variables also follows a χ^2 distribution.

5. (Central limit theorem) If $Y_1, Y_2, \ldots, Y_n$ are independent and identically distributed random variables with $E(Y_i) = \mu$ and $\text{Var}(Y_i) = \sigma^2 < \infty$, then

$$\frac{\bar{Y} - \mu}{\sigma/\sqrt{n}}$$

converges in distribution to a standard normal distribution as $n \to \infty$. The key point is that if n is sufficiently large, then $\bar{Y}$ approximately follows a normal distribution. What constitutes sufficiently large depends on the underlying distribution of the Y_i's.

6. If $Z \sim N(0, 1)$, $V \sim \chi_v^2$, and Z and V are independent, then

$$\frac{Z}{\sqrt{V/v}} \sim t_v$$

where t_v is the t distribution with v degrees of freedom.

7. Let $V \sim \chi_v^2$, and let $W \sim \chi_\eta^2$. If V and W are independent, then

$$\frac{V/v}{W/\eta} \sim F_{v,\eta}$$

where $F_{v,\eta}$ is the F distribution with v and η degrees of freedom. The key point is that the ratio of two independent χ^2 random variables, each divided by their respective degrees of freedom, follows an F distribution.

Noncentral Distributions

1. Let $Y \sim N(\delta, 1)$, and let $V \sim \chi^2_\nu$. If Y and V are independent, then

$$\frac{Y}{\sqrt{V/\nu}} \sim t'_{\nu,\delta}$$

where $t'_{\nu,\delta}$ is the noncentral t distribution with ν degrees of freedom and noncentrality parameter δ.

2. If $Y \sim N(\delta, 1)$, then

$$Y^2 \sim \chi^{2\prime}_{1,\delta^2}$$

where $\chi^{2\prime}_{1,\delta^2}$ is the noncentral χ^2 distribution with one degree of freedom and noncentrality parameter δ^2.

3. If $Y_1, Y_2, \ldots, Y_n$ are independent normally distributed random variables with $E(Y_i) = \delta_i$, and $\mathrm{Var}(Y_i) = 1$, then

$$\sum_{i=1}^{n} Y_i^2 \sim \chi^2_{n,\lambda}$$

where the noncentrality parameter, λ, is

$$\lambda = \sum_{i=1}^{n} \delta_i^2$$

4. Let $V \sim \chi^{2\prime}_{\nu,\lambda}$, and let $W \sim \chi^2_\eta$. If V and W are independent, then

$$\frac{V/\nu}{W/n} \sim F'_{\nu,\eta,\lambda}$$

where $F'_{\nu,\eta,\lambda}$ is a noncentral F distribution with ν and η degrees of freedom and noncentrality parameter λ.

APPENDIX A.2

Background from the Theory of Linear Models

Basic Definitions

1. (Rank of a matrix) The rank of a matrix, **A**, is the number of linearly independent columns. Equivalently, it is the number of linearly independent rows.
2. (Identity matrix) The identity matrix of order k, denoted by **I** or $\mathbf{I}_k$, is *a* $k \times k$ square matrix whose diagonal elements are 1's and whose nondiagonal elements are 0's; thus

$$\mathbf{I} = \begin{bmatrix} 1 & 0 & 0 & \cdots & 0 \\ 0 & 1 & 0 & \cdots & 0 \\ \vdots & & & & \\ 0 & 0 & 0 & \cdots & 1 \end{bmatrix}$$

3. (Inverse of a matrix) Let **A** be a $k \times k$ matrix. The inverse of **A**, denoted by $\mathbf{A}^{-1}$, is another $k \times k$ matrix such that

$$\mathbf{A}\mathbf{A}^{-1} = \mathbf{A}^{-1}\mathbf{A} = \mathrm{I}$$

   If the inverse exists, it is unique.
4. (Transpose of a matrix) Let **A** be a $n \times k$ matrix. The transpose of **A**, denoted by $\mathbf{A}'$ or $\mathbf{A}^T$, is a $k \times n$ matrix whose columns are the rows of **A**; thus if

$$\mathbf{A} = \begin{bmatrix} a_{11} & a_{12} & \dots & a_{1k} \\ a_{21} & a_{22} & \dots & a_{2k} \\ \vdots & & & \\ a_{n1} & a_{n2} & \dots & a_{nk} \end{bmatrix}$$

then

$$\mathbf{A}' = \begin{bmatrix} a_{11} & a_{21} & \dots & a_{n1} \\ a_{12} & a_{22} & \dots & a_{n2} \\ \vdots & & & \\ a_{1k} & a_{2k} & \dots & a_{nk} \end{bmatrix}$$

Note that if **A** is a $n \times m$ matrix and **B** is a $m \times p$ matrix, then

$$(\mathbf{AB})' = \mathrm{B}'\mathbf{A}'$$

5. (Symmetric matrix) Let **A** be a $k \times k$ matrix. **A** is said to be symmetric if $\mathbf{A} = \mathbf{A}'$.
6. (Idempotent matrix) Let **A** be a $k \times k$ matrix. **A** is called idempotent if

$$\mathbf{A} = \mathbf{AA}$$

If **A** is also symmetric, **A** is called symmetric idempotent. If **A** is symmetric idempotent, then $\mathbf{I} - \mathbf{A}$ is also symmetric idempotent.
7. (Orthonormal matrix) Let **A** be a $k \times k$ matrix. If **A** is an orthonormal matrix, then $\mathbf{A}'\,\mathbf{A} = \mathbf{I}$. As a consequence, if **A** is an orthonormal matrix, then $\mathbf{A}^{-1} = \mathbf{A}'$.
8. (Quadratic form) Let **y** be a $k \times 1$ vector, and let **A** be a $k \times k$ matrix. The function

$$\mathbf{y}'\mathbf{A}\mathbf{y} = \sum_{i=1}^{k} \sum_{j=1}^{k} a_{ij} y_i y_j$$

is called a quadratic form. **A** is called the matrix of the quadratic form.
9. (Positive definite and positive semidefinite matrices) Let **A** be a $k \times k$ matrix. **A** is said to be positive definite if the following conditions hold:
   (a) $\mathbf{A} = \mathbf{A}'$ (**A** is symmetric), and
   (b) $\mathbf{y}'\mathbf{A}\mathbf{y} > 0 \;\forall\; \mathbf{y} \in \mathrm{R}^k, \mathbf{y} \neq 0$.
   A is said to be positive semidefinite if the following conditions hold:
   (a) $\mathbf{A} = \mathbf{A}'$ (**A** is symmetric),

(b) $\mathbf{y}'\mathbf{A}\mathbf{y} \geq 0 \quad \forall \mathbf{y} \in R^k$, and

(c) $\mathbf{y}'\mathbf{A}\mathbf{y} = 0 \quad$ for some $\mathbf{y} \neq 0$.

10. (Trace of a matrix) Let $\mathbf{A}$ be a $k \times k$ matrix. The trace of $\mathbf{A}$, denoted by tr ($\mathbf{A}$), is the sum of the diagonal elements of $\mathbf{A}$; thus

$$\text{tr}(\mathbf{A}) = \sum_{i=1}^{k} a_{ii}$$

Note that

(a) If $\mathbf{A}$ is a $m \times n$ matrix and $\mathbf{B}$ is a $n \times m$ matrix, then

$$\text{tr}(\mathbf{AB}) = \text{tr}(\mathbf{BA})$$

(b) If the matrices are appropriately conformable, then

$$\text{tr}(\mathbf{ABC}) = \text{tr}(\mathbf{CAB})$$

(c) If $\mathbf{A}$ and $\mathbf{B}$ are $k \times k$ matrices and a and b are scalars, then

$$\text{tr}(a\mathbf{A} + b\mathbf{B}) = a\ \text{tr}(\mathbf{A}) + b\ \text{tr}(\mathbf{B})$$

11. (Rank of an idempotent matrix) Let $\mathbf{A}$ be an idempotent matrix. The rank of $\mathbf{A}$ is its trace.

12. (Important identity for a partitioned matrix) Let $\mathbf{X}$ be a $n \times p$ matrix partitioned such that

$$\mathbf{X} = [\mathbf{X}_1 \mathbf{X}_2]$$

We note that

$$\mathbf{X}(\mathbf{X}'\mathbf{X})^{-1}\mathbf{X}'\mathbf{X} = \mathbf{X}$$
$$\mathbf{X}(\mathbf{X}'\mathbf{X})^{-1}\mathbf{X}'[\mathbf{X}_1\mathbf{X}_2] = \mathbf{X}$$
$$\mathbf{X}(\mathbf{X}'\mathbf{X})^{-1}\mathbf{X}'[\mathbf{X}_1\mathbf{X}_2] = [\mathbf{X}_1\mathbf{X}_2]$$

Consequently,

$$\mathbf{X}(\mathbf{X}'\mathbf{X})^{-1}\mathbf{X}'\mathbf{X}_1 = \mathbf{X}_1$$
$$\mathbf{X}(\mathbf{X}'\mathbf{X})^{-1}\mathbf{X}'\mathbf{X}_2 = \mathbf{X}_2$$

Similarly,

$$\mathbf{X}_1'\mathbf{X}(\mathbf{X}'\mathbf{X})^{-1}\mathbf{X}' = \mathbf{X}_1'$$
$$\mathbf{X}_2'\mathbf{X}(\mathbf{X}'\mathbf{X})^{-1}\mathbf{X}' = \mathbf{X}_2'$$

Matrix Derivatives

Let $\mathbf{A}$ be a $k \times k$ matrix of constants, $\mathbf{a}$ be a $k \times 1$ vector of constants, and $\mathbf{y}$ be a $k \times 1$ vector of variables.

1. If $z = \mathbf{a}'\mathbf{y}$, then
$$\frac{\partial z}{\partial \mathbf{y}} = \frac{\partial \mathbf{a}'\mathbf{y}}{\partial \mathbf{y}} = \mathbf{a}$$
2. If $z = \mathbf{y}'\mathbf{y}$, then
$$\frac{\partial z}{\partial \mathbf{y}} = \frac{\partial \mathbf{y}'\mathbf{y}}{\partial \mathbf{y}} = 2\mathbf{y}$$
3. If $\mathbf{z} = \mathbf{a}'\mathbf{A}\mathbf{y}$, then
$$\frac{\partial z}{\partial \mathbf{y}} = \frac{\partial \mathbf{a}'\mathbf{A}\mathbf{y}}{\partial \mathbf{y}} = \mathbf{A}'\mathbf{a}$$
4. If $z = \mathbf{y}'\mathbf{A}\mathbf{y}$, then
$$\frac{\partial z}{\partial \mathbf{y}} = \frac{\partial \mathbf{y}'\mathbf{A}\mathbf{y}}{\partial \mathbf{y}} = \mathbf{A}\mathbf{y} + \mathbf{A}'\mathbf{y}$$
If $\mathbf{A}$ is symmetric, then
$$\frac{\partial \mathbf{y}'\mathbf{A}\mathbf{y}}{\partial \mathbf{y}} = 2\mathbf{A}\mathbf{y}$$

Expectations

Let $\mathbf{A}$ be a $k \times k$ matrix of constants, $\mathbf{a}$ be a $k \times 1$ vector of constants, and $\mathbf{y}$ be a $k \times 1$ random vector with mean $\boldsymbol{\mu}$ and nonsingular variance–covariance matrix $\mathbf{V}$.

1. $\text{E}(\mathbf{a}'\mathbf{y}) = \mathbf{a}'\boldsymbol{\mu}$.
2. $\text{E}(\mathbf{A}\mathbf{y}) = \mathbf{A}\boldsymbol{\mu}$.
3. $\text{Var}(\mathbf{a}'\mathbf{y}) = \mathbf{a}'\mathbf{V}\mathbf{a}$.
4. $\text{Var}(\mathbf{A}\mathbf{y}) = \mathbf{A}\mathbf{V}\mathbf{A}'$. Note that if $\mathbf{V} = \sigma^2\mathbf{I}$, then $\text{Var}(\mathbf{A}\mathbf{y}) = \sigma^2\mathbf{A}\mathbf{A}'$.
5. $\text{E}(\mathbf{y}'\mathbf{A}\mathbf{y}) = \text{tr}(\mathbf{A}\mathbf{V}) + \boldsymbol{\mu}'\mathbf{A}\boldsymbol{\mu}$. Note that if $\mathbf{V} = \sigma^2\mathbf{I}$, then $\text{E}(\mathbf{y}'\mathbf{A}\mathbf{y}) = \sigma^2\,\text{tr}(\mathbf{A}) + \boldsymbol{\mu}'\mathbf{A}\boldsymbol{\mu}$.

Distribution Theory

Let $\mathbf{A}$ be a $k \times k$ matrix of constants, and $\mathbf{y}$ be a $k \times 1$ multivariate normal random vector with mean $\boldsymbol{\mu}$ and nonsingular variance–covariance matrix $\mathbf{V}$; thus

$$\mathbf{y} \sim N(\boldsymbol{\mu}, \mathbf{V})$$

Let U be the quadratic form defined by $\boldsymbol{U} = \mathbf{y}'\mathbf{A}\mathbf{y}$.

1. If $\mathbf{A}\mathbf{V}$ or $\mathbf{V}\mathbf{A}$ is an idempotent matrix of rank p, then
$$U \sim \chi^{2'}_{p,\lambda}$$
where $\lambda = \boldsymbol{\mu}'\mathbf{A}\boldsymbol{\mu}$.

2. Let $\mathbf{V} = \sigma^2\mathbf{I}$, which is a typical assumption. If $\mathbf{A}$ is idempotent with rank p, then

$$\frac{U}{\sigma^2} \sim \chi^{2'}_{p,\lambda}$$

where $\lambda = \boldsymbol{\mu}' \mathbf{A}\boldsymbol{\mu}/\sigma^2$.

3. Let $\mathbf{B}$ be a $q \times k$ matrix, and let W be the linear form given by $W = \mathbf{By}$. The quadratic form $U = \mathbf{y}' \mathbf{Ay}$ and W are independent if

$$\mathbf{BVA} = \mathbf{0}$$

Note that if $\mathbf{V} = \sigma^2\mathbf{I}$, then U and W are independent if $\mathbf{BA} = \mathbf{0}$.

4. Let $\mathbf{B}$ be a $k \times k$ matrix. Let $V = \mathbf{y}'\mathbf{By}$. The two quadratic forms, U and V, are independent if

$$\mathbf{AVB} = \mathbf{0}$$

Note that if $\mathbf{V} = \sigma^2\mathbf{I}$, then U and V are independent if $\mathbf{AB} = \mathbf{0}$.

APPENDIX A.3

The Gauss–Markov Theorem, $\mathrm{Var}(\varepsilon) = \sigma^2 \mathbf{I}$

The Gauss–Markov theorem establishes that the ordinary least squares (OLS) estimator of $\boldsymbol{\beta}$, $\mathbf{b} = (\mathbf{X}'\mathbf{X})^{-1}\mathbf{X}'\mathbf{y}$, is BLUE (*best linear unbiased estimator*). By *best,* we mean that $\mathbf{b}$ has the smallest variance, in some meaningful sense, among the class of all unbiased estimators that are linear combinations of the data. One problem is that $\mathbf{b}$ is a vector; hence its variance is actually a matrix. Consequently, we seek to show that $\mathbf{b}$ minimizes the variance for any linear combination of the estimated coefficients, $\mathbf{l}'\mathbf{b}$. We note that

$$\begin{aligned}\mathrm{Var}(\mathbf{l}'\mathbf{b}) &= \mathbf{l}'\mathrm{Var}(\mathbf{b})\mathbf{l}\\ &= \mathbf{l}'\left[\sigma^2(\mathbf{X}'\mathbf{X})-1\right]\mathbf{l}\\ &= \sigma^2\mathbf{l}'(\mathbf{X}'\mathbf{X})^{-1}\mathbf{l}\end{aligned}$$

which is a scalar. Let $\tilde{\boldsymbol{\beta}}$ be another unbiased estimator of $\boldsymbol{\beta}$ that is a linear combination of the data. Our goal then is to show that

$$\mathrm{Var}(\mathbf{l}'\tilde{\boldsymbol{\beta}}) \geq \sigma^2\mathbf{l}'(\mathbf{X}'\mathbf{X})^{-1}\mathbf{l}$$

with at least one $\mathbf{l}$ such that

$$\mathrm{Var}(\mathbf{l}'\tilde{\boldsymbol{\beta}}) > \sigma^2\mathbf{l}'(\mathbf{X}'\mathbf{X})^{-1}\mathbf{l}$$

We first note that we can write any other estimator of $\boldsymbol{\beta}$ that is a linear combination of the data as

$$\mathbf{b} = \left[(\mathbf{X}'\mathbf{X})^{-1}\mathbf{X}' + \mathbf{B}\right]\mathbf{y} + \mathbf{b}_0$$

Generalized Linear Models, Second Edition, by Myers, Montgomery, Vining, and Robinson where $\mathbf{B}$ is a $p \times n$ matrix and $\mathbf{b}_0$ is a $p \times 1$ vector of constants that appropriately adjusts the OLS estimator to form the alternative estimate. We next note that if the model is correct, then

$$\begin{aligned} E(\tilde{\boldsymbol{\beta}}) &= E\left(\left[(\mathbf{X}'\mathbf{X})^{-1}\mathbf{X}' + \mathbf{B}\right]\mathbf{y} + \mathbf{b}_0\right) \\ &= \left[(\mathbf{X}'\mathbf{X})^{-1}\mathbf{X}' + \mathbf{B}\right]E(\mathbf{y}) + \mathbf{b}_0 \\ &= \left[(\mathbf{X}'\mathbf{X})^{-1}\mathbf{X}' + \mathbf{B}\right]\mathbf{X}\boldsymbol{\beta} + \mathbf{b}_0 \\ &= (\mathbf{X}'\mathbf{X})^{-1}\mathbf{X}'\mathbf{X}\boldsymbol{\beta} + \mathbf{B}\mathbf{X}\boldsymbol{\beta} + \mathbf{b}_0 \\ &= \boldsymbol{\beta} + \mathbf{B}\mathbf{X}\boldsymbol{\beta} + \mathbf{b}_0 \end{aligned}$$

Consequently, $\tilde{\boldsymbol{\beta}}$ is unbiased if and only if both $\mathbf{b}_0 = \mathbf{0}$ and $\mathbf{BX} = \mathbf{0}$. The variance of $\tilde{\boldsymbol{\beta}}$ is

$$\begin{aligned} \mathrm{Var}(\tilde{\boldsymbol{\beta}}) &= \mathrm{Var}\left(\left[(\mathbf{X}'\mathbf{X})^{-1}\mathbf{X}' + \mathbf{B}\right]\mathbf{y}\right) \\ &= \left[(\mathbf{X}'\mathbf{X})^{-1}\mathbf{X}' + \mathbf{B}\right]\mathrm{Var}(\mathbf{y})\left[(\mathbf{X}'\mathbf{X})^{-1}\mathbf{X}' + \mathbf{B}\right]' \\ &= \left[(\mathbf{X}'\mathbf{X})^{-1}\mathbf{X}' + \mathbf{B}\right]\sigma^2\mathbf{I}\left[(\mathbf{X}'\mathbf{X})^{-1}\mathbf{X}' + \mathbf{B}\right]' \\ &= \sigma^2\left[(\mathbf{X}'\mathbf{X})^{-1}\mathbf{X}' + \mathbf{B}\right]\left[\mathbf{X}(\mathbf{X}'\mathbf{X})^{-1} + \mathrm{B}'\right] \\ &= \sigma^2\left[(\mathbf{X}'\mathbf{X})^{-1} + \mathbf{BB}'\right] \end{aligned}$$

because $\mathbf{BX} = \mathbf{0}$, which in turn implies that $(\mathbf{BX})' = \mathbf{X}'\mathbf{B}' = \mathbf{0}$. As a result

$$\begin{aligned} \mathrm{Var}(\mathbf{l}'\tilde{\boldsymbol{\beta}}) &= \mathbf{l}'\mathrm{Var}(\tilde{\boldsymbol{\beta}})\mathbf{l} \\ &= \mathbf{l}'\left(\sigma^2\left[(\mathbf{X}'\mathbf{X})^{-1} + \mathbf{BB}'\right]\right)\mathbf{l} \\ &= \sigma^2\mathbf{l}'(\mathbf{X}'\mathbf{X})^{-1}\mathbf{l} + \sigma^2\mathbf{l}'\mathbf{BB}'\mathbf{l} \\ &= \mathrm{Var}(\mathbf{b}) + \sigma^2\mathbf{l}'\mathbf{BB}'\mathbf{l} \end{aligned}$$

We note that $\mathbf{BB}'$ is at least positive semidefinite matrix; hence $\sigma^2\mathbf{l}'\mathbf{BB}'\mathbf{l} \geq 0$. So we can define $\mathbf{1}^* = \mathbf{B}'\mathbf{l}$. As a result

$$\mathbf{l}'\mathbf{BB}'\mathbf{l} = \mathbf{l}^{*\prime}\mathbf{l}^* = \sum_{i=1}^{p} \mathbf{l}_i^{*2}$$

which must be strictly greater than 0 for some $\mathbf{l} \neq \mathbf{0}$ unless $\mathbf{B} = \mathbf{0}$. Thus the OLS estimate of $\boldsymbol{\beta}$ is the best linear unbiased estimator.

APPENDIX A.4

The Relationship Between Maximum Likelihood Estimation of the Logistic Regression Model and Weighted Least Squares

In Chapter 4, we showed that the maximum likelihood estimator solves

$$\mathbf{X}'(\mathbf{y} - \boldsymbol{\mu}) = \mathbf{0}$$

where $\mathbf{y}' = [y_1, y_2, \ldots, y_n]$ and $\boldsymbol{\mu}' = [n_1\pi_1, n_2\pi_2, \ldots, n_m\pi_m]$. This set of equations is often called the **maximum likelihood score equations.** The **Newton–Raphson** method is typically used to solve the score equations. This procedure observes that in the neighborhood of the solution, we can use a first-order Taylor series expansion to form the approximation

$$\frac{y_i}{n_i} - \pi_i \approx \left(\frac{\partial \pi_i}{\partial \boldsymbol{\beta}}\right)'(\boldsymbol{\beta}^* - \boldsymbol{\beta}) \tag{A.4.1}$$

where $\boldsymbol{\beta}^*$ is the value of $\boldsymbol{\beta}$ that solves the score equations. Now $\eta_i = \mathbf{x}'_i\boldsymbol{\beta}$, and

$$\frac{\partial \eta_i}{\partial \boldsymbol{\beta}} = x_i$$

so

$$\pi_i = \frac{\exp(\eta_i)}{1 + \exp(\eta_i)}$$

By the chain rule

$$\frac{\partial \pi_i}{\partial \boldsymbol{\beta}} = \frac{\partial \pi_i}{\partial \eta_i}\frac{\partial \eta_i}{\partial \boldsymbol{\beta}} = \frac{\partial \pi_i}{\partial \eta_i}\mathbf{x}_i$$

Therefore we can rewrite (A.4.1) as

$$\begin{aligned}
\frac{y_i}{n_i} - \pi_i &\approx \left(\frac{\partial \pi_i}{\partial \eta_i}\right)\mathbf{x}'_i(\boldsymbol{\beta}^* - \boldsymbol{\beta}) \\
\frac{y_i}{n_i} - \pi_i &\approx \left(\frac{\partial \pi_i}{\partial \eta_i}\right)(\mathbf{x}'_i\boldsymbol{\beta}^* - \mathbf{x}'_i\boldsymbol{\beta}) \qquad \text{(A.4.2)} \\
\frac{y_i}{n_i} - \pi_i &\approx \left(\frac{\partial \pi_i}{\partial \eta_i}\right)(\eta_i^* - \eta_i)
\end{aligned}$$

where η_i^* is the value of η_i evaluated at $\boldsymbol{\beta}^*$. We note that

$$\pi_i = \frac{\exp(\eta_i)}{1 + \exp(\eta_i)}$$

Thus we can write

$$\begin{aligned}
\frac{\partial \pi_i}{\partial \eta_i} &= \frac{\exp(\eta_i)}{1 + \exp(\eta_i)} - \left[\frac{\exp(\eta_i)}{1 + \exp(\eta_i)}\right]^2 \\
&= \pi_i(1 - \pi_i)
\end{aligned}$$

Consequently,

$$y_i - n_i\pi_i \approx [n_i\pi_i(1 - \pi_i)](\eta_i^* - \eta_i)$$

Now the variance of the linear predictor $\eta_i^* = \mathbf{x}_i'\boldsymbol{\beta}^*$ is, to a first approximation,

$$\text{Var}(\eta_i^*) \approx \frac{1}{n_i\pi_i(1 - \pi_i)} = \frac{1}{\text{Var}(y_i)}$$

Thus

$$y_i - n_i\pi_i \approx \text{Var}(y_i) \cdot (\eta_i^* - \eta_i)$$

and we may rewrite the score equations as

$$\sum_{i=1}^{m} \text{Var}(y_i) \cdot (\eta_i^* - \eta_i) = 0$$

or in matrix notation,

$$\mathbf{X}'\mathbf{V}(\boldsymbol{\eta}^* - \boldsymbol{\eta}) = \mathbf{0}$$

where $\mathbf{V}$ is a diagonal matrix of the weights formed from the variances of the y_i. Because $\boldsymbol{\eta} = \mathbf{X}\boldsymbol{\beta}$ we may write the score equations as

$$\mathbf{X}'\mathbf{Y}(\boldsymbol{\eta}^* - \mathbf{X}\boldsymbol{\beta}) = \mathbf{0}$$

and the maximum likelihood estimate of $\boldsymbol{\beta}$ is

$$\mathbf{b} = (\mathbf{X}'\mathbf{V}\mathbf{X})^{-1}\mathbf{X}'\mathbf{V}\boldsymbol{\eta}^*$$

However, there is a problem because we don't know $\boldsymbol{\eta}^*$. Our solution to this problem uses Equation (A.4.2):

$$\frac{y_i}{n_i} - \pi_i \approx \left(\frac{\partial \pi_i}{\partial \eta_i}\right)(\eta_i^* - \eta_i)$$

which we can solve for η_i^*,

$$\eta_i^* \approx \eta_i + \left(\frac{y_i}{n_i} - \pi_i\right)\frac{\partial \eta_i}{\partial \pi_i}$$

Let $z_i = \eta_i + (y_i/n_i - \pi_i)\partial\eta_i/\partial\pi_i$ and $\mathbf{z}' = [z_1, z_2, \ldots, z_n]$. Then the Newton–Raphson estimate of $\boldsymbol{\beta}$ is

$$\mathbf{b} = (\mathbf{X}'\mathbf{V}\mathbf{X})^{-1}\mathbf{X}'\mathbf{V}\mathbf{z}$$

In this case, $\mathbf{V} = \text{diag}\{n\pi_i(1-\pi_i)\}$. Thus the IRLS algorithm based on the Newton–Raphson method can be described as follows:

1. Use ordinary least squares to obtain an initial estimate of $\boldsymbol{\beta}$, say, $\mathbf{b}_0$.
2. Use $\mathbf{b}_0$ to estimate $\mathbf{V}$ and π.
3. Let $\boldsymbol{\eta}_0 = \mathbf{X}\mathbf{b}_0$.
4. Base $\mathbf{z}_1$ on $\boldsymbol{\eta}_0$.
5. Obtain a new estimate $\mathbf{b}_1$, and iterate until some suitable convergence criterion is satisfied.

If $\mathbf{b}$ is the final value that the above algorithm produces and if the model assumptions are correct, then we can show that asymptotically

$$E(\mathbf{b}) = \boldsymbol{\beta} \quad \text{and} \quad \text{Var}(\mathbf{b}) = (\mathbf{X}'\mathbf{V}^{-1}\mathbf{X})^{-1}$$

The fitted value of the logistic regression model is often written as

$$\hat{\pi}_i = \frac{\exp(\mathbf{x}_i'\mathbf{b})}{1 + \exp(\mathbf{x}_i'\mathbf{b})}$$
$$= \frac{1}{1 + \exp(-\mathbf{x}_i'\mathbf{b})}$$

APPENDIX A.5

Computational Details for GLMs for a Canonical Link

Recall that the log-likelihood function for GLMs is

$$L = \ln \mathscr{L}(\mathbf{y},\boldsymbol{\beta}) = \sum_{i=1}^{n} [y_i\theta_i - b(\theta_i)]/a(\phi) + c(y_i, \phi)$$

For the canonical link we have $\eta_i = g[E(y_i)] = g(\mu_i) = \mathbf{x}_i'\boldsymbol{\beta}$, and the score equations are

$$\frac{1}{a(\phi)} \sum_{i=1}^{n} (y_i - \mu_i)\mathbf{x}_i = \mathbf{0} \qquad \text{(A.5.1)}$$

In matrix form these equations are

$$\mathbf{X}'(\mathbf{y} - \boldsymbol{\mu}) = \mathbf{0}$$

where $\boldsymbol{\mu}' = [\mu_1, \mu_2, \ldots, \mu_n]$.

To solve the score equations, we can use iteratively reweighted least squares (IRLS), just as we did in the cases of logistic and Poisson regression. We start by finding a first-orderTaylor series approximation in the neighborhood of the solution, η_i^*, which is

$$y_i - \mu_i \approx \frac{d\mu_i}{d\eta_i}(\eta_i^* - \eta_i)$$

Now for a canonical link $\eta_i = \theta_i$, and

$$y_i - \mu_i \approx \frac{d\mu_i}{d\theta_i}(\eta_i^* - \eta_i)$$

Generalized Linear Models, Second Edition, by Myers, Montgomery, Vining, and Robinson

Therefore we have

$$\eta_i^* - \eta_i \approx (y_i - \mu_i)\frac{d\theta_i}{d\mu_i} \tag{A.5.2}$$

Recall that

$$\text{Var}_\mu = \frac{d\mu_i}{d\theta_i}$$

Thus we can reexpress Equation (A.5.2) as

$$\eta_i^* - \eta_i \approx \frac{(y_i - \mu_i)}{\text{Var}_\mu}$$

which implies that

$$y_i - \mu_i \approx \text{Var}_\mu \cdot (\eta_i^* - \eta) \tag{A.5.3}$$

Substituting Equation (A.5.3) into (A.5.1) yields

$$\frac{1}{a(\phi)}\sum_{i=1}^{n}(\eta_i^* - \eta_i)\cdot \text{Var}_\mu \cdot \mathbf{x}_i = \mathbf{0} \tag{A.5.4}$$

Let $\mathbf{V} = \text{diag}(\text{Var}_\mu\}$. In matrix form Equation (A.5.4) is

$$\mathbf{y} - \boldsymbol{\mu} \approx \frac{1}{a(\phi)}\mathbf{V}^{-1}\cdot(\boldsymbol{\eta}^* - \boldsymbol{\eta})$$

If $a(\phi)$ is constant, we rewrite the score equations as follows:

$$\mathbf{X}'(\mathbf{y} - \boldsymbol{\mu}) = \mathbf{0}$$
$$\mathbf{X}'\mathbf{V}^{-1}(\boldsymbol{\eta}^* - \boldsymbol{\eta}) = \mathbf{0}$$
$$\mathbf{X}'\mathbf{V}^{-1}(\boldsymbol{\eta}^* - \mathbf{X}\boldsymbol{\beta}) = \mathbf{0}$$

Thus the maximum likelihood estimate of $\boldsymbol{\beta}$ is

$$\mathbf{b} = (\mathbf{X}'\mathbf{V}^{-1}\mathbf{X})^{-1}\mathbf{X}'\mathbf{V}^{-1}\boldsymbol{\eta}^*$$

Now just as we saw in the logistic regression situation, we do not know $\boldsymbol{\eta}^*$, so we pursue an iterative scheme based on

$$z_i = \hat{\eta}_i + (y_i - \hat{\mu}_i)\frac{d\eta_i}{d\mu_i}$$

Using iteratively reweighted least squares with the Newton–Raphson method, we find the solution from

$$\mathbf{b} = (\mathbf{X}'\mathbf{V}^{-1}\mathbf{X})^{-1}\mathbf{X}'\mathbf{V}^{-1}\mathbf{z}$$

Asymptotically the random component of $\mathbf{z}$ comes from the observations y_i.

To take an example, consider the logistic regression case:

$$\begin{aligned}
\eta_i &= \ln\left(\frac{\pi_i}{1-\pi_i}\right) \\
\frac{d\eta_i}{d\mu_i} &= \frac{d\eta_i}{d\pi_i} = \frac{d\ln[\pi_i/(1-\pi)]}{d\pi_i} \\
&= \frac{1-\pi_i}{\pi_i}\left[\frac{\pi_i}{1-\pi_i} + \frac{\pi_i}{(1-\pi_i)^2}\right] \\
&= \frac{(1-\pi_i)}{\pi_i(1-\pi_i)}\left[1 + \frac{\pi_i}{1-\pi_i}\right] \\
&= \frac{1}{\pi_i}\left[\frac{1-\pi_i+\pi_i}{1-\pi_i}\right] \\
&= \frac{1}{\pi_i(1-\pi_i)}
\end{aligned}$$

Thus

$$z_i = \hat{\eta}_i + (p_i - \pi_i)\frac{d\eta_i}{d\pi_i} = \hat{\eta}_i + \frac{p_i - \pi_i}{\pi_i(1-\pi_i)}$$

and $\mathbf{V} = \text{diag}\left(\frac{1}{n_i\pi_i(1-\pi_i)}\right)$, which is exactly what we obtained previously. Therefore IRLS based on the Newton–Raphson method can be described as follows:

1. Obtain an initial estimate of $\boldsymbol{\beta}$, say, $\mathbf{b}_0$.
2. Use $\mathbf{b}_0$ to estimate $\mathbf{V}$ and $\boldsymbol{\mu}$.
3. Let $\boldsymbol{\eta}_0 = \mathbf{X}\mathbf{b}_0$.
4. Base $\mathbf{z}_1$ on $\boldsymbol{\eta}_0$.
5. Obtain a new estimate $\mathbf{b}_1$, and iterate until some suitable convergence criterion is satisfied.

APPENDIX A.6

Computational Details for GLMs for a Noncanonical Link

If we don't use the canonical link, then $\eta_i \neq \theta_i$, and the appropriate derivative of the log-likelihood is

$$\frac{\partial L}{\partial \boldsymbol{\beta}} = \frac{dL}{d\theta_i}\frac{d\theta_i}{d\mu_i}\frac{d\mu_i}{d\eta_i}\frac{\partial \eta_i}{\partial \boldsymbol{\beta}}$$

Note that

1. $\dfrac{dL}{d\theta_i} = \dfrac{1}{a(\phi)}\left[y_i \dfrac{db(\theta_i)}{d\theta_i}\right] = \dfrac{1}{a(\phi)}(y_i - \mu_i)$
2. $\dfrac{d\theta_i}{d\mu_i} = \dfrac{1}{\text{Var}_\mu}$
3. $\dfrac{\partial \eta_i}{\partial \boldsymbol{\beta}} = \mathbf{x}_i$

Putting all this together yields

$$\frac{\partial L}{\partial \boldsymbol{\beta}} = \sum_{i=1}^{n} \frac{y_i - \mu_i}{a(\phi)} \frac{1}{\text{Var}_\mu} \frac{d\mu_i}{d\eta_i} \mathbf{x}_i$$

Once again, we can use a Taylor series expansion to obtain

$$y_i - \mu_i \approx \frac{d\mu_i}{d\eta_i}(\eta_i^* - \eta_i)$$

The score equations then become

Generalized Linear Models, Second Edition, by Myers, Montgomery, Vining, and Robinson

$$\sum_{i=1}^{n} \frac{\eta_i^* - \eta_i}{a(\phi)} \frac{1}{\mathrm{Var}_\mu} \left(\frac{d\mu_i}{d\eta_i}\right)^2 \mathbf{x}_i = \mathbf{0} \qquad \text{(A.6.1)}$$

It is straightforward to show that

$$\mathrm{Var}(\eta_i^* - \eta_i) \approx a(\phi) \cdot \mathrm{Var}_\mu \cdot \left(\frac{d\eta_i}{d\mu_i}\right)^2$$

As a result we can reexpress Equation (A.6.1) as

$$\sum_{i=1}^{n} \frac{\eta_i^* - \eta_i}{\mathrm{Var}(\eta_i^* - \eta_i)} \mathbf{x}_i = \mathbf{0}$$

Let $\mathbf{V}_\eta = \mathrm{diag}\{\mathrm{Var}(\eta_i^* - \eta_i)\}$. The score equations in matrix form are

$$\mathbf{X}'\mathbf{V}_\eta^{-1}(\boldsymbol{\eta}^* - \boldsymbol{\eta}) = \mathbf{0}$$
$$\mathbf{X}'\mathbf{V}_\eta^{-1}(\boldsymbol{\eta}^* - \mathbf{X}\boldsymbol{\beta}) = \mathbf{0}$$
$$\mathbf{X}'\mathbf{V}_\eta^{-1}\boldsymbol{\eta}^* - \mathbf{X}'\mathbf{V}_\eta^{-1}\mathbf{X}\boldsymbol{\beta} = \mathbf{0}$$

The MLE of $\boldsymbol{\beta}$ is

$$\mathbf{b} = (\mathbf{X}'\mathbf{V}_\eta^{-1}\mathbf{X})^{-1}\mathbf{X}'\mathbf{V}_\eta^{-1}\boldsymbol{\eta}^*$$

which is generalized least squares on $\boldsymbol{\eta}^*$. Again, we do not know $\boldsymbol{\eta}^*$, so we pursue an iterative scheme based on

$$z_i = \hat{\eta}_i + (y_i - \hat{\mu}_i)\frac{d\eta_i}{d\mu_i}$$

Using iteratively reweighted least squares with the Newton–Raphson method, we find the solution from

$$\mathbf{b} = (\mathbf{X}'\mathbf{V}_\eta^{-1}\mathbf{X})^{-1}\mathbf{X}'\mathbf{V}_\eta^{-1}\mathbf{z}$$

Asymptotically the random component of $\mathbf{z}$ comes from the observations y_i. The use of a noncanonical link still leads to iteratively reweighted least squares; however, the weights are no longer $\mathrm{Var}(y_i)$.

Following an argument similar to that employed before, let Δ be a diagonal matrix whose diagonal elements are

$$\left[\frac{d\theta_i}{d\mu_i}\right]$$

We can show for a noncanonical link that the score equations become

$$\mathbf{X}'\boldsymbol{\Delta}(\mathbf{y}-\boldsymbol{\mu}) = \mathbf{0}$$

and eventually we can show that

$$\frac{\partial L}{\partial \boldsymbol{\beta}} = \sum_{i=1}^{n} \frac{\eta_i^* - \eta_i}{a(\phi)\mathrm{Var}(\eta_i)}\mathbf{x}_i$$

Equating this last expression to zero and writing it in matrix form, we obtain

$$\mathbf{X}'\mathbf{V}^{-1}(\boldsymbol{\eta}^* - \boldsymbol{\eta}) = \mathbf{0}$$

or since $\boldsymbol{\eta} = \mathbf{X}\boldsymbol{\beta}$,

$$\mathbf{X}'\mathbf{V}^{-1}(\boldsymbol{\eta}^* - \mathbf{X}\boldsymbol{\beta}) = \mathbf{0}$$

The Newton–Raphson solution is based on

$$\mathbf{b} = (\mathbf{X}'\mathbf{V}^{-1}\mathbf{X})^{-1}\mathbf{X}'\mathbf{V}^{-1}\mathbf{z}$$

where

$$z_i = \hat{\eta}_i + (y_i - \hat{\mu}_i)\frac{d\eta_i}{d\mu_i}$$

Just as in the case of the canonical link, the matrix $\mathbf{V}$ is a diagonal matrix formed from the variances of the estimated linear predictors, apart from $a\,(\varphi)$.

References

Abdelbasit, K.M. and Plackett, R.L. (1983). "Experimental Design for Binary Data," *Journal of the American Statistical Association*, 78, pp. 90–98.

Agresti, A. (1990). *Categorical Data Analysis*, Wiley, Hoboken, NJ.

Aitken, M. (1987). "Modeling Variance Heterogeneity in Normal Regression Using GLIM," *Applied Statistics*, 36, pp. 332–339.

Akaike, H. (1974). "A New Look at the Statistical Model Identification," *IEEE Transactions on Automatic Control*, 19, pp. 716–723.

Allen, D.M. (1971). "Mean Square Error of Prediction as a Criterion for Selecting Variables," *Technometrics*, 13, pp. 469–475.

Allen, D.M. (1974). "The Relationship Between Variable Selection and Data Augmentation and a Method for Prediction," *Technometrics*, 16, pp. 125–127.

Allison, P.D. (1999). *Logistic Regression Using the SAS System: Theory and Application*, SAS Institute, Cary, NC.

Anand, K.N. (1997). "Improving the Yield of Silica Gel in a Chemical Plant," *Quality Engineering*, 9, pp. 355–361.

Ashford, J.K. "An Approch to the Analysis of Data for Semi-Quantal Responses in Biological Assay," *Biometrics*, 15, pp. 573–581.

Bailer, A.J. and Piegorsch, W.W. (2000). "From Quantal Counts to Mechanisms and Systems: The Past, Present, and Future of Biometrics in Environmental Toxicology," *Biometrics*, 56, pp. 327–336.

Bast, R.C. Jr. Klug, T.L., St. John, E., Jenison, E., Niloff, J.M., Lazarus, H., Berkowitz, R.S., Leavitt, T., Griffiths, C.T., Parker, L., Zurawski, V.R. Jr., and Knapp, R.C. (1983). "Radioimmunoassay Using a Monoclonal Antibody to Monitor the Course of Epithelial Ovarian Cancer," *New England Journal of Medicine*, 309, pp. 883–887.

Bates, D.M. and Watts, D.G. (1988). *Nonlinear Regression Analysis and Its Applications*, Wiley, Hoboken, NJ.

Bickel, P.J. and Doksum, K.A. (2001). *Mathematical Statistics: Basic Ideas and Selected Topics*, 2nd ed., Prentice Hall, Upper Saddle River, NJ.

Bisgaard, S. and Fuller, H.T. (1994–1995). "Analysis of Factorial Experiments with Defects or Defectives as the Response," *Quality Engineering*, 7, pp. 429–443.

Bliss, C.I. (1940). "The Relation Between Exposure Time, Concentration, and Toxicity in Experiments on Insecticides," *Annals of the Entomological Society of America*, 33, pp. 721–766.

Box, G.E.P. (1982), "Choice of Response Surface Design and Alphabetic Optimality," *Utilitas Mathematica*, 21, pp. 11–55.

Box, G.E.P. and Behnken, D.W. (1960). "Some New Three-Level Designs for the Study of Quantitative Variables," *Technometrics*, 2, pp. 455–475.

Box, G.E.P. and Cox, D.R. (1964). "An Analysis of Transformations," *Journal of the Royal Statistical Society, Series B*, 26, pp. 211–252.

Box, G.E.P. and Draper, N.R. (1959). "A Basis for the Selection of a Response Surface Design," *Journal of the American Statistical Association*, 54, pp. 622–654.

Box, G.E.P. and Draper, N.R. (1987). *Empirical Model Building and Response Surfaces*, Wiley, Hoboken, NJ.

Box, G.E.P. and Hunter, J.S. (1961a). "The 2^{k-p} Fractional Factorial Designs, Part I," *Technometrics*, 2, pp. 311–352.

Box, G.E.P. and Hunter, J.S. (1961b). "The 2^{k-p} Fractional Factorial Designs,Part II," *Technometrics*, 2, pp. 449–458.

Box, G.E.P., Hunter, W.G., and Hunter, J.S. (2005). Statistics for Experimenters, 2nd ed., Wiley, Hoboken, NJ.

Box, G.E.P. and Lucas, H.L. (1959). "Design of Experiments in Non-linear Situations," *Biometrika*, 46, pp. 77–90.

Box, G.E.P. and Wilson, K.B. (1951). "On the Experimental Attainment of Optimum Conditions," *Journal of the Royal Statistical Society, Series B*, 13, pp. 1–45.

Branden, M.N., Vidmar, T.J., and McKean, J.W. (1988). *Drug Interaction and Lethality Analysis*, CRC Press, Boca Raton, FL.

Breslow, N.E. and Clayton, D.G. (1993). "Approximate Inference in Generalized Linear Mixed Models," *Journal of the American Statistical Association*, 88, pp. 9–25.

Breslow, N.E. and Lin, X. (1995). "Bias Correction Generalized Linear Mixed Models with a Single Component of Dispersion," *Biometrika*, 81, pp. 81–91.

Byrne, D.M. and Taguchi, S. (1987). "The Taguchi Approach to Parameter Design," *Quality Progress*, 21(12), pp. 19–26.

Carroll, R.J. and Ruppert, D. (1988). *Transformation and Weighting in Regression*, Chapman and Hall, London.

Casella, G. and George, E. (1992). "Explaining the Gibbs Sampler," *The American Statistician*, 46, pp. 167–174.

Chaloner, K. and Larntz, K. (1989). "Optimal Bayesian Design Applied to Logistic Regression Experiments," *Journal of Statistical Planning and Inference*, 21, pp. 191–208.

Chaloner, K. and Verdinelli, I. (1995). "Bayesian Experimental Design: A Review," *Statistical Science*, 10, pp. 273–304.

Chapman, R.E. (1997–98). "Degradation Study of a Photographic Developer to Determine Shelf Life," *Quality Engineering*, 10, pp. 137–140.

Chatterjee, S. and Hadi, A.S. (1988). *Sensitivity Analysis in Linear Regression*, Wiley, Hoboken, NJ.

Chib, S. and Greenberg, E. (1995), "Understanding the Metropolis–Hastings Algorithm," *The American Statistician*, 49, pp. 327–335.

Chowdhury, K.K., Gijo, E.V., and Raghavan, R. (2000). "Quality Improvements Through Design of Experiments: A Case Study," *Quality Engineering*, 12, pp. 407–416.

Collett, D. (1991). *Modelling Binary Data*, Chapman and Hall, London.

Condra, L.W. (1993). *Reliability Improvement with Design of Experiments*, Marcel Dekker, New York.

Cook, R.D. (1977). "Detection of Influential Observations in Linear Regression," *Technometrics*, 19, pp. 15–18.

Cook, R.D. (1979). "Influential Observations in Linear Regression," *Journal of the American Statistical Association*, 74, pp. 169–174.

Cornell, J.A. (2002). *Experiments in Mixtures*, 3rd ed. Wiley, Hoboken, NJ.

Cox, D.R. (1972). "Regression Models and Life Tables," *Journal of the Royal Statistical Society, Series B*, 74, pp. 187–220.

Daniel, C. (1976). *Applications of Statistics to Industrial Experimentation*, Wiley, Hoboken, NJ.

DeGroot, M.H. (1970). *Optimal Statistical Decisions*, McGraw-Hill, New York.

Dobson, A.J. (1990). *An Introduction to Generalized Linear Models*, Chapman and Hall, London.

Dror, H.A. and Steinberg, D.M. (2006). "Robust Experimental Design for Multivariate Generalized Linear Models," *Technometrics*, 48, pp. 520–529.

Engle, J. and Huele, A.F. (1996). "A Generalized Linear Modeling Approach to Robust Parameter Design," *Technometrics*, 38, pp. 365–373.

Fahrmeir, L. and Tutz, G. (1994). *Multivariate Statistical Modeling Based on Generalized Linear Models*, Springer-Verlag, New York.

Finney, D.J. (1950). *Probit Analysis*, Cambridge University Press, Cambridge, UK.

Ford, I., Kitsos, D., and Titterington, D.M. (1989). "Recent Advances in Norlinear Experimental Design," *Technometrics*, 31, pp. 49–60.

Gelman, A. (2006). "Prior Distributions for Variance Parameters in Hierarchical Models," *Bayesian Statistics*, 1, pp. 515–533.

Gelfand, A.E. and Smith, A.F.M. (1990). "Sampling-Based Approaches to Calculating Marginal Densities," *Journal of the American Statistical Association*, 85, pp. 398–409.

Gotwalt, C.M., Jones, B.A., and Steinberg, D.M. (2009). "Fast Computation of Designs Robust to Parameter Uncertainty for Nonlinear Settings," *Technometrics*, 51, pp. 66–95.

Gregoire, T.G., Schabenberger, O., and Barrett, J.P. (1995). "Linear Modeling of Irregularly Spaced, Unbalanced, Longitudinal Data from Permanent Plot Measurements," *Canadian Journal of Forest Research*. 25, pp. 137–156.

Gupta, A. and Das, A.K. (2000). "Improving Resistivity of UF Resin Through Setting of Process Parameters," *Quality Engineering*, 12, pp. 611–618.

Hamada, M. and Nelder, J.A. (1997). "Generalized Linear Models for Quality-Improvement Experiments," *Journal of Quality Technology*, 29, pp. 292–304.

Harvey, A.C. (1976). "Estimation of Regression Models with Multiplicated Heteroscedasticity," *Econometrica*, 44, pp. 461–475.

Hastie, T.J. and Tibshirani, R. (1990). *Generalized Additive Models*, Chapman and Hall, London.

Heise, M.A. and Myers, R.H. (1996). "Optimal Designs for Bivariate Logistic Regression," *Biometrika*, 52, pp. 613–624.

Henderson, R. and Shimakura S. (2003). "A Serially Correlated Gamma Frailty Model for Longitudinal Count Data," *Biometrika*, 90, pp. 355–366.

Hosmer, D.W. Jr. and Lemeshow, S. (2002). *Applied Logistic Regression*, 2nd ed., Wiley, Hoboken, NJ.

Jensen, W.A., Birch, J.B., and Woodall, W.W. (2008). "Monitoring Correlation Within Linear Profiles Using Mixed Models," *Journal of Quality Technology*, 40, pp. 167–183.

Jia, Y. and Myers, R.H. (2001). "Some Optimal and Robust Designs for the Two Variable Logistic Model," Technical Report No. 01–3, Department of Statistics, VPI & SU, Blacksburg, VA.

Kalish, L.A. and Rosenberger, J.L. (1978). "Optimal Designs and the Estimation of the Logistic Function," Technical Report No. 33, Department of Statistics, Pennsylvania State University, University Park, PA.

Kenward, M.G. and Roger, J.H. (1997). "Small Sample Inference for Fixed Effects from Restricted Maximum Likelihood," *Biometrics*, 53, pp. 983–997.

Kiefer, J. (1959). "Optimum Experimental Designs," *Journal of the Royal Statistical Society, Series B*, 21, pp. 272–304.

Kiefer, J. and Wolfowitz, J. (1959). "Optimum Designs in Regression Problems," *Annals of Mathematical Statistics*, 30, pp. 271–294.

Kleinbaum, D.G. (1994). *Logistic Regression: A Self-Learning Text*, Springer-Verlag, New York.

Khuri, A., Mukherjee, B., Sinh, B., and Ghosh, M. (2006). "Design Issues for Generalized Linear Models," *Statistical Science*, 21, pp. 376–399.

Kowalski, S.M., Parker, P.A., and Vining, G.G. (2007). "Tutorial: Industrial Split-Plot Experiments," *Quality Engineering*, 19, pp. 1–16.

Lee, Y. and Nelder, J.A. (1996). "Hierarchical Generalized Linear Models," *Journal of the Royal Statistical Society Series B*, 58, pp. 619–656.

Lee, Y. and Nelder, J.A. (1998a). Letter to the Editor: "Joint Modeling of Mean and Dispersion," *Technometrics*, 40, pp. 168–171.

Lee, Y. and Nelder, J.A. (1998b). "Generalized Linear Models for the Analysis of Quality-Improvement Experiments," *Canadian Journal of Statistics*, 26, pp. 95–105.

Lewis, S.L. (1998). "Analysis of Designed Experiments Using Generalized Linear Models," Ph.D. Dissertation, Department of Industrial Engineering, Arizona State University, Tempe, Arizona.

Lewis, S.L., Montgomery, D.C., and Myers, R.H. (2001a). "Examples of Designed Experiments with Nonnormal Responses," *Journal of Quality Technology*, 33, pp. 265–278.

Lewis, S.L., Montgomery, D.C., and Myers, R.H. (2001b). "Confidence Interval Coverage for Designated Experiments Analyzed with GLMs," *Journal of Quality Technology*, 33, pp. 279–292.

Liang, K.Y. and Zeger, S.L. (1986). "Longitudinal Analysis Using Generalized Linear Models," *Biometrika*, 73, pp. 13–22.

Lindsey, J.K. (1994). *Applying Generalized Linear Models*, Springer-Verlag, New York.

Lipsitz, S.R., Kim, K., and Zhao, L. (1994). "Analysis of Repeated Categorical Data Using Generalized Estimating Equations," *Statistics in Medicine*, 13, pp. 1149–1163.

Littell, R.C., Milliken, G.A Stroup, W.W., Wolfinger, R.D. and Schaben berger, O. (2006). *SAS for Mixed Models* 2nd ed. SAS Publishing, Caryl, N.C.

Mancl, L.A. and De Rouen, T.A. (2001). "A Covariance Estimator for GEE with Improved Small Sample Properties," *Biometrics*, 57, pp. 126–134.

Marquardt, D.W. (1963). "An Algorithm for Least Squares Estimation of Nonlinear Parameters," *Journal of the Society for Industrial Applications of Mathematics*, 2, pp. 431–441.

Maruthi, B.N. and Joseph, V.R. (1999–2000). "Improving the Yield of Printed Circuit Boards Using Design of Experiments," *Quality Engineering*, 12, pp. 259–265.

McCullagh, P. and Nelder, J.A. (1989). *Generalized Linear Models*, Chapman and Hall, London.

McKnight, B. and van den Eeden, S.K. (1993). "A Conditional Analysis for Two-Treatment Multiple-Period Crossover Designs with Binomial or Poisson Outcomes and Subjects Who Drop Out," *Statistics in Medicine*. 12, pp. 825–834.

Meyer, R. and Nachtsheim, C. (1995). "The Coordinate Exchange Algorithm for Constructing Exact Optimal Experimental Designs," *Technometrics*, 37, pp. 60–69.

Minkin, S. (1987). "Optimal Designs for Binary Data," *Journal of the American Statistical Association*, 82, pp. 1098–1103.

Minkin, S. (1993). "Experimental Designs for Clonogenic Assays in Chemotherapy," *Journal of the American Statistical Association*, 88, pp. 410–420.

Mitchell, T.J. (1974). "An Algorithm for the Construction of D-Optimal Exeperimental Designs," *Technometrics*, 16, pp. 203–210.

Montgomery, D.C. (2009). *Design and Analysis of Experiments*, 7th ed., Wiley, Hoboken, NJ.

Montgomery, D.C, Peck, E.A., and Vining, G.G. (2006). *Introduction to Linear Regression Analysis*, 4th ed., Wiley, Hoboken, NJ.

Myers, R.H. (1964). "Methods for Estimating the Composition of a Three Component Liquid Mixture," *Technometrics*, 6, pp. 343–356.

Myers, R.H. (1990). *Classical and Modem Regression with Applications*, 2nd ed., Duxbury Press, Boston.

Myers, R.H., Khuri, A.I., and Carter, W.H. (1989). "Response Surface Methodology: 1966–1988," *Technometrics*, 31, pp. 137–157.

Myers, R.H., Montgomery, D.C, and Anderson-Cook, C.M. (2009). *Response Surface Methodology* 3rd ed., Wiley, Hoboken, NJ.

Myers, R.H. and Montgomery, D.C. (1997). "A Tutorial on Generalized Linear Models," *Journal of Quality Technology*, 29, pp. 274–291.

Myers, W.R., Myers, R.H., and Carter, W.H. (1997). "Some Alphabetic Optimal Designs for Logistic Regression," *Journal of Statistical Planning and Inference*, 42, pp. 57–77.

Nelder, J.A. and Lee, Y. (1991). "Generalized Linear Models for the Analysis of Taguchi Type Experiments," *Applied Stochastic Models and Data Analysis*, 7, pp. 107–120.

Nelson, W. (1982). *Applied Life Data Analysis*, Wiley, Hoboken, NJ.

Patterson, H.D. and Thompson, R. (1971). "Recovery of Inter-Block Information when Block Sizes are Unequal," *Biometriko*, 58 pp. 545–554.

Pierce, D.A. and Schafer, D.W. (1986). "Residuals in Generalized Linear Models," *Journal of the American Statistical Association*, 81, pp. 977–986.

Pinheiro, J.C. and Chao, E.C. (2006). "Efficient Laplacian and Adaptive Gaussian Quadrature Algorithms for Multilevel Generalized Linear Mixed Models," *Journal of Computational Graphical Statistics* 15, pp. 58–81.

Potthoff, R.F. and Roy, S.N. (1964). "A Generalized Multivariate Analysis of Variance Model Useful Especially for Growth Curve Models," *Biometrika*, 51, pp. 665–680.

Pregibon, D. (1980). "Goodness of Link Tests for Generalized Linear Models," *Applied Statistics*, 29, pp. 15–24.

Pulkstenis, E.P. and Robinson T.J. (2004). "Goodness-of-Fit Tests for Ordinal Response Regression Models," *Statistics in Medicine*, 23, pp. 999–1014.

Pulkstenis E. and Robinson T.J. (2002). "Two Goodness-of-Fit Tests for Logistic Regression Models with Continuous Covariates," *Statistics in Medicine*, 21, pp. 79–93.

Reiczigel, J. (1999). "Analysis of Experimental Data with Repeated Measurements," *Biometrics*, 55, pp. 1059–1063.

Robinson, T. and Myers, R.H. (2001). "The Use of Generalized Estimating Equations and the Split-Plot Experiment," Technical Report No. 01–02, Department of Statistics, VP1 & SU, Blacksburg, VA.

Robinson, T.J., Myers, R.H., and Montgomery, D.C. (2004). "Analysis Considerations in Industrial Split-Plot Experiments with Non-normal Responses," *Journal of Quality Technology*, 36, pp. 180–192.

Robinson, T.J., Wulff, S.S., Montgomery, D.C., and Khuri, A.I. (2006). "Robust Parameter Design Using Generalized Linear Mixed Models," *Journal of Quality Technology*, 38, pp. 65–75.

Robinson, T.J., Anderson-Cook, C.M., and Hamada, M.S. (2009). "Bayesian Analysis of Split-Plot Experiments with Non-normal Responses for Evaluating Non-standard Performance Criteria," *Technometrics*, 51, pp. 56–65.

Schabenberger, O. (2007). "Growing Up Fast: SAS 9.2 Enhancements to the GLIMMIX Procedure," *SAS Globat Forum Proceedings Paper 177–2007*, pp. 1–19.

Schall, R. (1991). "Estimation in Generalized Linear Models with Random Effects," *Biometrika*, 78, pp. 719–727.

Schubert, K., Kerber, M.W., Schmidt, S.R., and Jones, S.E. (1992). "The Catapult Problem: Enhanced Engineering Modeling Using Experimental Design," *Quality Engineering*, 4, pp. 463–473.

Searle, S.R., and Casella, G., and McCulloch, C.E. (1992). *Variance Components*, Wiley, Hoboken, NJ.

Self, S.G. and Liang, K.Y. (1987). "Asymptotic Properties of Maximum Likelihood Estimators and Likelihood Ratio Tests Under Nonstandard Conditions," *Journal of the American Statistical Association*, 82, pp. 605–610.

Schwarz, G. (1978). "Estimating the Dimension of a Model," *The Annals of Statistics*, 6, pp. 461–464.

Sitter, R.R. (1992). "Robust Designs for Binary Data," *Biometrics*, 48, pp. 1145–1155.

Sitter, R.R. and Torsney, B. (1995). "D-Optimal Designs for Generalized Linear Models," *Advances in Model Oriented Data Analysis,* Heidelberg, pp. 87–102.

Slaton, T.L., Piegorsch, W.W., and Durham, S.D. (2000). "Estimation and Testing with Overdispersed Proportions Using the Beta-Logistic Regression Model of Heckman and Willis," *Biometrics*, 56, pp. 125–133.

Somner, A. (1982): *Nutritional Blindness: Xerophthalmia and Keratomalacia*, Oxford University Press, New York.

Spiegelhalter, D., Thomas, A., Best, N., and Lunn, D. (2004). *WinBUGS Version 1.4 User Manual.*

Stiger, T.R., Barnhart, H.X., and Williamson, J.M. (1999). "Testing Proportionality in the Proportional Odds Model Fitted with GEE," *Statistics in Medicine*, 18, pp. 1419–1433.

Stokes, M.E., Davis, C.S., and Koch, G. (1995). *Categorical Data Analysis Using the SAS System*, SAS Institute Inc., Cary, NC.

Thall, P.F. and Vail, S.C. (1990). "Some Covariance Models for Longitudinal Count Data with Overdispersion," *Biometrics*, 46, pp. 657–671.

Torres, V.A. (1993). "A Simple Analysis of Unreplicated Factorials with Possible Abnormalities," *Journal of Quality Technology*, 25, pp. 183–187.

U.S. Navy (1979). *Procedures and Analyses for Staffing Standards Development: Data/Regression Analysis Handbook*, Navy Manpower and Material Analysis Center, San Diego, CA.

Van Mullekom, J. and Myers, R.H. (2001). "Optimal Experimental Designs for Poisson Impaired Reproduction," Technical Report No. 01–1, Department of Statistics, VPI & SU, Blacksburg, VA.

Verbeke, G. and Molenberghs, G. (1997). *Linear Mixed Models in Practice: A SAS Oriented Approach.* Lecture Notes in Statistics. 126. Springer-Verlag, New York.

Verbeke, G. and Molenberghs, G. (2000). *Linear Mixed Models for Longitudinal Data*, Springer-Verlag, New York.

Verbeke, G. and Molenberghs, G. (2003). "The Use of Score Tests for Inference on Variance Components," *Biomertrics*, 59, pp. 254–262.

Vining, G.G. and Myers, R.H. (1990). "Combining Taguchi and Response Surface Philosophies: A Dual Response Approach," *Journal of Quality Technology*, 22, pp. 38–45.

Vining, G.G., Kowalski, S.M., and Montgomery, D.C. (2005). "Response Surface Designs within a Split-Plot Structure," *Journal of Quality Technology*, 37, pp. 115–129.

Wedderburn, R.W.M. (1974). "Quasi-Likelihood Functions, Generalized Linear Models and the Gauss Newton Method," *Biometrika*, 61, pp. 439–447.

Weiss, R.E. (2005). *Discrete Longitudinal Data*, Springer-Verlag, New York.

Westbrooke, I. and Robinson, T.J. (2009). "A Random-Coefficient Approach to the Analysis of Kukupa Counts," University of Wyoming Technical Report 1–09.

Wolfinger, R. and O' Connell, M. (1993). "Generalized Linear Mixed Models: A Pseudo-Likelihood Approach," *Journal of Statistical Computation and Simulation*, 48, pp. 233–243.

Woods, D.C., Lewis, S.M., Eccleston, J.A., and Russell, K.G. (2006). "Designs for Generalized Linear Models with Several Variables and Model Uncertainty," *Technometrics*, 48, pp. 287–292.

Wulff, S.S. and Robinson, T.J. (2009). "Assessing Uncertainty of Regression Estimates in a Response Surface Model for Repeated Measures," *Quality Technology and Quality Management*, 6, pp. 309–324.

Zeger, S.L. and Liang, K.Y. (1986). "Longitudinal Data Analysis for Discrete and Continuous Outcomes," *Biometrics*, 44, pp. 121–130.

Zhang, H. and Zelterman, D. (1999). "Binary Regression for Risks in Excess of Subject-Specific Thresholds," *Biometrics*, 55, pp. 1247–1251.

Index

WILEY SERIES IN PROBABILITY AND STATISTICS

ESTABLISHED BY WALTER A. SHEWHART AND SAMUEL S. WILKS

The ***Wiley Series in Probability and Statistics*** is well established and authoritative. It covers many topics of current research interest in both pure and applied statistics and probability theory. Written by leading statisticians and institutions, the titles span both state-of-the-art developments in the field and classical methods.

Reflecting the wide range of current research in statistics, the series encompasses applied, methodological and theoretical statistics, ranging from applications and new techniques made possible by advances in computerized practice to rigorous treatment of theoretical approaches.

This series provides essential and invaluable reading for all statisticians, whether in academia, industry, government, or research.

† ABRAHAM and LEDOLTER · Statistical Methods for Forecasting
AGRESTI · Analysis of Ordinal Categorical Data, *Second Edition*
AGRESTI · An Introduction to Categorical Data Analysis, *Second Edition*
AGRESTI · Categorical Data Analysis, *Second Edition*
ALTMAN, GILL, and McDONALD · Numerical Issues in Statistical Computing for the Social Scientist
AMARATUNGA and CABRERA · Exploration and Analysis of DNA Microarray and Protein Array Data
ANDĚL · Mathematics of Chance
ANDERSON · An Introduction to Multivariate Statistical Analysis, *Third Edition*
* ANDERSON · The Statistical Analysis of Time Series
ANDERSON, AUQUIER, HAUCK, OAKES, VANDAELE, and WEISBERG · Statistical Methods for Comparative Studies
ANDERSON and LOYNES · The Teaching of Practical Statistics
ARMITAGE and DAVID (editors) · Advances in Biometry
ARNOLD, BALAKRISHNAN, and NAGARAJA · Records
* ARTHANARI and DODGE · Mathematical Programming in Statistics
* BAILEY · The Elements of Stochastic Processes with Applications to the Natural Sciences
BALAKRISHNAN and KOUTRAS · Runs and Scans with Applications
BALAKRISHNAN and NG · Precedence-Type Tests and Applications
BARNETT · Comparative Statistical Inference, *Third Edition*
BARNETT · Environmental Statistics
BARNETT and LEWIS · Outliers in Statistical Data, *Third Edition*
BARTOSZYNSKI and NIEWIADOMSKA-BUGAJ · Probability and Statistical Inference
BASILEVSKY · Statistical Factor Analysis and Related Methods: Theory and Applications
BASU and RIGDON · Statistical Methods for the Reliability of Repairable Systems
BATES and WATTS · Nonlinear Regression Analysis and Its Applications
BECHHOFER, SANTNER, and GOLDSMAN · Design and Analysis of Experiments for Statistical Selection, Screening, and Multiple Comparisons

*Now available in a lower priced paperback edition in the Wiley Classics Library.
†Now available in a lower priced paperback edition in the Wiley–Interscience Paperback Series.

BELSLEY · Conditioning Diagnostics: Collinearity and Weak Data in Regression
† BELSLEY, KUH, and WELSCH · Regression Diagnostics: Identifying Influential Data and Sources of Collinearity
BENDAT and PIERSOL · Random Data: Analysis and Measurement Procedures, *Fourth Edition*
BERRY, CHALONER, and GEWEKE · Bayesian Analysis in Statistics and Econometrics: Essays in Honor of Arnold Zellner
BERNARDO and SMITH · Bayesian Theory
BHAT and MILLER · Elements of Applied Stochastic Processes, *Third Edition*
BHATTACHARYA and WAYMIRE · Stochastic Processes with Applications
BILLINGSLEY · Convergence of Probability Measures, *Second Edition*
BILLINGSLEY · Probability and Measure, *Third Edition*
BIRKES and DODGE · Alternative Methods of Regression
BISWAS, DATTA, FINE, and SEGAL · Statistical Advances in the Biomedical Sciences: Clinical Trials, Epidemiology, Survival Analysis, and Bioinformatics
BLISCHKE AND MURTHY (editors) · Case Studies in Reliability and Maintenance
BLISCHKE AND MURTHY · Reliability: Modeling, Prediction, and Optimization
BLOOMFIELD · Fourier Analysis of Time Series: An Introduction, *Second Edition*
BOLLEN · Structural Equations with Latent Variables
BOLLEN and CURRAN · Latent Curve Models: A Structural Equation Perspective
BOROVKOV · Ergodicity and Stability of Stochastic Processes
BOULEAU · Numerical Methods for Stochastic Processes
BOX · Bayesian Inference in Statistical Analysis
BOX · R. A. Fisher, the Life of a Scientist
BOX and DRAPER · Response Surfaces, Mixtures, and Ridge Analyses, *Second Edition*
* BOX and DRAPER · Evolutionary Operation: A Statistical Method for Process Improvement
BOX and FRIENDS · Improving Almost Anything, *Revised Edition*
BOX, HUNTER, and HUNTER · Statistics for Experimenters: Design, Innovation, and Discovery, *Second Editon*
BOX, JENKINS, and REINSEL · Time Series Analysis: Forcasting and Control, *Fourth Edition*
BOX, LUCEÑO, and PANIAGUA-QUIÑONES · Statistical Control by Monitoring and Adjustment, *Second Edition*
BRANDIMARTE · Numerical Methods in Finance: A MATLAB-Based Introduction
† BROWN and HOLLANDER · Statistics: A Biomedical Introduction
BRUNNER, DOMHOF, and LANGER · Nonparametric Analysis of Longitudinal Data in Factorial Experiments
BUCKLEW · Large Deviation Techniques in Decision, Simulation, and Estimation
CAIROLI and DALANG · Sequential Stochastic Optimization
CASTILLO, HADI, BALAKRISHNAN, and SARABIA · Extreme Value and Related Models with Applications in Engineering and Science
CHAN · Time Series: Applications to Finance
CHARALAMBIDES · Combinatorial Methods in Discrete Distributions
CHATTERJEE and HADI · Regression Analysis by Example, *Fourth Edition*
CHATTERJEE and HADI · Sensitivity Analysis in Linear Regression
CHERNICK · Bootstrap Methods: A Guide for Practitioners and Researchers, *Second Edition*
CHERNICK and FRIIS · Introductory Biostatistics for the Health Sciences
CHILÈS and DELFINER · Geostatistics: Modeling Spatial Uncertainty
CHOW and LIU · Design and Analysis of Clinical Trials: Concepts and Methodologies, *Second Edition*
CLARKE · Linear Models: The Theory and Application of Analysis of Variance

*Now available in a lower priced paperback edition in the Wiley Classics Library.
†Now available in a lower priced paperback edition in the Wiley–Interscience Paperback Series.

CLARKE and DISNEY · Probability and Random Processes: A First Course with Applications, *Second Edition*
* COCHRAN and COX · Experimental Designs, *Second Edition*
COLLINS and LANZA · Latent Class and Latent Transition Analysis: With Applications in the Social, Behavioral, and Health Sciences
CONGDON · Applied Bayesian Modelling
CONGDON · Bayesian Models for Categorical Data
CONGDON · Bayesian Statistical Modelling
CONOVER · Practical Nonparametric Statistics, *Third Edition*
COOK · Regression Graphics
COOK and WEISBERG · Applied Regression Including Computing and Graphics
COOK and WEISBERG · An Introduction to Regression Graphics
CORNELL · Experiments with Mixtures, Designs, Models, and the Analysis of Mixture Data, *Third Edition*
COVER and THOMAS · Elements of Information Theory
COX · A Handbook of Introductory Statistical Methods
* COX · Planning of Experiments
CRESSIE · Statistics for Spatial Data, *Revised Edition*
CSÖRGŐ and HORVÁTH · Limit Theorems in Change Point Analysis
DANIEL · Applications of Statistics to Industrial Experimentation
DANIEL · Biostatistics: A Foundation for Analysis in the Health Sciences, *Eighth Edition*
* DANIEL · Fitting Equations to Data: Computer Analysis of Multifactor Data, *Second Edition*
DASU and JOHNSON · Exploratory Data Mining and Data Cleaning
DAVID and NAGARAJA · Order Statistics, *Third Edition*
* DEGROOT, FIENBERG, and KADANE · Statistics and the Law
DEL CASTILLO · Statistical Process Adjustment for Quality Control
DEMARIS · Regression with Social Data: Modeling Continuous and Limited Response Variables
DEMIDENKO · Mixed Models: Theory and Applications
DENISON, HOLMES, MALLICK and SMITH · Bayesian Methods for Nonlinear Classification and Regression
DETTE and STUDDEN · The Theory of Canonical Moments with Applications in Statistics, Probability, and Analysis
DEY and MUKERJEE · Fractional Factorial Plans
DILLON and GOLDSTEIN · Multivariate Analysis: Methods and Applications
DODGE · Alternative Methods of Regression
* DODGE and ROMIG · Sampling Inspection Tables, *Second Edition*
* DOOB · Stochastic Processes
DOWDY, WEARDEN, and CHILKO · Statistics for Research, *Third Edition*
DRAPER and SMITH · Applied Regression Analysis, *Third Edition*
DRYDEN and MARDIA · Statistical Shape Analysis
DUDEWICZ and MISHRA · Modern Mathematical Statistics
DUNN and CLARK · Basic Statistics: A Primer for the Biomedical Sciences, *Third Edition*
DUPUIS and ELLIS · A Weak Convergence Approach to the Theory of Large Deviations
EDLER and KITSOS · Recent Advances in Quantitative Methods in Cancer and Human Health Risk Assessment
* ELANDT-JOHNSON and JOHNSON · Survival Models and Data Analysis
ENDERS · Applied Econometric Time Series
† ETHIER and KURTZ · Markov Processes: Characterization and Convergence
EVANS, HASTINGS, and PEACOCK · Statistical Distributions, *Third Edition*

*Now available in a lower priced paperback edition in the Wiley Classics Library.
†Now available in a lower priced paperback edition in the Wiley–Interscience Paperback Series.

FELLER · An Introduction to Probability Theory and Its Applications, Volume I, *Third Edition,* Revised; Volume II, *Second Edition*
FISHER and VAN BELLE · Biostatistics: A Methodology for the Health Sciences
FITZMAURICE, LAIRD, and WARE · Applied Longitudinal Analysis
* FLEISS · The Design and Analysis of Clinical Experiments
FLEISS · Statistical Methods for Rates and Proportions, *Third Edition*
† FLEMING and HARRINGTON · Counting Processes and Survival Analysis
FUJIKOSHI, ULYANOV, and SHIMIZU · Multivariate Statistics: High-Dimensional and Large-Sample Approximations
FULLER · Introduction to Statistical Time Series, *Second Edition*
† FULLER · Measurement Error Models
GALLANT · Nonlinear Statistical Models
GEISSER · Modes of Parametric Statistical Inference
GELMAN and MENG · Applied Bayesian Modeling and Causal Inference from Incomplete-Data Perspectives
GEWEKE · Contemporary Bayesian Econometrics and Statistics
GHOSH, MUKHOPADHYAY, and SEN · Sequential Estimation
GIESBRECHT and GUMPERTZ · Planning, Construction, and Statistical Analysis of Comparative Experiments
GIFI · Nonlinear Multivariate Analysis
GIVENS and HOETING · Computational Statistics
GLASSERMAN and YAO · Monotone Structure in Discrete-Event Systems
GNANADESIKAN · Methods for Statistical Data Analysis of Multivariate Observations, *Second Edition*
GOLDSTEIN and LEWIS · Assessment: Problems, Development, and Statistical Issues
GREENWOOD and NIKULIN · A Guide to Chi-Squared Testing
GROSS, SHORTLE, THOMPSON, and HARRIS · Fundamentals of Queueing Theory, *Fourth Edition*
GROSS, SHORTLE, THOMPSON, and HARRIS · Solutions Manual to Accompany Fundamentals of Queueing Theory, *Fourth Edition*
* HAHN and SHAPIRO · Statistical Models in Engineering
HAHN and MEEKER · Statistical Intervals: A Guide for Practitioners
HALD · A History of Probability and Statistics and their Applications Before 1750
HALD · A History of Mathematical Statistics from 1750 to 1930
† HAMPEL · Robust Statistics: The Approach Based on Influence Functions
HANNAN and DEISTLER · The Statistical Theory of Linear Systems
HARTUNG, KNAPP, and SINHA · Statistical Meta-Analysis with Applications
HEIBERGER · Computation for the Analysis of Designed Experiments
HEDAYAT and SINHA · Design and Inference in Finite Population Sampling
HEDEKER and GIBBONS · Longitudinal Data Analysis
HELLER · MACSYMA for Statisticians
HINKELMANN and KEMPTHORNE · Design and Analysis of Experiments, Volume 1: Introduction to Experimental Design, *Second Edition*
HINKELMANN and KEMPTHORNE · Design and Analysis of Experiments, Volume 2: Advanced Experimental Design
HOAGLIN, MOSTELLER, and TUKEY · Fundamentals of Exploratory Analysis of Variance
* HOAGLIN, MOSTELLER, and TUKEY · Exploring Data Tables, Trends and Shapes
* HOAGLIN, MOSTELLER, and TUKEY · Understanding Robust and Exploratory Data Analysis
HOCHBERG and TAMHANE · Multiple Comparison Procedures
HOCKING · Methods and Applications of Linear Models: Regression and the Analysis of Variance, *Second Edition*

*Now available in a lower priced paperback edition in the Wiley Classics Library.
†Now available in a lower priced paperback edition in the Wiley–Interscience Paperback Series.

HOEL · Introduction to Mathematical Statistics, *Fifth Edition*
HOGG and KLUGMAN · Loss Distributions
HOLLANDER and WOLFE · Nonparametric Statistical Methods, *Second Edition*
HOSMER and LEMESHOW · Applied Logistic Regression, *Second Edition*
HOSMER, LEMESHOW, and MAY · Applied Survival Analysis: Regression Modeling of Time-to-Event Data, *Second Edition*
† HUBER and RONCHETTI · Robust Statistics, *Second Edition*
HUBERTY · Applied Discriminant Analysis
HUBERTY and OLEJNIK · Applied MANOVA and Discriminant Analysis, *Second Edition*
HUNT and KENNEDY · Financial Derivatives in Theory and Practice, *Revised Edition*
HURD and MIAMEE · Periodically Correlated Random Sequences: Spectral Theory and Practice
HUSKOVA, BERAN, and DUPAC · Collected Works of Jaroslav Hajek—with Commentary
HUZURBAZAR · Flowgraph Models for Multistate Time-to-Event Data
IMAN and CONOVER · A Modern Approach to Statistics
† JACKSON · A User's Guide to Principle Components
JOHN · Statistical Methods in Engineering and Quality Assurance
JOHNSON · Multivariate Statistical Simulation
JOHNSON and BALAKRISHNAN · Advances in the Theory and Practice of Statistics: A Volume in Honor of Samuel Kotz
JOHNSON and BHATTACHARYYA · Statistics: Principles and Methods, *Fifth Edition*
JOHNSON and KOTZ · Distributions in Statistics
JOHNSON and KOTZ (editors) · Leading Personalities in Statistical Sciences: From the Seventeenth Century to the Present
JOHNSON, KOTZ, and BALAKRISHNAN · Continuous Univariate Distributions, Volume 1, *Second Edition*
JOHNSON, KOTZ, and BALAKRISHNAN · Continuous Univariate Distributions, Volume 2, *Second Edition*
JOHNSON, KOTZ, and BALAKRISHNAN · Discrete Multivariate Distributions
JOHNSON, KEMP, and KOTZ · Univariate Discrete Distributions, *Third Edition*
JUDGE, GRIFFITHS, HILL, LÜTKEPOHL, and LEE · The Theory and Practice of Econometrics, *Second Edition*
JUREČKOVÁ and SEN · Robust Statistical Procedures: Aymptotics and Interrelations
JUREK and MASON · Operator-Limit Distributions in Probability Theory
KADANE · Bayesian Methods and Ethics in a Clinical Trial Design
KADANE AND SCHUM · A Probabilistic Analysis of the Sacco and Vanzetti Evidence
KALBFLEISCH and PRENTICE · The Statistical Analysis of Failure Time Data, *Second Edition*
KARIYA and KURATA · Generalized Least Squares
KASS and VOS · Geometrical Foundations of Asymptotic Inference
† KAUFMAN and ROUSSEEUW · Finding Groups in Data: An Introduction to Cluster Analysis
KEDEM and FOKIANOS · Regression Models for Time Series Analysis
KENDALL, BARDEN, CARNE, and LE · Shape and Shape Theory
KHURI · Advanced Calculus with Applications in Statistics, *Second Edition*
KHURI, MATHEW, and SINHA · Statistical Tests for Mixed Linear Models
KLEIBER and KOTZ · Statistical Size Distributions in Economics and Actuarial Sciences
KLEMELÄ · Smoothing of Multivariate Data: Density Estimation and Visualization
KLUGMAN, PANJER, and WILLMOT · Loss Models: From Data to Decisions, *Third Edition*
KLUGMAN, PANJER, and WILLMOT · Solutions Manual to Accompany Loss Models: From Data to Decisions, *Third Edition*

*Now available in a lower priced paperback edition in the Wiley Classics Library.
†Now available in a lower priced paperback edition in the Wiley–Interscience Paperback Series.

KOTZ, BALAKRISHNAN, and JOHNSON · Continuous Multivariate Distributions, Volume 1, *Second Edition*
KOVALENKO, KUZNETZOV, and PEGG · Mathematical Theory of Reliability of Time-Dependent Systems with Practical Applications
KOWALSKI and TU · Modern Applied U-Statistics
KRISHNAMOORTHY and MATHEW · Statistical Tolerance Regions: Theory, Applications, and Computation
KROONENBERG · Applied Multiway Data Analysis
KVAM and VIDAKOVIC · Nonparametric Statistics with Applications to Science and Engineering
LACHIN · Biostatistical Methods: The Assessment of Relative Risks
LAD · Operational Subjective Statistical Methods: A Mathematical, Philosophical, and Historical Introduction
LAMPERTI · Probability: A Survey of the Mathematical Theory, *Second Edition*
LANGE, RYAN, BILLARD, BRILLINGER, CONQUEST, and GREENHOUSE · Case Studies in Biometry
LARSON · Introduction to Probability Theory and Statistical Inference, *Third Edition*
LAWLESS · Statistical Models and Methods for Lifetime Data, *Second Edition*
LAWSON · Statistical Methods in Spatial Epidemiology
LE · Applied Categorical Data Analysis
LE · Applied Survival Analysis
LEE and WANG · Statistical Methods for Survival Data Analysis, *Third Edition*
LEPAGE and BILLARD · Exploring the Limits of Bootstrap
LEYLAND and GOLDSTEIN (editors) · Multilevel Modelling of Health Statistics
LIAO · Statistical Group Comparison
LINDVALL · Lectures on the Coupling Method
LIN · Introductory Stochastic Analysis for Finance and Insurance
LINHART and ZUCCHINI · Model Selection
LITTLE and RUBIN · Statistical Analysis with Missing Data, *Second Edition*
LLOYD · The Statistical Analysis of Categorical Data
LOWEN and TEICH · Fractal-Based Point Processes
MAGNUS and NEUDECKER · Matrix Differential Calculus with Applications in Statistics and Econometrics, *Revised Edition*
MALLER and ZHOU · Survival Analysis with Long Term Survivors
MALLOWS · Design, Data, and Analysis by Some Friends of Cuthbert Daniel
MANN, SCHAFER, and SINGPURWALLA · Methods for Statistical Analysis of Reliability and Life Data
MANTON, WOODBURY, and TOLLEY · Statistical Applications Using Fuzzy Sets
MARCHETTE · Random Graphs for Statistical Pattern Recognition
MARDIA and JUPP · Directional Statistics
MASON, GUNST, and HESS · Statistical Design and Analysis of Experiments with Applications to Engineering and Science, *Second Edition*
McCULLOCH, SEARLE, and NEUHAUS · Generalized, Linear, and Mixed Models, *Second Edition*
McFADDEN · Management of Data in Clinical Trials, *Second Edition*
* McLACHLAN · Discriminant Analysis and Statistical Pattern Recognition
McLACHLAN, DO, and AMBROISE · Analyzing Microarray Gene Expression Data
McLACHLAN and KRISHNAN · The EM Algorithm and Extensions, *Second Edition*
McLACHLAN and PEEL · Finite Mixture Models
McNEIL · Epidemiological Research Methods
MEEKER and ESCOBAR · Statistical Methods for Reliability Data
MEERSCHAERT and SCHEFFLER · Limit Distributions for Sums of Independent Random Vectors: Heavy Tails in Theory and Practice

*Now available in a lower priced paperback edition in the Wiley Classics Library.
†Now available in a lower priced paperback edition in the Wiley–Interscience Paperback Series.

MICKEY, DUNN, and CLARK · Applied Statistics: Analysis of Variance and Regression, *Third Edition*
* MILLER · Survival Analysis, *Second Edition*
MONTGOMERY, JENNINGS, and KULAHCI · Introduction to Time Series Analysis and Forecasting
MONTGOMERY, PECK, and VINING · Introduction to Linear Regression Analysis, *Fourth Edition*
MORGENTHALER and TUKEY · Configural Polysampling: A Route to Practical Robustness
MUIRHEAD · Aspects of Multivariate Statistical Theory
MULLER and STOYAN · Comparison Methods for Stochastic Models and Risks
MURRAY · X-STAT 2.0 Statistical Experimentation, Design Data Analysis, and Nonlinear Optimization
MURTHY, XIE, and JIANG · Weibull Models
MYERS, MONTGOMERY, and ANDERSON-COOK · Response Surface Methodology: Process and Product Optimization Using Designed Experiments, *Third Edition*
MYERS, MONTGOMERY, VINING, and ROBINSON · Generalized Linear Models. With Applications in Engineering and the Sciences, *Second Edition*
† NELSON · Accelerated Testing, Statistical Models, Test Plans, and Data Analyses
† NELSON · Applied Life Data Analysis
NEWMAN · Biostatistical Methods in Epidemiology
OCHI · Applied Probability and Stochastic Processes in Engineering and Physical Sciences
OKABE, BOOTS, SUGIHARA, and CHIU · Spatial Tesselations: Concepts and Applications of Voronoi Diagrams, *Second Edition*
OLIVER and SMITH · Influence Diagrams, Belief Nets and Decision Analysis
PALTA · Quantitative Methods in Population Health: Extensions of Ordinary Regressions
PANJER · Operational Risk: Modeling and Analytics
PANKRATZ · Forecasting with Dynamic Regression Models
PANKRATZ · Forecasting with Univariate Box-Jenkins Models: Concepts and Cases
* PARZEN · Modern Probability Theory and Its Applications
PEÑA, TIAO, and TSAY · A Course in Time Series Analysis
PIANTADOSI · Clinical Trials: A Methodologic Perspective
PORT · Theoretical Probability for Applications
POURAHMADI · Foundations of Time Series Analysis and Prediction Theory
POWELL · Approximate Dynamic Programming: Solving the Curses of Dimensionality
PRESS · Bayesian Statistics: Principles, Models, and Applications
PRESS · Subjective and Objective Bayesian Statistics, *Second Edition*
PRESS and TANUR · The Subjectivity of Scientists and the Bayesian Approach
PUKELSHEIM · Optimal Experimental Design
PURI, VILAPLANA, and WERTZ · New Perspectives in Theoretical and Applied Statistics
† PUTERMAN · Markov Decision Processes: Discrete Stochastic Dynamic Programming
QIU · Image Processing and Jump Regression Analysis
* RAO · Linear Statistical Inference and Its Applications, *Second Edition*
RAUSAND and HØYLAND · System Reliability Theory: Models, Statistical Methods, and Applications, *Second Edition*
RENCHER · Linear Models in Statistics
RENCHER · Methods of Multivariate Analysis, *Second Edition*
RENCHER · Multivariate Statistical Inference with Applications
* RIPLEY · Spatial Statistics
* RIPLEY · Stochastic Simulation
ROBINSON · Practical Strategies for Experimenting

*Now available in a lower priced paperback edition in the Wiley Classics Library.
†Now available in a lower priced paperback edition in the Wiley–Interscience Paperback Series.

ROHATGI and SALEH · An Introduction to Probability and Statistics, *Second Edition*
ROLSKI, SCHMIDLI, SCHMIDT, and TEUGELS · Stochastic Processes for Insurance and Finance
ROSENBERGER and LACHIN · Randomization in Clinical Trials: Theory and Practice
ROSS · Introduction to Probability and Statistics for Engineers and Scientists
ROSSI, ALLENBY, and McCULLOCH · Bayesian Statistics and Marketing
† ROUSSEEUW and LEROY · Robust Regression and Outlier Detection
* RUBIN · Multiple Imputation for Nonresponse in Surveys
RUBINSTEIN and KROESE · Simulation and the Monte Carlo Method, *Second Edition*
RUBINSTEIN and MELAMED · Modern Simulation and Modeling
RYAN · Modern Engineering Statistics
RYAN · Modern Experimental Design
RYAN · Modern Regression Methods, *Second Edition*
RYAN · Statistical Methods for Quality Improvement, *Second Edition*
SALEH · Theory of Preliminary Test and Stein-Type Estimation with Applications
* SCHEFFE · The Analysis of Variance
SCHIMEK · Smoothing and Regression: Approaches, Computation, and Application
SCHOTT · Matrix Analysis for Statistics, *Second Edition*
SCHOUTENS · Levy Processes in Finance: Pricing Financial Derivatives
SCHUSS · Theory and Applications of Stochastic Differential Equations
SCOTT · Multivariate Density Estimation: Theory, Practice, and Visualization
† SEARLE · Linear Models for Unbalanced Data
† SEARLE · Matrix Algebra Useful for Statistics
† SEARLE, CASELLA, and McCULLOCH · Variance Components
SEARLE and WILLETT · Matrix Algebra for Applied Economics
SEBER · A Matrix Handbook For Statisticians
† SEBER · Multivariate Observations
SEBER and LEE · Linear Regression Analysis, *Second Edition*
† SEBER and WILD · Nonlinear Regression
SENNOTT · Stochastic Dynamic Programming and the Control of Queueing Systems
* SERFLING · Approximation Theorems of Mathematical Statistics
SHAFER and VOVK · Probability and Finance: It's Only a Game!
SILVAPULLE and SEN · Constrained Statistical Inference: Inequality, Order, and Shape Restrictions
SMALL and McLEISH · Hilbert Space Methods in Probability and Statistical Inference
SRIVASTAVA · Methods of Multivariate Statistics
STAPLETON · Linear Statistical Models, *Second Edition*
STAPLETON · Models for Probability and Statistical Inference: Theory and Applications
STAUDTE and SHEATHER · Robust Estimation and Testing
STOYAN, KENDALL, and MECKE · Stochastic Geometry and Its Applications, *Second Edition*
STOYAN and STOYAN · Fractals, Random Shapes and Point Fields: Methods of Geometrical Statistics
STREET and BURGESS · The Construction of Optimal Stated Choice Experiments: Theory and Methods
STYAN · The Collected Papers of T. W. Anderson: 1943–1985
SUTTON, ABRAMS, JONES, SHELDON, and SONG · Methods for Meta-Analysis in Medical Research
TAKEZAWA · Introduction to Nonparametric Regression
TAMHANE · Statistical Analysis of Designed Experiments: Theory and Applications
TANAKA · Time Series Analysis: Nonstationary and Noninvertible Distribution Theory
THOMPSON · Empirical Model Building
THOMPSON · Sampling, *Second Edition*
THOMPSON · Simulation: A Modeler's Approach

*Now available in a lower priced paperback edition in the Wiley Classics Library.
†Now available in a lower priced paperback edition in the Wiley–Interscience Paperback Series.

THOMPSON and SEBER · Adaptive Sampling
THOMPSON, WILLIAMS, and FINDLAY · Models for Investors in Real World Markets
TIAO, BISGAARD, HILL, PEÑA, and STIGLER (editors) · Box on Quality and Discovery: with Design, Control, and Robustness
TIERNEY · LISP-STAT: An Object-Oriented Environment for Statistical Computing and Dynamic Graphics
TSAY · Analysis of Financial Time Series, *Second Edition*
UPTON and FINGLETON · Spatial Data Analysis by Example, Volume II: Categorical and Directional Data
† VAN BELLE · Statistical Rules of Thumb, *Second Edition*
VAN BELLE, FISHER, HEAGERTY, and LUMLEY · Biostatistics: A Methodology for the Health Sciences, *Second Edition*
VESTRUP · The Theory of Measures and Integration
VIDAKOVIC · Statistical Modeling by Wavelets
VINOD and REAGLE · Preparing for the Worst: Incorporating Downside Risk in Stock Market Investments
WALLER and GOTWAY · Applied Spatial Statistics for Public Health Data
WEERAHANDI · Generalized Inference in Repeated Measures: Exact Methods in MANOVA and Mixed Models
WEISBERG · Applied Linear Regression, *Third Edition*
WELSH · Aspects of Statistical Inference
WESTFALL and YOUNG · Resampling-Based Multiple Testing: Examples and Methods for *p*-Value Adjustment
WHITTAKER · Graphical Models in Applied Multivariate Statistics
WINKER · Optimization Heuristics in Economics: Applications of Threshold Accepting
WONNACOTT and WONNACOTT · Econometrics, *Second Edition*
WOODING · Planning Pharmaceutical Clinical Trials: Basic Statistical Principles
WOODWORTH · Biostatistics: A Bayesian Introduction
WOOLSON and CLARKE · Statistical Methods for the Analysis of Biomedical Data, *Second Edition*
WU and HAMADA · Experiments: Planning, Analysis, and Parameter Design Optimization, *Second Edition*
WU and ZHANG · Nonparametric Regression Methods for Longitudinal Data Analysis
YANG · The Construction Theory of Denumerable Markov Processes
YOUNG, VALERO-MORA, and FRIENDLY · Visual Statistics: Seeing Data with Dynamic Interactive Graphics
ZACKS · Stage-Wise Adaptive Designs
ZELTERMAN · Discrete Distributions—Applications in the Health Sciences
* ZELLNER · An Introduction to Bayesian Inference in Econometrics
ZHOU, OBUCHOWSKI, and McCLISH · Statistical Methods in Diagnostic Medicine

*Now available in a lower priced paperback edition in the Wiley Classics Library.
†Now available in a lower priced paperback edition in the Wiley–Interscience Paperback Series.